Robotic Vision:
Technologies for Machine Learning and Vision Applications

José García-Rodríguez
University of Alicante, Spain

Miguel Cazorla
University of Alicante, Spain

Managing Director:	Lindsay Johnston
Editorial Director:	Joel Gamon
Book Production Manager:	Jennifer Yoder
Publishing Systems Analyst:	Adrienne Freeland
Development Editor:	Christine Smith
Assistant Acquisitions Editor:	Kayla Wolfe
Typesetter:	Erin O'Dea
Cover Design:	Nick Newcomer

Published in the United States of America by
Information Science Reference (an imprint of IGI Global)
701 E. Chocolate Avenue
Hershey PA 17033
Tel: 717-533-8845
Fax: 717-533-8661
E-mail: cust@igi-global.com
Web site: http://www.igi-global.com

Library of Congress Cataloging-in-Publication Data

Robotic vision: technologies for machine learning and vision applications / Jose Garcia-Rodriguez and Miguel A. Cazorla Quevedo, editors.
    pages cm
  Summary: "This book offers comprehensive coverage of the current research on the fields of robotics, machine vision, image processing and pattern recognition that is important to applying machine vision methods in the real world"-- Provided by publisher.
  Includes bibliographical references and index.
  ISBN 978-1-4666-2672-0 (hardcover) -- ISBN 978-1-4666-2703-1 (ebook) -- ISBN 978-1-4666-2734-5 (print & perpetual access) 1. Computer vision. 2. Pattern recognition systems. 3. Image processing. 4. Robotics--Human factors. I. Garcia-Rodriguez, Jose, 1970- II. Cazorla Quevedo, Miguel A., 1970-
  TA1634.R63 2013
  629.8'92637--dc23
                            2012029113

British Cataloguing in Publication Data
A Cataloguing in Publication record for this book is available from the British Library.

All work contributed to this book is new, previously-unpublished material. The views expressed in this book are those of the authors, but not necessarily of the publisher.

List of Reviewers

Raed Almomani, *Wayne State University, USA*
Cecilio Angulo, *CETpD-UPC, Universitat Politècnica de Catalunya, Spain*
Roseli Aparecida, *University of Sao Paulo, Brazil*
Jorge Azorín, *University of Alicante, Spain*
Antonio Bandera, *University of Málaga, Spain*
Juan Pedro Bandera, *University of Málaga, Spain*
Douglas Brooks, *Georgia Institute of Technology, USA*
Ivan Cabezas, *Universidad del Valle, Colombia*
José María Cañas, *Rey Juan Carlos University, Spain*
Ming Dong, *Wayne State University, USA*
Andrés Fuster, *University of Alicante, Spain*
Manuel Graña, *Basque Country University (UPV/EHU), Spain*
Juan Manuel Guede, *Basque Country University (UPV/EHU), Spain*
Ayanna M. Howard, *Georgia Institute of Technology, USA*
Laura Igual, *MAiA-UB, Spain*
Jim Juola, *University of Kansas, USA*
Vicente Matellán, *University of León, Spain*
Neils Meins, *University of Hamburg, Germany*
Vicente Morell, *University of Alicante, Spain*
Ramon Moreno, *University of Alicante, Spain*
Lazaros Nalpantidis, *Royal Institute of Technology – KTH, Sweden*
Sergio Orts, *Universitat Autonoma de Barcelona, Spain*
Chung Park, *Georgia Institute of Technology, USA*
Hae Wong Park, *Georgia Institute of Technology, USA*
Xavier Perez-Sala, *CETpD-UPC, Universitat Politècnica de Catalunya, Spain*
Domenec Puig, *University of Rovira I Virgili, Spain*
Renato Ramos, *University of Sao Paulo, Brazil*
J.A. Rodríguez, *University of Málaga, Spain*
Sreela Sasi, *Gannon University, USA*
Marcelo Saval, *University of Alicante, Spain*
Mohan Sridharan, *Texas Tech University, USA*
Elena Torta, *Eindhoven University of Technology, The Netherlands*
Diego Viejo, *University of Alicante, Spain*
Wenjie Yan, *University of Hamburg, Germany*

Table of Contents

Section 1
Computer Vision

Patrycia Barros de Lima Klavdianos, Université de Bourgogne, France
Lourdes Mattos Brasil, Universidade de Brasília (UnB), Brazil
Jairo Simão Santana Melo, Universidade de Brasília (UnB), Brazil

Xavier Perez-Sala, Universitat Politècnica de Catalunya, Spain
   & Computer Vision Center of Barcelona, Spain
Laura Igual, Universitat de Barcelona, Spain & Computer Vision Center of Barcelona, Spain
Sergio Escalera, Universitat de Barcelona, Spain & Computer Vision Center of Barcelona, Spain
Cecilio Angulo, Universitat Politècnica de Catalunya, Spain

Marcelo Saval-Calvo, University of Alicante, Spain
Jorge Azorín-López, University of Alicante, Spain
Andrés Fuster-Guilló, University of Alicante, Spain

Section 2
Computer Vision Applications

Section 3
3D Computer Vision and Robotics

Section 4
Social Robotics

Section 5
Vision Control

Section 6
Visual Attention

Detailed Table of Contents

Section 1
Computer Vision

Chapter 1

> *Patrycia Barros de Lima Klavdianos, Université de Bourgogne, France*
> *Lourdes Mattos Brasil, Universidade de Brasília (UnB), Brazil*
> *Jairo Simão Santana Melo, Universidade de Brasília (UnB), Brazil*

Recognition of human faces has been a fascinating subject in research field for many years. It is considered a multidisciplinary field because it includes understanding different domains such as psychology, neuro-science, computer vision, artificial intelligence, mathematics, and many others. Human face perception is intriguing and draws our attention because we accomplish the task so well that we hope to one day witness a machine performing the same task in a similar or better way. This chapter aims to provide a systematic and practical approach regarding to one of the most current techniques applied on face recognition, known as AAM (Active Appearance Model). AAM method is addressed considering 2D face processing only. This chapter doesn't cover the entire theme, but offers to the reader the necessary tools to construct a consistent and productive pathway toward this involving subject.

Chapter 2

> *Xavier Perez-Sala, Universitat Politècnica de Catalunya, Spain*
>   *& Computer Vision Center of Barcelona, Spain*
> *Laura Igual, Universitat de Barcelona, Spain & Computer Vision Center of Barcelona, Spain*
> *Sergio Escalera, Universitat de Barcelona, Spain & Computer Vision Center of Barcelona, Spain*
> *Cecilio Angulo, Universitat Politècnica de Catalunya, Spain*

Different methodologies of uniform sampling over the rotation group, SO(3), for building unbiased 2D shape models from 3D objects are introduced and reviewed in this chapter. State-of-the-art non uniform sampling approaches are discussed, and uniform sampling methods using Euler angles and quaternions are introduced. Moreover, since presented work is oriented to model building applications, it is not limited to general discrete methods to obtain uniform 3D rotations, but also from a continuous point of view in the case of Procrustes Analysis.

Chapter 3

In this chapter, a comparative analysis of basic segmentation methods of video sequences and their combinations is carried out. Analysis of different algorithms is based on the efficiency (true positive and false positive rates) and temporal cost to provide regions in the scene. These are two of the most important requirements of the design to provide to the tracking with segmentation in an efficient and timely manner constrained to the application. Specifically, methods using temporal information as Background Subtraction, Temporal Differencing, Optical Flow, and the four combinations of them have been analyzed. Experimentation has been done using image sequences of CAVIAR project database. Efficiency results show that Background Subtraction achieves the best individual result whereas the combination of the three basic methods is the best result in general. However, combinations with Optical Flow should be considered depending of application, because its temporal cost is too high with respect to efficiency provided to the combination.

Section 2
Computer Vision Applications

Computer vision plays a significant role in a wide range of homeland security applications. The homeland security applications include: port security (cargo inspection), facility security (embassy, power plant, bank), and surveillance (military or civilian), et cetera. Video surveillance cameras are placed in offices, hospitals, banks, ports, parking lots, parks, stadiums, malls, train stations, airports, et cetera. The challenge is not for acquiring surveillance data from these video cameras, but for identifying what is valuable, what can be ignored, and what demands immediate attention. Computer vision systems attempt to construct meaningful and explicit descriptions of the environment or scene captured in an image. A few Computer Vision based security applications are presented here for securing building facility, railroad (Objects on railroad, and red signal detection), and roads.

Chapter 5

In this chapter, a system to identify the different elements of a Linked Multi-Component Robotic System (L-MCRS) is specified, designed, and implemented. A L-MCRS is composed of several independent robots and a linking element between them which provide a greater complexity to these systems. The identification system is used to model each component of the L-MCRS using very basic information about each of the individual components. So, different state models that have been used in several works of the literature that have been reviewed can be covered. The chapter explains the design of the system and shows its frontend. This work is the first step towards a realistic implementation of L-MCRS.

Video tracking systems are increasingly used day in and day out in various applications such as surveillance, security, monitoring, and robotic vision. In this chapter, the authors propose a novel multiple objects tracking system in video sequences that deals with occlusion issues. The proposed system is composed of two components: An improved KLT tracker, and a Kalman filter. The improved KLT tracker uses the basic KLT tracker and an appearance model to track objects from one frame to another and deal with partial occlusion. In partial occlusion, the appearance model (e.g., a RGB color histogram) is used to determine an object's KLT features, and the authors use these features for accurate and robust tracking. In full occlusion, a Kalman filter is used to predict the object's new location and connect the trajectory parts. The system is evaluated on different videos and compared with a common tracking system.

The representation of the RGB color space points in spherical coordinates allows to retain the chromatic components of image pixel colors, pulling apart easily the intensity component. This representation allows the definition of a chromatic distance and a hybrid gradient with good properties of perceptual color constancy. In this chapter, the authors present a watershed based image segmentation method using this hybrid gradient. Oversegmentation is solved by applying a region merging strategy based on the chromatic distance defined on the spherical coordinate representation. The chapter shows the robustness and performance of the approach on well known test images and the Berkeley benchmarking image database and on images taken with a NAO robot.

This chapter aims to address the ability of self-organizing neural network models to manage video and image processing in real-time. The Growing Neural Gas networks (GNG) with its attributes of growth, flexibility, rapid adaptation, and excellent quality representation of the input space makes it a suitable model for real time applications. A number of applications are presented, including: image compression, hand and medical image contours representation, surveillance systems, hand gesture recognition systems, and 3D data reconstruction.

Section 3
3D Computer Vision and Robotics

Chapter 9

Vicente Morell-Gimenez, University of Alicante, Spain
Sergio Orts-Escolano, University of Alicante, Spain
José García-Rodríguez, University of Alicante, Spain
Miguel Cazorla, University of Alicante, Spain
Diego Viejo, University of Alicante, Spain

The task of registering three dimensional data sets with rigid motions is a fundamental problem in many areas as computer vision, medical images, mobile robotic, arising whenever two or more 3D data sets must be aligned in a common coordinate system. In this chapter, the authors review registration methods. Focusing on mobile robots area, this chapter reviews the main registration methods in the literature. A possible classification could be distance-based and feature-based methods. The distance based methods, from which the classical Iterative Closest Point (ICP) is the most representative, have a lot of variations which obtain better results in situations where noise, time, or accuracy conditions are present. Feature based methods try to reduce the great number or points given by the current sensors using a combination of feature detector and descriptor which can be used to compute the final transformation with a method like RANSAC or Genetic Algorithms.

Chapter 10

Ivan Cabezas, Universidad del Valle, Colombia
Maria Trujillo, Universidad del Valle, Colombia

The use of disparity estimation algorithms is required in the 3D recovery process from stereo images. These algorithms tackle the correspondence problem by computing a disparity map. The accuracy assessment of a disparity estimation process has multiple applications such as comparing among different algorithms' performance, tuning algorithm's parameters within a particular context, and determining the impact of components, among others. Disparity estimation algorithms can be assessed by following an evaluation methodology. This chapter is dedicated to present and discuss methodologies for evaluating disparity estimation algorithms. The discussion begins with a review of the state-of-the-art. The constitutive components of a methodology are analysed. Finally, advantages and drawbacks of existing methodologies are presented.

Chapter 11

Ashwin P. Dani, University of Florida, USA
Zhen Kan, University of Florida, USA
Nic Fischer, University of Florida, USA
Warren E. Dixon, University of Florida, USA

In this chapter, an online method is developed for estimating 3D structure (with proper scale) of moving objects seen by a moving camera. In contrast to traditionally developed batch solutions for this problem, a nonlinear unknown input observer strategy is used where the object's velocity is considered as an unknown input to the perspective dynamical system. The estimator is exponentially stable, and hence, provides robustness against modeling uncertainties and measurement noise from the camera. The developed method provides first causal, observer based structure estimation algorithm for a moving camera viewing a moving object with unknown time-varying object velocities.

Chapter 12

Traversability estimation is the process of assessing whether a robot is able to move across a specific area. Autonomous robots need to have such an ability to automatically detect and avoid non-traversable areas and, thus, stereo vision is commonly used towards this end constituting a reliable solution under a variety of circumstances. This chapter discusses two different intelligent approaches to assess the traversability of the terrain in front of a stereo vision-equipped robot. First, an approach based on a fuzzy inference system is examined and then another approach is considered, which extracts geometrical descriptions of the scene depth distribution and uses a trained support vector machine (SVM) to assess the traversability. The two methods are presented and discussed in detail.

Section 4
Social Robotics

Learning by imitation allows people to teach social robots new tasks using natural and intuitive interaction channels. Vision is the main of these channels. This chapter describes a learning-by-imitation architecture that uses stereo vision to perceive, recognize, learn, and imitate social gestures. This description is based on the identification of a set of generic components, which can be found in any learning by imitation architecture. It highlights the main contribution of the proposed architecture: the use of an inner human model to help perceiving, recognizing and learning human gestures. This allows different robots to share the same perceptual and knowledge modules. Experimental results show that the proposed architecture is able to meet the requirements of learning by imitation scenarios. It can also be integrated in complete software structures for social robots, which involve complex attention mechanisms and decision layers.

Computer vision is essential to develop a social robotic system capable to interact with humans. It is responsible to extract and represent the information around the robot. Furthermore, a learning mechanism, to select correctly an action to be executed in the environment, pro-active mechanism, to engage in an interaction, and voice mechanism, are indispensable to develop a social robot. All these mechanisms together provide a robot emulate some human behavior, like shared attention. Then, this chapter presents a robotic architecture that is composed with such mechanisms to make possible interactions between a robotic head with a caregiver, through of the shared attention learning with identification of some objects.

This chapter presents an overview of a typical scenario of Ambient Assisted Living (AAL) in which a robot navigates to a person for conveying information. Indoor robot navigation is a challenging task due to the complexity of real-home environments and the need of online learning abilities to adjust for dynamic conditions. A comparison between systems with different sensor typologies shows that vision-based systems promise to provide good performance and a wide scope of usage at reasonable cost. Moreover, vision-based systems can perform different tasks simultaneously by applying different algorithms to the input data stream thus enhancing the flexibility of the system. The authors introduce the state of the art of several computer vision methods for realizing indoor robotic navigation to a person and human-robot interaction. A case study has been conducted in which a robot, which is part of an AAL system, navigates to a person and interacts with her. The authors evaluate this test case and give an outlook on the potential of learning robot vision in ambient homes.

Developments in sensor technology and sensory input processing algorithms have enabled the use of mobile robots in real-world domains. As they are increasingly deployed to interact with humans in our homes and offices, robots need the ability to operate autonomously based on sensory cues and high-level feedback from non-expert human participants. Towards this objective, this chapter describes an integrated framework that jointly addresses the learning, adaptation, and interaction challenges associated with robust human-robot interaction in real-world application domains. The novel probabilistic framework consists of: (a) a bootstrap learning algorithm that enables a robot to learn layered graphical models of environmental objects and adapt to unforeseen dynamic changes; (b) a hierarchical planning algorithm based on partially observable Markov decision processes (POMDPs) that enables the robot to reliably and efficiently tailor learning, sensing, and processing to the task at hand; and (c) an augmented reinforcement learning algorithm that enables the robot to acquire limited high-level feedback from non-expert human participants, and merge human feedback with the information extracted from sensory cues. Instances of these algorithms are implemented and fully evaluated on mobile robots and in simulated domains using vision as the primary source of information in conjunction with range data and simplistic verbal inputs. Furthermore, a strategy is outlined to integrate these components to achieve robust human-robot interaction in real-world application domains.

Section 5
Vision Control

Chapter 17

Domenec Puig, Rovira i Virgili University, Spain

This chapter focuses on the study of SLAM taking into account different strategies for modeling unknown environments, with the goal of comparing several methodologies and test them in real robots even if they are heterogeneous. The purpose is to combine them in order to reduce the exploration time. Indubitably, it is not an easy work because it is important to take into account the problem of integrating the information related with the changes into the map. In this way, it is necessary to obtain a representation of the surrounding in an efficiently way. Furthermore, the author is interested in the collaboration between robots, because it is well-known that a team of robots is capable of completing a given task faster than a single robot. This assumption will be checked by using both simulations and real robots in different experiments. In addition, the author combines the benefits of both vision-based and laser-based systems in the integration of the algorithms.

Chapter 18

P. Cavestany Olivares, University of Murcia, Spain

D. Herrero-Pérez, University of Murcia, Spain

J. J. Alcaraz Jiménez, University of Murcia, Spain

H. Martínez Barberá, University of Murcia, Spain

In this chapter, the authors describe their vision system used in the Standard Platform League (SPL), one of the official leagues in RoboCup competition. The characteristics of SPL are very demanding, as all the processing must be done on board, and the changeable environment requires powerful methods for extracting information and robust filters. The purpose is to show a vision system that meets these goals. The chapter describes the architecture of the authors' system as well as the flowchart of the image process, which is designed in such a manner that allows a rapid and reliable calibration. The authors deal with field features detection by finding intersections between field lines at frame rate, using a fuzzy-Markov localisation technique. Also, the methods implemented to recognise the ball and goals are explained.

Chapter 19

L. M. Alkurdi, University of Edinburgh, UK

R. B. Fisher, University of Edinburgh, UK

The problem of visual control of an autonomous indoor blimp is investigated in this chapter. Autonomous aerial vehicles have been an attractive platform for a wide range of applications, especially since they don't have the terrain limitations the autonomous ground vehicles face. They have been used for advertisements, terrain mapping, surveillance, and environmental research. Blimps are a special kind of autonomous aerial vehicles; they are wingless and have the ability to hover. This makes them overcome the maneuverability constraints winged aerial vehicles and helicopters suffer from. The authors' blimp platform also provides an exciting platform for the application and testing of control algorithms. This is because blimps are notorious for the uncertainties within their mathematical model and their susceptibility for environmental disturbances such as wind gusts. The authors have successfully applied visual control by using a fuzzy logic controller on the robotic blimp to achieve autonomous waypoint tracking.

Section 6
Visual Attention

Chapter 20

Juan F. García, Universidad de León, Spain

Francisco J. Rodríguez, Universidad de León, Spain

Vicente Matellán, Universidad de León, Spain

The purpose of this chapter is both to review some of the most representative visual attention models, both theoretical and practical, that have been proposed to date, and to introduce the authors' attention model, which has been successfully used as part of the control system of a robotic platform. The chapter has three sections: in the first section, an introduction to visual attention is given. In the second section, relevant state of art in visual attention is reviewed. This review is organised in three areas: psychological based models, connectionist models, and features-based models. In the last section, the authors' attention model is presented.

Chapter 21

Julio Vega, Rey Juan Carlos University, Spain

Eduardo Perdices, Rey Juan Carlos University, Spain

José María Cañas, Rey Juan Carlos University, Spain

Cameras are one of the most relevant sensors in autonomous robots. Two challenges with them are to extract useful information from captured images and to manage the small field of view of regular cameras. This chapter proposes a visual perceptive system for a robot with a mobile camera on board that cope with these two issues. The system is composed of a dynamic visual memory that stores the information gathered from images, an attention system that continuously chooses where to look at, and a visual evolutionary localization algorithm that uses the visual memory as input. The visual memory is a collection of relevant task-oriented objects and 3D segments. Its scope and persistence is wider than the camera field of view and so provides more information about robot surroundings and more robustness to occlusions than current image. The control software takes its contents into account when making behavior or navigation decisions. The attention system considers the need of reobserving objects already stored, of exploring new areas and of testing hypothesis about objects in the robot surroundings. A robust evolutionary localization algorithm has been developed that can use both the current instantaneous images or the visual memory. The system has been programmed and several experiments have been carried out both with simulated and real robots (wheeled Pioneer and Nao humanoid) to validate it.

Chapter 22

E. Antúnez, Universidad de Málaga, Spain

Y. Haxhimusa, Vienna University of Technology, Austria

R. Marfil, Universidad de Málaga, Spain

W. G. Kropatsch, Vienna University of Technology, Austria

A. Bandera, Universidad de Málaga, Spain

Computer vision systems have to deal with thousands, sometimes millions of pixel values from each frame, and the computational complexity of many problems related to the interpretation of image data is very high. The task becomes especially difficult if a system has to operate in real-time. Within the Combinatorial Pyramid framework, the proposed computational model of attention integrates bottom-up and top-down factors for attention. Neurophysiologic studies have shown that, in humans, these two factors are the main responsible ones to drive attention. Bottom-up factors emanate from the scene and focus attention on regions whose features are sufficiently discriminative with respect to the features of their surroundings. On the other hand, top-down factors are derived from cognitive issues, such as knowledge about the current task. Specifically, the authors only consider in this model the knowledge of a given target to drive attention to specific regions of the image. With respect to previous approaches, their model takes into consideration not only geometrical properties and appearance information, but also internal topological layout. Once the focus of attention has been fixed to a region of the scene, the model evaluates if the focus is correctly located over the desired target. This recognition algorithm considers topological features provided by the pre-attentive stage. Thus, attention and recognition are tied together, sharing the same image descriptors.

Foreword

The application of vision to robotics has seen an enormous progress in the last decade with the introduction on new algorithms and very powerful computer hardware. This progress has also been extended to a number of very dissimilar areas such as automation, medicine, and surveillance to name a few. Computer vision has seen very successful application in field and service robots, in particular in autonomous machines and automotive applications. This has been possible not only by the improvement of computer hardware but from the development of new very efficient algorithms.

Until a few years ago, fundamental problems affected computer vision that makes most algorithms not viable for real time application. This has started to change dramatically. Over the last few years, we have seen an enormous growth of very successful practical implementation of computer vision for robotics. Furthermore, some of them have exploited mass production of proprietary hardware to make the deployment of impressive applications at very reduced costs. This has also been possible due to a number of significant breakthroughs in the underlying algorithms and techniques, including feature detectors, classifiers and a large variety of very efficient machine learning algorithms.

This book presents a comprehensive introduction and the latest development to the fields of computer vision and applications to robotics, social robotics, visual control, and visual attention. The material is organized in various sections with a number of contributions from world experts in the different areas.

The target audience of this book includes robotics scientist, engineers, and students interested in getting a comprehensive background in the rapidly developing field of robotics and computer vision.

It is impossible to select a number of papers to cover all the recent progress in computer vision. Nevertheless, the editors have chosen a number of fundamental aspects of robotic vision that are addressed in a very comprehensive manner in this book. The material presented is intended to be a fundamental first step towards understanding the main challenges involved in robotic vision application.

Eduardo Nebot
University of Sydney & Australian Centre for Field Robotics, Australia

Eduardo Nebot, *BSc. EE, UNS, Argentina and MS and PhD CSU, USA, is a Professor at the University of Sydney and the Director of the Australian Centre for Field Robotics. His main research areas are in field robotics automation. The major impact of his fundamental research is in autonomous system, navigation, and mining safety.*

Preface

Computer vision and robotics connections have grown dramatically in the last years. Any robotic system includes object or scene recognition, vision-based motion control, vision-based mapping, and dense range sensing. Developments in hardware and sensing permit most vision algorithms to work in real-time and using cheap and flexible sensors.

"Robotic Vision: Technologies for Machine Learning and Vision Applications" is an edited collection of contributed chapters of interest for both researchers and practitioners in the fields of computer vision and robotics.

Written by leading researchers in the field, the chapters are organized into six sections. The first two sections deal with computer vision basics and computer vision applications. Section 3 is devoted to 3D data processing applied to robotics. In section 4, some works describing social robotics systems are presented, while section 5 presents works related with vision control, and section 6 introduces some research in visual attention.

SECTION 1: COMPUTER VISION

Barros de Lima Klavdianos, Mattos Brasil, and Simão Santana Melo propose a systematic and practical approach regarding to one of the most current techniques applied on face recognition, known as AAM (Active Appearance Model). Different methodologies of uniform sampling over the 3D rotation group, SO(3), for building unbiased 2D shape models from 3D objects are introduced and reviewed by Perez-Sala, Igual, Escalera, and Angulo. A comparative analysis of basic segmentation methods of video sequences and their combinations is presented by Saval-Calvo, Azorín-López, and Fuster-Guilló.

SECTION 2: COMPUTER VISION APPLICATIONS

Sasi presents a system for identifying what are valuable, what can be ignored, and what demands immediate attention, in a vision security system. In the chapter "Visual Detection in Linked Multi-Component Robotic Systems" by Lopez-Guede, Fernandez-Gauna, Moreno, and Graña, a system to identify the different elements of a Linked Multi-Component Robotic System (L-MCRS) is specified, designed, and implemented. Almomani and Dong propose a novel multiple objects tracking system in video sequences that deals with occlusion issues. The proposed system is composed of two components: An improved KLT tracker, and a Kalman filter. Moreno, Graña, and Madani introduce a watershed and region merging segmentation algorithm based on the zenithal and azimuthal angles of the spherical representation of colors in the RGB space. Garcia-Rodriguez et al. demonstrate the capacity of self-organizing neural networks to solve some computer vision an image processing tasks presenting different examples like image segmentation and compression, tracking, or 3D reconstruction.

SECTION 3: 3D COMPUTER VISION AND ROBOTICS

In chapter "A Review of Registration Methods on Mobile Robots" by Morell-Gimenez, Orts-Escolano, García-Rodríguez, Cazorla, and Viejo, the authors provide a review of the main registration methods in the literature, where registration is a process to find the transformation between two consecutive poses, from 3D data. In "Methodologies for Evaluating Disparity Estimation Algorithms" by Cabezas and Trujillo, the chapter is dedicated to present and discuss methodologies for evaluating disparity estimation algorithms. An online method for estimating 3D structure (with proper scale) of moving objects seen by a moving camera is developed by Dani, Kan, Fischer, and Dixon. Two different intelligent approaches to assess the traversability of the terrain in front of a stereo vision-equipped robot are presented by Nalpantidis, Kostavelis, and Gasterato.

SECTION 4: SOCIAL ROBOTICS

Bandera, Rodríguez, Molina-Tanco and Bandera describe a learning by imitation architecture that uses stereo vision to perceive, recognize, learn and imitate social gestures. An overview of a typical scenario of Ambient Assisted Living (AAL) in which a robot navigates to a person for conveying information is presented by Yan, Torta, van der Pol, Meins, Weber, Cuijpers and Wermter. Da Silva and Romero deals with Computer Vision for learning to interact socially with humans presenting a robotic architecture for a simple interaction between a caregiver and a robot face. Sridharan describes an integrated framework that jointly addresses the learning, adaptation, and interaction challenges associated with robust human-robot interaction in real-world application domains.

SECTION 5: VISION CONTROL

The chapter by Puig and Aviles presents a framework for simultaneous localization and mapping based on an active coordination of a team of robots. Cavestany Olivares, Herrero-Pérez, Alcaraz Jiménez, and Martínez Barberá describes their vision system used in the Standard Platform League (SPL), one of the official leagues in RoboCup competition. Alkurdi and Fisher applied visual control by using a fuzzy logic controller on the robotic blimp to achieve autonomous waypoint tracking.

SECTION 6: VISUAL ATTENTION

García, Rodríguez, and Matellán make a review of some of the most representative visual attention models, which can be used for reducing the time to process images by a robot. Vega, Perdices, and Cañas propose a visual perceptive system for a robot with a mobile camera on board that copes with two challenges arising when using cameras: to extract useful information from captured images and to manage the small field of view of regular cameras. The chapter by Antúnez, Haxhimusa, Marfil, Kropatsch, and A. Bandera proposes a visual attention model using a hierarchical grouping process that encodes the input image into a Combinatorial Pyramid.

José García-Rodríguez
University of Alicante, Spain

Miguel Cazorla
University of Alicante, Spain

Acknowledgment

The authors want to acknowledge the valuable help of reviewers and colleagues that helped to evaluate the project.

The work has been also supported by grant DPI2009-07144 from Ministerio de Ciencia e Innovacion of the Spanish Government, by the University of Alicante projects GRE09-16 and GRE10-35, and Valencian Government project GV/2011/034.

Section 1
Computer Vision

Chapter 1
Face Recognition with Active Appearance Model (AAM)

Patrycia Barros de Lima Klavdianos
Université de Bourgogne, France

Lourdes Mattos Brasil
Universidade de Brasília (UnB), Brazil

Jairo Simão Santana Melo
Universidade de Brasília (UnB), Brazil

ABSTRACT

Recognition of human faces has been a fascinating subject in research field for many years. It is considered a multidisciplinary field because it includes understanding different domains such as psychology, neuroscience, computer vision, artificial intelligence, mathematics, and many others. Human face perception is intriguing and draws our attention because we accomplish the task so well that we hope to one day witness a machine performing the same task in a similar or better way. This chapter aims to provide a systematic and practical approach regarding to one of the most current techniques applied on face recognition, known as AAM (Active Appearance Model). AAM method is addressed considering 2D face processing only. This chapter doesn't cover the entire theme, but offers to the reader the necessary tools to construct a consistent and productive pathway toward this involving subject.

INTRODUCTION

Computer Vision is a relatively new research field since it was not before the late 70's that it was widely recognized by the scientific community. This was due mainly to the lack of proper computational resources to process images and data

on an acceptable rate. Most of these limitations have been overcome and now is possible to take advantage of complex algorithms running on fast microprocessors in order to make machines "see" what it was only seen by humans.

Actually, nowadays there are machines able to see beyond of what humans are capable of. The

DOI: 10.4018/978-1-4666-2672-0.ch001

hardware used for image acquisition has taken a real technological and innovative leap for the past 20 years. As a result of this, we can easily add smart cameras, depth cameras, IR (infrared) cameras and many other types of input devices and modern sensors in our projects. The new challenge, however, is to make the machines not only see, but also understand and interpret the scene.

New methods for image processing and object recognition are emerging every year. One of the most prominent areas of research in object recognition and pattern recognition is, undoubtedly, face processing. According to Zhao and Chellappa (2003, 2006) this is due to the high demand for commercial and law-enforcement application associated, of course, with the current status of our technological resources to attend such requests. One can say that several years of improvements in hardware and software have contributed positively for the growth in this research field. In addition to that is essential to acknowledge all the effort and endeavor of many researchers working in the areas of psychology, neuroscience, computer vision, artificial intelligence, mathematics and many others.

Another aspect which contributed for the high interest in face recognition is mentioned by Zhao and Chellappa (2003) and is related to the nature of method itself. Face recognition is different from other human identification methods such as biometrics because it doesn't need the cooperation of the person being detected or identified. Therefore, you can think of face recognition process as a passive method which relies only on the camera acquisition and the processing mechanism in order to generate results. This is also a controversial theme since people may start questioning about privacy policy. Does the Government and the so-called Special Agencies have the right on spying on you?

To answer this and other questions related to face recognition there are many conferences and workshops around the world. It is worth to mention the followings: IEEE Conference on Face and Gesture Recognition, International Conference on Pattern Recognition (ICPR), Structural and Syntactic Pattern Recognition (SSPR), International Conference on Image Analysis and Recognition (ICIAR) and International Conference on Machine Learning and Data Mining.

As you may notice recognizing a human face is not a trivial task and demands plenty of work from several experts. Nowadays, for instance, computer vision, artificial intelligence and pattern recognition practitioners work side by side to create models and improve techniques for the purpose of recognizing human faces in many situations. Researchers from physics have been collaborated in full extension with technological breakthrough in the production of sensors, cameras and different types of vision systems for robotics.

Besides, since the recent spate of terrorist attacks such as September 11 in 2001 and London bombings in 2005, face recognition and facial reconstruction have won a spot on the list of priority systems necessary to promote social well-fare. Despite of security systems and surveillance, a typical list of applications for face recognition are: games, human-robot interaction, photo management, smart cards and psychological studies of human behavior (expressions, mood and appearance).

The next sections explore the field of face recognition by focusing on the AAM technique. First we present the workflow of face recognition by considering it as a problem in the pattern recognition domain. Subsequently, AAM method is described in detail including key points regarding to its implementation in Matlab. Then, a small guide toward open field subjects, usage of tools and frameworks are provided in order to capture the interest of future researchers. Finally, we conclude the work by presenting our opinion and summarizing the topics discussed along this chapter.

UNDERSTANDING FACE RECOGNITION PROCESS

Face Processing as a Pattern Recognition Problem

The first step on face recognition is to understand the problem statement so that goals and problem boundaries are well identified. With this in mind, it is important to realize that face recognition belongs to a class of problems in the pattern recognition field. Therefore, as such types of problems its main goal is to recognize or, in a more formal way, to classify an object according to certain rules and constraints. In the case of face recognition the main objective, of course, is to classify human faces.

Another important aspect in face processing is to distinguish two types of procedures: (a) the verification of a given face and (b) the identification of a given face. In the former case, the main goal of the system is to confirm or deny a claimed identity. For the latter case, the system aims to identify a given input through a comparison process of known individuals taken from a database (Zhao & Chellappa, 2006).

The verification case is a simplified version of face recognition problem and, thus, the processing time required for the task is smaller as well. The database used for computation is also different. This problem requires only a database containing a set of variations for those faces which will be recognized. The identification problem, on the other hand, demands that a much larger database is used so that the system can recognize a great number of people.

Although both processes are different in many aspects, they present similarities in terms of technique and because of that, they can be solved the same way as another problem in pattern recognition domain. Consequently, understanding the two most important concepts related to pattern recognition are of fundamental importance. These concepts are: features and classifiers.

Features are measurements taken from an object and used to infer unique characteristics or properties so that this object can be easily distinguished from other types of objects. When a set of features are organized together they comprise a "feature vector" (Theodoridis & Koutroumbas, 2009). In mathematical terms you can think of all feature vectors laying down on a space which delimits the problem boundary. The problem space, thus, is referred as the object feature space because it contains all the unique characteristics for the object in study.

Classifiers are processes constructed under certain rules and constraints with the objective of dividing the object feature space into regions (Theodoridis & Koutroumbas, 2009). These multiple regions represent all classes which the classifier is able to identify. As a result, a good classifier for face identification, for instance, creates an abstract decision line in the feature space which separates those faces resembled to the model being identified from other models in the database of faces.

Another way of understanding these concepts in face recognition domain is to consider features as some special locations which uniquely describe a human face. The same way, you can comprehend a classifier as the technique applied to solve the problem of recognizing a given face. Good locations used to identify human faces are the positions of eyes, nose, ears and mouth. Regarding to classifiers, there are many techniques available nowadays and this chapter addresses AAM which is based on the shape and appearance of human faces. More information about AAM operation is provided in subsequent sections.

Now that features and classifiers were described, a general pattern recognition problem can be formalized in a five stage process shown in Figure 1.

The first stage involves capturing the object through a sensor, for instance, a camera or video device. The second and third stages correspond

Figure 1. A generalized pattern recognition workflow (adapted from Theodoridis & Koutroumbas, 2009)

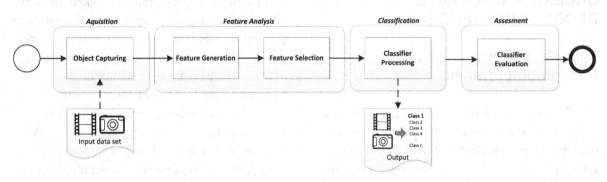

to feature generation and feature selection, respectively. Feature generation performs an automatic or manual identification of all features which uniquely describe the object being recognized. The feature selection process performs a reduction on the number of features. This step is needed because the computational resources available are limited and a better approach is to consider only those features which contribute for solving the problem. Therefore, the main goal of feature selection stage is to discard the redundant and useless features. After feature selection, the next step implies the classifier processing which effectively solve the problem by recognizing the object. In Figure 1 the additional stage called "Classifier Evaluation" is responsible for computing the accuracy level of the whole process. This is an important stage in all pattern recognition problems because a classifier needs to prove being efficient in solving a problem. A good classifier solves the problem for 80% of the cases and one can use this estimate to identify when the classifier design should be improved or not (Theodoridis & Koutroumbas, 2009).

Face Recognition Process

Being the face recognition process a special case of pattern recognition, its workflow can be derived from the generalized workflow shown in Figure 1. Figure 2 shows up this extended process.

Three general phases are distinguished in face recognition: training, testing and evaluation. In training phase a database of uniquely detected features representing human faces is created. One can think about training as a preparation phase for face processing. Testing phase, on the other hand, is when the face verification or face identification tasks take place. Finally, the evaluation phase measures the performance of the entire system by calculating metrics which verify among other things the classifier accuracy.

The next sections provide more details about each one of these phases. It is important to mention, however, that our reference to training, testing and evaluation as phases in face recognition attends uniquely to the purpose of organizing ideas and may not be part of a common vocabulary in this field.

First Phase: Training

The training phase starts by receiving a set of faces recorded in image or video format. The input data set may contain labels which describe the person's identity and a feature vector representing uniquely each face.

In cases where labels are missing, the system provides the functionalities for detecting faces, annotating them and generating the feature vectors. For those cases where labels are present the system provides a validation algorithm so that

Figure 2. Face recognition workflow

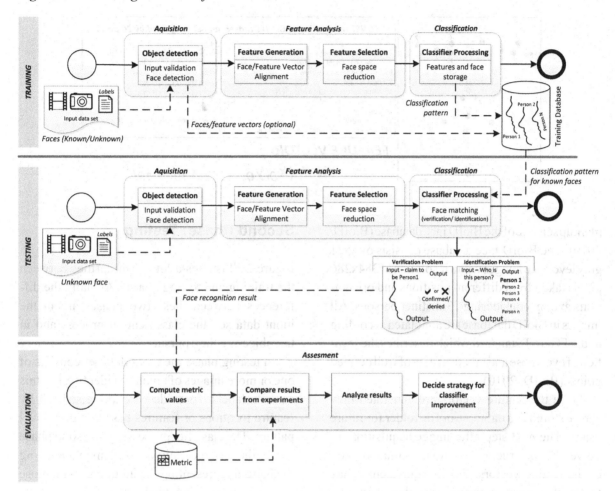

feature vectors can be verified. In face recognition, a feature vector represents the face shape or appearance. When a given face is represented by its shape the feature vector is comprised by locations of eyes, nose and mouth. When the representation is for face appearance, a set of grey level pattern defines the feature vector. There are many ways to construct a feature vector for shape representation. For instance, one can use boundary models (points, curves and surfaces), interior models (medial and solid meshes) and transformation models based on splines curves (Bookstein, 1997). Considering these possibilities for input, a well-designed system must provide a validation process so that the compliance to one annotation standard can be confirmed.

When we consider face recognition in two dimension the most common model to represent a feature vector based on shape is the point approach also known as landmarks. Good landmarks models contain well defined corners, 'T' junctions points and are generally based on biological studies (Cootes, 2004). Additional points are placed between corners and 'T' junctions so that a final shape model is represented by a continuous and equally spaced set of marks (Cootes, 2004).

One example of landmarks is shown in Figure 3 which also provides a representation of the correspondent feature vector.

Other examples of landmarks models which consider 22 points and 68 points are provided by the University of Manchester under the FGNET

Figure 3. Landmarks and feature vector representation

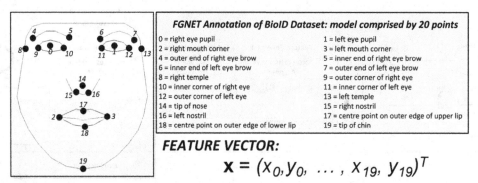

FGNET Annotation of BioID Dataset: model comprised by 20 points

0 = right eye pupil	1 = left eye pupil
2 = right mouth corner	3 = left mouth corner
4 = outer end of right eye brow	5 = inner end of right eye brow
6 = inner end of left eye brow	7 = outer end of left eye brow
8 = right temple	9 = outer corner of right eye
10 = inner corner of right eye	11 = inner corner of left eye
12 = outer corner of left eye	13 = left temple
14 = tip of nose	15 = right nostril
16 = left nostril	17 = centre point on outer edge of upper lip
18 = centre point on outer edge of lower lip	19 = tip of chin

FEATURE VECTOR:

$$\mathbf{x} = (x_0, y_0, \ldots, x_{19}, y_{19})^T$$

markup scheme of the BioID face database (BioID, 2010). The BioID face database consists of 1521 gray level face images with a resolution of 384x286 pixels taken from different positions and orientations and by considering 23 distinct persons. All images in BioID database are annotated according to the FGNET standard which describes the location of eyes, nose and mouth as a set of well defined points (BioID, 2010).

Once the feature vectors are generated they can be stored in a database or file folder for future usage. The next step after image acquisition involves the construction of a more accurate version of the feature vectors. This is equivalent to say that each feature vector must be alignment with one another so that differences in scale, position and rotation in the data set are corrected. Every face recognition process should take into account that the acquisition process may result in images containing slightly differences or pose variances.

Subsequently to the feature alignment, the feature vector space or simply the face space should be reduced so that only the most important features are considered by the classifier. This process is known as feature selection, the same feature selection mentioned previously for generalized pattern recognition workflow. The final process is the classifier operation which, in case of training stage, generates as a result a pattern to be used in testing stage for localizing and recognizing the given face.

Second Phase: Testing

Figure 2 illustrates a lot of similarities between the training and testing phases. Actually the differences between these two phases are in the input data set, the classification process and in the objective they pursue.

In testing phase the input data set consists of one or more images of unknown identity. For this case, the labels, if provided, are comprised only by feature locations or features based on grey level pattern. The classifier processing in testing phase solves the face recognition problem by matching the pattern representing the input face with those stored in the training database.

As you can see the main goal associated to the training and testing phases are quite different. While the former aims to construct a database of classification patterns, the latter intends to use this database to perform face verification or face identification tasks.

One last observation about training and testing phases is that sometimes they are implemented in such a way that is almost impossible to distinguish one from the other. Therefore, remember that Figure 2 gives you a didactic and comprehensive way of thinking about face recognition as a whole. In practice, however, the proposed workflow can varies depending on the system architecture, resources and algorithms applied to perform each task.

Third Phase: Evaluation

Regarding to evaluation phase, it can be implemented in different ways and by following a variety number of methodologies. Due to the extension of this subject this chapter will not describe this phase. Therefore, we provided only a brief idea about this process on Figure 2 for a matter of completeness.

Besides, regarding to evaluation stage just keep in mind that the main idea is to compute some metric so that the system performance can be verified. In face recognition field, though, when we mention the word "performance" it means that we must measure the efficiency of the system considering three aspects: the rate of success, the rate of errors and the response time under a user point of view.

CLASSIFIER BASED ON ACTIVE APPEARANCE MODEL

Models for Face Recognition

The literature survey of Zhao, Chellappa and Rosenfeld (2003) depict very well all stages of evolution in face recognition.

The earliest techniques were based on deformable template models known as Active Contour Model or simply Snakes. This technique was first proposed by Kass, Witkin and Terzopoulos in 1988 and gained respect along the years. Actually, this method is still being used for the purpose of segmentation and object recognition. The basic idea of snakes is to represent an object as a set of outline landmarks upon which a correlation structure is forced to constrain local shape changes (Kass, Witkin & Terzopoulos, 1988). Changing a set of parameters through an optimization process the shape of the object is modified until it fits a desirable and localized form.

The next stage consisted on bringing some prior knowledge to the deformable template method, for instance, the shape, size and color of the object being recognized. Following this path, is worth mentioning the work of Cootes, T. F. et al. (1995) whom introduce the concept of Active Shape Models (ASM). Their work followed the Procrustes Analysis and Principal Component Analysis, two well-known techniques at that time.

Subsequently, Cootes, Edwards and Taylor (1998) published an extension for ASM which they called AAM (Active Appearance Model). While ASM considered only the shape of a human face, AAM takes into account also the appearance of the face. Many other publications followed these two initial papers which qualified these techniques known as "statistical models" as promising methods for object recognition field.

Other quite similar methods appeared in the same period. For instance, Sclaroff and Isidoro (1998) proposed the Active Blob approach which consisted of a real-time technique for detecting shape information from a prototype image.

In recent years, the model-based approach towards image interpretation has proven very successful. This is especially true in the case of images containing objects with large variability Cootes (2004). Therefore, understanding and capturing the variances of objects have become a true challenge.

Statistical Shape Model (SSM)

As stated previously several techniques for face modeling have been developed along the years with the purpose of representing faces in a computational way so that variances can be studied, simulated and recognized.

In order to detect the many variations found in human faces we first need to understand and represent their shapes. A good definition for shapes is provided by Dryden & Mardia (1998) whom states that "Shape is all the geometrical information that remains when location, scale and rotational effects are filtered out from an object" (p.1).

Another way of understanding shapes is to think about an object which is invariant to Euclidean transformation in such a way that changes in position, size and orientation (these parameters are referred as "POSE") doesn't modify the form of the object. Other point of interest in shapes involves its representation which as stated in previous section, can be done by using the landmarks technique and the feature vectors scheme.

Now that we know the concept of shapes and how to represent them, let's understand the main goal of SSM. Statistical models are based on a study performed upon a training data set. Images patterns stored in the training database give clues not only about the shape of a given face but also how this shape varies in different situations. The training phase in this case consists of applying SSM to compute and store these many patterns representing the shape and its variations Cootes (2004).

Therefore, SSM can be understood as a technique which takes a set of face images represented by its feature vectors and gives, as a result, statistical models representing the shapes changes in many situations. It is also important to keep in mind that statistical models generated through SSM should be generic and compact enough in order to represent all possible variations but in a way that further computation can be done without compromising system performance.

According to the workflow described in Figure 2, training phase starts with the object detection which can be performed manually, on a semi-automatic way or even through an entirely automatic process. This chapter doesn't go further in the techniques to automate this process because we understand that this is a subject related to the field of object detection and tracking which is beyond the scope of this work. If you are an avid researcher for such topic we recommend reading the work of Kim, M. et al. (2008) which gives insights regarding the problematic of detecting

and tracking human faces in video sequences. Another reference is the paper of Yan, T. et al. (2009) which describes a method for finding landmarks automatically in human faces.

The work presented in this chapter assumes that we already have a set of points for each input data set representing a face. For a complete reference of databases containing annotated faces read the section "Applications and Tools" further on this chapter.

Since we already have a set of annotated faces, the next step consists of generating a consolidated version of the feature vectors. By "consolidated version" we mean that all feature vectors representing the input data set should be aligned properly so that one unique coordinate reference system is established for all the shapes we have. The alignment is needed because images are taken from different POSE and not correcting these differences leads to models which don't represent the face and its variances (Cootes, 2004).

A classical solution for the alignment problem is based on the Procrustes Analysis which aims to minimize the sum of distances from each shape to a mean shape during an iterative process (Cootes, 2004). In order words, Procrustes Analysis results in shapes with identical centre of gravity and approximately the same scale and orientation. A detailed explanation of Procrustes Analysis can be found in Bookstein (1997), Cootes (2004), Dryden and Mardia (1998) and Goodall (1991).

One common misunderstanding regarding Procrustes Analysis is that it comes in two versions, the classical method which performs the alignment between two objects and the so-called "Generalized Procrustes Analysis" or GPA which provides an alignment among many objects on an iterative and complete process (Bookstein, 1997, Cootes, 2004). Let's understand the simpler case and, then, move on to the generalized version. The algorithm for performing the classical version of Procrustes Analysis is given in Table 1.

The first step in Procrustes Analysis is to remove the translation from each shape. This can

Table 1. Procrustes analysis

1. Compute the centroid for each shape.
2. Remove the translation from each shape.
3. Remove the scale from each shape.
4. Remove the rotation from each shape.

be done simply by translating the shapes to its centre of mass. Then, for n feature vectors (or shapes) comprised by k landmark locations (x, y) we can define Equation (1) for translation removal:

$$\hat{x}_i^{\,j} = x_i^{\,j} - \overline{X}^{\,j}$$
$$\scriptstyle j=1...n \atop \scriptstyle i=1...k$$

and

$$\hat{y}_i^{\,j} = y_i^{\,j} - \overline{Y}^{\,j} \qquad (1)$$
$$\scriptstyle j=1...n \atop \scriptstyle i=1...k$$

From Equation (1), $\overline{X}^{\,j}$ and $\overline{Y}^{\,j}$ are the centroid computed for each shape according to Equation (2).

$$\overline{X}^{\,j} = \frac{1}{n}\sum_{i=1}^{k} x_i^{\,j}$$
$$\scriptstyle j=1...n$$

and

$$\overline{Y}^{\,j} = \frac{1}{n}\sum_{i=1}^{k} y_i^{\,j} \qquad (2)$$
$$\scriptstyle j=1...n$$

After removing translation, the next step in Procrustes Analysis includes re-scaling each shape so that they have equal size. This can be done by applying any scale metric methods such as the

Root-Mean-Square Distance (RMSD) (Dryden & Mardia, 1998). Equation (3) describes the formula to apply this method. The main goal at this point is to find a scale factor S^j for each shape so that their sizes can be changed to be compliant to a unit scale.

$$S^{j} = \sqrt{\frac{1}{k}\sum_{i=1}^{k}\left(\hat{x}_i^{\,j} - \overline{X}^{\,j}\right)^2 + \left(\hat{y}_i^{\,j} - \overline{Y}^{\,j}\right)^2}$$
$$\scriptstyle j=1...n$$
$$(3)$$

Now that the scale factors were found, all the shapes can be re-scaled according to Equation (4).

$$\tilde{x}_i^{\,j} = \frac{\hat{x}_i^{\,j} - \overline{X}^{\,j}}{S}$$
$$\scriptstyle j=1...n \atop \scriptstyle i=1...k$$

and

$$\tilde{y}_i^{\,j} = \frac{\hat{y}_i^{\,j} - \overline{Y}^{\,j}}{S} \qquad (4)$$
$$\scriptstyle j=1...n \atop \scriptstyle i=1...k$$

Up to this point we have the shapes aligned with respect to translation and scale. The last step in Procrustes Analysis is to perform the alignment considering the differences in rotation among the shapes. There are many methods to do this. In our experiments we've applied the Singular Value Decomposition (SVD) method so that the system of equations representing the rotation problem can be solved.

To understand this let's consider the rotation alignment by taken two shapes only. After aligning these two shapes according to translation and scale we end up with two feature matrices Z_1 and Z_2 which represent both shapes. Equation (5) formalizes the construction of the feature matrix for these two shapes.

$$Z_1 \underset{i=1...k}{=} \left[\tilde{x}_i^1 \mid \tilde{y}_i^1 \right]_{kx2}$$

and

$$Z_2 \underset{i=1...k}{=} \left[\tilde{x}_i^2 \mid \tilde{y}_i^2 \right]_{kx2} \qquad (5)$$

From Equation (5), Bookstein (1997) suggests applying SVD on the system of equation represented by $\left[Z_2 \quad Z_1 \right]$ which results in a rotation matrix $R = UV^T$ where the columns of U and V represent a set of orthonormal vectors also known as basis vectors. Through the decomposition of the matrix $\left[Z_2 \quad Z_1 \right]$ one can find the rotation matrix R and then apply it to Z_2 in order align this shape to Z_1. Equations (6) and (7) formalize this final step of Procrustes Analysis.

$$svd(shape_j \mid shape_{ref})_{j=1..n} = U_j S_j V_j^T \qquad (6)$$

$$R_{j/ref} \underset{j=1...n}{=} U_j V_j^T \qquad (7)$$

Bookstein (1997), Cootes (2004), Dryden and Mardia (1998) and Goodall (1991) suggest many other methods to perform shape alignment. Cootes (2004), in special, suggest some interesting optimization methods from which the scale and rotation alignment are performed at once, in a single algorithm.

The Procrustes Analysis explained until now only describes the general operation of the algorithm, that is, it doesn't take into account the alignment of several shapes as for the case of our training data set. Therefore, we need to apply the concepts learnt so far using the Generalized

Procrustes Analysis mentioned previously. Table 2 fully describes this method.

The main idea behind GPA is to perform the alignment of all shapes compared to a reference shape so that the process is repeated as many times as needed. During the iteration process the reference shape is updated as well as the transformation matrices used for aligning each shape. According to experiments performed by Bookstein (1997) the process should converge after two iterations which make it suitable to restriction in terms of performance. After applying GPA it is interesting to plot the results of all aligned faces so that a visual inspection can be made. The plot() function in Matlab can help you on this task.

From Procrustes Analysis we move on to the next step in the face workflow for the training phase which corresponds to feature selection. Feature selection involves the analysis of the feature vectors or feature matrices regarding to three aspects: relevance, redundancy and noise. What we want performing a feature selection process is to maximize the relevance and minimize the redundancy and noise (Yan et al., 2009). Feature selection can be performed in many different ways, for instance, by visual inspection or applying some automatic technique such as filtering, wrapper functions, embedded methods and feature combination methods.

For the purpose of SSM, feature selection comprises basically of a reduction in the face space so that only those features which really represent the basis for our training data set are effectively selected. There are many mathematical techniques for space reduction; however, the classical version of SSM relies on the PCA to perform this task.

Table 2. Generalized Procrustes analysis

1. Choose a reference shape as an estimate of the mean shape.
2. Align the other shapes according to the reference shape.
3. Calculate the mean shape from the aligned shapes.
4. Go to step 2 if the mean shape has changed.

PCA is a linear technique which aims to find the highest variations on a given data set. The main idea of PCA is to project the data into a new reference system so that only the components representing the highest variances in the sample are kept while the others are discarded. The components kept from the sample data set are called "Principal Components" because they represent, in fact, the uncorrelated version of the data (Jolliffe, 2010). Table 3 describes the step-by-step procedure necessary to implement PCA algorithm.

PCA method starts from the mean shape calculated according Equation (8). The mean shape is derived from the cloud of aligned shapes computed during the Procrustes Analysis. Therefore, X_j in Equation 8 represents each aligned shape in a matrix notation. Keep in mind that the coordinates x and y for each shape are represented on the matrix as columns vectors. This is the most common arrangement when implementing such method.

$$\overline{X} = \frac{1}{n} \sum_{j=1}^{n} X_j \tag{8}$$

PCA algorithm uses the mean shape to compute the covariance of the data as described in Equation (9).

$$Cov = \frac{1}{n-1} \sum_{j=1}^{n} \left(X_j - \overline{X} \right) \left(X_j - \overline{X} \right)^T \tag{9}$$

Table 3. Principal component analysis

1. Compute the mean shape from the aligned set of shapes.
2. Compute the covariance matrix.
3. Compute the eigenvectors and eigenvalues from the covariance matrix.
4. Sort the eigenvectors and eigenvalues and extract the highest ones.

Since the covariance matrix contains the statistic measurement of how much two variables change one in relation to the other, the computation time for such method is a matter of concern. Therefore, a well known trick to compute efficiently the covariance matrix is to subtract the mean shape from each aligned shape (Equation 10) and then, apply the formula shown in Equation (11). Notice that the parameter X in Equation 10 is a matrix built by all aligned shapes after the subtraction from the mean shape. A usual way to create matrix X when implementing PCA is to consider its columns as representing each aligned shape and half of the rows as the x coordinates and the other half as the y coordinates of the landmarks. One problem faced during implementation is that depending on the number of shapes you are using, the total number of columns may be larger than the number of rows. If this is the case, Equation (11.a) should be applied; otherwise Equation (11.b) is more appropriated.

$$X_j = X_j - \overline{X}$$
$$\scriptstyle j=1...n$$

and

$$X = \left[X_1, ..., X_j \right] \tag{10}$$

$$Cov = XX^T \text{ (a)}$$

and

$$Cov = X^T X \text{ (b)} \tag{11}$$

From the covariance matrix, the eigenvectors and their eigenvalues can be calculated. In Matlab the function eig() do this work for you. The first result value from eig() function is the eigenvectors while the second value represents the eigenvalues (Equation 12).

$$\Psi = \left\{\psi_1, ..., \psi_n\right\} \text{ (eigenvectors)}$$

and

$$\Lambda = \left\{\lambda_1, ..., \lambda_n\right\} \text{ (eigenvalues)} \tag{12}$$

Another step needed if you are using Equation (11.b) to compute the covariance is to adjust the eigenvectors values which can be done by applying Equation (13).

$$\psi_i = \frac{1}{\sqrt{\lambda_i}} \psi_i \tag{13}$$
$$\scriptstyle i=1...n$$

The next step on PCA is to sort the eigenvectors and its eigenvalues in descendent order so that only the highest values are kept and the space dimension is finally reduced. In Matlab the function sort() is in hand to arrange the eigenvectors and eigenvalues in descendent order. Regarding to the number of eigenvectors and eigenvalues which should be kept, Equation (14) can be used to compute the new dimension size k of our data. The constant c_p indicates the proportion of the total variation which should be kept. Usually this number is around 0.98.

$$\sum_{i=1}^{k} \lambda_i = c_p \sum_{i=1}^{n} \lambda_i \tag{14}$$

The eigenvectors and eigenvalues representing the highest variations of the training data set is used, then, to construct statistical shape models which simulates many variations applied to the mean shape. Equation (15) illustrates the formula needed to project the new shape model parameter b into the face space. Another way to understand this is to think that new faces can be created by varying the shape model parameter b since shapes in SSM are linear combinations of eigenvectors.

$$SSM_i = \overline{X} + \psi_i b_i$$
$$\scriptstyle i=1...l$$

and

$$b_i = \pm c_v \sqrt{\lambda_i} \tag{15}$$
$$\scriptstyle i=1..l$$

In Equation (15) the parameters $\overline{X}$, ψ_i and b_i are respectively the mean shape, some selected eigenvectors and the shape model parameter which is based on eigenvalues and a constant value c_v representing a variation for the mean face. Note from Equation (15) that not all the highest eigenvectors and eigenvalues are used. Generally a common approach is to use three to six components. Figure 4 gives you an example of how to compute shapes by using different values for b_i.

Statistical Models of Appearance

Previous section depicted how to construct realistic synthetic faces from a training data set through SSM process. SSM is necessary because we need to express the face to be recognized in many different ways or variances so that we avoid the system being deceived when changes occur to the images. This process of finding shape models is also known as synthesis. Although synthesis is an important step on face recognition, it isn't sufficient to solve the entire problem, mainly because it only gives information about face structure. No information regarding to the appearance is gathered from that which means to say that no pixel information inside the shape structure is considered. As a conclusion, to improve the results of face recognition is indispensable to construct statistical models of appearance by collecting texture information between landmarks in a mean shape structure (Cootes, 2004).

Figure 4. Configurations for SSM computation

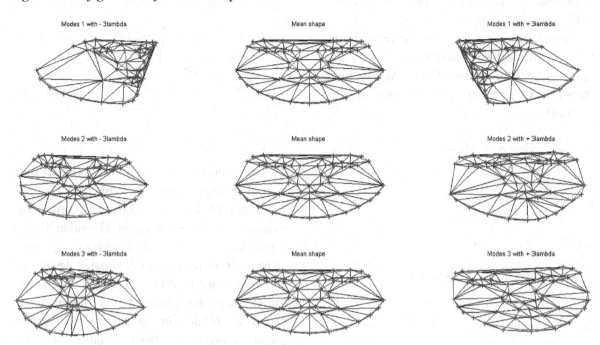

The main idea behind appearance models is that there is a correlation between variations on shape and texture models. Therefore, an appearance model is just a combination of the mean shape model with texture information. Cootes (2004) defines texture "as a pattern of intensities or colors across the image patch". He also states that the appearance model should be obtained from the same training set of labeled images which generated the mean shape model in SSM process.

The first step in appearance models is to build texture models from the mean shape models. This can be accomplished by a method known as "image warping" proposed by Lanitis, Taylor and Cootes (1995). Warping each training image to the mean shape will result in free patches images (Figure 6). The texture models, thus, can be built based on the variations of these free patches.

Another observation about warping technique is that one can be tempted to get the texture models directly from eigenvector decomposition, a well known solution. However, Cootes (2004) affirms that warping each image to its mean shape will

remove spurious texture variations and this aspect increases the accuracy of the face representation. Consequently, the accuracy of a recognition process which uses this method is also increased.

Warping technique involves, thus, a mapping from the landmarks of our training data set to the landmarks of the mean shape computed during SSM process. The direct correspondence between landmarks is not sufficient to create an efficient warping function for building texture models. This fact is true simply because what is needed is the mapping of a 2D surface represented by the inside pixels of the landmark contour and not a simple one-to-one correspondence between points. Then, another important step is to generate meshes among landmarks, for both training set and the mean shape structure. A well recommended way to do this is to apply Delaunay triangulation which in Matlab can be solved by delaunay() function.

After Delaunay triangulation the warping procedure can be applied by mapping each pixel of each triangular mesh from the training set to each pixel of each triangular mesh on the mean shape structure. The schematic procedure is shown in Figure 5.

From Figure 5 we can construct relationships between the mesh surface in training set *I(x)* and mesh surface in mean shape *I'(x')*. These relationships are expressed through Equations (16), (17) and (18).

$$x = a + \beta(b - a) + \gamma(c - a)$$
$$x = (1 - \beta - \gamma)a + \beta b + \gamma c$$
$$\boxed{x = \alpha a + \beta b + \gamma c}$$

and

$$x' = a' + \beta(b' - a') + \gamma(c' - a')$$
$$x' = (1 - \beta - \gamma)a' + \beta b' + \gamma c' \qquad (16)$$
$$\boxed{x' = \alpha a' + \beta b' + \gamma c'}$$

$$\alpha + \beta + \gamma = 1$$

and x is inside the triangular mesh if

$$0 \le \alpha, \beta, \gamma \le 1 \qquad (17)$$

Warping function is given by

$$f(x) = x' = \alpha x_1' + \beta x_2' + \gamma x_3' \qquad (18)$$

Equations (19), (20) and (21) describe the formulas to compute α, β and γ, respectively.

$$\alpha = 1 - (\beta + \gamma) \qquad (19)$$

$$\beta = \frac{yx_3 - x_1y - x_3y_1 - y_3x + x_1y_3 + xy_1}{-x_2y_3 + x_2y_1 + x_1y_3 + x_3y_2 - x_3y_1 - x_1y_2} \qquad (20)$$

$$\gamma = \frac{xy_2 - xy_1 - x_1y_2 - x_2y + x_2y_1 + x_1y}{-x_2y_3 + x_2y_1 + x_1y_3 + x_3y_2 - x_3y_1 - x_1y_2} \qquad (21)$$

Figure 5. Warping function scheme

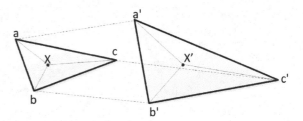

The complete algorithm for implementing warping method is given in Table 4.

From Table 4, another problem to be overcome when implementing warping algorithm is that Equation (18) doesn't guarantee the mapping for all pixels in the result free image patch. Therefore, an interpolation algorithm is needed to complete the texture model. The most used solution nowadays is the bilinear interpolation method which predicts a pixel value from its four neighboring pixels. Another alternative method is to use a second order approximation using bicubic interpolation considering 16 neighboring pixels. Our experience has shown that interpolation mechanisms generally demand high processing time and the choice of a good method signifies an increase on the overall system performance.

Another important aspect to be considered after applying warping algorithm and building the texture vector is to consider global changes in illumination. If we consider that these changes are linear, we can solve the problem by applying a scale and offset value to adjust the texture samples (Cootes, Edwards & Taylor, 1998). Equation (22) explains the normalization of the texture samples. In Equation (22) $g_{texture}$ represents the texture sample in grey level after executing warping method, g_{norm} is its normalized version and β and α are the offset and scaling factors which can be computed, respectively, by taking the mean and the standard deviation of $g_{texture}$.

$$g_{norm} = (g_{texture} - \beta) / \alpha \qquad (22)$$

Table 4. Warping algorithm

Table 4. Warping algorithm

1. For each pixel *x* inside the shape model
2. Determine the triangle *t* that *x* belongs to.
3. Find the position of *x* inside the triangle *t*. 4. Find the position of *x* inside the triangle *t'*. 5. Define *I'(x')* as *I(f(x))* where *f(x)* is given by Equation (18). 6. End

After texture models normalization the final statistical appearance models can be generated. The idea behind appearance models is the same as for shape models. However, in this case PCA is applied upon the normalized version of the texture models so that eigenvectors and eigenvalues can be extracted. The eigenvectors and eigenvalues are, then, used to get the variation models for the face appearance.

PCA method starts from the mean texture model calculated according Equation (23). The mean texture model is derived from the normalized version of texture models.

$$\overline{g_{norm}} = \frac{1}{n} \sum_{j=1}^{n} g_{j_{norm}} \qquad (23)$$

The mean texture model is used afterwards to compute the covariance of the data as described in Equation (24).

$$Cov = \frac{1}{n-1} \sum_{j=1}^{n} \left(g_{j_{norm}} - \overline{g_{norm}} \right) \left(g_{j_{norm}} - \overline{g_{norm}} \right)^{T} \qquad (24)$$

The same trick used to compute the covariance matrix for SSM can be applied here so that the computation time is reduced. From the covariance matrix, the eigenvectors and their eigenvalues are calculated. As stated before, you can use the function eig() in Matlab. The first result value from eig() function is the eigenvectors while the second

value represents the eigenvalues (Equation 12). As explained before for SSM, depending on the method used in the covariance matrix computation, a adjustment on the eigenvectors are needed (Equation 13).

Up to this point, the eigenvectors and its eigenvalues are sorted in descendent order so that only the highest values which represent the highest variance are kept and the space dimension is finally reduced. Return to the Equation (14) for this step.

The eigenvectors and eigenvalues representing the highest variations of the training data set is used, then, to construct statistical appearance models which simulates many variations of the face in terms of texture. Equation (25) summarizes the formula needed to project the new appearance model parameter b_i into the face appearance space. Another way to understand this process is to think that new faces can be created by varying appearance model parameter b since appearances are just linear combinations of eigenvectors.

$$MA_i = \overline{g_{norm}} + \psi_i b_i$$
$$\scriptstyle i=1...l$$

and

$$b_i = \pm c_v \sqrt{\lambda_i} \qquad (25)$$
$$\scriptstyle i=1..l$$

In Equation (25) the parameters $\overline{g_{norm}}$, ψ_i and b_i are respectively the mean texture model, some selected eigenvectors and the appearance model parameter which is based on eigenvalues, and a constant c_v representing a variation for the mean appearance. Note from Equation (25) that not all the highest eigenvectors and eigenvalues are used. Generally a common approach is to use three to six components as in SSM.

Combined Model: Shape and Appearance

The two previous sections described methods to construct shape and appearance models from a set of training images. Since shape and appearance models generally don't change in a detached way, there may be some correlation between them. Therefore, it is necessary to combine both shape and appearance models in order to remove the correlations and obtain a more compact and unique face models. These new models are simply referred as Combined Appearance Models (Cootes, 2004).

The new appearance models are generated using PCA on the combined vector of shape and appearance models. Let's understand this process by considering b_s and b_g as the shape and appearance parameter vectors respectively. These values were calculated previously by Equation (15) and (25). Therefore, the integrated vector b is defined based on their combination (Equation 26).

$$b = \begin{bmatrix} W_s b_s \\ b_g \end{bmatrix} = \begin{bmatrix} W_s P_s^T (x - \bar{x}) \\ P_g^T (g - \bar{g}) \end{bmatrix} \tag{26}$$

In Equation (26), P_s^T and P_g^T are the shape and appearance eigenvectors. The parameters x and g are the shape and appearance data while $\bar{x}$ and $\bar{g}$ are the associated mean. The parameter referred as W_s is a diagonal matrix of weights for each shape parameters (Cootes, 2004). This matrix is required because b_s and b_g represent different parameters. While the appearance model is based on intensity variations, the shape model is in accordance to distance unit. In order to combine these two models, analysis of variations of b_s on sample g must be considered (Cootes, 2004). The value W_s can be obtained, then, by considering the element displacement of b_s with reference to

its optimum value in each training sample. The RMS metric can be used to measure the change in g per unit for each change in the shape parameters b_s. At the end, the mean change of each training set will give the appropriate W_s.

After constructing the integrated vector b, PCA is applied on it by preserving about 99% of the eigenvectors and their respective eigenvalues. Another way of understanding this method is to consider that the integrated b is comprised of two elements: P_c element which is the eigenvectors and c element which represents the final appearance parameter that control both variations in shape and texture (Equation 27).

$$b = P_c c$$

and

$$P_c = \begin{pmatrix} P_{cs} \\ P_{cg} \end{pmatrix} \tag{27}$$

Therefore, in order to obtain new appearance models which represent the possible variations in a human face, we can use the eigenvectors (P_c) while promoting variations on the values of c. Equations (28) and (29) formalize this process.

$$c_{i=1...n} = P_c^T (b_i - \bar{b}) \tag{28}$$

$$shape_{new} = \bar{x} + P_s W_s^{-1} P_{cs} c$$

and

$$appearance_{new} = \bar{g} + P_g P_{cg} c \tag{29}$$

This concludes the generation of a general model which combines shape (structure) and appearance (texture) representing a human face.

Active Appearance Model (AAM)

Until now, we've seen how to build the shape and appearance models from a set of training images. These two processes, although different in many ways, are performed during training phase. According to the face recognition workflow depicted in Figure 2, we still have to deal with the testing phase which comprises of a recognition process. This means that for a given unknown image face, the system must recognize it by executing a comparison or classification process. This phase is known as Active Appearance Model or simply AAM.

Therefore, the question now is how we can use the statistical appearance models constructed from a training set to interpret new images given as input by the user. The simplest approach is to use the iterative method of matching appearance models to the input image (Cootes, 2004). The idea is quite simply; appearance models are projected on the new image and the differences are calculated (Equation 30).

$$\delta M = M_{training} - M_{new} \qquad (30)$$

As you may notice the fitting problem is actually an optimization problem since what we want is to find a set of parameters that minimize the residual difference between the reference and the training models. Table 5 describe step-by-step the AMM algorithm for the iterative approach.

Other approaches used to solve AAM problem are the followings: multiresolution search, shape AAM based on the work of Sclaroff and Isidoro on Active Blobs, compositional approach and direct AAM.

APPLICATIONS AND TOOLS

Three academic centers are reference in the area of face recognition using AAM: the Robot Institute of Carnegie Mellon University, the Technical University of Denmark (DTU) and the Research Group in Imaging Sciences of Manchester University.

Carnegie Mellon University deserves attention because of the several works being done in face processing such as: face detection, face recognition, face expression analysis, techniques for face enhancement and 3D head motion recovery in real time.

The Technical University of Denmark (DTU) is recognized by the outstanding work of Mikkel B. Stegmann who developed AAM-API. This framework comprises of a set of functions implemented in C++ for face recognition based on AAM technique. Stegmann provides the source code and documentation of AAM library as well as another tool called AAM Explorer which allows the user to perform experimentations by changing the parameters of shape and appearance models through a very intuitive graphical interface. Additionally, Stegmann also makes available data sets

Table 5. AAM iterative algorithm

1. Execute SSM and Appearance Model to build the training database (shape and appearance models - $M_{training}$)
2. Compute the combined shape and appearance model for the new face $\left(M_{new} \right)$
3. Compute the residual (Equation 30)

containing images of the hand, face and medical images for testing AAM.

Talking about data set, it is very important to have different images to test your AAM application. Some of useful sources are the followings: Facial Recognition Technology (FERET), BioID database, XM2VTS, AR face database and Talking face video. The last four sources are related to the members of Research Group in Imaging Sciences of Manchester University from which Tim Cootes is the main reference in AAM.

From a more generic and commercial point of view in face recognition world, we would like to mention the work of face.com which offers a platform for detecting and recognizing faces. The same is said about the Luxand FaceSDK and Falcet SDK from Bayometric Corporation. We also have seen the emergence of face recognition in the consumer market through products such as Picasa from Google and iPhoto from Apple. Actually, people are expecting face recognition capabilities to be included on the next version of iPhone5.

FUTURE RESEARCH DIRECTIONS

With all the advances in face recognition there are many open research fields to be explored yet. Face recognition systems must improve in their capabilities, for instance. It is also needed to improve in the automatic task of detecting and recognizing faces. Nowadays, most of the systems require the user to perform some tiresome tasks in order to set up the system and keep the training database updated. Another issue that deserves major criticisms on the current applications concerns to the poor performance of the system and the low-quality images generated by the sensors, especially in portable devices.

Mass market applications will demand a lot of work on this field, mainly to attend mobile users, social networks and search engine services. However, in our opinion the largest market share

will still be concentrated on security applications. These, will still continue to motivate a lot of face recognition researches, particularly for people recognition in non-ideal situations meaning that the acquisition process is uncontrolled or poor in quality. In the same direction, another promising field is face recognition in video surveillance. This field has been of strong interest from the Government and from private companies and has also proven to be a very challenging and encouraging subject.

CONCLUSION

This chapter has focused on the AAM technique for face recognition. This technique is part of the so-called "Statistical Models" and aims to build synthetic models representing the shape and appearance of human faces. These models can be doted of a wide range of variation modes which constitute the possible changes regarding to expression, pose and illumination. Therefore, such models can be used by matching or classification systems to detect similarities among faces or, even, to confirm or detect the identity of a person.

This work presented in detail the process for building the shape and appearance models as well as the iterative method for applying AAM to recognize a given face. The problem of fitting the training models and the new images remains an open research field. Recently, advanced techniques using Support Vector Machines (SVN), Artificial Neural Networks and LDA (Linear Discriminant Analysis) have been applied in order to improve the recognition and the performance of the system. Despite the efforts, many questions remain unanswered such as those related to models correspondences, the better usage and organization of the training database, the size of the training input set, the automatic detection of landmarks, the mechanism for fitting the shape and appearance models, etc.

According to Cootes, Edwards and Taylor (1998), AAM is also recommended to tracking objects through images sequences or to locate structures in 3D datasets including MR (medical resonance) images.

In summary, AAM is an interesting and very versatile technique which has many aspects to be improved, but which has been proved to be efficient in real-world applications.

REFERENCES

Bio, I. D. GmbH. (2010). *The BioID face database*. Retrieved November 30, 2011, from http://support.bioid.com/downloads/facedb/index.php

Bookstein, F. L. (1997). Landmark methods for forms without landmarks: Localizing group differences in outline shape. *Medical Image Analysis*, *1*(3), 225–244. doi:10.1016/S1361-8415(97)85012-8

Cootes, T. F. (2004). *Statistical models of appearance for computer vision*. Online Technical Report. Retrieved December 3, 2011, from http://www.isbe.man.ac.uk/~bim/refs.html

Cootes, T. F. (1995). Active shape models - Their training and application. *Computer Vision and Image Understanding*, *61*(1), 38–59. doi:10.1006/cviu.1995.1004

Cootes, T. F., Edwards, G. J., & Taylor, C. J. (1998). Active appearance models. In H. Burkhardt & B. Neumann (Ed.), *European Conference on Computer Vision*, Vol. 2, (pp. 484-498). Springer.

Cootes, T. F., Edwards, G. J., & Taylor, C. J. (1998). A comparative evaluation of active appearance models algorithms. *British Machine Vision Conference*, Vol. 2, (pp. 680-689). BMVA Press.

Cootes, T. F., Edwards, G. J., & Taylor, C. J. (2001). Active appearance models. *IEEE Transactions on Pattern Analysis and Machine Intelligence*, *23*(6), 681–685. doi:10.1109/34.927467

Cootes, T. F., & Kittipanyangam, P. (2002). Comparing variations on the active appearance model algorithm. *British Machine Vision Conference*, Vol. 2, (pp. 837-846).

Dryden, I. L., & Mardia, K. V. (1998). *Statistical shape analysis*. London, UK: John Wiley & Sons.

Edwards, G. J., Taylor, C. J., & Cootes, T. F. (1998). Interpreting face images using active appearance models. *International Conference on Automatic Face and Gesture Recognition*, (pp. 300-305).

Goodall, C. (1991). Procrustes methods in the statistical analysis of shape. *Royal Statistical Society, Series B*, *53*, 285–339.

Jolliffe, I. T. (Ed.). (2010). *Principal component analysis*. New York, NY: Springer-Verlag.

Kass, M., Witkin, A., & Terzopoulos, D. (1988). Snakes: Active contour models. *International Journal of Computer Vision*, *8*(2), 321–331. doi:10.1007/BF00133570

Kim, M. (2008). Face tracking and recognition with visual constraints in real-world videos. *Computer Vision and Pattern Recognition, CVPR*, *2008*, 1–8.

Lanitis, A., Taylor, C. J., & Cootes, T. F. (1995). Automatic face identification system using flexible appearance models. *Image and Vision Computing*, *13*(5), 393–401. doi:10.1016/0262-8856(95)99726-H

Lanitis, A., Taylor, C. J., & Cootes, T. F. (1997). Automatic interpretation and coding of face images using flexible models. *IEEE Transactions on Pattern Analysis and Machine Intelligence, 19*(7), 742–756. doi:10.1109/34.598231

Matthews, I., & Baker, S. (2004). Active appearance models revisited. *International Journal of Computer Vision, 60*(2), 135-164. Retrieved December 1, 2011, from http://www.ri.cmu.edu/publication_view.html?pub_id=4601

Sclaroff, S., & Isidoro, J. (1998). Active blobs. *IEEE International Conference on Computer Vision*, (pp. 1146-1153).

Sclaroff, S., & Isidoro, J. (2003). Active blobs: Region-based, deformable appearance models. *Computer Vision and Image Understanding, 89*(2/3), 197–225. doi:10.1016/S1077-3142(03)00003-1

Theodoridis, S., & Koutroumbas, K. (Eds.). (2009). *Pattern recognition*. Burlington, MA: Academic Press, Elsevier.

Yan, T. (2009). Automatic facial landmark labeling with minimal supervision. *Computer Vision and Pattern Recognition, CVPR, 2009*, 2097–2104.

Zhao, W., & Chellappa, R. (2006). *Face processing: Advanced modeling and methods*. Burlington, MA: Academic Press, Elsevier.

Zhao, W., Chellappa, R., & Rosenfeld, A. (2003). Face recognition: A literature survey. *ACM Computing Surveys, 35*(4), 399–458. doi:10.1145/954339.954342

ADDITIONAL READING

Baker, S., & Matthews, I. (2001). Equivalence and efficiency of image alignment algorithms. *Computer Vision and Pattern Recognition, 1*, 1090–1097.

Baker, S., & Matthews, I. (2004). Lucas-Kanade 20 years on: A unifying framework part 1: The quantity approximated, the warp update rule, and the gradient descent approximation. *International Journal of Computer Vision, 56*(3).

Batur, A. U., & Hayes, M. H. (2003). A novel convergence scheme for active appearance models. *Computer Vision and Pattern Recognition, 1*, 359–366.

Benson, P., & Perrett, D. (1991). Synthesizing continuous-tone caricatures. *Image and Vision Computing, 9*, 123–129. doi:10.1016/0262-8856(91)90022-H

Bergen, J., et al. (1992). Hierarchical model-based motion estimation. *European Conference on Computer Vision*, (pp. 237-252).

Beveridge, J. R., et al. (2001). Givens: A nonparametric statistical comparison of principal component and linear discriminant subspaces for face recognition. *IEEE Conference on Computer Vision and Pattern Recognition*, (pp. 535-542).

Black, M. J., & Yacoob, Y. (1995). Recognizing facial expressions under rigid and non-rigid facial motions. *International Workshop on Automatic Face and Gesture Recognition 1995*, (pp. 12–17).

Bookstein, F. (1989). Principal warps: Thin-plate splines and the decomposition of deformations. *IEEE Transactions on Pattern Analysis and Machine Intelligence, 11*(6), 567–585. doi:10.1109/34.24792

Covell, M. (1996). Eigen-points: Control-point location using principal component analysis. In *International Workshop on Automatic Face and Gesture Recognition 1996,* (pp. 122–127).

Ezzat, T., & Poggio, T. (1996). Facial analysis and synthesis using image-based models. *International Workshop on Automatic Face and Gesture Recognition 1996,* (pp. 116–121).

Ginneken, B. (2002). Active shape model segmentation with optimal features. *IEEE-TMI, 21,* 924–933.

Hou, X. (2001). Direct appearance models. *Computer Vision and Pattern Recognition, 1,* 828–833.

Jones, M. J., & Poggio, T. (1998). Multidimensional morphable models: A framework for representing and matching object classes. *International Journal of Computer Vision, 2*(29), 107–131. doi:10.1023/A:1008074226832

Lucas, B., & Kanade, T. (1981). An iterative image registration technique with an application to stereo vision. *International Joint Conference on Artificial Intelligence,* (pp. 674-679).

Matas, K. J. J., & Kittler, J. (1997). *Fast face localisation and verification.* In British Machine Vision Conference 1997.

Nastar, C., Moghaddam, B., & Pentland, A. (1996). Generalized image matching: Statistical learning of physically-based deformations. *European Conference on Computer Vision,* Vol. 1, (pp. 589–598).

Romdhani, S., Gong, S., & Psarrou, A. (1999). A multi-view non-linear active shape model using kernel PCA. *British Machine Vision Conference,* Vol. 2, (pp. 483-492).

Stegmann, M. B., Ersboll, B. K., & Larsen, R. (2003). FAME - A flexible appearance modeling environment. *IEEE Transactions on Medical Imaging, 22*(10). doi:10.1109/TMI.2003.817780

Stegmann, M. B., & Larsen, R. (2003). Multiband modeling of appearance. *Image and Vision Computing, 21*(1), 66–67. doi:10.1016/S0262-8856(02)00126-9

Turk, M., & Pentland, A. (1991). Eigenfaces for recognition. *Journal of Cognitive Neuroscience, 3*(1), 71–86. doi:10.1162/jocn.1991.3.1.71

Viola, P. (1995). Alignment by maximization of mutual information. *International Conference on Computer Vision,* (pp. 16–23).

KEY TERMS AND DEFINITIONS

Active Appearance Model (AAM): Describes an algorithm that matches statiscal models of the object shape and appearance to a new image.

Active Shape Model (ASM): Describes an algorithm that matches statiscal models of the object shape to a new image by interactively promoting many deformations to it.

Classifier Process: Is a set of tasks constructed under certain rules and constraints which aims to classify an object or group of objects for verification and/or identification purposes.

Face Identification: Is a face recognition process which aims to recognize a give face through a comparison process of known individuals taken from a training database.

Face Processing: Is a generic term used to describe the processes associated to face recognition or face perception.

Face Verification: Is a face recognition process which aims to confirm or deny a claimed identity of a person or group of persons.

Features: Are measurements taken from an object and used to infer unique characteristics or properties of an object. A set of features comprise a "feature vector".

Pattern Recognition Process: Comprises a set of tasks which aims to recognize/classify an object or to predict a value given an input data set. In the context of this chapter, only problems related to the recognition/classification of faces will be considered.

Statistical Shape Model (SSM): Describes an algorithm that aims to building statistical shape models from objects represented, for instance, by landmark points.

Training Database: Is a storage unit comprised by feature vectors whose true class is known (labeled sets) and which can be used for classification purposes.

Chapter 2
Uniform Sampling of Rotations for Discrete and Continuous Learning of 2D Shape Models

Xavier Perez-Sala
Universitat Politècnica de Catalunya, Spain & Computer Vision Center of Barcelona, Spain

Laura Igual
Universitat de Barcelona, Spain & Computer Vision Center of Barcelona, Spain

Sergio Escalera
Universitat de Barcelona, Spain & Computer Vision Center of Barcelona, Spain

Cecilio Angulo
Universitat Politècnica de Catalunya, Spain

ABSTRACT

Different methodologies of uniform sampling over the rotation group, SO(3), for building unbiased 2D shape models from 3D objects are introduced and reviewed in this chapter. State-of-the-art non uniform sampling approaches are discussed, and uniform sampling methods using Euler angles and quaternions are introduced. Moreover, since presented work is oriented to model building applications, it is not limited to general discrete methods to obtain uniform 3D rotations, but also from a continuous point of view in the case of Procrustes Analysis.

INTRODUCTION

Two-dimensional shape models are able to change their shape according to a labeled training set. A shape is composed by a finite set of landmarks whose geometrical information remains unchanged when the shape suffers from rigid transformations. Common 2D shape models (e.g.

Point Distribution Models, Active Shape Models) have been successfully applied to solve several problems in Computer Vision, such as face tracking, object recognition, and image segmentation. Usually, these models are learned from a discrete set of 2D shapes once the rigid transformations are removed by aligning the training set, i.e., applying Procrustes Analysis (PA). However, PA is

DOI: 10.4018/978-1-4666-2672-0.ch002

sensible to incomplete and biased set of views of the objects in the training set. In order to solve this problem, examples of 3D objects can be used in two ways: on the one hand, 3D objects are used to extract uniform 2D views to be aligned by standard PA; on the other hand, Continuous Procrustes Analysis (CPA) is used to learn all 3D rigid transformations directly from the 3D examples. In the past, such techniques could only be applied to a limited number of objects, since the most part of databases were storing two-dimensional information. However, recently many 3D databases, as well as 2D databases with three-dimensional information, have become available because of the market release of low cost depth cameras. It is illustrated in Figure 1 how 2D data could be extracted from 3D information provided for this kind of cameras. Different approaches to achieve non biased 2D shape models from 3D data will be explained and discussed along this Chapter.

Uniform sampling of 3D objects is considered a key step when building unbiased 2D shape models. Frequently, Euler angles are used to define three dimensional rotations; however, Euler angles suffer from known problems like gimbal lock or non-uniform rotations (Kuffner, 2004). The main part of this Chapter will be devoted to discuss different configurations to parameterize 3D rotations: usual non-uniform rotations and uniform sampling alternatives using quaternions and Euler angles.

2D shape models are also able to modify their shape in a non-rigid mode, consistent with shape deformations in the training set. The extraction process for non-rigid variations is outlined in the next section; however, deformable models are not addressed in this research, though they could be a direct extension.

BACKGROUND

Building 2D shape models will be the main goal when generating and analyzing uniform rotations. From this perspective, construction of statistical models will be introduced and, specifically, the Procrustes Analysis (PA) technique is described. PA is an important step in model building, as well as it is closely related with the continuous approach of uniform sampling.

2D Shape Models

Two-dimensional shape models are statistical models which are able to modify their shape according to the different transformations present

Figure 1. (left) Image sequence obtained from a depth camera; (middle) 3D skeletons extracted using Kinect© for Windows SDK; (right) 5 cameras, randomly distributed around the 3D skeleton, are displayed.

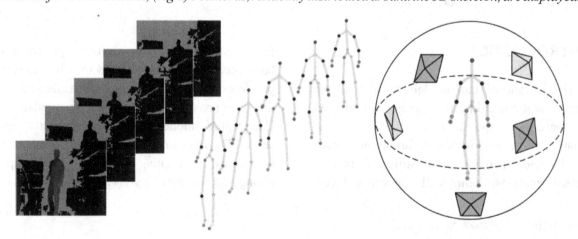

in a training set (Cootes & Taylor, 2001). The training set should be labeled with landmark correspondences across all the training shapes, where a shape is defined as a finite set of landmarks whose geometrical information remains unchanged when the shape suffers rigid transformations.

The building process for 2D shape models requires a set of training shapes composed by a number l of landmarks, where each landmark should be consistently labeled, representing the same anatomical location across all training shapes. The building process starts by removing rigid transformations in the training set using Procrustes Analysis (PA), which minimizes the least squares error between landmarks from training shapes and the mean shape m, which is also computed.

After aligning the shapes, each d-dimensional training shape is represented as a point $s \in \mathbb{R}^{dl}$, as well as the mean shape $m \in \mathbb{R}^{dl}$. In the case of three dimensional shapes, each point has the form $s = (x_1, y_1, z_1, ..., x_l, y_l, z_l)$. In the second step of model building, it is assumed that all points represented in the space $\mathbb{R}^{dl}$ follow a Gaussian probability density function. Eigenvectors and eigenvalues are computed using Principal Component Analysis (PCA), and the set of most informative eigenvectors is used as the basis of a subspace B, which summarizes the non-rigid variations among all the training shapes. The percentage of information explained using the subspace B is selected by choosing the number r of ordered basis. Columns in matrix $B \in \mathbb{R}^{dl \times r}$ are the most informative eigenvectors, each one describing a principal mode of variation in the training set.

The last step of the process is shape representation using the computed information. Following the Point Distribution Model (PDM) approach, new shapes can be represented as $s = m + Bc$; where $c \in \mathbb{R}^{l \times r}$ is a vector weighting each basis over the shape s. Hence, an infinite number of shapes can be created by modifying parameters in c. Moreover, limits over c guarantee that new shapes will follow the variations present in the training set. Usually, c parameters are limited such as $|c_i| < 3\sqrt{\lambda_i}$, where λ_i is the i-th eigenvalue, which means that c parameters are limited within ± 3 their corresponding standard deviation. Further details are specified in (Cootes & Taylor, 2001).

The most popular techniques using 2D shape models are PDM and Active Shape Models (ASM) (Cootes & Taylor, 2001). Many computer vision problems have been successfully treated by their application: image segmentation (Osher & Paragios, 2003; Mumford & Shah, 1989), object recognition (Ullman & Basri, 1991; Jones & Poggio, 1989), and face tracking (Baker et al., 2004; Dela Torre & Nguyen, 2008; Blanz & Vetter, 1999; Cootes et al., 2001; Gong et al., 2000), among others.

Procrustes Analysis

Rigid registration among different shapes, composed by labeled landmarks, is usually addressed by Procrustes Analysis (PA) (Goodall, 1991). More precisely, it is called Generalized Procrustes analysis (GPA) when more than two shapes are aligned (see Figure 2). GPA minimizes the least squares error between the landmarks of each training shape and a mean shape, while this mean shape is as well estimated:

$$E_1(M, A_1, ..., A_n) = \sum_{i=1}^{n} \left\| A_i D_i - M \right\|_F^2 .$$

PA is used to compute the mean shape $M \in \mathbb{R}^{2 \times l}$ of the training set and the n rigid transformations $A_i \in \mathbb{R}^{2 \times 2}$ between the mean shape and each training sample $D_i \in \mathbb{R}^{2 \times 1}$, where n is the number of training shapes, l is the number of landmarks that compose each shape, and $\left\| X \right\|_F^2 = tr\left(X^T X \right)$ designates the square of the

Figure 2. (left) Training set after translation removal; (right) aligned training set (blue points) and the mean shape (red line) using Procrustes analysis

Frobenius norm of a matrix. PA is usually expressed as a vector (Cootes & Taylor, 2001), however E_i is formulated using matrices for easy comparison with Continuous Procrustes Analysis (CPA), later introduced.

Though analytic solutions exist for training set alignment (Horn, 1987), iterative approaches are commonly used (Eggert et al., 1997; Cootes & Taylor, 2001) because of their intuitiveness and fast convergence. A standard iterative approach proceeds as follows. First of all, shape translations are removed by displacing the centers of gravity of all shapes to the origin. The second step of GPA consists on choosing one training shape as an initial estimate of the mean shape. In the third step, all training shapes are aligned with the estimation of the mean shape. Finally, the mean shape M is re-estimated from the aligned shapes. The algorithm iterates from the third step until convergence of M.

It is known that neither iterative methods nor analytic approaches guarantee the convergence to the global optimal solution. Recent works have proposed more accurate solutions (Bartoli et al., 2010) and new optimization procedures for finding the global optimal solution (Pizarro & Bartoli, 2011).

UNIFORM ROTATIONS

Building 2D shape models from 3D objects databases can be addressed by applying 2D techniques over 2D views, sampled from 3D training examples. Since 2D views from training set can bias the estimation of the shape model, an uniform coverage of all 3D transformations of the objects is needed, which can be addressed from both, discrete and continuous formulations.

Issues, Controversies, and Problems

The final goal of uniform rotations in our domain is obtaining a set of 2D object views, uniformly distributed along all possible rotations of a 3D object. Given a 3D shape and a virtual camera with a fixed point of view, it could be achieved by rotating the three-dimensional object over the rotation group SO(3) in a uniform way.

The Special Orthogonal group in three dimensions, SO(3), forms a group with its action being the composition of rotations. Each rotation is a linear transformation that preserves vectors length and space orientation. SO(3) is not only a group, but also a manifold, which makes it a Lie group. Moreover, this manifold has the topology of real 3-dimensional projective space $\mathbb{RP}^3$.

In order to appreciate the transformations suffered for a 3D object by the rotations, its minimal representation is the composition of two unit 3D vectors from the origin (see Figure 3). Our problem about uniform sampling of rotations, therefore, can be understood as the uniform distribution of couples of unit vectors into the unit 2-sphere, so that both vectors follow a uniform density function over the sphere surface, and the angle represented by the union of both vectors in the sphere surface follows a uniform distribution.

There exist several representations for 3D rotations, the most common of them Euler angles and quaternions. In the following sections different parameterizations for Euler angles and quaternions are presented. Moreover, in order to achieve uniform rotations with the final goal of building 2D shape models, a continuous methodology is also described in the scope of Procrustes Analysis.

Euler Angles

Euler angles represent orientations in the rotation group SO(3) through the composition of three

Figure 3. Rotation of ψ angle between y-axis (vertical arrow) and the union of the couple of unit vectors (dashed), i.e. the shape, in the sphere surface

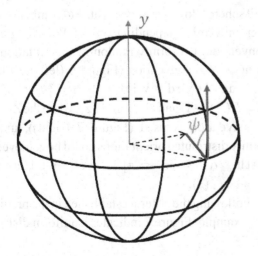

rotations (α, β, γ), each one around a single axis of a basis. The final rotation is obtained multiplying three rotation matrices:

$$R_{xyz} = R_z(\gamma)R_y(\beta)R_x(\alpha)$$

where:

$$R_z(\gamma)\begin{bmatrix} \cos\gamma & -\sin\gamma & 0 \\ \sin\gamma & \cos\gamma & 0 \\ 0 & 0 & 1 \end{bmatrix}$$

$$R_y(\beta)\begin{bmatrix} \cos\beta & 0 & \sin\beta \\ 0 & 1 & 0 \\ -\sin\beta & 0 & \cos\beta \end{bmatrix}$$

$$R_x(\alpha)\begin{bmatrix} 1 & 0 & 0 \\ 0 & \cos\alpha & -\sin\alpha \\ 0 & \sin\alpha & \cos\alpha \end{bmatrix}.$$

Since matrix product is not commutative, different permutations of the axes result in different orientations. There are at least 24 conventions for Euler angles, depending on the chosen Cartesian axes and which order is applied: 3 (*first axis: x, y, z*) × 2 (*second axis*) × 2 (*first axis repetition, or not*) × 2 (*static or rotating axes*) = 3 × 2 × 2× 2 = 24 possibilities.

Nevertheless, all of them suffer the same issues: non-uniform distribution of orientations, singularities, and the gimbal lock problem.

The gimbal lock problem for rotations in the three-dimensional space appears when two out of the three axes are parallel. One degree of freedom is lost and, therefore, only rotations in two-dimensional space are performed. An easy example to understand this issue appears when using the convention Z-Y-Z, i.e., first, a rotation on the Z-axis by α angle, followed by a turn on the rotated *Y-axis* of β angle and, finally, a rotation by γ angle on the new *Z-axis*. If $\beta = 0$, it produces a rotation by $\delta_1 = \alpha + \gamma$ angle, only on *Z-axis*. In this case, the system loses a degree of freedom and it is "locked" rotating in a degenerate two-dimensional space. Of course, the same situation occurs when $\beta = \pi$, with a final rotation of $\delta_2 = \alpha - \gamma$ angle around *Z-axis*.

It is a clear example of singularity on Euler angles, where different rotations in SO(3) are mapped onto a single rotation in the Euler representation. In the previous example, the final rotation described by $\beta = 0$ and $\delta_1 = \alpha + \gamma$ could be achieved by any different combination of α and γ, as well as infinity. Other simple example to visualize singularities is the Mercator projection of the globe. In this case, lines representing North and South poles are mapped as single points in the globe. Singularities usually lead to representation problems around its influence area, because small changes in one representation may lead to large changes in other representation.

Hence, points uniformly distributed near the pole lines in the Mercator projection, representing large distances in this representation, will be overrepresented near the North pole or South pole points of the globe. Returning to the Euler angles example, rotations with $\beta \approx 0$ or $\beta \approx \pi$ will be overrepresented, resulting in a non-uniform distribution of 3D orientations.

Perhaps other Euler angles standards, like X-Y-Z convention, make more difficult to guess the final orientations. However, for the X-Y-Z convention example (shown in Figure 4(left)) the distribution of orientations are not uniform, focusing rotations in the poles of the unit sphere. Angle ranges chosen to perform the experiment in Figure 4(left) are the domains for Euler angles: $\alpha, \gamma = \mathcal{U}(-\pi, \pi]$ and $\beta = \mathcal{U}\left[-\dfrac{\pi}{2}, \dfrac{\pi}{2}\right]$, where $\mathcal{U}$ means that angles can take values uniformly distributed on the interval.

It was previously stated that a couple of vectors is the minimum expression of a shape where 3D rotations can be appreciated. Hence, the perfor-

mance of different approaches when sampling rotations over SO(3) will be checked by observing their application over a couple of unit vectors (i.e. the shape). Our goal is to obtain uniform distribution of shapes into the unit 2-sphere.

Shoemake (1992) stated that coordinates $\{x, y, z\}$ of a vector uniformly distributed on a sphere are also uniformly distributed between their limits ($x, y, z \in [-1, +1]$ in the case of the unit sphere). Therefore, 2D shape rotations over the unit sphere surface are uniformly distributed while one of the vectors that compose the shape has its components uniformly distributed between the limits of the unit sphere.

The third rotation of the shape is referred to the relative change of orientation between the pair of vectors that compose the shape. 3D shape rotations are uniformly distributed while the ψ angle between the Y-axis and the line painted by the couple of vectors in the sphere surface (illustrated in Figure 3) follows as well a uniform distribution.

It is shown in Figure 5 (lower row) that angle follows a uniform distribution. However, Euler angles do not perform a uniform distribution of rotations because $\{x, y, z\}$ components of rotated vectors have a non-uniform distribution (Figure 5, upper row). Qualitative results (Figure 4(left)) also indicated a non-uniform distribution of the vectors, since the number of rotations over the poles is visually higher than for the rest of the unit sphere. However, it is possible to compensate non-uniformity depending on the Euler angles convention. Uniformly randomized orientations using X-Y-Z convention (Figure 6 (left)) could be achieved with $\alpha, \gamma = \mathcal{U}(-\pi, \pi]$, $z = \mathcal{U}(-1 + 1)$ and $\beta = \sin^{-1}(z)$; when the Z-Y-Z convention is used (Figure 7 (right)), uniformly distributed orientations could be achieved with $\alpha, \gamma = \mathcal{U}(-\pi, \pi]$, $z = \mathcal{U}(-1 + 1)$ and $\beta = \cos^{-1}(z)$.

Following the criteria established in the previous example, Figure 7 indicates a uniform distri-

Figure 4. Sampling of 5000 rotations of a couple of unit vectors: (left) uniform sampling of Euler angles following the x-y-z convention; (right) uniform sampling of quaternion parameters using the method presented in (Shoemake, 1992)

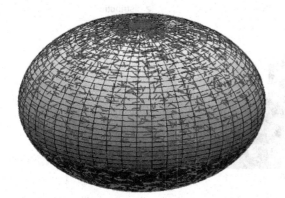 

bution over the sphere because each coordinate of the rotated vectors is uniformly distributed (Figure 7, upper row); moreover, angle (Figure 7, lower row) shows that relative rotations between two vectors painted in the sphere are uniform as well.

Quaternions

Quaternions were conceived by Hamilton (1853) as extended complex numbers $q = [a, bi, cj, dk]$. Each unit quaternion can be interpreted as a point in the unit 3-sphere $S^3 \in \mathbb{R}^4$, which represents a rotation on SO(3). For any unit quaternion $q = q_0 + q = \cos(\theta / 2) + \hat{u} \sin(\theta / 2)$ and for any vector $v \in \mathbb{R}^3$, the action of the triple product $q_{v'} = q q_v q^*$ may be interpreted geometrically as a rotation of the vector v through an angle θ around $\hat{u}$ as the axis of rotation (Kuipers, 1999), where a unit quaternion is defined as $|q| = |q^*| = \sqrt{q^* q} = 1$, q^* is the quaternion conjugate, and $q_v = [0, v]$ is the vector v expressed as a pure quaternion, i.e. a quaternion whose scalar part is 0.

Following quaternion algebra (Lerios, 1995), an equivalent rotation can be represented in matrix formulation as $q_{v'} = q q_v q^* = R_q q_v$, with:

$$R_q = \begin{pmatrix} a^2 + b^2 + c^2 + d^2 & 0 & 0 & 0 \\ 0 & a^2 + b^2 - c^2 - d^2 & 2bc - 2ad & 2bd + 2ac \\ 0 & 2bc + 2ad & a^2 - b^2 + c^2 - d^2 & 2cd - 2ad \\ 0 & 2bd - 2ac & 2cd + 2ab & a^2 - b^2 - c^2 + d^2 \end{pmatrix}.$$

Figure 5. Non-uniform distribution of the rotated vector parameters (up) with the mean of all distributions (horizontal line) and distribution of spherical coordinates (down) with the mean of ψ distribution (horizontal line), using Euler angles following the x-y-z convention.

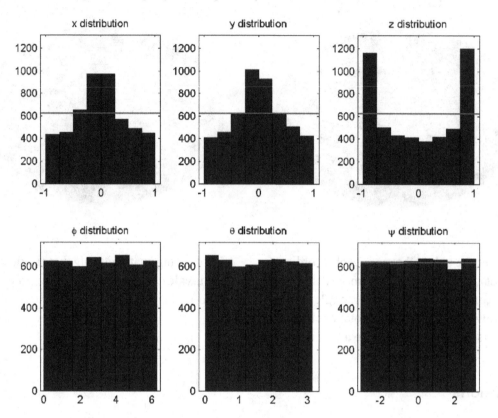

Since $q = [a, b, c, d]$ is a unit quaternion such that $\|q\| = q^*q = a^2 + b^2 + c^2 + d^2 = 1$, R_q can be simplified to:

$$R_q = \begin{vmatrix} |q|^2 & 0 & & \\ 0 & |q|^2 - 2(c^2 + d^2) & & \\ 0 & 2(bc + ad) & & \\ 0 & 2(bd \text{ - } ac) & & \end{vmatrix}$$

$$\begin{matrix} 0 & 0 \\ 2(bc - ad) & 2(bd + ac) \\ |q|^2 + 2(b^2 + d^2) & 2(cd - ab) \\ 2(cd + ab) & |q|^2 - 2(b^2 + c^2) \end{matrix} \Bigg] ==$$

$$\begin{vmatrix} 1 & 0 & 0 \\ 0 & 1 - 2(c^2 + d^2) & 2(bc - ad) \\ 0 & 2(bc + ad) & 1 - 2(b^2 + d^2) \\ 0 & 2(bd - ac) & 2(cd + ab) \end{vmatrix}$$

$$\begin{matrix} 0 \\ 2(bd + ac) \\ 2(cd - ab) \\ 1 - 2(b^2 + c^2) \end{matrix} \Bigg) = \begin{pmatrix} 1 & 0 \\ 0 & Q \end{pmatrix},$$

which satisfies the properties of rotation matrices (Lerios, 1995). An equivalent matrix is presented in (Shoemake, 1991) using a different order of the quaternion components

Figure 6. Sampling of 5000 rotations of a couple of unit vectors by non-uniform sampling of Euler angles: (left) following x-y-z convention to achieve uniform rotations; and (right) following the z-y-z convention

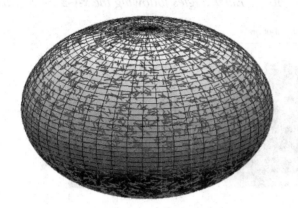 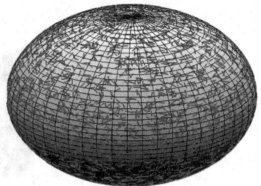

$q = [x, y, z, w] = [b, c, d, a]$. Moreover, to avoid unnecessary calculus and vector-quaternion conversions, vector rotations using quaternions could be calculated by the product $v' = Qv$, where $Q \in \mathbb{R}^{3\times3}$ and $v' \in \mathbb{R}^{3\times1}$ is the rotated vector.

The axis-angle interpretation shows that quaternions are composed by only one rotation and, therefore, they do not suffer from the gimbal lock problem, though opposite quaternions $q = -q$ represent the same rotation. However, not all parameterizations perform uniform rotations despite of the use of quaternions. In the following paragraphs three different parameterizations are considered.

Euler Angles Conversion

Given the difficulty linked to quaternions, the first naïve attempt to present intuitive rotations lies on using a unit quaternion expressed in Euler angles. Nevertheless, a uniform sampling of Euler angles does not perform uniform rotations despite of quaternion conversion, and gimbal lock problem persists.

Given the three Euler angles (α, β, γ), three independent quaternions can be formed:

$$q_x = \left[\cos\left(\frac{\alpha}{2}\right), \left(\sin\left(\frac{\alpha}{2}\right), 0, 0\right)\right],$$

$$q_y = \left[\cos\left(\frac{\beta}{2}\right), \left(0, \sin\left(\frac{\beta}{2}\right), 0\right)\right],$$

$$q_z = \left[\cos\left(\frac{\gamma}{2}\right), \left(0, 0, \sin\left(\frac{\gamma}{2}\right)\right)\right].$$

Similarly to rotation matrices, the joint rotation is represented by the final quaternion

$$q_{xyz} = q_z q_y q_x = \left[a_{xyz}, b_{xyz}, c_{xyz}, d_{xyz}\right],$$

where:

$$a_{xyz} = \cos\left(\frac{\alpha}{2}\right)\cos\left(\frac{\beta}{2}\right)\cos\left(\frac{\gamma}{2}\right) + \sin\left(\frac{\alpha}{2}\right)\sin\left(\frac{\beta}{2}\right)\sin\left(\frac{\gamma}{2}\right),$$

Figure 7. Uniform distribution of the rotated vector parameters (up) with the mean of all distributions (horizontal line) and distribution of spherical coordinates (down) with the mean of ψ distribution (horizontal line), using a uniform sampling distribution of Euler angles following the x-y-z convention

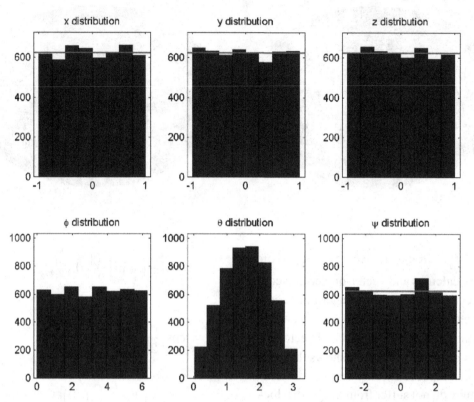

$$b_{xyz} = \sin\left(\frac{\alpha}{2}\right)\cos\left(\frac{\beta}{2}\right)\cos\left(\frac{\gamma}{2}\right) -$$
$$\cos\left(\frac{\alpha}{2}\right)\sin\left(\frac{\beta}{2}\right)\sin\left(\frac{\gamma}{2}\right),$$

$$c_{xyz} = \cos\left(\frac{\alpha}{2}\right)\sin\left(\frac{\beta}{2}\right)\cos\left(\frac{\gamma}{2}\right) +$$
$$\sin\left(\frac{\alpha}{2}\right)\cos\left(\frac{\beta}{2}\right)\sin\left(\frac{\gamma}{2}\right),$$

$$d_{xyz} = \cos\left(\frac{\alpha}{2}\right)\cos\left(\frac{\beta}{2}\right)\sin\left(\frac{\gamma}{2}\right) -$$
$$\sin\left(\frac{\alpha}{2}\right)\sin\left(\frac{\beta}{2}\right)\cos\left(\frac{\gamma}{2}\right).$$

If q_{xyz} is not a unit vector due to numerical drift, it should be normalized such that $\hat{q}_{xyz} = q_{xyz} / \left|q_{xyz}\right|$. Given a uniform sampling of the three parameters, such that $\alpha, \gamma = \mathcal{U}(-\pi, \pi]$ and $\beta = \mathcal{U}\left[-\frac{\pi}{2}, \frac{\pi}{2}\right]$, rotations obtained follow the same distribution as orientations performed using rotation matrices in the previous section. Qualitative results comparing both approaches are presented in Figure 8, where rotations are concentrated onto the poles of the 2-sphere.

As it was performed in previous sections, the uniform distribution of couples of unit vectors onto the unit 2-sphere is checked for the uniformity of $\{x,y,z\}$ coordinates of unit vectors and, in addition, for the uniform distribution of the angle (shown in Figure 2) between the Y-axis and the union joining both vectors.

Figure 8. Sampling of 5000 rotations of a couple of unit vectors by uniform sampling of Euler angles following x-y-z convention, (left) using rotation matrices and (right) using quaternions, based on Euler angles parameterization

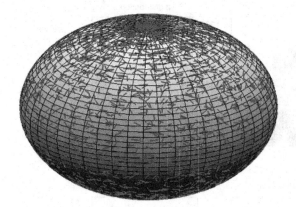
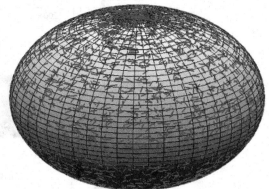

Similarly to Figure 5, rotations performed with quaternions parameterized by Euler angles result in non-uniform rotations. Figure 9 (lower row) shows a uniform distribution over ψ angle, though the vector distribution is not uniform around the sphere S^2 because vector coordinates are not uniformly distributed (Figure 9, upper row). In addition, a parameterization based on Euler angles conversion does not solve the gimbal lock problem.

Axis-Angle Parameterization

In order to avoid previous problems, it is necessary to work with other parameters closely related to quaternion rotations. As it was explained in the beginning of this section, a unit quaternion $q = \boldsymbol{q}_0 + q = \cos(\theta/2) + \hat{\boldsymbol{u}}\ \sin(\theta/2)$ can be expressed as a rotation by θ angle around $\hat{\boldsymbol{u}} -$ axis. Since the uniform sampling of becomes in uniform rotations over $\hat{\boldsymbol{u}} -$ axis, the intuitive attempt is to perform these 2D rotations over a set of uniformly distributed $\hat{\boldsymbol{u}}$ on the unit sphere S^2. The process considered to obtain a uniform distribution of unit vectors $\hat{\boldsymbol{u}} = \begin{bmatrix} u_x, u_y, u_z \end{bmatrix}$ in the unit 2-sphere (Shoemake, 1992) is illustrated in Figure 10.

Given three random variables θ, u_z, ρ, rotations are performed by the unit quaternion q, where $\theta = \mathcal{U}(0, 2\pi)$ and $\hat{u}$ are defined as:

$$u_z = \mathcal{U}(-1 + 1),$$

$$\rho = \mathcal{U}(0, 2\pi),$$

$$z^2 + r^2 = 1 \rightarrow r = \sqrt{1 - z^2},$$

$$u_x = r \cos \rho,$$

$$u_x^2 + u_y^2 = r^2 \rightarrow r^2((\sin \rho)^2 + (\cos \rho)^2) = r^2 \rightarrow (r \sin \rho)^2 + (r \cos \rho)^2 = r^2,$$

$$u_y = \sqrt{(r \sin \rho)^2} = \pm r \sin \rho.$$

Therefore, two quaternions are created with opposite $\hat{u}$ axis for each three random values. Qualitative results in Figure 11 show that the composition of uniform rotation axis $\hat{u}$ and uniform angle θ does not perform a uniform distribution of unit quaternions.

Figure 9. Non-uniform distribution of the rotated vector parameters (up) with the mean of all distributions (horizontal line) and distribution of spherical coordinates (down) with the mean of ψ distribution (horizontal line), using Euler angles-to-quaternion conversion

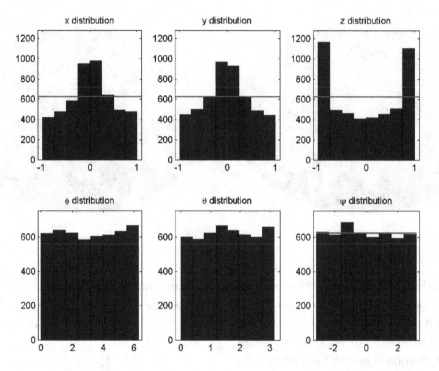

Figure 10. To achieve uniformly distributed unit vectors $\hat{u}$ in the unit 2-sphere, u_z takes values along the diameter following a uniform distribution, $u_z = \mathcal{U}(-1+1)$; and u_x, u_y are distributed on the circle (green) of radius r (red), which cuts the sphere by u_z (blue)

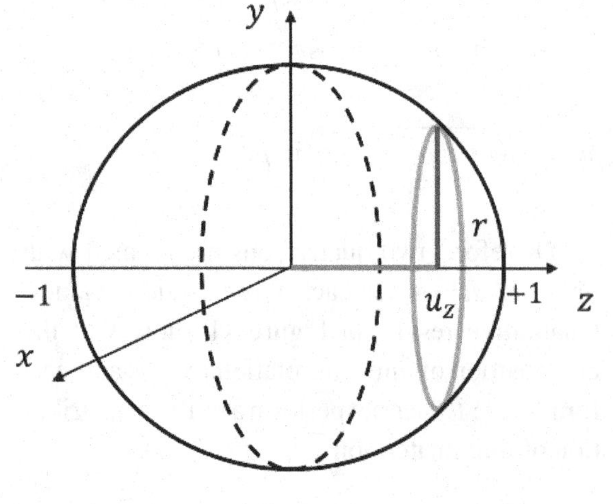

The distribution of vector components (Figure 12, upper row) when rotated using the previously created random quaternions are non-uniform, which means that neither rotations are uniform onto the unit 2-sphere, nor orientations over the sphere surface measured by angle (Figure 12, lower row).

Parameters Defined in (Shoemake, 1992)

Finally, taking inspiration as well from a composition of rotations, a direct method is presented in (Shoemake, 1992), where the four quaternion parameters are calculated through the use of three random variables. Figure 4 (right) shows the correctness of this method, obtaining uniform three-dimensional orientations. Similarly to the method proposed in (Yershova et al., 2010), Shoemake's approach computes unit random quaternions addressing the problem from the subgroup algorithm.

Figure 11. Sampling of 5000 rotations of a couple of unit vectors using random quaternions created from a uniform sampling of θ and uniformly distributed $\hat{\mathbf{u}} - axis$

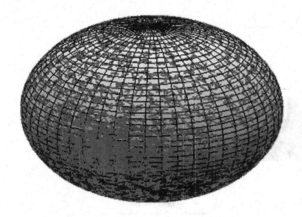

If all 3D rotations in the space compose a group, a subgroup $q = (c, 0, 0, s)$ of this group is constituted by planar rotations around the Z-axis, and cosets $q = (w, x, y, 0)$ of this subgroup are rotations pointing Z-axis in different directions. Following the subgroup algorithm (Diaconis & Shahshahani, 1987), a uniformly distributed element of the complete group can be achieved by the multiplication of a uniformly distributed element from the subgroup with a uniformly distributed *coset*:

$$\mathcal{U}([c, 0, 0, s])\mathcal{U}([w, x, y, 0]) =$$
$$\mathcal{U}\big[[cw, cx + sy, -sx + cy, sw]\big].$$

Given three independent random variables $X_0, X_1, X_2 \in \mathcal{U}(0, 1)$, random unit quaternions $q = \big[\cos\theta_2 r_2, \sin\theta_1 r_1, \cos\theta_1 r_1, \sin\theta_2 r_2\big]$ are computed, where:

$$\theta_1 = 2\pi X_1,$$

$$\theta_2 = 2\pi X_2,$$

$$r_1 = \sqrt{1 - X_0},$$

$$r_2 = \sqrt{X_0}.$$

Results exposed in Figure 13 indicate a uniform distribution over the sphere because each coordinate $\{x, y, z\}$ of the rotated vector is uniformly distributed (Figure 13, upper row). Moreover, ψ angle distribution (Figure 13, lower row) shows that relative rotations between the Y-axis and the union of the couple of vectors in the sphere follow a uniform distribution as well.

Continuous Procrustes Analysis

In the particular field of learning 2D shape models, an alternative to discrete model building is proposed in (Igual & De la Torre, 2010), where 2D discrete techniques are extended to be applied over 3D models in a continuous form. Continuous Procrustes Analysis (CPA) is presented as a competitive alternative to Generalized Procrustes Analysis (GPA) when 3D models are available.

Shape models learned using GPA (Figure 14, left) could be biased by a non-uniform sampling of the 2D views of the 3D examples. This issue, the computational cost, and large amount of data associated to the uniform sampling of 3D transformations are solved by CPA (Figure 14, right). CPA gives a closed form solution for the learning of 2D models directly from a 3D training set, incorporating the information of all 3D rigid transformations.

CPA minimizes the least-squared error between the 2D projections of 3D landmarks of each training shape and a 2D mean shape, while this mean shape is also estimated. Let $\Omega = \big\{\omega = (\phi, \theta, \psi)\big\} \in \mathbb{R}^3$ be the set of 3D rotations, where ω are the Euler angles and the Haar measure is $d\omega = \sin\theta d\phi d\theta d\psi$. CPA minimizes the following energy functional:

Figure 12. Non-uniform distribution of the rotated vector parameters (up) with the mean of all distributions (horizontal line) and distribution of spherical coordinates (down) with the mean of ψ distribution (horizontal line), using random quaternions composed by random uniform rotation axis $\hat{u}$ and uniform angle θ

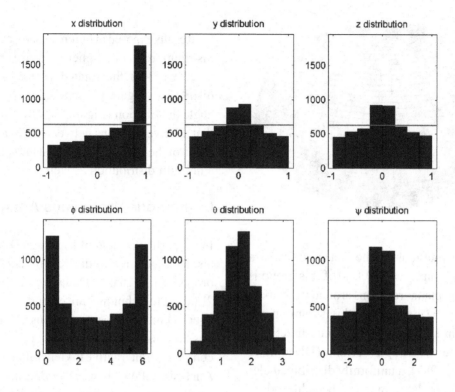

$$E_2(M, A_1, ..., A_n) =$$

$$\sum_{i=1}^{n} \int_{\Omega} F_2(M, A_i(\omega)) d\omega =$$

$$\sum_{i=1}^{n} \int_{\Omega} \left\| PR(\omega)D_i - A_i(\omega)M \right\|_F^2 d\omega.$$

$$P = \begin{pmatrix} 1 & 0 & 0 \\ 0 & 1 & 0 \end{pmatrix}.$$

CPA computes the two-dimensional mean shape $M \in \mathbb{R}^{2 \times 1}$ of the training set and the n rigid transformations $A_i(\omega) \in \mathbb{R}^{2 \times 2}$ between the mean shape and the orthographic projection $P \in \mathbb{R}^{2 \times 3}$ of each training sample $D_i \in \mathbb{R}^{3 \times 1}$, which is rotated over all possible Euler angles of the domain Ω, by the rotation matrix $R \in \mathbb{R}^{3 \times 3}$ As in the previous section, n is the number of training shapes and l is the number of landmarks that form each shape. The matrix P describes the orthographic projection onto the plane Z=0, defined as:

The CPA functional $E_2(M, A_1, ..., A_n)$ is similar to the energy function $E_1(M, A_1, ..., A_n)$ from GPA; however, there are three main differences: first, E_2 is a continuous formulation where discrete sums are extended to integrals; the second difference relies on the 2D views used in E_2, which depend directly on the 3D structure of the training examples D_i and the 3D transformation parameters $R(\omega)$; and the third difference is that, A_i in E_1 are variables, whereas in E_2 are functions depending on Euler angles ω.

Note that uniform rotations are achieved by $R(\omega)$. Since using Haar measure $d\omega$, an invariant integral for functions on the rotation group SO(3)

Figure 13. Uniform distribution of the rotated vector parameters (top) with the mean of all distributions (horizontal line) and distribution of spherical coordinates (down) with the mean of ψ distribution (horizontal line), using the methodology proposed in (Shoemake 1992)

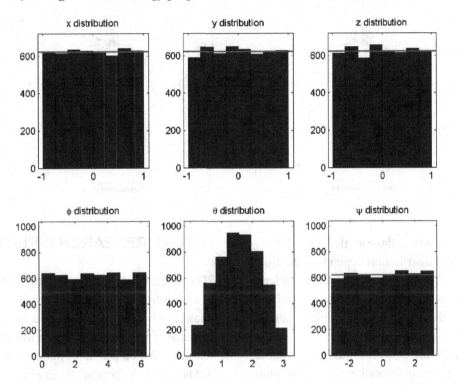

is obtained (Naimark, 1964), despite of the use of Euler angles. Therefore, the problem of discrete non-uniform distribution discussed above is avoided in the integral definition.

SOLUTIONS AND RECOMMENDATIONS

The final goal to achieve uniform rotations is obtaining a set of 2D object views uniformly distributed along all possible rigid transformations of a 3D object. To test the presented methodologies, a couple of unit vectors is rotated into the unit 2-sphere. Distribution of rotations is considered uniform when each rotated vector follows a uniform distribution over the sphere surface and, moreover, a uniform distribution of the angle between the vertical axis and the union of the

couple of vectors in the sphere surface. Since Shoemake (1992) stated that $\{x,y,z\}$ coordinates of a vector uniformly distributed on a sphere are also uniformly distributed between their limits (into the unit sphere: $x, y, z \in [-1, +1]$), the correct distribution of vectors over the sphere surface is checked by the distribution of its components. In addition, distribution function of the angle between both vectors onto the sphere surface is tested by the uniformity of ψ angle, as illustrated in Figure 2.

Results presented in the previous section show, first of all, that uniform sampling of Euler angles does not result in uniform rotations when three Euler angles take values between the common limits: $\alpha, \gamma = \mathcal{U}(-\pi, \pi]$ and $\beta = \mathcal{U}\left[-\dfrac{\pi}{2}, \dfrac{\pi}{2}\right]$. A uniform distribution of the rotations can be obtained with Euler angles, when a non-uniform

Figure 14. Illustration of 2D shape model building using (left) Procrustes analysis (PA) and (right) Continuous Procrustes analysis (CPA)

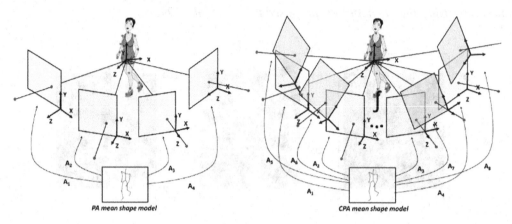

sampling of one of the three angles is used. However, angle and distribution depend on the Euler angles convention chosen, and there exist, at least, 24 conventions. In addition, despite of the uniform distribution, the gimbal lock problem persists while Euler angles are used.

Euler angles are also presented as an alternative to perform uniform rotations over continuous domains. Using the Haar measure into the integral definition the problem of discrete non-uniform distribution is avoided.

The use of quaternions can solve the previous problems while an appropriate parameterization is chosen. Presented results illustrate that Euler-to-quaternion conversion neither perform uniform distribution nor solves gimbal lock problem, despite of the use of quaternions. Axis-angle parameterization is also used to create unit quaternions, from uniformly distributed rotation axis and uniform rotation angle. However, this approach does not perform uniform rotations either.

Finally, it is checked that the method proposed in (Shoemake, 1992) results in uniform rotations. In this method, uniform unit quaternions are computed from three random variables. Therefore, this solution solves uniformity as well singularities.

FUTURE RESEARCH DIRECTIONS

2D shape models were introduced in the Chapter as state-of-the-art methods for object detection, image segmentation, and face tracking. The future research of non-biased 2D shape models is closely related with their possible applications in research fields such as improving Active Shape Models in the case of face detection.

Procrustes Analysis (PA) technique was also discussed. It is a state-of-the-art research area because of recent works about global optimization of PA. Hence, interesting research would be on the direction of global optimization of the, as well described, Continuous Procrustes Analysis (CPA). Other promising extensions of CPA would be a continuous formulation of PA using quaternions instead of Euler angles, and, independently of the parameterization of rotations, incorporating a subspace into the CPA formulation.

Finally, an additional research focus in the field of uniform sampling of SO(3) would be the uniform sampling of rotations in a limited part of the two-sphere using quaternions, since vectors rotated in the sphere are used to illustrate the orientations of a rotated object. The research on this subject would be useful to limit the rotations according to the problem at hands, avoiding memory resources and computational time.

CONCLUSION

The main goal of this chapter has been the study of the uniform sampling of rotations of 3D objects to be used on building unbiased 2D shape models. Along this work, standard non uniform rotations techniques over rotation group SO(3) were discussed, as well as uniform methods using both, Euler angles and quaternions, were presented. Moreover, the problem was addressed from discrete and continuous points of view. For the problem at hands, uniformity was checked trough the distribution of rotated couples of unit vectors into the unit 2-sphere S^2.

The concise review of the non-uniform sampling approaches discussed in this Chapter is that, on the one hand, Euler angles do not perform uniform rotations from a uniform sampling of the three angles and, in addition it is a parameterization that suffers singularities, i.e. the gimbal lock problem. On the other hand, these problems can be solved applying quaternions when correct parameterization is used; intuitive creation of unit quaternions as Euler angles conversion or angle-axis approach do not perform uniformly distributed rotations.

Three methodologies were presented to perform uniform rotations. The first solution to the problem was to use Euler angles. Uniformity is obtained for a certain distribution of one of the three angles, which depends on the convention that is being used; however, despite of uniform rotations, singularities are not solved with this method. The second approach consisted on using quaternions, following a methodology which generates uniform random unit quaternions from three random variables. In this case, uniform rotations are achieved, as well as problems related to singularities are solved. Finally, instead of general rotation methodologies, the third approach relied on the use of Continuous Procrustes Analysis (CPA), a particular method that builds unbiased 2D shape models from 3D objects. A continuous formulation of Euler angles was used in order to avoid non uniformity of rotations. This method does not solve problems related with Euler angles singularities. However, uniform rotations are performed and, in addition, CPA allows the possibility of limiting the rotations into a specific domain of the problem.

Each one of the three previous approaches performs uniform rotations. However, it is recommended to use each one of them according to the case to be applied. Uniform Euler angles are useful while intuitive rotations are needed, according to the axis of a basis. Otherwise, it is recommended to use other alternatives in order to avoid problems related with singularities and the rotation order, which varies depending on the convention of Euler angles. For the purpose of building unbiased 2D shape models from 3D objects, CPA presents similar performance that discrete Procrustes Analysis (PA), using a 2D dataset obtained by the use of uniform unit quaternions. The main differences are that CPA allows the use of limits over the sampling domain and unit quaternions solve not only uniform sampling, but also singularities.

ACKNOWLEDGMENT

This work is partly supported by the Spanish Ministry of Science and Innovation (projects TIN2011-28854-C03-01, TIN2009-14404-C02, CONSOLIDER INGENIO CSD 2007-00018) and the Comissionat per a Universitats i Recerca del Departament d'Innovació, Universitats i Empresa de la Generalitat de Catalunya.

REFERENCES

Baker, S., Matthews, I., & Schneider, J. (2004). Automatic construction of active appearance models as an image coding problem. *IEEE Transactions on Pattern Analysis and Machine Intelligence, 26*(10), 1380–1384. doi:10.1109/TPAMI.2004.77

Bartoli, A., Pizarro, D., & Loog, M. (2010, August). *Stratified generalized procrustes analysis.* In British Machine and Vision Conference, Aberystwyth, UK.

Blanz, V., & Vetter, T. (1999, August). *A morphable model for the synthesis of 3D faces.* In International Conference and Exhibition on Computer Graphics and Interactive Techniques, Los Angeles, CA.

Cootes, T., Edwards, G., & Taylor, C. (2001). Active appearance models. *IEEE Transactions on Pattern Analysis and Machine Intelligence, 23*(6), 681–685. doi:10.1109/34.927467

Cootes, T. F., & Taylor, C. J. (2001). *Statistical models of appearance for computer vision.* Retrieved from http://www.itu.dk/stud/projects_f2004/handtracking/referencer/Cootes%20den%20lange%20-%20app_model.pdf

De la Torre, F., & Nguyen, M. (2008, June). *Parameterized kernel principal component analysis: Theory and applications to supervised and unsupervised image alignment.* In IEEE Computer Vision and Pattern Recognition, Anchorage, AK.

Diaconis, P., & Shahshahani, M. (1987). The subgroup algorithm for generating uniform random variables. *Probability in the Engineering and Informational Sciences, 1*(1), 15–32. doi:10.1017/S0269964800000255

Eggert, D., Lorusso, A., & Fisher, R. (1997). Estimating 3-D rigid body transformations: A comparison of four major algorithms. *Machine Vision and Applications, 9*(5), 272–290. doi:10.1007/s001380050048

Gong, S., McKenna, S. J., & Psarrou, A. (Eds.). (2000). *Dynamic vision: From images to face recognition.* London, UK: Imperial College Press. doi:10.1142/p155

Goodall, C. (1991). Procrustes methods in the statistical analysis of shape. *Journal of the Royal Statistical Society. Series A, (Statistics in Society), 53*(2), 285–339.

Hamilton, W. R. S. (Ed.). (1853). *Lectures on quaternions.* Dublin, Ireland: Hodges and Smith.

Horn, B. K. P. (1987). Closed-form solution of absolute orientation using unit quaternions. *Journal of the Optical Society of America, 4*(4), 629–642. doi:10.1364/JOSAA.4.000629

Igual, L., & De la Torre, F. (2010, June). *Continuous Procrustes analysis to learn 2D Shape models from 3D objects.* In Computer Vision and Pattern Recognition Workshops, San Francisco, CA.

Jones, M. J., & Poggio, T. (1989). Multidimensional morphable models. In *International Conference on Computer Vision*, Vol. 29, (pp. 683-688). Springer.

Kuffner, J. J. (2004). Effective sampling and distance metrics for 3D rigid body path planning. In *Proceedings of IEEE International Conference on Robotics and Automation, Vol. 4*, (pp. 3993-3998).

Kuipers, J. B. (Ed.). (1999). *Quaternions and rotation sequences: A primer with applications to orbits, aerospace and virtual reality.* Princeton, NJ: Princeton University Press.

Lerios, A. (1995). *Rotations and quaternions.* Technical Report. Retrieved from http://server2.phys.uniroma1.it/doc/giansanti/COMP_BIOPHYS_2008/LECTURES/LECT_2/Lerios1995.pdf

Mumford, D., & Shah, J. (1989). Optimal approximations by piecewise smooth functions and associated variational problems. *Communications on Pure and Applied Mathematics, 42*(5), 577–685. doi:10.1002/cpa.3160420503

Naimark, M. A. (Ed.). (1964). *Linear representation of the Lorentz Group.* New York, NY: Macmillan.

Osher, S., & Paragios, N. (Eds.). (2003). *Geometric level set methods in imaging, vision, and graphics.* New York, NY: Springer.

Pizarro, D., & Bartoli, A. (2011). Global optimization for optimal generalized Procrustes analysis. In *IEEE Conference on Computer Vision and Pattern Recognition.* (pp. 2409-2415).

Shoemake, K. (1991). *Quaternions.* Tech Report. Retrieved from http://campar.in.tum.de/twiki/pub/Chair/DwarfTutorial/quatut.pdf

Shoemake, K. (1992). Uniform random rotations. In Kirk, D. (Ed.), *Graphics Gems III* (pp. 124–132). San Francisco, CA: Morgan Kaufmann.

Ullman, S., & Basri, R. (1991). Recognition by linear combinations of models. *IEEE Transactions on Pattern Analysis and Machine Intelligence, 13*(10), 992–1006. doi:10.1109/34.99234

Yershova, A., Jain, S., LaValle, S. M., & Mitchell, J. C. (2010). Generating uniform incremental grids on SO (3) using the Hopf fibration. *The International Journal of Robotics Research, 29*(7), 801–812. doi:10.1177/0278364909352700

ADDITIONAL READING

Arvo, J. (1992). Fast random rotation matrices. In Kirk, D. (Ed.), *Graphics Gems III* (pp. 117–120). San Francisco, CA: Morgan Kaufmann.

Dela Torre, F., & Black, M. J. (2003). Robust parameterized component analysis: Theory and applications to 2D facial appearance models. *Computer Vision and Image Understanding, 91*(1), 53–71. doi:10.1016/S1077-3142(03)00076-6

Goodall, C. (1991). Procrustes methods in the statistical analysis of shape. *Journal of the Royal Statistical Society. Series A, (Statistics in Society), 53*(2), 285–339.

Górski, K. M., Hivon, E., Banday, A. J., Wandelt, B. D., Hansen, F. K., Reinecke, M., & Bartelmann, M. (2005). HEALPix: A framework for high-resolution discretization and fast analysis of data distributed on the sphere. *The Astrophysical Journal, 622,* 759–771. doi:10.1086/427976

Hernández, A., Reyes, M., Escalera, S., & Radeva, P. (2010). *Spatio-temporal GrabCut human segmentation for face and pose recovery.* Computer Vision and Pattern Recognition, 2010, San Francisco, CA.

Kookinos, I., & Yuille, A. (2007, October) *Unsupervised learning of object deformation models.* In International Conference on Computer Vision, Rio de Janeiro, Brazil.

Miles, R. E. (1965). On random rotations in R^3. *Biometrika, 52*(3-4), 636–639. doi:10.1093/biomet/52.3-4.636

Murnaghan, F. D. (1962). *The unitary and rotation groups. Lectures on applied mathematics.* Washington, DC: Spartan Books.

Pham, H. L., Perdereau, V., Adorno, B. V., & Fraisse, P. (2010). Position and orientation control of robot manipulators using dual quaternion feedback. In *Proceedings of IEEE/RSJ International Conference on Intelligent Robots and Systems* (pp. 658-663).

Ramamoorthy, S., Rajagopal, R., Ruan, Q., & Wenzel, L. (2006) Low-discrepancy curves and efficient coverage of space. In *Proceedings of Workshop on Algorithmic Foundations of Robotics* (pp. 203-218). Springer.

Rovira, J., Wonka, P., Castro, F., & Sbert, M. (2005). Point sampling with uniformly distributed lines. In *Proceedings of Point-Based Graphics: Eurographics/IEEE VGTC Symposium Proceedings* (pp. 109-118).

Sahu, S., Biswal, B. B., & Subudhi, B. (2008). A novel method for representing robot kinematics using quaternion theory. In *Proceedings of IEEE Sponsored Conference on Computational Intelligence, Control And Computer Vision In Robotics & Automation*.

Tomasi, C., & Kanade, T. (1992). Shape and motion from image streams under orthography: A factorization method. *International Journal of Computer Vision, 9*(2), 137–154. doi:10.1007/BF00129684

KEY TERMS AND DEFINITIONS

Euler Angles: Representation of orientations in SO(3) through the composition of three rotations, each one around a single axis of a basis.

Landmark: Consistently labeled keypoint which represents an anatomical location of the model at hands.

Procrustes Analysis: Typical method to remove rigid transformations between shapes, composed by labeled landmarks, which minimizes the least-squared error between landmarks from training shapes.

Rigid Transformation: Transformation that preserves isometry, i.e. distances between every pair of points are preserved. Typical rigid transformations are rotation, translation, and isomorphic scaling.

Shape: Finite set of landmarks whose geometrical information remains unchanged when the shape suffers rigid transformations.

Statistical Shape Models: Models which are able to modify their shape according to the different transformations present in a training set.

Unit Quaternion: Point in the unit 3-sphere $S^3 \in \mathbb{R}^4$, which represents a rotation in SO(3).

Chapter 3
Comparative Analysis of Temporal Segmentation Methods of Video Sequences

Marcelo Saval-Calvo
University of Alicante, Spain

Jorge Azorín-López
University of Alicante, Spain

Andrés Fuster-Guilló
University of Alicante, Spain

ABSTRACT

In this chapter, a comparative analysis of basic segmentation methods of video sequences and their combinations is carried out. Analysis of different algorithms is based on the efficiency (true positive and false positive rates) and temporal cost to provide regions in the scene. These are two of the most important requirements of the design to provide to the tracking with segmentation in an efficient and timely manner constrained to the application. Specifically, methods using temporal information as Background Subtraction, Temporal Differencing, Optical Flow, and the four combinations of them have been analyzed. Experimentation has been done using image sequences of CAVIAR project database. Efficiency results show that Background Subtraction achieves the best individual result whereas the combination of the three basic methods is the best result in general. However, combinations with Optical Flow should be considered depending of application, because its temporal cost is too high with respect to efficiency provided to the combination.

INTRODUCTION

Nowadays, analysis of behavior in video sequences is one of the most popular topics in the field of computer vision. Video surveillance, ambient intelligence, economization of space, urban planning, robot control (swarm robots, safety in industrial environments) and others examples of applications in which more and more an automated behavioral analysis is needed. To carry out this task is necessary to process the sequence of images previously to the cognitive analysis of the

DOI: 10.4018/978-1-4666-2672-0.ch003

scene. The process steps are usually: segmentation and tracking. The former extracts the region of interest of each frame. The latter analyses which elements of a frame correspond to the same in the next, that is, following a region of interest along the sequence.

This chapter focuses on the first step: segmentation. The aim is to study methods of video segmentation to determine the best one that fulfills requirements of efficiency and time that can be given by an application in which the processing is finally embedded.

Methods can be classified as temporal and spatial segmentation. The former segment regions of interest by using the temporal information of sequences extracted from different frames in a given time interval. Among these the most used, and the base for much other variations, are: Background Subtraction (BG), Temporal Differencing (TD, also known as Interframe) and Optical Flow (OF). Spatial methods are those that divide the image space into regions based on certain features (color, shape, etc.). Currently, methods combining basic segmentation techniques have been developed in order to improve efficiency given by individual methods (Hu, 2010; Velastin, 2005). This chapter is focused at temporal segmentation methods and their combinations.

CAVIAR project sequences (PETS04 (Fisher, 2004)) are used for experimentation. This is a public and a well-known database containing a quite precise ground truth.

BACKGROUND

Specifically among the works directly related to this chapter, analysis of efficiency of segmentation methods has been carried out recently. A comparative analysis of BG methods and its variations are frequent in the literature. It is worth mentioning the works of El Baf et al. (El Baf, 2007) and Hall et al. (Hall, 2005) in which a comparison of BG Simple Gaussian, Mixture of Gaussians, Kernel Density Estimation, W4 (Haritaoglu, 2000) and LOTS

method (based on different background models) can be found. Also, Benezeth et al. (Benezeth, 2010) perform a wide comparison with an extensive database of sequences and methods based on the BG. Interesting analysis of contour based methods has been carried out by VenuGopal et al. (VenuGolap, 2011) and Arbelaez et al. (Arbelaez, 2009). These works compare segmentation based on border extraction methods including Canny, Sobel and Laplacian of Gaussians. A comparison of parametric and non-parametric methods was proposed by Herrero et al. (Herrero, 2009). Basic methods (Temporal differencing, Median filter), parametric (Simple, Mixture of Gaussians and Gamma algorithm), and non-parametric methods (Histogram-based approach, Kernel Density Estimation) were analyzed concluding that parametric methods have the best results but they have problems to properly adjust the parameters. Finally, it is interesting to mention the comparative of region and contour based methods proposed by Zhang (Zhang, 1997).

To the best of our knowledge, although segmentation methods have been combined to improve efficiency in specific applications (Hu, 2010; Velastin, 2005), no evaluations of algorithms in combination have been performed. Therefore, the objective of this chapter is to study the combination of temporal methods most commonly used to decide which algorithm best suits the application requirements. The evaluation of the algorithm is based on the efficiency (true positive and false positive rates) and time cost of the system. These are two of the most important requirements of the design to provide to the next step (tracking) with segmentation in an efficient and timely manner constrained to the application.

SEGMENTATION OF VIDEO SEQUENCES

When segmentation is mentioned, it is usual to include different steps in the term such as conditioning of captured images, segmentation method processing itself, filtering of results and object detection (Figure 1).

Pre-segmentation step groups all methods that are applied to frame or frames of the sequence before they are processed. Usually it has not got information of the scene. Spatial filters are included in this step to condition image for method used. For example, smooth images taken by the camera, apply resizes, changes of resolution and other transformations of them. They use the information of the whole frame, being mean and median the most common. With this, it is possible to make images more homogeneous minimizing the noise for the next step.

In the present chapter, a filter to normalize each frame with the background has been done. This process smooth changes of environment lighting and other possible changes that could cause errors while frames are treated. Normalization is carried out obtaining the deviation mean between frame and background image, applying this factor to the whole frame approaching then common parts. After that a Wiener filter, based on statistic estimations of pixel neighborhood, is used.

Segmentation methods represent the core of the process. In this step, filtered images are divided in meaningful regions that make easier their analysis. In order to do this, different algorithms have been proposed that use different features of each frame, or the whole sequence, to find and label important regions.

The methods and their specific implementation are explained in next subsections.

Background Subtraction

Background subtraction method is a widely used algorithm in image segmentation for moving regions. Each frame, I, is compared with a model of the background scene (MBG). Different modeling have been proposed such as static images of the background, Gaussian models, histograms, W4 of Haritaoglu et al. (Haritaoglu, 2000), etc. All of them use noise thresholds as static values or standard deviations to reduce camera intrinsic errors not filtered in previous step. Moving regions result of this process, called foreground (FG), corresponding to those pixels of I which difference with MBG is greater than the noise threshold, n.

$$FG(x,y) = |MBG(x,y) - I(x,y)| > n$$

Specifically a Gaussian model has been implemented for this work. Each pixel is modeled as the mean of values of the same pixel in a sequence of images in the empty environment (MeBG, mean of background). In this case, threshold n is substituted by the standard deviation of each pixel, being a matrix of values called SBG (standard deviation of background). The form is now the next:

Figure 1. Segmentation schema with the different steps included on it

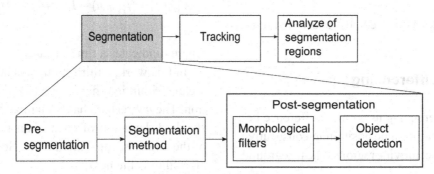

Figure 2. Complete process of segmentation and combination: (a) is a background model used to BG; (b) is a frame (TD and OF use two more frames); (c) is the same frame after pre-segmentation is applied; (d), (e), and (f) are results of each basic method; (g) is a combination of all of basic methods, (h) is the result of applying morphological post-segmentation filters, and (i) is the final result after application of small area algorithm

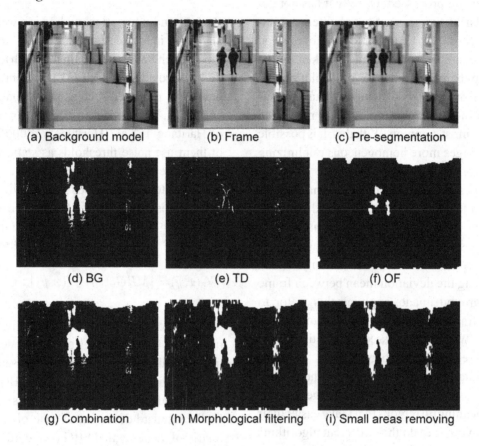

$$BG(x,y) = |MeBG(x,y) - I(x,y)| > c * SBG(x,y),$$

where c is a factor that indicates how many times the SGB has to be exceeded to be considered as foreground.

An example of this method is shown in Figure 2d.

Temporal Differencing

Temporal differencing uses value difference between pixels in the same position in consecutive frames to extract moving regions. The foreground (FG) corresponds to those parts of the image that have changed more than the rest of the frame. Here a noise threshold, n, is also used to reduce false positive errors. The general formula is:

$$FG(x,y) = |I_t(x,y) - I_{t+1}(x,y)| > n,$$

where t indicates a time instant.

In this work, a third frame has been added to the algorithm in order to enhance the segmentation. The n threshold has been modeled using the standard deviation of each pixel in a sequence of the empty scene, called STD. So the specific algorithm is the next:

$$TD(x,y) = |I_t(x,y) - I_{t+1}(x,y) - I_{t+2}(x,y)| > c * STD(x,y),$$

being c a factor that indicates how many times the SGB has to be exceeded to be considered as foreground.

Figure 2e shows an example of TD method in a specific sequence.

Optical Flow

Optical flow, OF, is a method of moving extraction based on relative local moving between two observations of an object. In Barron et al. (Barron, 2005) a more detailed analyses of OF is shown. Moreover, Barron et al. (Barron, 1994) made a review of different approaches of this method. Other applications are proposed in (Moeslund, 2006; Hu, 2004).

The proposal for this work uses an algorithm that uses different iterations changing the size of local areas of search to accurate the result.

Moreover a third frame has been added in order to enhance the results, applying first the method to images t and $t+1$ and after $t+1$ and $t+2$. Once those results are extracted, an addition of themselves is made to obtain the full segmentation. Algorithm returns a displacement vectors matrix of the whole image (MOF, movement of Optical Flow), being necessary distinguish those parts with more movement than the rest. In order to do this, a median is extracted of vectors length and their standard deviation as well (SOF(I_1,I_2), standard deviation of OF). FG would be the parts that movement is more significant:

$$OF(x,y) = \begin{cases} MOF(I_t, I_{t+1}) > 2 * SOF(I_t, I_{t+1})) \\ OR \\ MOF(I_{t+1}, I_{t+2}) > 2 * SOF(I_{t+1}, I_{t+2})) \end{cases}$$

An example of the result is shown in Figure 2f.

Figure 3. Four different instants of sequence S1. During the whole sequence a couple of people are passing through the corridor, starting near the camera and going away. Other people are crossing in different moments producing occlusions.

Table 1. Summary table of area, element to element (B2B) and time values. Values are shown for all the sequences and divided by the different algorithms and combinations.

Area													
	BG		TD		OF		BGTD		BGOF		TDOF		BGTDOF	
	TPR	FPR	TPR	FPR	TPR	FPR	TPR	FPR	TPR	FPR	TPR	FPR	TPR	FPR
S1	90,2	8,8	37,9	3,1	52,1	5,8	96,1	12,2	97,3	15,7	65,0	10,4	**98,4**	19,8
S2	88,0	7,0	37,7	3,2	57,1	6,1	92,1	11,1	94,8	14,1	69,0	11,6	**96,1**	19,4
S3	95,6	10,4	38,0	3,8	42,1	6,6	97,1	15,0	96,9	17,7	67,8	12,8	**97,7**	23,4
S4	90,2	8,8	37,9	3,1	52,1	5,8	96,1	12,2	97,3	15,7	65,0	10,4	**98,4**	19,8
Mean	91,0	8,7	37,9	3,3	50,5	6,1	95,3	12,5	**96,6**	15,7	66,7	11,3	**97,6**	20,5
B2B													
	BG		TD		OF		BGTD		BGOF		TDOF		BGTDOF	
	TPR	FPR	TPR	FPR	TPR	FPR	TPR	FPR	TPR	FPR	TPR	FPR	TPR	FPR
S1	76,7	4,5	32,2	0,4	30,5	1,3	84,8	5,1	85,2	6,0	51,7	1,8	**87,8**	6,4
S2	75,9	5,8	28,1	0,7	43,3	2,0	81,7	6,8	84,8	8,1	54,5	2,9	**86,9**	8,7
S3	90,0	9,3	24,2	0,7	24,8	1,3	**91,1**	10,0	90,5	11,3	44,7	2,6	89,0	11,1
S4	76,7	4,5	32,2	0,4	30,5	1,3	84,8	5,1	85,2	6,0	51,7	1,8	**87,8**	6,4
Mean	79,6	5,7	29,0	0,5	31,6	1,4	85,5	6,5	**86,4**	7,6	50,5	2,2	**87,9**	7,9
Time (seconds/frame)													
	BG		TD		OF		BGTD		BGOF		TDOF		BGTDOF	
S1	0,0320		0,0137		24,6949		0,0457		24,7269		24,7086		24,7406	
S2	0,0301		0,0107		21,5668		0,0407		21,5969		21,5775		21,6075	
S3	0,0537		0,0114		21,3137		0,0650		21,3673		21,3250		21,3787	
S4	0,0323		0,0140		25,5024		0,0463		25,5347		25,5164		25,5487	
Mean	0,0359		0,0123		23,1957		0,0487		23,2333		23,2081		23,2456	

Combinations

A very important part of this study is combination of basic methods of segmentation to observe how better or worse are those combinations. In order to do this, the whole possibilities have been implemented including the next exposed:

- $BGTD = BG(x,y) + TD(x,y)$
- $BGOF = BG(x,y) + OF(x,y)$
- $TDOF = TD(x,y) + OF(x,y)$
- $BGTDOF = BG(x,y) + TD(x,y) + OF(x,y)$

Combinations have been done adding sequentially the results of basic methods previously to use morphological filters in post-segmentation step (View Figure 1). BG returns the body of the person segmented (in this particular application), however TD and OF segment the contour of people. Hence, adding BG with the others a more complete segmentation is achieved. Another thing to take into account is that adding segmentation, wrong labeled parts will be added as well, then error is incremented. An example of this addition is possible to view in Figure 2.

Post-Segmentation

The last two steps of the segmentation process are morphological filtering and object detection (view Figure 1), allowing results to be refined. Usually they are implemented in this kind of systems and are grouped as post-segmentation. It is necessary to have some knowledge of the scene and objects that are going to be detected because filters and detection algorithms depend on them.

On the one hand, morphological filters are used to fill holes in an object, join divided areas, filter small areas, etc. This kind of filters uses mathematical morphology in order to smooth frame objects based on their own shape. Basic methods are dilate and erode, and from them derive opening (erode and dilate) and close (dilate and erode). Furthermore, the filter might be applied with different shapes (linear, circular, square, diamond, etc.) to fit better the image. Foundations of this kind of operators are explained in (Dougherty, 2003).

On the other hand object detection allows the system discriminate objects extracted as result of previous steps. There is no general method, it depends on the objects are going to be analyzed. Features as colors, shapes, size (height and width) and position in the image or combination of some of them are used to distinguish different elements. Position allows false positives be eliminated such as areas similar to people located on the ceiling or on the water. Moreover, it is possible to use this feature in order to determinate different situations, but this goal is not studied in segmentation phase. For example Velastin et al. (Velastin, 2005) uses position to detect people in unsafely or forbidden regions in underground stations.

In this work, dilate and close morphological filters with a linear shape have been implemented, as well as an open filter with a circular shape. These have been applied sequentially and selected experimentally. After that, an algorithm to eliminate small size areas have been used to erase those parts that previous filters could not eliminate. Shapes of those areas are not considered because it is known that in this scene small segmented areas are errors independently of how they look like. People are analyzed in this system, therefore the object detection is optimized to this purpose. Thresholds of maximal size have been used, however

minimum have not because in segmentation step objects might be divided in smaller areas than the normal size of a person. Those errors and others like union of regions near each other's, can be corrected in following phases such as tracking, detecting patterns of movement, etc.

Complete Process

See Figure 2 for the complete process of segmentation and combination.

Solutions and Recommendations

Evaluation of methods has been done using image sequences from the database of the project CAVIAR (Fisher, 2004). Results shown in this section correspond to different sequences representing different situation with specific features. Those sequences are found in: http://groups.inf. ed.ac.uk/vision/CAVIAR/CAVIARDATA1/, and will be referred in the text as: *S1* - EnterExit-CrossingPaths1cor; *S2* - OneShopOneWait1cor; *S3* - ThreePastShop2cor; *S4* - TwoLeaveShop-2cor. In order to analyze methods quantitatively, pixels rightly labeled are taken into account (TP, true positive), and incorrectly labeled as part of foreground being background (FP, false positive). Using both values, the next rates are obtained and evaluated:

- TPR (True Positive Rate) = TP/T, being T the positive labeling in Ground truth.
- FPR (False Positive Rate) = FP/N, being N the negative labeling in Ground truth.

Receiver Operating Characteristics space is used in some cases to represent these values. ROC space allows both rates be evaluated in the same graph. It is described in vertical axis with TPR and in horizontal one with FPR.

Segmentation methods have been implemented in Matlab R2009a being BG and TD written specifically for this work and OF obtained from the

Figure 4. Results of methods and its combinations. Slim lines represent TPR and thickers FPR. (a) shows the best result for individual method. (g) is the best combination having a good TPR and acceptable FPR.

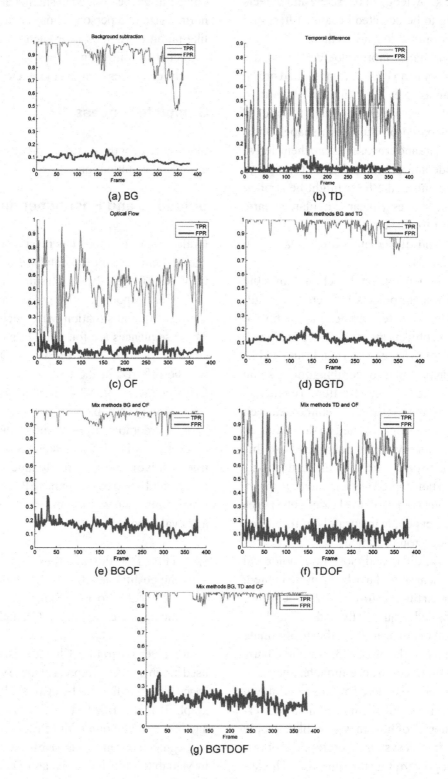

official website of Mathworks[1]. Programs have been tested using Matlab as well and executed in a personal computer with Pentium Dual-Core 2.20Ghz and a RAM memory of 4GB.

This section is divided in different subsections to evaluate algorithms from different points of view. On the one hand efficiency is evaluated using TPR, FPR and ROC space. This evaluation is also separated in two parts, one comparing resulted areas with the labeling of the ground truth (section Area comparison), and the other one compared element to element individually (section Element to element comparison). On the other hand a temporal cost evaluation has been done, showing its results in section Temporal evaluation.

First of all, a summary table is presented showing both values of area and elements (B2B) comparison. Values in the table represent mean of TPR and FPR for each sequence for the different algorithms implemented. Those values that represent the best TPR are marked as bold, showing that combining the three methods is the best result achieved.

Area Comparison

In this section, analyze of result obtained compared to those in the ground truth (GT) is done. This comparison returns TPR and FPR values to evaluate methods and their characteristics in different situations. In this section areas are compared not taking into account the elements individually. Therefore, situations such as occlusions are not relevant because the whole frame is compared with its own result in GT.

A representative case is explained in detail, corresponding to *S1* (Figure 3), where two people are walking across a corridor and other persons crossing them during the sequence. With it, summary table results (Table 1) are analyzed and understood.

Specifically to this sequence, Figure 4 shows TPR and FPR results. According to basic methods

BG obtains the best results with a TPR over 70% in the most of the sequence. At the end of *S1* the couple of women is at the end of corridor, this method has worse segmentation because all the shapes are joined forming a big and non-person shape. OF obtains a low TPR at the beginning of the sequence due to a person is crossing the couple causing difficulties to find movement vectors. If there are areas with similar features, OF algorithm needs more iterations to find correspondences.

Analyzing combination of basic methods, the best result is achieved by joining the three algorithms. Its values are similar to the BG and OF union, but sequences where there are big crowd of moving people, TD provides better results (view Figure 5). On the other hand, it is important to notice that TD and OF algorithms obtain mainly borders, however BG returns the body, in this case, of objects. That is the reason because combining BG with one of the others is better than TD and OF together.

TPR shows how many pixels labeled as a foreground match with the same label of the ground truth, therefore FPR is also an important value to take into account due it shows how many pixels tagged as foreground are not that in GT. Combining algorithms always the FPR grows, but this increasing is worth the augmented of the TPR. This work studies the first step of the system, so it is important to balance how much error is possible to assume and provide to the next step and be corrected as much as possible then.

Element to Element Comparison (B2B)

In this section TPR and FPR values are shown to compare each object of each image with their respective ones in the ground truth. This comparative allows seeing how good methods are segmenting individual elements. Supposing a frame where two persons are joined after post-segmentation step, it is one person for this analyze but in the ground

Figure 5. Different examples of how combining BG, TD and OF is better than only BG and OF. Circle shows those parts where it is possible to notice the improvement.

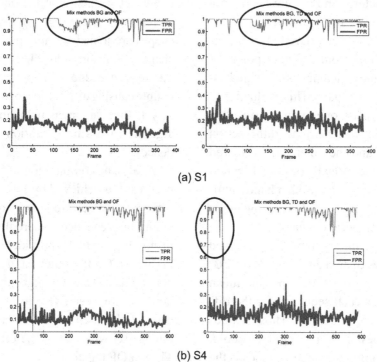

(a) S1

(b) S4

truth there are two, so in both comparison of each actual persons there will be part of TPR but also FPR due joining to the other person. Moreover it is important to say that false positives as shines, reflections, etc. are not taken into account.

As in the previous section same representative case is presented (S1, Figure 3) to analyze better the results and compare easily with those observed in section Area comparison. Concretely results are exposed using the ROC Space which permits see the relationship TPR/FPR of each method.

Graphs shown in Fig 6 confirm that the best result is achieved combining all the basic methods. These results are represented using dots that correspond to each frame of the sequence, and a bigger diamond that is the mean of TPR and FPR. Focusing in results, BG has the best rates individually with over 70% of TPR and fewer than 10% of FPR. The standard deviation, represented with an

ellipse, shows a value of 10% in the vertical axis meaning that it is a good reference and showing that the worst case, the norm is still upper than 50%. Combined method using all the basics is the one that achieve best results in the average of TPR and FPR, being 87.79% and 6.44% respectively. Values of BG and OF combined are similar, but slightly lowers in TPR (TPR = 85.18% and FPR = 6.00%).

On another way, a study of results depending on number of actual blobs in the scene has taken into account. This allows comparing results depending on the crowd people. In Figure 7, mean values of results separated with 1, 2, 3 and 4 or more elements in the frame could be appreciated. Combining all the algorithms a 94.48% of TPR and 1.25%FPR for 2 persons has been obtained. For 3 people, 90.12% has reached for TPR and 2.61% for FPR, and a TPR of 88.56% and FPR

Figure 6. Results of element to element comparison of basic methods and its combinations are shown. Graphs have diamonds representing mean values, and ellipses of standard deviation. BGTDOF has the best result with high TPR and low FPR as well as a good standard deviation of those values.

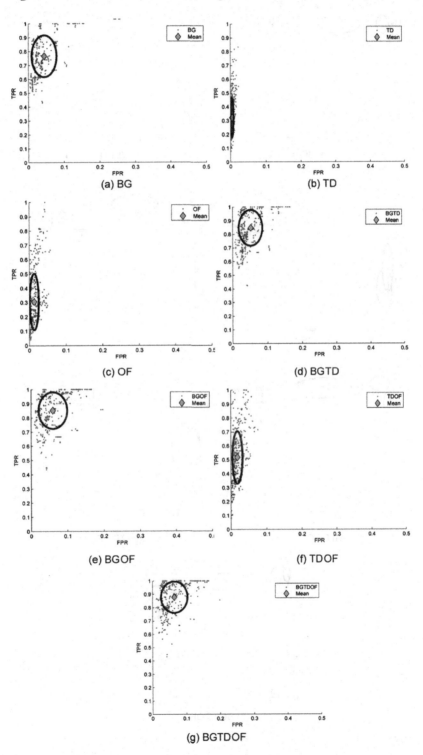

Figure 7. Results of B2B depending the number of blobs in the scene is presented. Means for different range of blobs are presented showing how FPR grows according to the number of blobs.

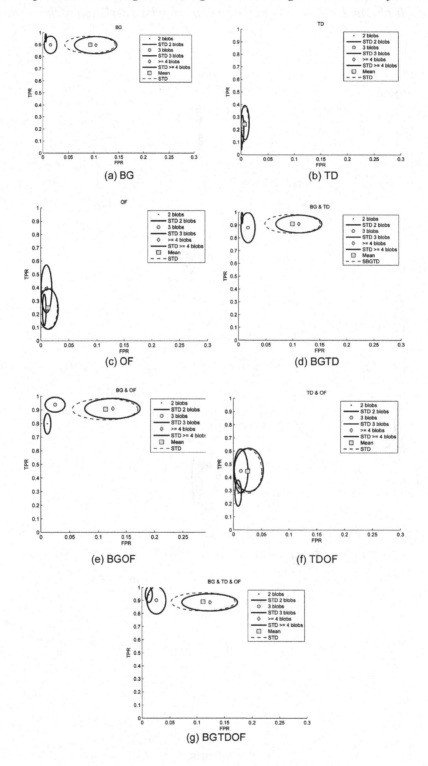

of 12.37% for 4 or more people. According to those results, more people appear in scene, worse TPR and FPR are obtained.

Temporal Evaluation

Computing time is an important variable to take into account. In this chapter, it is not important the accurate time of each method, but the relation between them and if are depending on the scene or not. Table 1 shows in the time part those values obtained for all the basic methods as the average of the time for all frames of sequences *S1*, *S2*, *S3* and *S4*. Combined algorithms time are the addition of each method, due combination have been done sequentially. BG and TD times are similar, but OF time is over 400% slower due its nature. Moreover, times of faster methods are almost the same on each sequence, but OF is not. With these values it is possible to decide which combination or individual method is appropriated for an application, depending the time requirements and segmentation precision.

BG and TD methods are not dependents of the scene because they are based on subtractions, but OF is because its nature. For example, sequence *S2* has a first part of empty scene and this produce a monotone time of processing and after time starts to have up and down values. It is produced due the OF depends on the scene and its changes during the sequence to find correspondences and movement parts.

FUTURE RESEARCH DIRECTIONS

Future research is opened with this study. Short-term study pretends to add more segmentation methods, including spatial as Mean Shift. Medium-term objectives are: on the one hand continue with other steps of segmentation, as tracking; and on the other hand, improve performance using embedded systems such as FPGAs or GPGPUS (General-Purpose Computing on Graphics Processing Units).

CONCLUSION

A comparative study of segmentation methods for video sequences is presented in this chapter. True positive and false positive rates as well as computing time have been used to evaluate efficiency and computational cost. Moreover, combination of basic methods has been evaluated in order to propose an improved method. Concretely, basic temporal segmentation methods explored have been Background subtraction, Temporal differencing and Optical flow. Using a ground truth, evaluation has been done from three points of view. First compare segmented areas, achieving the best individual result BG method, and the combination of the three the best result in general. On the other hand, element to element comparison has been evaluated obtaining same results than areas study, but with slightly worse results. Furthermore, number of elements in the scene has been taken into account. Results show that as more people are in the scene, worse results are reached.

Time cost has been evaluated in order to analyze relative differences between methods and their dependence to the sequence features. BG and TD have similar times, but OF is over 400% slower, and this last method is also depending on the characteristics of homogeneity and variation between frames.

ACKNOWLEDGEMENT

This work has been partially supported by the University of Alicante project GRE11-01.

REFERENCES

Arbelaez, P., Maire, M., Fowlkes, C., & Malik, J. (2009). *From contours to regions: An empirical evaluation* (pp. 2294–2301). IEEE CSC on CVPR. doi:10.1109/CVPR.2009.5206707

Barron, J., & Thacker, N. (2005). *Tutorial: Computing 2D and 3D optical flow.* (2004-012). Tina memo.

Barron, J. L., Fleet, D. J., & Beauchemin, S. S. (1994). Performance of optical flow techniques. *International Journal of Computer Vision, 12,* 43-77. Retrieved from http://dx.doi.org/10.1007/BF01420984

Benezeth, Y., Jodoin, P. M., Emile, B., Laurent, H., & Rosenberger, C. (2010). Comparative study of background subtraction algorithms. *Journal of Electronic Imaging, 19*(3). doi:10.1117/1.3456695

Dougherty, E. R., & Lotufo, R. A. (2003). *Hands-on morphological image processing* (*Vol. 59*). Bellingham, WA: The International Society for Optical Engineering, ETATS-UNIS. doi:10.1117/3.501104

El Baf, F., Bouwmans, T., & Vachon, B. (2007). Comparison of background subtraction methods for a multimedia application. In *Systems, Signals and Image Processing, 2007 and 6th EURASIP Conference focused on Speech and Image Processing, Multimedia Communications and Services* (pp. 385-388).

Fisher, R. B. (2004). PETS04 surveillance ground truth data set. In *Sixth IEEE International Workshop on Performance Evaluation of Tracking and Surveillance* (pp. 1-5).

Hall, D., Nascimento, J., Ribeiro, P., Andrade, E., Moreno, P., & Pesnel, S. … Crowley, J. (2005). Comparison of target detection algorithms using adaptive background models. *Joint IEEE International Workshop on Visual Surveillance and Performance Evaluation of Tracking and Surveillance,* (pp. 113-120).

Haritaoglu, I., Harwood, D., & Davis, L. S. (2000). W4: Real-time surveillance of people and their activities. *IEEE Transactions on Pattern Analysis and Machine Intelligence, 22,* 809–830. doi:10.1109/34.868683

Herrero, S., & Bescós, J. (2009). Background subtraction techniques: Systematic evaluation and comparative analysis. In *Proceedings of the 11th International Conference on Advanced Concepts for Intelligent Vision Systems* (pp. 33-42). Springer-Verlag.

Hu, Q., Li, S., He, K., & Lin, H. (2010). A robust fusion method for vehicle detection in road traffic surveillance. In Huang, D.-S., Zhang, X., Reyes Garcia, C., & Zhang, L. (Eds.), *Advanced Intelligent Computing Theories and Applications with Aspects of Artificial Intelligence* (*Vol. 6216*, pp. 180–187). Berlin, Germany: Springer. doi:10.1007/978-3-642-14932-0_23

Hu, W., Tan, T., Wang, L., & Maybank, S. (2004). A survey on visual surveillance of object motion and behaviors. *Pattern Recognition, 34,* 334–352.

Moeslund, T. B., Hilton, A., & Krüger, V. (2006). A survey of advances in vision-based human motion capture and analysis. *Computer Vision and Image Understanding, 104*(2-3), 90–126. doi:10.1016/j.cviu.2006.08.002

Velastin, S. A., Boghossian, B. A., Ping, B., Lo, L., Sun, J., & Vicencio-Silva, M. A. (2005). PRISMATICA: Toward ambient intelligence in public transport environments. In *Good Practice for the Management and Operation of Town Centre CCTV, European Conference on Security and Detection,* Vol. 35 (pp. 164-182)

VenuGopal, T., & Naik, P. P. S. (2011). Image segmentation and comparative analysis of edge detection algorithms. *International Journal of Electrical. Electronics & Computing Technology, 1*(3), 38–42.

Zhang, Y. (1997). Evaluation and comparison of different segmentation algorithms. *Pattern Recognition Letters, 18*(10), 963–974. doi:10.1016/S0167-8655(97)00083-4

ADDITIONAL READING

Avidan, S. (2005). Ensemble tracking. *IEEE Computer Society Conference on Computer Vision and Pattern Recognition*, Vol. 2, (pp. 494-501).

Baumann, A., Boltz, M., Ebling, J., Koenig, M., Loos, H. S., & Merkel, M. (2008). … Yu, J. (2008). A review and comparison of measures for automatic video surveillance systems. *EURASIP Journal on Image and Video Processing*, (n.d.), 30.

Beleznai, C., Fruhstuck, B., & Bischof, H. (2005). Tracking multiple humans using fast mean shift mode seeking. In *Workshop on Performance Evaluation of Tracking and Surveillance, Breckenridge*.

Collins, R. T. (2003). *Mean-shift blob tracking through scale space* (*Vol. 2*, p. 234). IEEE CSC on CVPR.

Comaniciu, D., & Meer, P. (2002). Mean shift: A robust approach toward feature space analysis. *IEEE Transactions on Pattern Analysis and Machine Intelligence, 24*, 603–619. doi:10.1109/34.1000236

Comaniciu, D., Ramesh, V., & Meer, P. (2000). *Real-time tracking of non-rigid objects using mean shift* (*Vol. 2*, p. 2142). IEEE CSC on CVPR. doi:10.1109/CVPR.2000.854761

Erdem, C., Murat Tekalp, A., & Sankur, B. (2001). Metrics for performance evaluation of video object segmentation and tracking without ground-truth. In *Proceedings 2001 International Conference on Image Processing*, Vol. 2 (pp. 69 -72).

Estrada, F., & Jepson, A. (2009). Benchmarking image segmentation algorithms. *International Journal of Computer Vision, 85*, 167–181. doi:10.1007/s11263-009-0251-z

Gelasca, E., Ebrahimi, T., Farias, M., Carli, M., & Mitra, S. (2004). Towards perceptually driven segmentation evaluation metrics. In *Conference on Computer Vision and Pattern Recognition Workshop, CVPRW '04* (p. 52).

Kang, W.-X., Yang, Q.-Q., & Liang, R.-P. (2009). The comparative research on image segmentation algorithms. In *First International Workshop on Education Technology and Computer Science, ETCS '09*, Vol. 2 (pp. 703 -707).

Littmann, E., & Ritter, H. (1997). Adaptive color segmentation-a comparison of neural and statistical methods. *IEEE Transactions on Neural Networks, 8*(1), 175–185. doi:10.1109/72.554203

McGuinness, K., & O'Connor, N. E. (2010). A comparative evaluation of interactive segmentation algorithms. *Pattern Recognition, 43*(2), 434–444. doi:10.1016/j.patcog.2009.03.008

McGuinness, K., & O'Connor, N. E. (2011). Toward automated evaluation of interactive segmentation. *Computer Vision and Image Understanding, 115*(6), 868–884. doi:10.1016/j.cviu.2011.02.011

Mignotte, M. (2008). Segmentation by fusion of histogram-based -means clusters in different color spaces. *IEEE Transactions on Image Processing, 17*(5), 780–787. doi:10.1109/TIP.2008.920761

Pantofaru, C. (2005). A comparison of image segmentation algorithms. *Robotics, 18*, 123–130.

Parks, D. H., & Fels, S. S. (2008). Evaluation of background subtraction algorithms with post-processing. *IEEE Conference on Advanced Video and Signal Based Surveillance*, (pp. 192-199).

Phung, S. L., Bouzerdoum, A., & Chai, D. (2005). Skin segmentation using color pixel classification: Analysis and comparison. *IEEE Transactions on Pattern Analysis and Machine Intelligence, 27*, 148–154. doi:10.1109/TPAMI.2005.17

Piccardi, M. (2004). Background subtraction techniques: A review. In *2004 IEEE International Conference on Systems, Man and Cybernetics*, Vol. 4 (pp. 3099-3104).

SanMiguel, J. C., & Martinez, J. M. (2010). On the evaluation of background subtraction algorithms without ground-truth. *IEEE Conference on Advanced Video and Signal Based Surveillance,* (pp. 180-187).

Schlogl, T., Beleznai, C., Winter, M., & Bischof, H. (2004). Performance evaluation metrics for motion detection and tracking. In *Proceedings of the 17th International Conference on Pattern Recognition, ICPR 2004,* Vol. 4 (pp. 519 - 522).

Sfikas, G., Nikou, C., & Galatsanos, N. (2008). Edge preserving spatially varying mixtures for image segmentation. *IEEE Computer Society Conference on Computer Vision and Pattern Recognition,* (pp. 1-7).

Sikka, K., & Deserno, T. M. (2010). Comparison of algorithms for ultrasound image segmentation without ground truth. In *Society of Photo-Optical Instrumentation Engineers (SPIE) Conference Series, Vol. 7627.*

Sturm, P., & Maybank, S. (1999). On plane-based camera calibration: A general algorithm, singularities, applications. In *IEEE Computer Society Conference on Computer Vision and Pattern Recognition,* Vol. 1 (p. 2).

Unnikrishnan, R., Pantofaru, C., & Hebert, M. (2005). A measure for objective evaluation of image segmentation algorithms. *Computer Vision and Pattern Recognition Workshop,* (p. 34).

Unnikrishnan, R., Pantofaru, C., & Hebert, M. (2007). Toward objective evaluation of image segmentation algorithms. *IEEE Transactions on Pattern Analysis and Machine Intelligence, 29,* 929–944. doi:10.1109/TPAMI.2007.1046

Wang, L., Hu, W., & Tan, T. (2003). Recent developments in human motion analysis. *Pattern Recognition, 36*(3), 585–601. doi:10.1016/S0031-3203(02)00100-0

Zhang, H., Fritts, J. E., & Goldman, S. A. (2008). Image segmentation evaluation: A survey of unsupervised methods. *Computer Vision and Image Understanding, 110*(2), 260–280. doi:10.1016/j.cviu.2007.08.003

KEY TERMS AND DEFINITIONS

Background Subtraction: Segmentation method that uses a model of the scene to compare the images extracting the non-background part.

Behavior Analysis: Analyze of objects behavior during a sequence of images, such as displacement, change of shapes and other possible movements.

Computer Vision: Area of computer intelligence that study how to provide a computer the ability of vision.

Image Segmentation: Process to extract different regions of images, in order to make easier their analysis.

Optical Flow: Segmentation method that uses relative local moving between two observations of an object to extract moving parts.

Spatial Segmentation: Image segmentation using the information of a single image, such as color, borders, etc.

Temporal Differencing: Segmentation method that uses the differences between frames of a sequence in different moments to extract moving parts.

Temporal Segmentation: Image segmentation using the information of a sequence of images.

ENDNOTES

1. http://www.mathworks.com/matlabcentral/fileexchange/17500 (last visit: 13/12/2011)

Section 2
Computer Vision Applications

Chapter 4
Security Applications Using Computer Vision

Sreela Sasi
Gannon University, USA

ABSTRACT

Computer vision plays a significant role in a wide range of homeland security applications. The homeland security applications include: port security (cargo inspection), facility security (embassy, power plant, bank), and surveillance (military or civilian), et cetera. Video surveillance cameras are placed in offices, hospitals, banks, ports, parking lots, parks, stadiums, malls, train stations, airports, et cetera. The challenge is not for acquiring surveillance data from these video cameras, but for identifying what is valuable, what can be ignored, and what demands immediate attention. Computer vision systems attempt to construct meaningful and explicit descriptions of the environment or scene captured in an image. A few Computer Vision based security applications are presented here for securing building facility, railroad (Objects on railroad, and red signal detection), and roads.

INTRODUCTION

Securing Building Facility

Homeland security functions focus on intelligence and warning, protecting critical infrastructure and domestic counterterrorism. Biometric is a reliable way to authenticate the identity of a living person based on the physiological or behavioral characteristics. Gait of a person is a non-invasive biometric that can be used for recognition at a greater distance without the knowledge or cooperation of the person being recognized. Body weight, limb length, habitual posture, bone structure, and age influence the gait of a person. It has applications in visual surveillance, aware-spaces, and intelligent human-computer interfaces. These factors give each person a distinctive gait, which can be used as a biometric. The non-linear characteristics associated with gait pose a major challenge for research in this area. In this research, three methods are devised and evaluated for performance for the recognition of static postures in gait by combining Hidden Markov model with Visual Hull technique by Gomatam A.M., & Sasi S. (2004), Stereovision with 3D Template Matching by Gomatam A.M.,

DOI: 10.4018/978-1-4666-2672-0.ch004

& Sasi S. (2004) and Isoluminance lines with 3D Template Matching (TM) by Gomatam, A. M. & Sasi. S. (2005). These methods were tested on silhouettes of different person that are extracted from Carnegie Melon University's Motion of Body (MoBo) database (2004) and performances were compared.

Railroad and Road Safety

Securing Railroad

Rail accidents pose a major threat in terms of lives and cost. The widespread of concerns for the nation's railroad have grown direr since the attacks of September 11, 2001. While the Federal Government has implemented extensive safety and security measures in the aviation industry, it has left railroad security entirely up to rail corporations. Statistics collected by Operation Lifesaver show that fifty percent of the rail accidents occur at rail crossings equipped with flashing lights, barrier gates and warning bells. Though railroad crashes are rare, when they do occur, lead to massively destructive and deadly railroad crashes. According to the National Transportation Safety 60% of all crossing fatalities occur at unprotected crossings, and approximately 80% of all public railroad crossings are not protected by lights and safety gates. Collisions with other trains, derailment, and collisions with passenger vehicles are the common types of railroad accidents.

Here are some statistics from CNN News (2005) regarding the railroad accidents:

- Every 90 minutes there is a train collision or derailment.
- A train carrying hazardous material goes off the tracks approximately every 2 weeks in the United States.
- More than 50% of all railroad accidents occur at unprotected crossings.

These kinds of massive destructions could occur due to various reasons. Train accidents are caused due to mechanical failures, communication failures, railroad crossings littered with debris or a simple human error by an individual employee such as a locomotive engineer, driver, train conductor, rail inspector or railroad maintenance mechanic. The people responsible for the smooth operation of current rail system are listed below:

- **Locomotive Engineer:** Controls the locomotive
- **Driver:** Assists Locomotive engineer
- **Rail Inspector:** Inspects signals and track wiring
- **Train Conductor:** Deals with emergency situations
- **Railroad Maintenance Mechanic:** Repairs damaged tracks

There is a horn that can be honked to warn any vehicles on the rail track. But any common human error can lead to a train crash. There is no automated warning signal to alert the locomotive engineer about a possible threatening object or vehicle on the rail track in the current locomotive engine cabins. Whenever a train accident occurs, there will be serious personal injuries and extensive economical loss. The economical loss is due to both damaged property and the huge compensation paid to the victims. Lately, the locomotives are equipped with camera and a microphone as investigation aids. These are mounted in the locomotive engine cabin and a live video and audio are recorded continuously from the perspective of a locomotive engineer while the train is moving. Apart from this, they can aid in the recording of gate crossing incidents, near misses, or other operating incidents. Digital video recordings provide clear and detailed evidence that are more reliable than an eye witness's accounts in case of accidents. Also, digital recordings are legally admissible in court if needed. These video recordings can also be used for monitoring obstacles present on railroad using digital image processing techniques such as path tracking, edge detection, object recognition, and red signal detection.

Object Detection on Railroads

An automated computer "Vision-based Real-time Smart system to Prevent Railroad Accidents (VRSPRA)" that analyses individual frames in the video stream and generates a warning signal for the locomotive driver is presented. Abilash Sanam who was a graduate student at Gannon University in 2004 has done this research and devised an application for locomotive industry to save human life as well as major economic loss caused by the rail accidents.

Tracking Red Signal Lights near Railroads

Color of signal lights and position of signal poles is a major concern when designing an automatic signal detection system. The investigation of accidents completely depends on the unadulterated information gathered at an accident zone. Limited information available from the accident zone, such as eyewitness and physical evidence, causes several problems for investigations. These investigations indirectly affect the organizations which are depending on them. Generally, this kind of accident occurs at rail-road crossings, intersections of the road, and highway crossings. According to the investigators from the National Transportation Safety Board, an accident between an Amtrak train and a tractor-trailer outside of Chicago caused the death of at least 11 people. Both the locomotive engineer and the truck driver have given totally different accounts of what had happened prior to the crash (Train Accident Report, 2004).

According to the US Department of Transportation, human fault causes 70 to 80 percent of the transportation accidents. Human fatigue is also playing an important role in these accidents which are not understood by normal investigations. This kind of accidents causes much loss to the organizations, both economically and on reputation. The examples of human fatigue that caused losses according to Rail Accident Report (2004) and Rail Employee Fatigue (Amtrak 2004) report.

- A $52 million judgment was awarded against Conrail in an accident involving a railroad worker who was crushed by a train controlled by a sleep deprived driver on double shifts.
- A $4 million judgment was awarded against Nabors Drilling for an accident where an employee driving home after working long hours fell asleep behind the wheel.

The survey conducted by the Farmers Insurance Group of Companies showed that more than 36% of motorists admitted to running a red light in the past year, which is one of the leading causes of crashes in urban areas. The statistics gathered by the Insurance Institute for Highway Safety (IIHS) shows red light running crashes cause nearly 1,000 deaths and more than 200, 000 injuries each year. The main reason is due to the automobile driver failing to stop at the red signal, and runs over other road users. This happens when the driver fails to see that the signal is red due to his negligence after being drunk, or if the signal is invisible because of the bad climatic conditions.

A Vehicle Mounted Recording Systems (VMRS) can be used to provide evidence for organizations such as law enforcement, insurance agencies, and transportation. This recording system continuously monitors and records all the events. In the case of any accident these event recorded videos are used as evidence for an investigation. These recorded videos will provide complete information about an accident with all the causes behind it unaltered. The insurance industry will benefit from vehicle Mounted Video Recording Systems since they provide accurate evidence against the fraudulent claim losses as mentioned by Trax (2004) and National Transportation Safety Board (2004).

Though VMRS records continuously the events such as speed, time, location, transmission position, and heading direction of the vehicle etc.; the position of signal lights and the color associated with these lights are not automatically detected by the current system. A computer vision-based system is designed and implemented to track the position and color of signal lights. The position and the color of the signal are written to a log file that can be used as concrete evidence while investigating the causes of accidents. The "Color-based Signal light Tracking in Real-time Video (CSTRV)" system is an intelligent system using La*b* color model in combination with contour tracking by Yelal et al. (2006). This method analyzes each frame of the video sequentially, and detects the presence of signal lights and its color.

Ensuring Road Safety

Red light running is a leading cause of urban crashes that often results in injury and death. Total road deaths in USA for the year 2004 were 42,636. A survey conducted during 1999-2000 revealed that 20% of vehicles involved in road accidents did not obey the signal. Each year "red" light running causes nearly 200,000 accidents resulting in above 800 deaths and 180,000 injuries according to Drive and Safety Alive, Inc. (2006) and Department for Transport - UK (2004). Signal lights on the road intersection are for controlling traffic. Some people do not abide by the traffic rules and cross the intersection when the signal light is 'red'. In order to reduce the accident rate at the intersections, busy and accident prone intersections should be monitored. The authorities may not be able to monitor all the intersections continuously round the clock on all days. This demands a cost effective and automated system for continuously monitoring all the intersections, and to penalize the people who would violate the traffic rules.

Automatic License Plate Recognition (ALPR) systems have been developed and discussed in Wikipedia for Number Plate (2005), Motorola Solutions for Government (2005), CCTV Info – UK (2005), and License Plate recognition (2005). In ALPR systems for monitoring intersections, a still camera is placed adjacent to the signal lights for capturing the license plate of the car at the intersection. Sensors are located on the road to detect the presence of a vehicle at the intersection. When the signal light is 'red' and the sensors are active then a still photograph is taken which is used for issuing penalty ticket by the law enforcement authorities. The camera is equipped with a bright flash light for helping to capture quality image, and for cautioning the driver for his/her violation. The ALPR system is not a foolproof system because the license plate can be tampered or the plate might be stolen from another car or the license plate would not be visible due to bad weather conditions, or the sensors on the road might be tampered.

In this research an expert system that would capture the Vehicle Identification Number (VIN) and the License plate of the vehicle crossing the intersection on 'red' signal is presented. Using the VIN, it is possible to find the owner of the car, insurance details and the car facts report. No sensors are needed for this Vision-based Monitoring System for Detecting Red signal crossing (VMSDR) that captures and recognizes both VIN and License plate of the vehicle running 'red' signal light at the intersections presented by the research conducted by Sharma, R. and Sasi, S. (2007). VMSDR system needs two video cameras out of which one is placed on the sidewalk and the other is placed on the pole above the intersection adjacent to the signal lights. The video camera placed on the side walk captures the license plate and the video camera placed on the pole along with the 'signal light' captures the VIN. This research is intended to provide a support system for the law enforcement agencies to proactively ensure that intersections are engineered to discourage red light running.

ARCHITECTURE AND SIMULATION RESULTS

Securing Building Facility

As a newly emergent biometric, gait recognition aims at discriminating individuals by the way they walk. Gait has the advantages of being non-invasive and difficult to conceal, and is also the only perceivable biometric at a distance. The need for automated person identification is growing in many applications such as surveillance, access control and smart interfaces. It is well known that biometrics is a powerful tool for reliable and automated person identification system. In the study of gait, recognition is performed with a technique called an activity-specific static biometric. The advantage of measuring a static property is that it is amenable to being done from multiple viewpoints. In this research, recognition of gait characteristics is performed using static postures. Here, three different techniques are used to identify the gait characteristic from the given static postures. They are "Enhanced Gait Recognition Using HMM and Visual Hull Techniques," "Multimodal Gait Recognition based on Stereo Vision and 3D Template Matching," and "Gait Recognition Based on Isoluminance Line and 3D Template matching."

Gait Recognition Using Hidden Markov Model and Visual Hull Technique

A combination of Hidden Markov Model (HMM) and Visual Hull (VH) techniques is utilized to recognize the gait characteristics from the static postures are used in this method. Initially, the static pictures of various individual are taken from the right and left side using a camera. The silhouette is extracted for each of the pictures because the 3D information provided by the cameras is not enough to get a precise representation and also to provide a limit. This limit is the outer contour for the fitting process so that the model should not overpass that silhouette. Body silhouette extraction is achieved by a simple background subtraction and thresholding followed by applying a 3x3 median filter operator to suppress isolated pixels.

The concept of visual hull is to characterize the best geometric approximation that can be achieved using a shape-from-silhouette reconstruction method. This method exploits the fact that the silhouette of an object in an arbitrary view re-projects onto a generalized polyhedral Visual Hull. After the background subtraction is performed, the system approximates the visual hull in the form of a polyhedral volume. Without loss of generality it is presumed that the XZ-plane of coordinate system is the ground plane, and the Y-axis is the normal to the ground. The four different gait characteristics identified in the obtained polyhedral volume are the head section, pelvic or the centroid, foot section and the two-foot regions. The location of the centroid of the subject is estimated by taking the center of gravity of the VH. For the polyhedral VH, it is the centroid of the polyhedral model that can be computed while building the model. For the sampled silhouettes, one estimates the VH by integrating the volume enclosed within the endpoints of the ray intervals. The HMM is applied to the obtained polyhedral volume and all the points corresponding to the gait character is identified. The set of points are then stored in a gait database. This database contains values of several gait postures of different samples. A gait sample, which needs to be identified and authenticated, is subjected to the above-mentioned procedure and sets of points are obtained. The obtained points are then compared against each corresponding set of points in the database. An architecture is presented in Figure 1 using these techniques for authentication purposes.

Gait Recognition Based on Stereo Vision and 3D Template Matching

Stereovision technique is combined with the 3D template matching technique for effective iden-

Figure 1. Architecture for enhanced gait recognition using HMM and Visual Hull Technique

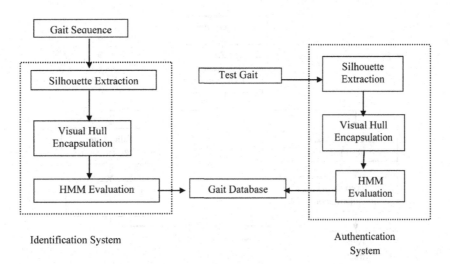

tification and authentication of a person in this method. An HMM technique is also implemented in parallel to ensure accuracy of recognition as shown in Figure 2. The silhouettes are extracted from the given pictures using a simple background subtraction and thresholding followed by applying a 3x3 median filter operator to suppress isolated pixels. The visual hull technique is used as an approximate geometric model of the objects in the scene. This system approximates the visual hull in the form of a polyhedral volume. This is made use by both the systems. The polyhedral volume is then subjected to stereovision technique for the purpose of reconstruction. Here, the structural based technique is made use for reconstructing the 3D image. The image is converted into the boundary representation, and the correspondence candidates are found from the epipolar condition, intensity, and the shape. The connectivity of the segments is evaluated according to less distance, same intensity, and same angle. Based on the similarity, the correspondence between the left and right segments is found, and then the 3D information is reconstructed. From the 3D image thus obtained, a 3D template is extracted using the segmentation-based technique. This is used to find the precise position and orientation

of the target object from depth data by projecting the corresponding 3D model. The extracted 3D template is stored in a database. A sample gait, which needs to be identified, is subjected to the above-mentioned procedure and is compared with the templates in the database. Parallel to this process, values are obtained from the silhouettes of image using HMM analysis. This is also used to identify and authenticate the person.

Gait Recognition Based on Isoluminance Line and 3D Template Matching

The Isoluminance lines for stereovision technique are combined with 3D template matching technique for effective identification and authentication in this method. This method is similar to the second method except that isoluminance line based stereovision is used to reconstruct the image and to find depth information. Initially, the visual hull technique is used as an approximate geometric model of the objects in the scene. After the background subtraction and thresholding, the system approximates the visual hull in the form of a polyhedral volume. Then the Isoluminance line based stereovision technique is applied to

Figure 2. Architecture for multimodal gait recognition based on stereovision and 3D template matching

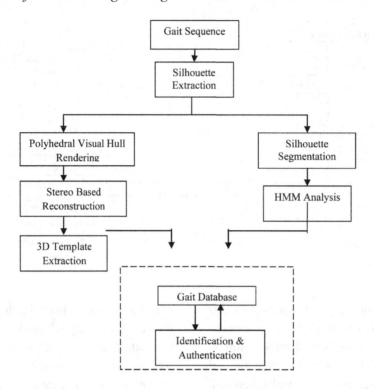

the polyhedral volume. The volume is segmented into various levels of black and white areas by setting a threshold. Black areas smaller than a certain threshold are deleted. A 3D partial image is thus reconstructed from the isoluminance lines. The partial images are merged together into one final image, which gives the depth information. From this 3D image, a 3D template is extracted using the segment-based 3D template matching technique. This is used to find the precise position and orientation of the target object from depth data by projecting the corresponding 3D model. The extracted 3D template is stored in a database. A sample gait, which needs to be identified, is subjected to the above-mentioned procedure and is compared with the templates in the database. The architecture is shown in Figure 3.

The architecture shown in Figure 1, Figure 2, and Figure 3 correspond to Gait Recognition Using HMM and Visual Hull Technique, Gait Recognition based on Stereo Vision and 3D Template Matching, and Gait Recognition Based on Isolu-

minance Line and 3D Template matching respectively are simulated using 40 test samples of silhouettes of different persons. The methods were simulated using Image Processing Toolbox in MATLAB. The silhouettes are extracted using the photo samples stored in the Carnegie Melon University database.

Simulation: Gait Recognition Using HMM and Visual Hull Technique

The silhouettes are extracted using static images and VH encapsulation is applied on them. HMM is used to find the maximum likelihood detection using the values obtained from the VH encapsulation for identification purpose and these values are stored in a sample database. Testing is conducted with a new set of test sample against the ones stored in the database. The simulation results are shown below using two sample test cases. For the first case, the samples used are from two different persons for the same posture. The pictures used

Figure 3. Architecture of gait recognition system

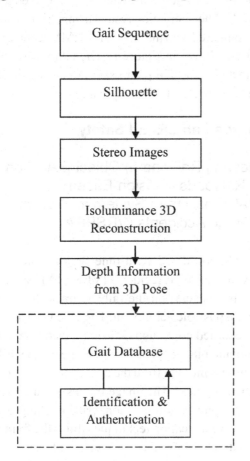

Figure 4. VH simulation

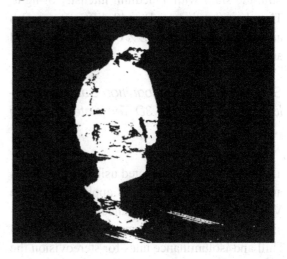

Simulation: Gait Recognition based on Stereo Vision and 3D Template Matching

in this research are obtained from a controlled environment, i.e. laboratory.

In the second test case, the samples are of the same person but for different postures. Initially, using the VH technique both cases are simulated to extract the features from silhouettes. Secondly four points are identified using HMM technique, and the four values corresponding to the centroid, distance between centroid and shoulders, and distance between the centroid and the foot are recorded. The values measured for case1 and case2 are given in Table 1 & 2. The values in Table 2 clearly demonstrate that the two samples are not identical and hence are in compliance with the actual data. Table 2 depicts a match greater than 86% from which an authentication can be concluded.

The silhouettes are extracted using static images, and then VH encapsulation is applied. The depth of the image is obtained using structural analysis method of stereovision. The combined application of Visual Hull and Stereo Vision increases the accuracy of the depth information of the image. The depth information of the pre-identified points corresponding to centroid, shoulder points and the leg points are obtained for this structure. These values are stored in a database and a 3D model is constructed. From this 3D templates are extracted and are compared with the templates stored in the database for authentication. An HMM analysis is performed in parallel to locate the centroid of the silhouette, and is stored along with four different values corresponding to the regions identified. Validation is conducted with a characteristic feature of a new person against the ones stored in the database. Authentication is valid if the feature matches with more than 87% of the values of the corresponding image in the database. The silhouettes for this research are extracted using the photo samples stored in the Carnegie Melon University database. The algorithm supports a variation of 45 degrees from the front posture on both the sides with a medium intensity of light. The algorithm is tested on 40 different samples and achieved a result of 89% of identification and authentication.

Simulation: Gait Recognition Based on Iso-luminance Line and 3D Template Matching (3DTM)

The silhouettes are extracted using static images and the VH encapsulation is applied. Once the Visual Hull is built, isoluminance line for stereovision is applied. With the combination of Visual Hull and isoluminance lines for stereovision the depth information of the image is obtained accurately. The depth information is obtained for the pre-identified points of the structure. For this model 3D templates are extracted and are compared with the templates stored in the database for authentication. Testing is conducted with a characteristic feature of a new person against the one stored in the database. The simulation results for the architecture using isoluminance line and 3DTM are shown in Figure 4 and Figure 5. Authentication is valid if the feature matches with

most of the values in the database for a given test sample. The simulation results using this method are compared with the ones using HMM technique and are shown in Figure 6. The silhouettes are extracted using the photo samples stored in the Carnegie Melon University database.

Railroad and Road Safety

Securing Railroad:(i) Object Detection on Railroads - "Vision-Based Real-Time Smart System to Prevent Railroad Accidents (VRSPRA)"

The "Vision-based Real-time Smart system to Prevent Railroad Accidents (VRSPRA)" system detects obstacles on the railway track that may cause a possible derailment. The video and audio are recorded continuously using a loco cam that is mounted in the locomotive engine cabin while the train is moving from the perspective of a locomotive engineer. The image frames are analyzed by applying edge detection, path tracking and object identification techniques. Initially, frames are extracted from the video data and processed to monitor the railway track for gaps and for any

Figure 5. Template matching

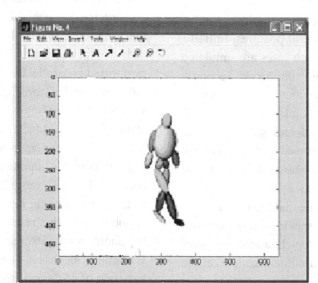

Figure 6. Comparison graph of the two methods

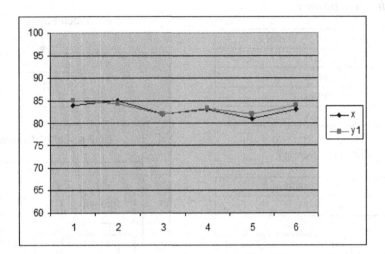

obstacles. If there are any sudden variations in the number of edges in the current region of interest then a warning signal is generated for the locomotive engineer. The VRSPRA system architecture is shown in Figure 7.

A customized sample video stream collected during normal operation of the locomotive from GE Transportation Systems in Erie, Pennsylvania is used for testing the VRSPRA system. Initially during preprocessing, audio-video interleaved (avi) data is extracted from the customized data stream. This avi file is encoded using MPEG-4 compression format. Hence 'DivX' codec is used to extract the sample I-Frames. Obstacles of different size are introduced in these frames and combined together to form a new video sequence. This video sequence is used to identify the objects as a possible threat and to generate an alarm signal. The image frames are analyzed by applying edge detection, path tracking and object identification techniques. The video is recorded at a speed of 30 frames per second. According to the research carried out at Defense Advanced Research Projects Agency (DARPA) in (2004), the normal speed of a locomotive is less than 100MPH. The distance traveled in one frame time is approximately 1.5 yards. The visibility range

of the camera used is 1.5 miles. The distance covered before activating the alarm signal is 0.012 miles. The breaking distance for the train with a usual force of 40Newton is less than a mile. Hence, there is enough time for the driver to take proper action after receiving the warning alarm. The timing computation for alarm signal generation is given below:

Let Train Speed be 100 MPH

Let Frame Rate be 30 FPS

Distance Covered in 1 Frame time = ~1.5 yards

Visibility range of camera used = 1.5 miles

Distance covered before alerting = 0.012 miles

Breaking distance for the train with an usual force of 40 Newton < 1 Mile

Figure 7. VRSPRA system architecture

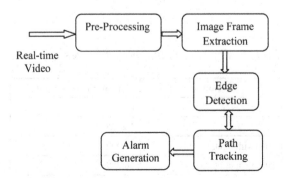

Table 1. Sample for different persons for the same posture and with different posture

Person 1	Person 2
490.00	201.00
317.00	251.50
307.00	240.00
220.00	263.65

Table 2. Test sample for same person

Posture 1	Posture 2
561.50	560.45
507.45	507.35
480.12	481.54
390.56	390.56

Table 3. List of objects and respective edge count

Object	No. Of Edges	Alarm
SUV	225	Yes
Bus	269	Yes
Locomotive Cabin	311	Yes
Log of wood	160	No

Table 4. The log file of signal lights

Frame Numbers	Presence of signal Lights	Color of signal light
1-30	No	No color
31-60	No	No color
61-90	Yes	Red
91-120	No	No color

Table 5. The execution times of frames

Frames	Frame rate (seconds)
1	0.1406
10	1.2438
100 (with lights)	11.999
100(without lights)	10.799

Figure 8. CSTRV system architecture

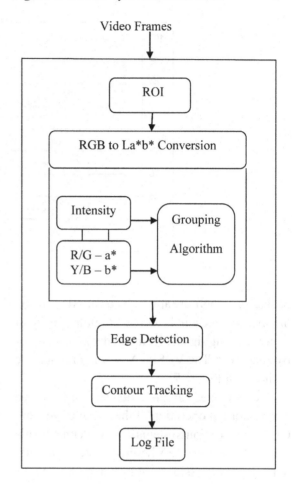

Figure 9. The image frame from a video

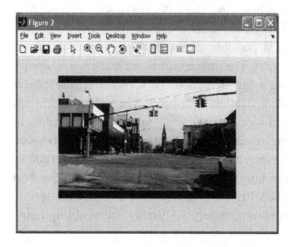

Hence it is possible to apply the brakes and stop the train in case any threatening object is identified on the tracks. Image frames with a sport utility vehicle (SUV), bus, locomotive cabin and a log of wood are inserted in the video stream collected from the locomotive cabin. The number of edges with and without the objects parked on railway track is counted and is given in Table 3.

Color-Based Signal Light Tracking in Real-Time Video (CSRTV)

The "Color-based Signal light Tracking in Real-time Video (CSTRV)" system uses the luminous values of the glowing points combined with color values present in the image frame of a real time video. The color-based prediction finds application in the field of transportation for detecting the color of signal lights. This system uses the luminous values of the glowing points combined with color values present in the image frame of a real time video. La*b* color model is used to extract the luminous values in the image frame, and contour tracking is used for shape detection of the signal lights. The architecture for CSTRV system is shown in Figure 8.

The image frames extracted from the real-time video are subjected to region segmentation to extract the Region of Interest (ROI). The ROI is extracted by analyzing the position of the signal lights in an image frame based on road tracking. This road tracking technique uses the edges of a road as reference lines. Based on these reference lines the image frame is segmented into different regions. The ROI is selected in such a way that the probability of presence of signal lights is maximal. Now the system needs to detect the presence of signal lights and then color. This can be done by converting the ROI image to La*b* color space and by applying contour tracking for shape recognition of signal lights.

The Commission Internationale de l'Eclairage (CIE) La*b* is the most complete color model used conventionally to describe all the colors visible to the human eye. CIELAB allows the specification of color perceptions in terms of a three-dimensional space in which L-axis deals with the lightness which extends from black to white, the a* deals with red and green colors and the b* deals with yellow and blue colors. The L, a*, and b* are calculated from the tristimulus values using following equations.

$$L=116(Y/Yn)1/3-16$$

$$a*=500[(X/Xn)1/3-(Y/Yn)1/3]$$

Figure 10. The image frame with ROI

Figure 11. The ColorValueSet associated with contour

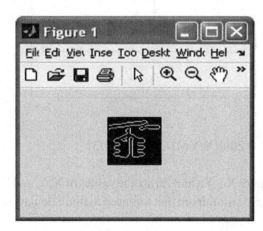

Figure 12. High level VMSDR architecture

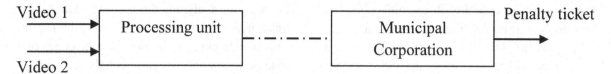

Figure 13. Detailed VMSDR architecture

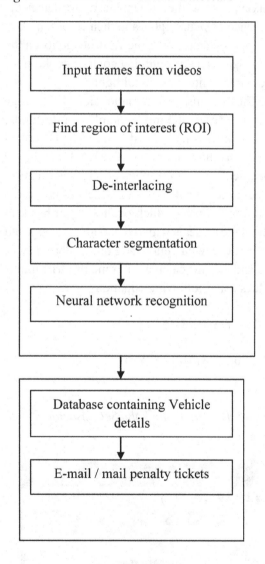

$$b* = 200[(Y/Yn)1/3 - (Z/Zn)1/3]$$

where Xn, Yn and Zn are the values of X, Y, and Z for the illuminant that was used for the calculation

of X, Y, and Z of the sample, and the quotients X/Xn, Y/Yn, and Z/Zn are all greater than 0.008856

The simulation is done using MATLAB 7.0.4 version. For simulation, several videos are taken in which some videos are with signal lights and others are without signal lights. The videos which are used for the simulation are taken from the computer-vision class of Gannon University database. This database consists of videos taken from an automobile moving at a speed of 25, 30, and 35 miles/hour. Sample videos taken while running at these speeds are tested using this algorithm.

Region segmentation is performed on the transformed binary image of the original RGB image extracted from the input video. The binary image is a matrix of 0's and 1's. The 1's in binary image represents either the sharp or blunt edges, or other structures containing sharp or blunt edges in the image. The tracing of the sharp edges is done using threshold values. The edges which lie below threshold values are eliminated and the values that lie above the threshold values are stored in a dataset. The dataset is subjected to a line tracing function to check the parallelism and continuity of the edges in the binary image. Line tracing function uses the adjacent pixel value method, in which it looks for the position of (coordinates) 1's in all the directions and stores them in a database. The resultant coordinates of the edges found in the line tracing are used as reference coordinates for tracing the signal lights.

The ColorValueSet needs to be tracked for the shape of the pole; i.e., whether the ColorValueSet lies inside the signal light pole or not. Tracing the shape of the signal light around the ColorValueSet will lead to the detection of the exact signal light

Figure 14. Region of interest that includes the VIN on the metallic plate

with color using this contour or shape tracing. The shape tracing function takes the center point [X, Y] of the ColorValueSet as reference points, with width +X and height +Y as boundary points. It checks for the presence of horizontal and vertical lines of the pole based on these reference points and the corresponding boundary points. The presence of the horizontal and vertical lines is treated as presence of a signal light pole. The simulation produced effective results in detecting the color associated with signal lights. A log file will be written with frame number and presence or absence of color associated with signal lights as shown in Table 4. The execution times for various numbers of frames with and without signal lights are shown in Table 5.

Figure 9 shows the original frame extracted from the video in which the signal light is present, and Figure 10 shows the image with ROI used to extract ColorValueSet i.e., the group of glowing

Figure 15. Region of interest (ROI)

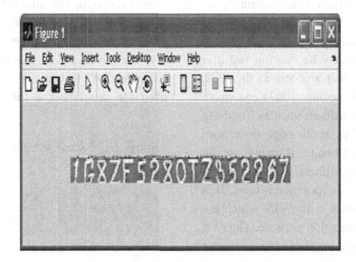

pixels (LEPs) associated with either red, yellow, or green colors. Figure 11 shows the image with ColorValueSet surrounded by contour that is used to predict the shape of the pole.

Ensuring Road Safety: Architecture and Simulation

A high level architecture for the VMSDR system is presented, and is shown in Figure 12.

A detailed diagram is shown in Figure 13.

For simulating the VMSDR architecture a video camera is placed at a height of 10 feet above the ground on a fixed pole in the parking lot at Gannon University, and another camera on the sidewalk. The timer of the processing unit is used for synchronization of videos from both the video cameras. The video captures 25 frames per second. These are used to identify the vehicle identification number (VIN) and the license plate number. As an example a single frame is used to explain the steps for detecting the VIN.

Initially the region of interest (ROI) that includes the VIN is cropped from the video frame and is shown in the Figure 14.

The ROI is further narrowed down to only the VIN and is used for further processing. A fine cropping is performed that includes only the VIN of size 24x135. Matlab 7.04 is used for simulating the algorithms. This code can be embedded in the processing unit attached to the video camera unit. The ROI consisting of only the VIN is shown in Figure 15 inside a Matlab window.

Since the vehicles may be moving fast it is likely to have interlacing artifacts in the ROI. Interlacing (from Wikipedia - 2007) can cause the image to have the artifacts such as Combing effect, Mice teeth, Saw tooth edge distortion, Interlaced lines, Ghost image, Blurring, etc. In order to remove these artifacts, a de-interlacing technique using linear interpolation is used. This will refine the characters in the VIN which are used for character recognition purpose. The ROI obtained after de-interlacing is used for character

recognition. The characters are segmented into a 7x5 matrix and are stored in arrays. Figure 16 shows some of these character images.

Each character array is fed to a neural network which recognizes the characters of the VIN. The neural network is trained using the method given at University of Florida website for Neural Network Training (2007). These were for 26 alphabets and 10 digits for both noise and without noise, and used backpropagation algorithm to recognize the array. Figure 17 shows the result of feeding the image consisting of "1" to the neural network.

All the images are fed to the neural network one after another and the complete VIN is recognized. The procedure is repeated for at least 10 frames for each vehicle.

The license plate is recognized using the algorithms given in Parker, J. R. (1994), and Lotufo, R. A., Morgan, A. D. & A.S. Johnson, A. S. (1990). This is sent along with the recognized VIN and the 10th frame from video camera as a proof of the vehicle in the middle of intersection on red light, to the test database management system which had data for 20 people.

The VIN and the license plate numbers are verified using this database containing the vehicle's details such as VIN, License plate number, owners name and address, insurance details, tickets issued in the current year. Using the address of the owner, a penalty ticket is issued based on his/ her previous driving records. A time period is given to the owner to contest/challenge in case someone else was driving during the ticketed time. In this way the driver is penalized instead of the owner.

Instead of a test database management system used for simulation, a fully pledged Database

Figure 16. Character images sent to neural network

Management System can be used at the municipal corporation side and penalty tickets can be issued automatically through e-mail or using ordinary mail. In case more details are needed, car facts of the vehicle can be checked using the VIN.

CONCLUSION

The first method for Gait Recognition uses Hidden Markov Model, which is a stochastic process, is quite efficient in identification of gait characteristics for static postures. The images considered for this method are all 2D, which makes the process robust and less time consuming. Since the processing is done on 2D images, the recognition rate is not quite high compared to the methods using Stereovision and Isoluminance lines with 3D Template Matching. The two methods using 3DTM have the distinct advantage of having higher recognition rate since they use 3D images. But the drawback of this method is the higher computing time that makes the process slower. Also the luminous intensity of the pictures plays an important role.

An automated computer "Vision-based Real-time Smart system to Prevent Railroad Accidents (VRSPRA)" that analyses individual frames in the video stream and generates a warning signal for

Figure 17. Character images recognized by neural network algorithm

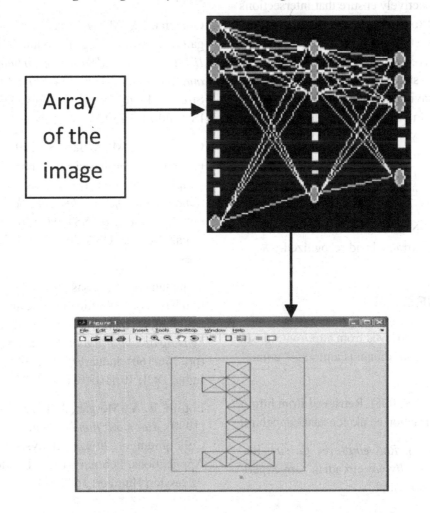

the locomotive driver for possible stopping. This system is simulated and was noted to be successful.

The "Color-based Signal light Tracking in Real-time Video (CSTRV)" system is an intelligent system using La*b* color model in combination with contour tracking. This method analyzes each frame of the video sequentially, and detects the presence of signal lights and its color. This finds application in tracking the color of signal lights on railroads that could be used for railroad accident investigation.

Vision-based Monitoring System is for Detecting Red signal crossing (VMSDR) that captures and recognizes both VIN and License plate of the vehicle running 'red' signal light at the intersections presented. This research is intended to provide a support system for the law enforcement agencies to proactively ensure that intersections are engineered to discourage red light running. The VMSDR system can be extended to include the details of drivers who rent vehicles from rental services from in state or out of state. Future work will focus on having access to the VIN details of all the vehicles in 50 states which are distributed across the country. This system can monitor tickets issued in another state to the same driver. The limitations of this system are the extreme weather conditions. If there is snow on the windscreen right above the VIN or if there is a sticker which obstructs the VIN as seen from the camera, the VIN cannot be extracted and recognized.

REFERENCES

Amtrak. (2004). Retrieved from http://www.ntsb. gov/events/2002/bourbonnais/amtrak59_anim. htm

CCTV Info – UK. (2005). Retrieved from http:// www.cctv-information.co.uk/constant3/anpr.html

Circadian. (2004). *Rail employee fatigue.* Retrieved from http://www.circadian.com/expert/ fatigue_inattention.html

Defense Advanced Research Projects Agency (DARPA). (2005). *DARPA grand challenge.* Retrieved March 14, 2005, from http://www.darpa. mil/grandchallenge

Department for Transport – UK. (2004). Retrieved in 2004 from http://www.dft.gov.uk/stellent/groups/dft_control/documents/homepage/ dft_home_page.hcsp

Drive and Safety Alive, Inc. (2006). *Key annual statistics for the USA, 2006.* Retrieved from http:// www.driveandstayalive.com/info%20section/ statistics/stats-usa.htm

Gomatam, A. M. (2004). *Non-invasive multimodal biometric recognition techniques.* Unpublished MS Thesis from Gannon University, Erie, PA, USA.

Gomatam, A. M., & Sasi, S. (2004). Enhanced gait recognition using HMM and VH techniques. *IEEE International Workshop on Imaging Systems and Techniques,* (pp. 144-147). 14 May 2004, Stresa - Lago Maggiore, Italy. DOI: 10.1109/ IST.2004.1397302

Gomatam, A. M., & Sasi, S. (2004). Multimodal gait recognition based on stereo vision and 3D template matching. *Proceedings of the International Conference on Imaging Science, Systems and Technology* (CISST'04) (pp. 405-410). Las Vegas, Nevada, USA, June 21-24, 2004, CSREA Press.

Gomatam, A. M., & Sasi, S. (2005). Gait recognition based on isoluminance line and 3D template matching. *International Conference on Intelligent Sensing and Information Processing -ICISIP '05,* (pp. 156-160). January 04-07, 2005, IIT Chennai, India. DOI: 10.1109/ICISIP.2005.1529440

Lotufo, R. A., Morgan, A. D., & Johnson, A. S. (1990). *Automatic number plate recognition.* IEE Colloquium on Image Analysis for Transport Applications, February 1990, London, INSPEC Accession Number: 3649590.

MoBo. (2004). *The CMU motion of body (MoBo) database*. Retrieved from http://www.ri.cmu.edu/publication_view.html?pub_id=3904

Motorola Solutions for Government. (2005). *Government and enterprise North America*. Retrieved from http://www.motorola.com/governmentandenterprise/northamerica/en-us/solution.aspx?navigationpath=id_801i/id_826i/id_2694i/id_2695i

National Transportation Safety Board. (2004). *Symposium Proceedings 2004*. Retrieved from http://www.ntsb.gov/Events/symp_rec/proceedings/authors/scaman.htm

News, C. N. N. (2005). *Train collision near Los Angeles kills 11*. Retrieved from http://www.cnn.com/2005/US/01/26/train.derailment/

Office of Law Enforcement Technology Commercialization. (2005). *License plate recognition*. Retrieved from http://www.oletc.org/oletctoday/0415_licplate.pdf#search=%22automatic%20license%20plate%20recognition%22

Parker, J. R. (1994). *Practical computer vision using C*. New York, NY: Wiley.

Sharma, R., & Sasi, S. (2007). *Vision-based monitoring system for detecting red signal crossing. Innovations and Advanced Techniques in Computer and Information Sciences and Engineering* (pp. 29–33). Springer.

Train Accident Report. (2004). Retrieved from http://www.visualexpert.com/Resoures/trainaccidents.html

Trax. (2004). Retrieved from http://www.avtangeltrax.com/digital.htm

University of Florida website. (2007). *Neural network training*. Retrieved from http://www.math.ufl.edu/help/matlab/ReferenceTOC.html

Wikipedia. (2005). *Automatic number plate recognition*. Retrieved from http://en.wikipedia.org/wiki/Automatic_number_plate_recognition

Wikipedia. (2007). *Interlacing*. Retrieved from http://en.wikipedia.org/wiki/Interlacing

Yelal, M. R., Sasi, S., Shaffer, G. R., & Kumar, A. K. (2006). *Color-based signal light tracking in real-time video*. IEEE International Conference on Advanced Video and Signal Based Surveillance, November 22-24, 2006, Sydney, Australia. DOI: 10.1109/AVSS.2006.34

Chapter 5
Visual Detection in Linked Multi-Component Robotic Systems

Jose Manuel Lopez-Guede
Basque Country University, Spain

Ramon Moreno
Basque Country University, Spain

Borja Fernandez-Gauna
Basque Country University, Spain

Manuel Graña
Basque Country University, Spain

ABSTRACT

In this chapter, a system to identify the different elements of a Linked Multi-Component Robotic System (L-MCRS) is specified, designed, and implemented. A L-MCRS is composed of several independent robots and a linking element between them which provide a greater complexity to these systems. The identification system is used to model each component of the L-MCRS using very basic information about each of the individual components. So, different state models that have been used in several works of the literature that have been reviewed can be covered. The chapter explains the design of the system and shows its frontend. This work is the first step towards a realistic implementation of L-MCRS.

INTRODUCTION

This chapter deals with the practical problem of parts identification and segmentation of the single elements that compose any instance of a Linked Multi-Component Robotic System (L-MCRS) (Duro, Graña, & de Lope, 2010). L-MRCS are composed of a number of mobile robotic elements that are linked by any flexible unidimensional

element, and the interaction between the passive, flexible element and the robots introduces highly non-linear effects in the system's dynamics. As it is an open issue, we are interested in autonomous behavior learning in L-MRCS.

Several algorithms and techniques to achieve this objective have been tested and validated through computer simulations using accurate geometrical and dynamical models based on sev-

DOI: 10.4018/978-1-4666-2672-0.ch005

eral computational tools, as it will be referenced later. Then, they have to be verified in real world systems to ensure that they maintain the expected performance under real world circumstances, ensuring this through several realistic experiments.

Each one of the elements which compose the L-MCRS is characterized by several attributes, and obtaining those object attributes of each element is relatively easy in the computer simulation environment. Once they have been obtained, other attributes that describe the relation between the mobile robots and the flexible element can be derived from them. To be able to reproduce the results obtained previously through computer simulations (Fernandez-Gauna, Lopez-Guede, Zulueta, & Graña, 2010), any kind of real-time perception system is required. We have addressed the possibility of doing this by means of a computer based vision system that fulfills several requirements, which will be exposed later. This vision system must give support to the decision algorithms that have been designed and trained assuming that they could know in any moment several relevant properties about the different elements of the L-MCRS.

The chapter is structured as follows: first, a background section introduces the L-MCRS. Later, a main section, where we describe the identification system is composed of several subsections devoted to its specifications, to the global pipeline and to a detailed explanation of the pipeline illustrated with several figures showing all process. Finally, we address future research directions and expose our conclusions.

BACKGROUND

In this section we are going to introduce the Linked Multi-Component Robotic Systems (L-MCRS) through a short review of the literature that shows several works that have been done up to now.

Linked Multi-Component Robotic Systems

Linked Multi-Component Robotic Systems (L-MCRS) are categorized by (Duro, Graña, & de Lope, 2010) as a collection of autonomous robots linked by a non-rigid physical link which must be modelled precisely because it is the source of strong non-linearities in the system dynamics. Multiple models proposed for uni-dimensional objects are reviewed in (Echegoyen, 2009) (Echegoyen, Villaverde, Moreno, Graña, & d'Anjou, 2010): differential equations (Pai, 2002), rigid element chains (Hergenrother & Dhne, 2000), spring mass systems (Gregoire, & Schomer, 2007), combinations of spline geometrical models and physical dynamical models (Qin, & Terzopoulos, 1996), and combinations of spline models and the Cosserat rod theory (Theetten, Grisoni, Andriot, & Barsky, 2008). We have done several computer simulations based on building our system dynamics model on a combination of spline models and the Cosserat rod theory to perform our simulations, because it improves the geometrical spline representation by adding force and torques (Antman, 1995) (Rubin, 2000) allowing to model the twisting of the hose. This approach, known as Geometrically Exact Dynamic Splines (GEDS), represents the control points of the splines by the three Cartesian coordinates plus a fourth coordinate representing the twisting state of the hose.

More recently, a new approach based on Reinforcement Learning (RL) (Sutton, & Barto, 1998) has been carried out by our group to learn autonomously behaviors because RL techniques have been used successfully in several areas of robotics. These areas are navigation (Duan, & Hexu, 2005), indoor navigation (Chen, Yang, Zhou, & Dong, 2008), cooperative navigation task (Melo, & Ribeiro, 2008) and automatic path search (Miyata, Nakamura, Yanou, & Takehara, 2009).

As a first step towards the general application of RL techniques to L-MCRS control, we have chosen the hose transportation problem for a single robot: from a predefined configuration of a hose, a robot fixed to the hose tip is expected to carry it to a given position. Computer simulations have been carried out to provide the information for the Q-Learning algorithm to avoid the main drawback of the experience-based learning algorithms: the huge amount of time required to realize the experiments in the real world. We have developed two works where this idea has been followed. First of them is (Fernandez-Gauna, Lopez-Guede, Zulueta, & Graña, 2010), where only one state model and one reward function have been used. Later, in (Lopez-Guede, Fernandez-Gauna, Graña, & Zulueta, 2011) the same approximation is followed, but testing more state models reward functions, with the aim of studying the effect of these elements in the performance of the learning algorithm.

All these works have obtained promising results, but all of them have been developed under simulation conditions and a visual detection and identification system is needed to implement them and test if such advances are suitable to a real environment.

ROBOTIC VISION SYSTEM

Specifications of the Robotic Vision System

In this subsection we are going to describe the input and the output specifications that our system of visual detection must fulfill to be useful for monitoring a L-MCRS during physical experiments similar to those that we have done using computer simulations. We have made some reasonable hypothesis about those specifications.

Regarding to the characteristics of the identification and segmentation system that we will design, we must take into account the specificities discussed in the section devoted to L-MCRS and

that the capability of adapting to different realistic environments is also required. So, regarding to the inputs, we can enumerate the following specifications:

- The L-MCRS that we want to monitorize can be in an unstructured and non controlled environment, so we cannot assume any particular condition about the light sources. The monitorization system must deal with different and unstructured light sources, and it must not be limited to work properly with a concrete illumination. In any case, the working area must be sufficiently enlightened to the human eye to understand the scene.

- The system that we want to build must be flexible enough to allow that the environment had any size. That means that it is reasonable to use it in an environment that is graspable with sufficient precision by the camera that is installed in the system. In any case, the system must not be limited to a fixed working area, and it is allowed that a human operator bounds the working area only one time, at the beginning of the monitorization process.

- Once a L-MCRS has been configured, its number of elements does not change. In the previous section devoted to describe these systems, only works with one robot were referenced. However, the monitoring system must be able to work with L-MCRS consisting of a flexible number of robots. The system must not be limited to recognize only a prefixed number of robots, and it is allowed that a human operator supply the number of robots of the L-MCRS only one time, at the beginning of the monitorization process.

- The robots that compose the L-MCRS can be of different colors. These colors must be sufficiently different one from another to distinguish between them. The system must not be limited to recognize only a prefixed set of colors, and it is allowed that a human operator supply the colors of each robot only one time, at the beginning of the monitorization process.

- The flexible element that links the elements that compose the L-MCRS must be identified by the identification system. The only hypothesis that can be accepted on its shape and arrangement is that it is an unidimensional object, and its behavior corresponds to this kind of objects.

- The flexible element of the L-MCRS can be of any color, provided that it was different from the color of all robots. The system must not be limited to recognize only a prefixed set of colors, and it is allowed that a human operator supply the color of the linking element only one time, at the beginning of the monitorization process.

So far, only the input specifications have been enumerated. Regarding to the outputs that the identification system must obtain, we can enumerate only one, but very ambitious specification:

- In the previous section devoted to describe the L-MCRS systems, several works with different models of state for the RL algorithm were referenced. The identification system must be able to compose these different state models, taking into account that some attributes of the state are derived from more basic attributes of the robots or the linking element. Obviously, this flexible specification needs some ad hoc code in each case, but the point in which this code has to be implemented must be very clear and must be isolated from the rest of the code.

Although in principle a system can be specified completely determining its inputs and its outputs, additional specifications can be imposed from the point of view of the internal process that is being carried out by the system. So, we enumerate two final and desirable specifications about the internal process:

- Regarding to the device that will be used to take the images, it is desirable that it was so simple and inexpensive as possible. In this way, it will not need calibration or complex processes and it will allow a fast and intuitive use by non experimented users.

- Finally, it is very desirable that the identification system has real time performance. In this context, real time means that the control system had information enough and with a frequency that allows him to take the most suitable actions in each moment of the control loop. I.e., it is assumed that it is sufficient to sample several times per second the complete L-MCRS.

Pipeline

Taking into account the previously exposed specifications of the identification system, we have designed the pipeline that is shown in the Figure 1. The square boxes correspond to defined and specific processes. The rhomboidal boxes correspond to well defined inputs to the system (acquired in an automatic way or by means a human operator), or they correspond to outputs that the system generates. As a summary, before the deep exposition of each well defined part, we can highlight schematically the following well defined parts:

- The first part corresponds to the image acquisition of the real world by means of the

Figure 1. Pipeline of the process to the visual identification of the different components of the L–MCRS

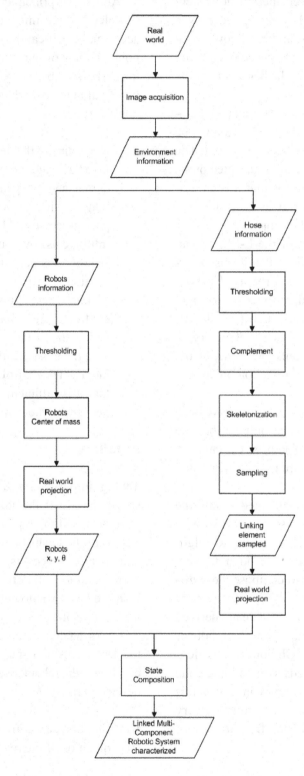

Figure 2. Virtual simulation of a very simple L-MCRS, where only one robot is attached to the linking element. This scene corresponds to the initial situation of the L-MCRS before a transporting task started.

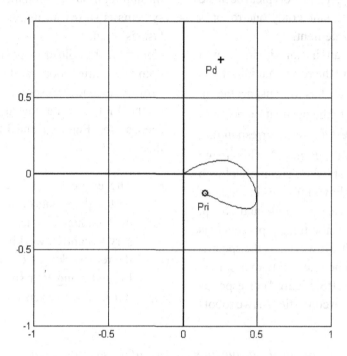

image capture device. In this part some information about the environment is provided by a human operator.

- The left branch corresponds to the segmentation and the identification of each robot with their characterizing attributes.
- The right branch corresponds to the segmentation of the linking element of the L-MCRS. It could be executed independently from the left branch, but normally it will be executed after the robot detecting branch.
- The final part of the pipeline corresponds to the fusion of the information that is produced by the two previously explained branches.

In the following subsections, each identified process and each input/output of information of the pipeline will be explained more deeply.

Image Acquisition

This is the main process of the initial common branch at the beginning of the pipeline. The system starts considering the real world as a generic input, and the image acquisition process involves taking only one image and showing it to the human operator.

Following the specifications of the system given in a previous section, the captured image must contain the working area of the L-MCRS with sufficient precision and with sufficient extension to contain all of its elements, i.e., all the robots and the linking element completely, without leaving any part of them out of the image.

Since the identification system must be equivalent to the monitorization parts of the L-MCRS that were used in the works referenced in the background section, we have to take into account how were these systems in the computer simulations that were carried out. We show one prototypical L-MCRS in the Figure 2. It is prototypical because it is one of the simplest, since only one robot is attached to the linking element.

This scene shows an initial situation of a transporting task, where the robot that initially is placed in the position P_{ri} has to transport the tip of the linking element to the destination position P_d. Figure 3 shows the real scene corresponding to the virtual scene of the Figure 2. This image has been taken using a standard commercial web cam that was placed above the working area of the L-MCRS. This device is simple and cheap, and neither calibration nor tuning process has been carried out with it. However, as the specifications stated, it could be interesting to monitorize more complex L-MCRS. Figure 4 corresponds to a more complex system consisting of two robots

and the linking element, that was generated through computer simulations. In this scene the systems is shown in its initial position, and the task is again to transport the tip of the linking element to the bold cross position, but in this case, both robots must cooperate to do it. Figure 5 shows to the real scene corresponding to the virtual scene of the Figure 4. Paying attention to both real scenes, i.e., Figure 3 and Figure 5, we can see that:

- The entire scene is captured, leaving no part of the robots or the linking element out of the image. In fact, to get this it is necessary to capture working area in excess, and the real working area of the L-MCRS is delimited using four small boxes that delimiter its four corners.

Figure 3. Real image captured by the acquisition camera of the identification system. It is the physical implementation of the very simple L-MCRS corresponding to the figure 2.

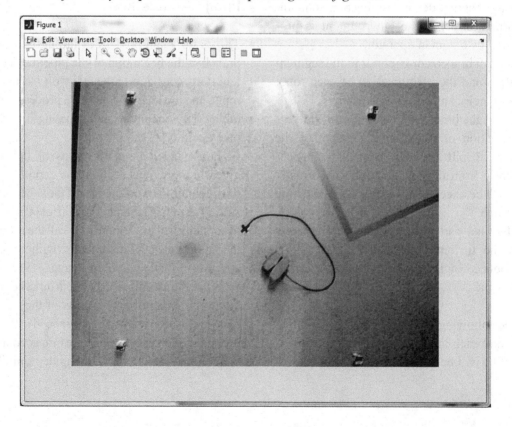

Figure 4. Virtual simulation of a very simple L-MCRS, where two robots are attached to the linking element. This scene corresponds to the initial situation of the L-MCRS before a transporting task started.

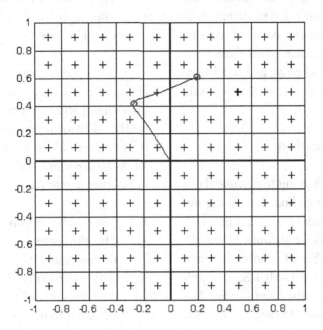

- In the images, the working areas are misaligned with respect to the image limits. It is because it is very difficult to have a precise alignment, and besides, we have not done any work to correct this effect, considering that the system must deal with these circumstances.
- The source of light is not structured and irregular in both scenes.
- A considerable brightness is present in the superior part of both scenes, and it is particularly relevant in the Figure 5, where the first robot of the L-MCRS is inside that brightness. Besides, on the right of the scenes, there is placed a long shadow that crosses the scene from top to bottom.
- There are several stains on the floor. There are mate stains, i.e., without brightness, like a small circle on the third quadrant of the working area. On the other hand, there are also bright stains, like an angular trajectory that was placed on the first and the fourth quadrants using duct tape.

- In the parts of the floor that are clean, its coloration is not uniform since the pavement of which it is made has a lot of small dots and speckles.

Once the real image of the scene has been acquired, it is shown to the human operator, and he is asked to introduce some information regarding to the environment. More concretely, this information is regarding to the size of the working area, and he is asked to determine the X coordinate of the upper left and the X coordinate of the upper right corners of the working area, which is assumed to be perfectly square. This process of supplying information by the human operator about the working area is carried out only one time, at the beginning of the identification process, and the human operator can use the tool "Data cursor" that the programming environment offers, as is shown in the Figure 6. By clicking on the upper left and on the upper right small boxes that delimiter the working area, the human operator can know their X coordinates, and introduce them when it is asked to do it, as we can see in the Figure 7, where only the X coordinate of the upper left corner is determined. A similar process would be carried out to determine the X coordinate of the upper right corner. From these data, and assuming that it has a known and fixed area, some equivalence can be done to transform pixels into meters.

Robots Characterization

In this subsection we are going to discuss about the left branch of the pipeline. Following the specifications of the system given in a previous section, the system must be flexible enough to identify a variable number of robots contained in the working area of the L-MCRS, although this number is fixed once the identification has started. It has been considered that for each robot, the following information must be obtained to consider it monitorized:

- The X coordinate in the working area,
- The Y coordinate in the working area,
- The angle θ, i.e., the orientation in the working area.

This branch starts considering the image acquired in the previous common branch as the main input, and based on it, the human operator will be asked to introduce some information about the robots to help to identify and to follow them while the experiment takes place. The information to be entered into the system is:

- The number of robots to monitorize,
- The chromatic tolerance that will be used for the recognition of those robots, since it will be carried out by means of their colors,
- And for each of the robots to monitorize, since the RGB color space to be used:
  ○ The R, G and B components of the front of the robot,

Figure 5. Real image captured by the acquisition camera of the identification system. It is the physical implementation of the L-MCRS corresponding to the figure 4.

Figure 6. Real image of the scene where the human operator can determine the X coordinate of the upper left and the X coordinate of the upper right corners of the working area using the "data cursor" tool (only one determination is shown).

○ The R, G and B components of the back of the robot.

Supplying this information to the system, it will be able to deal with a variable number of robots and will admit different and unknown in advance color patterns for each robot. Once the system shows the image acquired in the previous common branch to the human operator, he easily determines the number of robots involved in the L-MCRS, and besides, he determines heuristically the chromatic tolerance that will be used for the recognition of that number of robots. This parameter is determined taking into account the light of the scene, the quality of the image acquisition device and subjective impressions based on the uniformity that could be seen in each color of each robot in the image. So, this parameter will measure the difficult that the system will have to identify each color, and the more difficult the scene is, the higher its value will be.

In the Figure 8, we can see how the human operator can determine the R, G and B components corresponding to the front of the first robot of the L-MCRS. The system helps him by means of the "Data cursor" tool provided by the programming environment, which supplies those values of a concrete point of the image by clicking on it. For space reasons, we have omitted to show through figures the remaining parts of this process, but the human operator has to supply the R, G and B components corresponding to the back of the first robot, and the front and back color information of each of remaining robots, in this case, of the second robot. In the Figure 9 we show the screen where the human operator introduces the number of robots to be considered, the tolerance to use in the colors detection and the characteristic color of each part of each robot, after having used the "Data cursor" tool, as is shown in the Figure 8.

Once these chromatic values have been introduced, for each of two parts of each robot, the system carries out a thresholding process based on the chromatic distance of each pixel of the original image to the values that have been introduced for each part of each robot. The distance obtained in each case is softened by means of the tolerance value T that has been introduced by the human operator. Based on this distance, a binarized image is created for each case, where each pixel activated if the correspondent pixel of the original image is chromatically near of the part to detect, otherwise it is deactivated. The criterion that is used is shown in the Equation 1:

$$\forall i \in \left[1, n\right], \forall j \in \left[1, m\right], C\left(i, j\right) = 1 \Leftrightarrow$$

$$\sqrt{\begin{array}{l} \left(A\left(i, j\right)_R - R\right)^2 + \left(A\left(i, j\right)_G - G\right)^2 \\ + \left(A\left(i, j\right)_B - B\right)^2 \end{array}} < T$$

(1)

where:

- n : The number of rows of the original image,
- m : The columns of rows of the original image,
- C : The binarized image that is created,
- A : The original image,
- $A\left(i, j\right)_R, A\left(i, j\right)_G, A\left(i, j\right)_B$: The R, G and B components of the pixel of coordinates $\left(i, j\right)$ of the original image,
- R, G, B : The chromatic components of the part to detect,
- T : The tolerance value inserted by the human operator.

In the Figure 10, a real image of the scene where the human operator can determine the R, G and B components of the back of the first

Figure 7. Screen where the human operator introduces the relevant coordinates to delimit the working area. That coordinates have been determined using the tool shown in the figure 6.

Figure 8. Real image of the scene where the human operator can determine the R, G and B components of the front of the first robot. He uses the "data cursor" tool. Only one determination is shown, but the process is similar to the back of the first robot, and analogue to the remaining robots.

robot using the "Data cursor" tool is shown. Besides, Figure 10 shows the binarized image corresponding to the front of the first robot, where only those pixels are activated. Based on those activated pixels, the coordinates (X, Y) of the center of mass of the part to detect is calculated following the Equations 2 and 3:

$$X = \frac{\sum_{i=1}^{n} \sum_{j=1}^{m} C(i,j) * j}{\sum_{i=1}^{n} \sum_{j=1}^{m} C(i,j)} \quad (2)$$

$$Y = \frac{\sum_{i=1}^{n} \sum_{j=1}^{m} C(i,j) * i}{\sum_{i=1}^{n} \sum_{j=1}^{m} C(i,j)} \quad (3)$$

where:

- n : The number of rows of the original image,
- m : The number of columns of the original image,
- C : The binarized image.

In the Figure 10, on the left hand, a small white cross on the front of the first robot (green) is drawn to indicate where the system has estimated that the center of mass of this front part is placed, once this procedure has been followed. When the center of mass of the front and the back part of each robot has been calculated, the general center of mass of the robot is calculated through the mean of the two previous. To calculate the angle θ of each robot in the working area, the system makes use of the two centers of mass of the robot, as the Equation 4 shows:

$$\theta = \arctan\left(\frac{Y_F - Y_B}{X_F - X_B}\right) \quad (4)$$

where:

- X_F, Y_F : The coordinates of the center of mass of the front of the robot,
- X_B, Y_B : The coordinates of the center of mass of the back of the robot.

Regarding to the robots identification, the last step is to project the position of the global center of mass of each robot to the real world. Once their positions have been calculated in terms of pixels of the image, they are projected by means a rule of three, assuming a fixed size of the working area and taking into account its limits, which were introduced by the human operator in a previous step.

Linking Element Characterization

In this subsection we are going to discuss about the right branch of the pipeline. Following the specifications of the system given in a previous section, the system must be flexible enough to identify an uni-dimensional flexible linking element contained in the working area of the L-MCRS. Although its length and its shape are not known a priori, the length is fixed once the identification has started. However, the shape could change as the linked element move through the working area. It has been considered that for the linking element, a number of sampled points must be obtained to consider it monitorized.

This branch starts like the branch devoted to robots identification, i. e., considering the image acquired in the previous common branch as the main input, and based on it, the human operator will be asked to introduce some information about the linking element to help to identify and to follow it while the experiment takes place. The information to be entered into the system is:

- The chromatic tolerance that will be used for the recognition of the linking element,

Figure 9. Screen where the human operator introduces the relevant information to monitorize the robots. Some of that information has been determined using the tool shown in the figure 8.

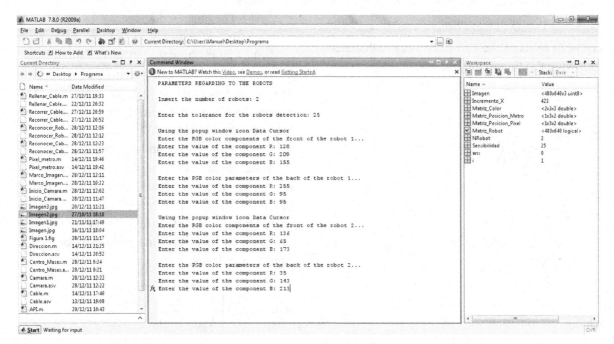

Figure 10. On the left: real image of the scene where the human operator can determine the R, G and B components of the back of the first robot using the "data cursor" tool. On the right: the binarized image where only the pixels corresponding to the front of the first robot are activated.

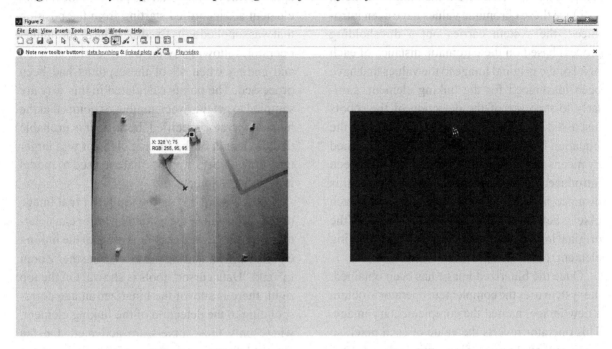

since it will be carried out by means of its colors,

- The position of the fixed extreme of the linking element,
- Since the RGB color space to be used to monitorize it:
  - The R, G and B components of the linking element.

Supplying this information to the system, it will be able to deal with a linking element of variable length and shape and will admit different and unknown in advance color patterns for it. Once the system shows the image acquired in the previous common branch to the human operator, he determines heuristically the chromatic tolerance that will be used for the recognition of the linking element. This parameter is determined again as in the robot identification branch, i.e., taking into account the light of the scene, the quality of the

image acquisition device and subjective impressions based on the uniformity that could be seen in each color of each robot in the image. So, this parameter will measure the difficult that the system will have to identify each color, and the more difficult the scene is, the higher its value will be.

In the Figure 11, we can see how the human operator can determine the R, G and B components corresponding to the linking element and the position of its fixed extreme. The system helps him by means of the "Zoom in" and "Data cursor" tools provided by the programming environment, which supplies those values of a concrete point of the image by clicking on it.

In the Figure 12 we show the screen where the human operator introduces the tolerance to use in the colors detection and the characteristic color of the linking element, after having used the "Zoom In" and "Data cursor" tools, as is shown in the Figure 11. Regarding to the auxiliary information that the systems needs to generate the output, the operator introduces the number of samples that are taken from the linking element.

Once these chromatic values have been introduced, the system carries out a thresholding process based on the chromatic distance of each pixel of the original image to the values that have been introduced for the linking element, similarly to the case of the detection of the robots discussed in a previous subsection, using the Equation 1. The distance obtained is also softened by means of the tolerance value T that has been introduced by the human operator. Based on this distance, a binarized image is created, where each pixel is activated if the correspondent pixel of the original image is chromatically near of the linking element, otherwise it is deactivated.

Once the binarized image has been obtained, the system uses the complement operator to obtain a new image, named the complementary image. This operator reverts the value of each pixel, so that it activated a pixel if is deactivated, and deactivates it if it is activated. This image is obtained to get an appropriate input to the skeletonization

process, which removes pixels on the boundaries of objects but it does not allow objects to break apart. As the final step, all the points of the linking element are identified starting from the position that was supplied as its fixed extreme by the human operator, following by the points of the skeleton and ending when all of these points had been processed. The points calculated in this way are sampled to get the exact number of points that the human operator specified, because it is probable that the length of the linking element was larger than the number of points that are used to model it in the experiments.

In the Figure 13, on the top left a real image of the scene where the human operator can determine the R, G and B components of the linking element and its starting position using the "Zoom In" and "Data cursor" tools is shown. On the top right, there is shown the binarized image corresponding to the detection of the linking element, where only those pixels are activated. On the bottom left, the complement of the binarized image is shown. On the bottom right, the skeletonized image based on the complement image is shown.

Figure 11. Real image of the scene where the human operator can determine the R, G and B components of the linking element. He also determines the position of its fixed extreme. He uses the "zoom in" and "data cursor" tools.

Figure 12. Screen where the human operator introduces the relevant information to monitorize the linking element. Some of that information has been determined using the tools shown in the figure 11.

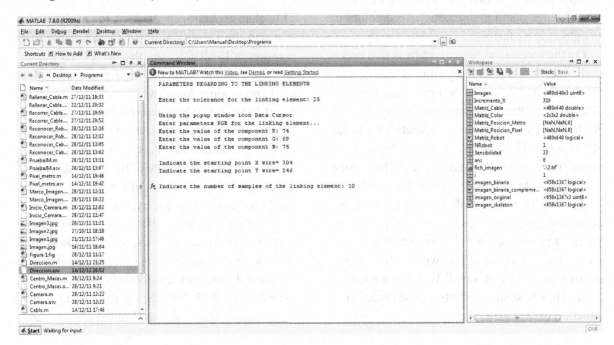

Regarding to the linking element identification, the last step is to project the position of each sampled point to the real world. Once their positions have been calculated in terms of pixels of the image, they are projected by means a rule of three, assuming a fixed size of the working area and taking into account its limits, which were introduced by the human operator in a previous step. From these data, and assuming that it has a known and fixed area, the transformation from pixels into meters can be done.

RL State Composition

This is the main process of the final common branch at the end of the pipeline. Following the left branch, the system has characterized all the robots implied in the L-MCRS, and following the right branch, it has sampled the linking element. I.e., individual components of the L-MCRS are identified, but the L-MCRS is not characterized.

Following the specifications of the system given in a previous section, we have to remember that in the section devoted to the L-MCRS background, several works with different models of state for the RL algorithm were referenced. The identification system must be able to compose and return the different state models, taking into account that some attributes of the state are derived from more basic attributes of the robots or the linking element. It was said that we assume that this flexible specification needs some ad hoc code in each case, and this final process is the point in which that code has to be implemented, where it is perfectly isolated from the rest of the code that carries out the identification of the robots and the linking element.

FUTURE RESEARCH DIRECTIONS

As a minor challenge, we address an issue regarding to the level of interaction of the human operator with the system in its initial stage, always guided to reduce it. More efforts should be made in order to detect the robots and the linking element of the L-MCRS with less collaboration of the human operator, so that the system could locate these elements and their characteristic colors in an autonomous fashion.

On the other hand, we can address two major challenges. The former is regarding to the sensibility of the actual system to the brightness. It would be interesting that the system was more robust to brightness, exploiting some chromatic properties of the scene. In this sense, it seems appropriate to use reflectance analysis methods to reduce sensitivity to brightness (specular free methods).

The second major challenge is regarding the general objective where this work is embedded. We must remember that as was said in previous sections, the purpose of this work is to develop the first step towards a realistic implementation of a L-MCRS, so more efforts will be made in the direction of that general objective.

CONCLUSION

This chapter has been devoted to the localization and the identification of the elements that compose a L-MCRS, i.e., the individual robots and the linking element. We have started this work reviewing some works of the literature about L-MCRS, where we can see that different models of the state were used. This was taken as a main specification to the system, which should be able to express the state of the L-MCRS using different models. Besides, many other requirements were added to the system specifications. A global pipeline that differentiates clearly four parts was proposed: the first, capture the image and retrieves some information of the system; the second, obtains the information regarding to the robots of the L-MCRS; the third, obtains the information regarding to the linking element, and

Figure 13. On the top left: real image of the scene where the human operator can determine the R, G and B components and the starting point of the linking element using the "data cursor" tool. On the top right: the binarized image where only the pixels that are chromatically near of the linking element are activated. On the bottom left: the complementary of the binarized image. On the bottom right: the skeletonized image based in the complementary image.

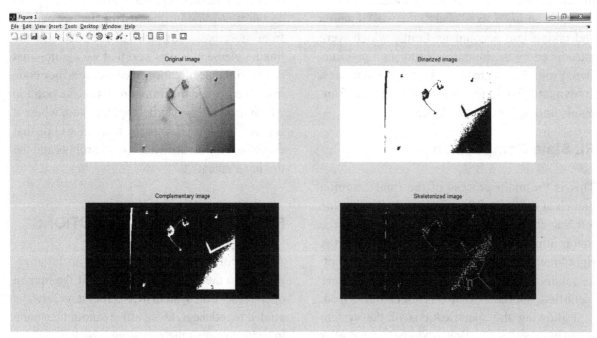

finally, the fourth part, composes the model of the state of the L-MCRS.

As has been shown in the main section of the chapter, the specifications of the system have been fulfilled, although some information must be provided by the human operator at the beginning of the process.

In summary, the work explained in this chapter is the first step towards a realistic implementation of the L-MCRS.

REFERENCES

Antman, S. (1995). *Nonlinear problems of elasticity*. Maryland, MD: Springer-Verlag.

Chen, C., Yang, P., Zhou, X., & Dong, D. (2008). A quantum-inspired qlearning algorithm for indoor robot navigation. In *IEEE International Conference on Networking, Sensing and Control (ICNSC 2008)* (pp. 1599 –1603).

Duan, Y., & Hexu, X. (2005). Fuzzy reinforcement learning and its application in robot navigation. In *Proceedings of 2005 International Conference on Machine Learning and Cybernetics,* (pp. 899 –904)

Duro, R., Graña, M., & de Lope, J. (2010). On the potential contributions of hybrid intelligent approaches to multicomponent robotic system development. *Information Sciences, 180*(14), 2635–2648. doi:10.1016/j.ins.2010.02.005

Echegoyen, Z. (2009). *Contributions to visual servoing for legged and linked multicomponent robots*. Ph.D. dissertation. San Sebastian, Spain: UPV/EHU.

Echegoyen, Z., Villaverde, I., Moreno, R., Graña, M., & d'Anjou, A. (2010). Linked multi-component mobile robots: Modeling, simulation and control. *Robotics and Autonomous Systems, 58*(12), 1292–1305. doi:10.1016/j.robot.2010.08.008

Fernandez-Gauna, B., Lopez-Guede, J. M., Zulueta, E., & Graña, M. (2010). Learning hose transport control with q-learning. *Neural Network World, 20*(7), 913–923.

Gregoire, M., & Schomer, E. (2007). Interactive simulation of one-dimensional flexible parts. *Computer Aided Design, 39*(8), 694–707. doi:10.1016/j.cad.2007.05.005

Hergenrother, E., & Dhne, P. (2000) Real-time virtual cables based on kinematic simulation. In Proceedings of the International Conference in Central Europe on Computer Graphics, Visualization and Computer Vision.

Lopez-Guede, J. M., Fernandez-Gauna, B., Graña, M., & Zulueta, E. (2011). Empirical study of q-learning based elemental hose transport control. In Corchado, E., Kurzynski, M., & Wozniak, M. (Eds.), *Hybrid Artificial Intelligent Systems* (*Vol. 6679*, pp. 455–462). Berlin, Germany: Springer. doi:10.1007/978-3-642-21222-2_55

Melo, F., & Ribeiro, M. (2008). Reinforcement learning with function approximation for cooperative navigation tasks. In *IEEE International Conference on Robotics and Automation,* (pp. 3321 –3327)

Miyata, S., Nakamura, H., Yanou, A., & Takehara, S. (2009). Automatic path search for roving robot using reinforcement learning. In *Fourth International Conference on Innovative Computing, Information and Control,* (pp. 169 –172).

Pai, D. (2002). Strands: Interactive simulation of thin solids using cosserat models. *Computer Graphics Forum, 21*(3), 347–352. doi:10.1111/1467-8659.00594

Qin, H., & Terzopoulos, D. (1996). *D-nurbs: A physics-based framework for geometric design*. Los Alamitos, CA. USA, Tech. Rep.

Rubin, M. (2000). *Cosserat theories: Shells, rods and points*. Norwell, MA: Kluwer.

Sutton, R., & Barto, A. (1998). *Reinforcement learning: An introduction.* Cambridge, MA: MIT Press.

Theetten, A., Grisoni, L., Andriot, C., & Barsky, B. (2008). Geometrically exact dynamic splines. *Computer Aided Design, 40*(1), 35–48. doi:10.1016/j.cad.2007.05.008

ADDITIONAL READING

Bartels, R. H., Beatty, J. C., & Barsky, B. A. (1987). *An introduction to splines for use in computer graphics & geometric modeling.* San Francisco, CA, USA: Morgan Kaufmann Publishers Inc.

Bellman, R. (1957). A Markovian decision process. *Indiana University Mathematics Journal, 6,* 679–684. doi:10.1512/iumj.1957.6.06038

Boor, C. D. (1994). *A practical guide to splines.* Springer.

Cao, Y., Fukunaga, A., & Kahng, A. (1997). Cooperative mobile robotics: Antecedents and directions. *Autonomous Robots, 4*(1), 7–27. doi:10.1023/A:1008855018923

Dudek, G., Jenkin, M. R. M., Milios, E., & Wilkes, D. (1996). A taxonomy for multi-agent robotics. *Autonomous Robots, 3*(4), 375–397. doi:10.1007/BF00240651

Farinelli, A., Iocchi, L., & Nardi, D. (2004). Multirobot systems: A classification focused on coordination. *IEEE Transactions on Systems, Man, and Cybernetics. Part B, Cybernetics, 34*(5), 2015–2028. doi:10.1109/TSMCB.2004.832155

Ferber, J. (1999). *Multi-agent systems: An introduction to distributed artificial intelligence.* Addison-Wesley.

Gadanho, S., & Hallam, J. (2001). Emotion-triggered learning in autonomous robot control. *Cybernetics and Systems, 32*(5), 531–559. doi:10.1080/019697201750257766

Huang, L. (2005). Speed control of differentially driven wheeled mobile robots: Model-based adaptive approach. *Journal of Robotic Systems, 22*(6), 323–332. doi:10.1002/rob.20068

Kala, R., Shukla, A., & Tiwari, R. (2010). Dynamic environment robot path planning using hierarchical evolutionary algorithms. *Cybernetics and Systems, 41*(6), 435–454. doi:10.1080/0196 9722.2010.500800

Kiguchi, K., Watanabe, K., Izumi, K., & Fukuda, T. (2003). A humanlike grasping force planner for object manipulation by robot manipulators. *Cybernetics and Systems, 34*(8), 645–662. doi:10.1080/716100282

Klancar, G., & Skrjanc, I. (2007). Tracking-error model-based predictive control for mobile robots in real time. *Robotics and Autonomous Systems, 55,* 460–469. doi:10.1016/j.robot.2007.01.002

Koh, K. C., & Cho, H. S. (1999). A smooth path tracking algorithm for wheeled mobile robots with dynamic constraints. *Journal of Intelligent & Robotic Systems, 24,* 367–385. doi:10.1023/A:1008045202113

McLain, T. W., & Beard, R. W. (2005). Coordination variables, coordination functions, and cooperative timing missions. *AIAA Journal of Guidance, Control, &. Dynamics (Pembroke, Ont.), 28*(1), 150–161.

Mehrabian, A. R., Lucas, C., & Roshanian, J. (2008). Design of an aerospace launch vehicle autopilot based on optimized emotional learning algorithm. *Cybernetics and Systems, 39*(3), 284–303. doi:10.1080/01969720801944364

Nuseirat, A., & Abu-Zitar, R. (2003). Hybrid trajectory planning using reinforcement and backpropagation through time techniques. *Cybernetics and Systems, 34*(8), 747–765. doi:10.1080/716100275

Oriolo, G., De Luca, A., & Vendittelli, M. (2002). WMR control via dynamic feedback linearization: Design, implementation, and experimental validation. *IEEE Transactions on Control Systems Technology, 10*(6), 835–852. doi:10.1109/TCST.2002.804116

Raimondi, F., & Melluso, M. (2005). A new fuzzy robust dynamic controller for autonomous vehicles with nonholonomic constraints. *Robotics and Autonomous Systems, 52*, 115–131. doi:10.1016/j.robot.2005.04.006

Ren, W., & Beard, R. W. (2007). *Distributed consensus in multi-vehicle cooperative control: Theory and applications*. Springer Publishing Company, Incorporated.

Shi-Cai, L., Da-Long, T., & Guang-Jun, L. (2007). Formation control of mobile robots with active obstacle avoidance. *Acta Automatica Sinica, 33*(5), 529–535.

Tan, R. T., Nishino, K., & Ikeuchi, K. (2004). Separating reflection components based on chromaticity and noise analysis. *IEEE Transactions on Pattern Analysis and Machine Intelligence, 26*(10), 1373–1379. doi:10.1109/TPAMI.2004.90

Tijms, H. C. (2004). *Discrete-time Markov decision processes* (pp. 233-277). John Wiley & Sons, Ltd. Retrieved from http://dx.doi.org/10.1002/047001363X.ch6

Villaverde, I. (2009). *On computational intelligence tools for vision based navigation of mobile robots*. San Sebastian, Spain: UPV/EHU.

Villaverde, I., Echegoyen, Z., Moreno, R., & Graña, M. (2010). Experiments on robotic multiagent system for hose deployment and transportation. In P. Pawlewski, et al., (Eds.), *Trends in Practical Applications of Agents and Multiagent Systems, Vol. 71 of Advances in Intelligent and Soft Computing* (pp. 573–580). Springer.

Watkins, C., & Dayan, P. (1992). Q-learning. *Machine Learning, 8*(3-4), 279–292. doi:10.1007/BF00992698

KEY TERMS AND DEFINITIONS

L-MCRS: Linked Multi-Component Robotic System.

Pipeline: General schema of the process that is carried out by the system.

Q-Learning: A concrete Reinforcement Learning algorithm.

Reinforcement: A signal that indicates to an agent how good an action that has been carried out is (good, bad or indifferent).

Reinforcement Learning: A learning paradigm based on giving reinforcement signals to an agent after doing any action.

Spline: A mathematical model of a curve.

State: Perception of an agent of its environment.

Chapter 6
Building a Multiple Object Tracking System with Occlusion Handling in Surveillance Videos

Raed Almomani
Wayne State University, USA

Ming Dong
Wayne State University, USA

ABSTRACT

Video tracking systems are increasingly used day in and day out in various applications such as surveillance, security, monitoring, and robotic vision. In this chapter, the authors propose a novel multiple objects tracking system in video sequences that deals with occlusion issues. The proposed system is composed of two components: An improved KLT tracker, and a Kalman filter. The improved KLT tracker uses the basic KLT tracker and an appearance model to track objects from one frame to another and deal with partial occlusion. In partial occlusion, the appearance model (e.g., a RGB color histogram) is used to determine an object's KLT features, and the authors use these features for accurate and robust tracking. In full occlusion, a Kalman filter is used to predict the object's new location and connect the trajectory parts. The system is evaluated on different videos and compared with a common tracking system.

INTRODUCTION

Object Tracking is the process of locating an object in every frame of the video frames and using its locations to generate a trajectory. The efficient object tracking is important for many computer vision applications, such as surveillance, security, monitoring, robotic vision, etc. Although many object tracking systems have been proposed, tracking is still one of the most challenging research topics in computer vision. Tracking systems face a big challenge arises from dealing with abrupt object motion, illumination variations, occlusions, scale variations, and various types of objects

DOI: 10.4018/978-1-4666-2672-0.ch006

(Park, Makhmalbaf, & Brilakis, 2011). So, in most cases, a proposed system shows best results under specific situation.

In complex scenes, overlapping of moving objects greatly affects the accuracy and robustness of tracking. Thus, occlusion handling plays a central role in building a tracking system. In this chapter, we describe a system to automatically track multiple objects which are partially or fully occluded in a moving or standing pose.

Recently, there has been an increasing interest in tracking objects in moving cameras, such as cell phones, vehicles and robots. Tracking objects in moving cameras are more complicated than stationary cameras where the image motion is induced by both the camera motion and the object motion. Different techniques are suggested, such as homography-based motion detection (Kim & Kweon, 2011), Pattern Matching (Reddy, 2012), background subtraction (Sheikh, Javed & Kanade, 2009) and contour (Yilmaz, Li & shah, 2004). In this chapter, our system uses the videos that are captured by using a stationary camera. In future work, we are going to develop our system to deal with objects in moving cameras.

BACKGROUND

Generally, Object tracking can be divided into three major categories (Yilmaz, Li, & Shah, 2004): Correspondence-based object tracking, transformation-based object tracking and contour-based object tracking.

Correspondence-Based Object Tracking

Tracking is performed by collecting the object information during tracking and using this information to predict and verify the object new location. The researchers used different object previous information such as object state (velocity and acceleration) and regional information (color,

texture, area and shape) (McKenna, Raja, & Gong, 1999; Zhang, & Freedman, 2005). Different filtering techniques are suggested to model object information such as Kalman filtering (Stauffer, & Grimson, 2000) and particle filtering (Rittscher, Kato, Joga, & Blake, 2000).

Transformation-Based Object Tracking

Tracking is performed by using the object information in consecutive frames to estimate the motion of the object. The most common transformation based trackers are ''template matching'' (Lipton, Fujiyoshi, & Patil, 1998), ''Mean shift'' (Comaniciu, Ramesh, & Meer, 2003) and KLT (Kanade-Lucas-Tomasi) (Shi, & Tomasi, 1994). Template matching tracker is implemented by searching the whole image for a similar template. The template matching tracker is challenged by the heavy computation required and the sensitivity to illumination variation. Mean shift, on the other hand, tracks the distribution of an object in real time with robust tracking performance. However, Mean shift tracker does not deal with deformable objects and the appearance variations decreases the tracker performance (Allili, & Ziou, 2008). KLT tracker finds features that are optimal for tracking first, and then computes the translation of these features and the quality of each tracked path (Zhou, Yuan, & Shi, 2008).

Contour-Based Object Tracking

Tracking is performed by evolving the contour of the target object in the current frame and finding its new position in the next frame. Snake (Sun, Haynor, & Kim, 2003) and level set (Paragios, & Deriche 2005) are mainly used to track object contours. Snake, active contour, uses the contour in the previous frame as an initial contour in current frame, then uses external and internal energies to fix the contour. Level set is used to represent contour as zero level set where the zero set function equals zero for each point on the contour (Yilmaz, Li, & Shah, 2004).

The main challenge that faces the researchers when they build an object tracking system is occlusion. The occlusion could be a full or partial occlusion. A common approach to handle full occlusion is to use the object previous information to predict the object new location in next frame by using linear or nonlinear motion model, such as Kalman filter that is used for predicting the location and motion of objects and a particle filter that is used for state estimation. Partial occlusion is more complex than full occlusion since it is difficult to separate between objects during occlusion. Appearance models such as color histogram and mixture of Gaussians are used to separate objects during partial occlusion (Haritao, & Flickner, 2001; McKenna, Jabri, Duran, & Wechsler 2000; Roh, Kang, & Lee, 2000; Senior, 2002; Senior, Hampapur, Tian, Brown, Pankanti, & Bolle, 2001; Senior et al., 2004). In addition to using appearance model, some researchers add the position of merged objects to detect and solve partial occlusion (Elgammal, Duraiswami, Harwood, & Anddavis, 2002; Maccormick, & Blake, 2000). Silhouette-based approaches and contour-based approaches are common approaches too (Chen, Sun, Heng, & Xia, 2010; Cucchiara, Grana, Tardini, & Vezzani, 2000; Eng, Wang, Kam, & Yau, 2004). Generally the silhouette-based approaches are more stable in noisy images than contour-based approaches. More complicated tracking systems assume that each person is a connected set of blobs, such as a person's shirt and pants, and track each part individually (Elgammal, & Davis, 2001; Khan, & Shah, 2000). The object motion is also used to build tracking systems (Jiang, Huynh, Morany, Challay, & Spadaccini, 2010). Tracking during full or partial occlusion in complex scene is still very far from the complete solution. As a result, some systems do not address the occlusion at all (Azarbayerjani, & Wren, 1997). Other systems minimize the occlusion issues by using multiple camera inputs (Dockstader, & Tekalp, 2001) or

selecting an appropriate position for cameras (Bobick et al., 1999).

MAIN FOCUS OF THE CHAPTER

Issues, Controversies, Problems

Object tracking system should be a real time system and dealing with occlusion issues. Generally, Fast tracking methods are very important. The correspondence-based object trackers face a challenge because of noise, illuminations changes and self-shadowing, which may decrease the efficiency. Contour-based object trackers have very expensive calculations. However, some of transformation-based object tracker works effectively and fast in most circumstances, such as KLT tracker.

While the problem of robust object tracking in the presence of occlusion has been studied in literature, to the best of our knowledge none of these methods provide accurate tracking of the occluding objects. In this chapter, we propose a tracking system for automatically tracking multiple objects and dealing with particularly occlusion issues. The proposed system is composed of two components: An improved KLT tracker, and a Kalman filter. The improved KLT tracker uses the basic KLT tracker and an appearance model to track objects from one frame to another and deal with partial occlusion. In partial occlusion, the appearance model (e.g., a RGB color histogram) is used to determine an object's KLT features during partial occlusion, and we use these features for accurate and robust tracking. In full occlusion, a Kalman filter is used to predict the object's new location and connect the trajectory parts.

Proposed System

In the proposed system, we automatically search for tracking multiple objects and dealing with

Figure 1. Motion histogram is used to estimate the number of objects in blobs: (a) the original frame with the estimated number of objects in each blob; (b) the KLT features of comparing the frame in (a) with the previous frame; (c) the histogram of the left blob that has two objects (the man and the lady); (d) the histogram of the right blob that has one object (the man).

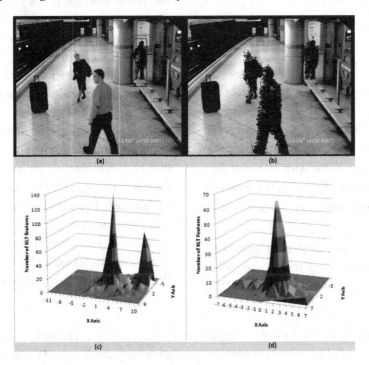

occlusion issues. The whole proposed tracking system is described in detail as follow.

The Improved KLT Tracker

The first step in tracking objects is to separate the objects from the background. Background subtraction (Musa, & Watada 2008; Wren, Azarbayejani, & Pentland, 1997) is a straightforward and widely used method. Background subtraction is performed by finding the difference between the current frame and an image of the statistical background image. The statistical background image can be built by using a single Gaussian kernel with YUV color space (Wren, Azarbayejani, & Pentland, 1997) or a Gaussian mixture model with RGB color space (Stauffer, & Grimson 2000) that is used in our system. After removing shadows and reflections (KadewTraKuPong & Bowden,

2001) then small blobs, a set of foreground blobs will be the result of our background subtraction system where each blob is one object or overlapped objects.

The foreground blobs are tracked by comparing the previous frame with the current frame and finding the KLT features. The goal of finding the KLT features is to determine distance (d) between the KLT feature at location (x) in the first frame (I) and the new location $(x + d)$ in the second frame (J) that minimizes dissimilarity ($\in$) where the dissimilarity computes from the Equation (1) and the residual is minimized by solving the the Equation (2) (Ghiasi, Nguyen & Sarrafzadeh, 2003):

$$\in = \iint_w [J(x + d) - I(x)]^2 \, dx \qquad (1)$$

$$Zd = e \qquad (2)$$

where:

$$Z = \iint_w \begin{bmatrix} g_x^2 & g_x g_y \\ g_x g_y & g_y^2 \end{bmatrix} dx,$$

$$e = \iint_w [I(x) - J(x)] \begin{bmatrix} g_x \\ g_y \end{bmatrix} dx,$$

$I(x)$ denotes the intensity of the feature point $x = $ [x y] in image (I), $J(x + d)$ denotes the intensity of the feature point with constant displacement (d) in image (J), w is the window size, g_x and g_y are the intensity gradients for x and y directions.

Some KLT features are not accurate and could have an effect on the tracking results. The KLT features that are longer than the max speed of the objects in the previous frame are deleted. The system uses the remaining KLT features to find the relationship between the blobs in the previous frame with the blobs in the current frame by counting the number of KLT features and connect between blobs that have the max number of features. For each blob in the current frame, there are four cases: New blob, existing blob, splitting blob and merging blob. These cases are described in detail as follow.

New blob: There is no match between the blob in the current frame with any blob in the database that has all the active blobs. We add the blob to the database as a new blob and extract four types of features: Location, area, variance of motion direction and centroid.

Exiting blob: There is a match between the blob in the current frame and a blob in the database. We update the blob information in the database.

Figure 2. Histogram projection is used to estimate the number of objects in blobs: a) original frame with the estimated number of objects in each blob; b) the foreground/background of the image in (a) where two objects are in the left blob and one object is in right blob; c) the histogram of the left blob that has two objects; d) the histogram of the right blob that has one object.

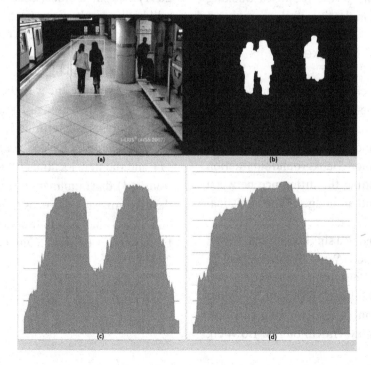

Figure 3. Tracking multiple objects with occlusion handling. (a) The foreground/background image of frame 1436 shows the occlusion. (b) The result of our tracking system of frame 1436. (c) The foreground background image of frame 4484 shows the occlusion. (b) The result of our tracking system of frame 4484.

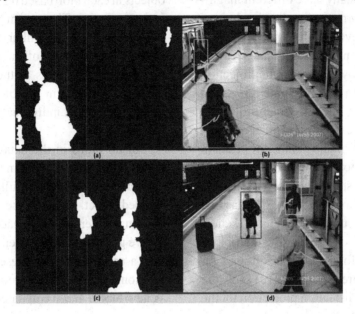

Splitting blob: There are more than one new blob in the current frame that are matched to one blob in the database. We add the new blobs to the database as new blobs.

Merging blob: There is one new blob in the current fame that is matched to more than one blob in the database. The basic KLT tracker alone is not enough in this case to keep tracking each blob individually because it cannot separate each blob's KLT features from the other merged blob features.

The improved KLT tracker is used to keep tracking partial occlusion blobs by building a RGB color histogram for each overlapped blob by using the information of the blobs before they merge. Then, the KLT features in the overlapping area are classified and assigned to each blob based on the histograms and the RGB color of the KLT features. The KLT features of each blob are used to determine the blob's bounding box. Finally, KLT features in each bounding box are used to match the blob with an existing blob in the database.

For each blob that is added to the database as a new blob, our system runs two methods, mo-

tion histogram and histogram projection (Shan, Sawhney, Matei, & Kumar, 2006), respectively, to estimate the number of objects in the blob. Motion histogram is built by using KLT features since the magnitude and direction of KLT features of an object are mostly the same. The number of objects in the blob can be estimated by the number of peaks in the motion histogram. The motion histogram gives a correct estimation when the objects in the blob have different magnitude or direction and separates them in different bounding boxes. Each bounding box is built to include all the KLT features of a peak in the motion histogram. Figurer 1 (a) shows the original image with the estimated number of objects in each blob and the corresponding KLT features are shown in Figurer 1 (b). Figure 1 (c) shows the motion histogram of the left blob that has two objects and Figure 1 (d) shows the motion histogram of the right blob that has one object. Clearing, in this case the motion histogram correctly estimates the number of object where the number of peaks refers to the number of objects.

When objects are located different distances away from the camera (i.e., in the Y direction), their KLT features usually have different magnitude, resulting in a correct estimation by the motion histogram. However, this may not be the case for the objects with similar Y coordinates. To this end, we further employ the histogram projection in the X direction for a more accurate estimation in each bounding box. The histogram projection is built by counting the foreground pixels on each point in the X direction. The number of peaks in the histogram refers to the number of objects in the blob. Figure 2 (a) shows the result of running the histogram projection on a frame that has two blobs. The left blob has two objects (the two ladies) and the right blob has one object (the man). Figure 2 (b) shows the foreground/background for the image in (a). Figures 2 (c) and 2 (d) show the result of building the histogram projection for the

two blobs. Clearly, in this case the histogram projection can correctly estimate the number of objects in each blob based on the number of peaks in the projection. The system repeats the examination of the blob for a number of successive frames and uses the average of the results to represent the number of objects in the blob. The system tracks the objects in the blob as one blob and tracks each object individually when it separates from the blob.

For all blobs in the database that cannot be matched in the current frame, there are two scenarios: The blob is either fully occluded or it is a stopped object. For the first case: Our system runs a Kalman filter to predict each blob's position as we are going to explain later. The system tries to find a match between these positions and the positions of the blobs that are added to the database as new blobs, and we update the active blob da-

Figure 4. The result of our tracking system for multiple objects in frames 1337, 1347, 1357, 1367, 1377, and 1387 of AVSS 2007 video. The original frames have the tracking results and foreground background images show the partial occlusion of one car by another.

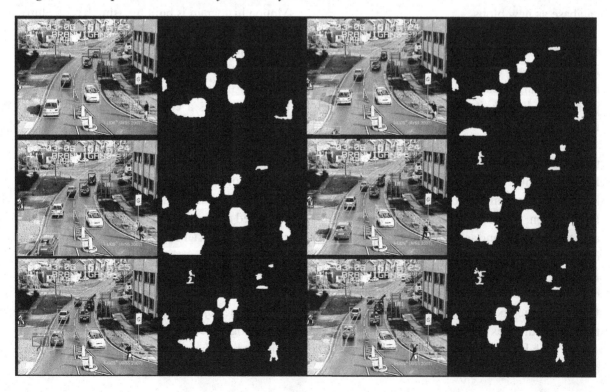

Figure 5. The result of our tracking system for multiple cars

tabase for each match. For the second case: Our system checks the centroid of the blob in the last few frames. If there is a little change, it is considered as a stopped object and marked accordingly in the database. Finally, a blob is deleted from the database when it has no match and the prediction of the Kalman filter is out of the video frame.

The Kalman Filter to Predict Object Position

The Kalman filter (Kalman, 1960) is a mathematical method that uses the previous object information to predict the state of the object in the next frames. In this chapter, a Kalman filter is used to predict the object location after the object is full occluded or has no KLT features. Let the state vector is X=[x,y,Sx,Sy,Ax,Ay], where [x,y] is the object location, [Sx,Sy] is the object speed in the x and y directions, and [Ax,Ay] is the object area. So, the Kalman filter system model and Measurement model are:

$$x_k = Fx_{k-1} + w_k \qquad (3)$$

$$z_k = Hx_k + v_k \qquad (4)$$

where:

$$F = \begin{pmatrix} 1 & 0 & 1 & 0 & 0 & 0 \\ 0 & 1 & 0 & 1 & 0 & 0 \\ 0 & 0 & 1 & 0 & 1 & 0 \\ 0 & 0 & 0 & 1 & 0 & 1 \\ 0 & 0 & 0 & 0 & 1 & 0 \\ 0 & 0 & 0 & 0 & 0 & 1 \end{pmatrix},$$

$$H = \begin{pmatrix} 1 & 0 & 0 & 0 & 0 & 0 \\ 0 & 1 & 0 & 0 & 0 & 0 \end{pmatrix}$$

w_k is the process noise where $w_k \sim N(0, Cov_1)$, and v_k is the observation noise where $v_k \sim N(0, Cov_2)$.

Experimental Results

The proposed system can track multiple objects in real-time and efficiently deal with partial and full occlusions. Our system draws a bounding box with different color to each tracked object in the scene. A line that has the same color of the object's bounding box is used to show the object's trajectory. The two numbers at the top of the bounding box are used to provide the blob ID and the estimated number of objects in the blob, respectively.

We have tested the system on a computer that has AMD Sempron 2.10 GHz processor and 2.00 GB RAM. Four publicly videos are used to evaluate our proposed system. These videos consists of indoors/outdoors and one object/multiple objects testing environments.

Figure 3 shows an example of tracking multiple objects with partial and complete occlusions. The video is from the AVSS 2007 dataset. The AVSS 2007 dataset is provided by the 2007 IEEE International Conference on Advanced Video and Signal based Surveillance. Each video is digitized with a frame size 720 by 520 and rate of 25 fps. The video includes moving persons and trains. Figure 3(a) shows the foreground/background of frame 1436 in Figure 3(b), and three objects are merged together. Cleary, our system tracks the two persons after they disappear behind the pole (full occlusion) and show up again. In addition, the three objects -the two guys and the lady- are tracked during the partial occlusion as shown in Figure 3 (b). Figure 3 (c) shows the foreground/background of frame 4484 from the same video sequences and the two guys are merged in one blob. Each object of the three objects in the frame 4484 is tracked individually without any effect to partial occlusion as shown in Figure 3 (d). So, a correct segmentation is made during the partial occlusion and each object is tracked individually.

Figure 4 shows an example of tracking multiple objects during partial occlusion. The video is from the AVSS 2007 dataset. The video includes moving persons and cars. Figure 4 shows the result of tracking multiple cars where one car is partially occluded by another one for about 50 frames as the foreground background images

show. Our system smoothly and nicely tracks the multiple objects with and without occlusion.

Figure 5 shows the result of tracking multiple objects in video sequences that are selected from are real surveillance videos that are taken on intersections during daytime. The moving cars are tracked nicely as shown by the different colors for the bounding boxes and trajectories as Figures 5 (a) and 5(c) show. For a more complex scene, Figure 5 (b) shows multiple stop-and-go cars, in which some cars stop for a short time before the traffic light and then start moving again. Our system keeps tracking them and maintains a bounding box around each car.

We also compared our system with a state-of-the-art tracking system that is called Predator (Kalal, Z., Matas, J. & Mikolayczyk, K., 2010) as Figure 6 shows. The first row in Figure 6 shows the results of using Predator to track an object in a subway and the second row shows the results of tracking the same object by using our system. The object is tracked nicely in our system where Predator failed.

FUTURE RESEARCH DIRECTIONS

Robotic vision is one of the most important research areas. Accurate tracking systems are useful in robotic vision applications. Many video tracking systems have been proposed, but no tracking system can deal with all tracking issues yet.

Figure 6. The result of tracking an object by using Predator (first row) and our system (second row)

In the future, our research will attempt to develop the system to deal with objects in the videos of moving cameras by using homography-based motion detection (Kim & Kweon, 2011) or Pattern Matching (Reddy, 2012). In addition, evolving the object contours during partial occlusion to enhance the efficiency of the proposed system. On the other hand, this proposed system needs to be comprehensively evaluated in wider video sets.

CONCLUSION

In this chapter, we have proposed a novel tracking system for effectively tracking objects in surveillance videos. The proposed system is composed of two components: An improved KLT tracker, and a Kalman filter. The improved KLT tracker uses the basic KLT tracker and an appearance model to track objects from one frame to another and deal with partial occlusion. In partial occlusion, the appearance model (e.g., a RGB color histogram) is used to determine an object's KLT features during partial occlusion, and we use these features for accurate and robust tracking. In full occlusion, a Kalman filter is used to predict the object's new location and connect the trajectory parts. The experimental results demonstrated that our system successfully tracks multiple objects with partial or full occlusions.

REFERENCES

Allili, M., & Ziou, D. (2008). Object tracking in videos using adaptive mixture models and active contours. *Neurocomputing, 71*(10-12), 2001–2011. doi:10.1016/j.neucom.2007.10.019

Azarbayerjani, A., & Wren, C. (1997). Real-time 3D tracking of the human body. *Proceedings of Image'com.*

Bobick, A., Intille, S., Davis, J., Baird, F., Pinhanez, C., & Campbell, L. (1999). The KidsRoom: A perceptually based interactive and immersive story environment. *Teleoperators and Virtual Environment, 8*(4), 367–391.

Chen, Q., Sun, Q., Ann, P. & Xia, D. (2010). Two-stage object tracking method based on kernel and active contour. *IEEE Transactions on Circuits and System for Video Technology, 20.*

Comaniciu, D., Ramesh, V., & Meer, P. (2003). Kernel-based object tracking. *IEEE Transactions on Pattern Analysis and Machine Intelligence, 25*, 564–577. doi:10.1109/TPAMI.2003.1195991

Cucchiara, R., Grana, C., & Vezzani, R. (2000). Probabilistic people tracking for occlusion handling. *International Conference on Production Research, 39*, 57-71.

Dockstader, S., & Tekalp, A. (2001). Multiple camera tracking of interacting and occluded human motion. *Proceedings of the IEEE, 89*, 1441–1455. doi:10.1109/5.959340

Elgammal, A., & Davis, L. (2001). Probabilistic framework for segmenting people under occlusion. *Eighth International Conference on Computer Vision*, Vol. 2, (pp. 145-152).

Elgammal, A., Duraiswami, R., Harwood, D., & Anddavis, L. (2002). Background and foreground modeling using nonparametric kernel density estimation for visual surveillance. *Institute of Electrical and Electronics Engineers, 7*, 1151–1163.

Eng, H., Wang, J., Kam, A., & Yau, W. (2004). A bayesian framework for robust human detection and occlusion handling using human shape model. *International Conference on Production Research*, Vol. 2, (pp. 257-260).

Ghiasi, S., Nguyen, K., & Sarrafzadeh, M. (2003). Profiling accuracy-latency characteristics of collaborative object tracking applications. *International Conference on Parallel and Distributed Computing and Systems,* (pp. 694-701).

Haritao-Glue, I., & Flickner, M. (2001). Detection and tracking of shopping groups in stores. *IEEE Computer Vision and Pattern Recognition, 1,* 431–438.

Jiang, Z., Huynh, D. Q., Morany, W., Challay, S., & Spadaccini, N. (2010). Multiple pedestrian tracking using colour and motion models. *Digital Image Computing: Techniques and Applications,* (pp. 328-334).

KadewTraKuPong. P., & Bowden, R. (2001). An improved adaptive background mixture model for real-time tracking with shadow detection. *Proceedings of the 2nd European Workshop on Advanced Video-Based Surveillance Systems.*

Kalal, Z., Matas, J., & Mikolajczyk, K. (2010). *P-N learning: Bootstrapping binary classifiers by structural constraints* (pp. 49–56). IEEE Computer Vision and Pattern Recognition. doi:10.1109/CVPR.2010.5540231

Kalman, R. (1960). A new approach to linear filtering and prediction problems. *American Society of Mechanical Engineers, 82,* 35–45.

Khan, S., & Shah, M. (2000). Tracking people in presence of occlusion. *Asian Conference on Computer Vision.*

Kim, W., & Kweon, I. (2011). Moving object detection and tracking from moving camera, *The 8th International Conference on Ubiquitous Robots and Ambient Intelligence,* Nov. 23-26.

Lipton, A., Fujiyoshi, H., & Patil, R. (1998). *Moving target classification and tracking from real-time video.* DARPA Image Understanding Workshop.

Maccormick, J., & Blake, A. (2000). Probabilistic exclusion and partitioned sampling for multiple object tracking. *International Journal of Computer Vision, 39,* 57–71. doi:10.1023/A:1008122218374

McKenna, S., Jabri, J., Duran, Z., & Wechsler, H. (2000). Tracking interacting people. *International Workshop on Face and Gesture Recognition,* (pp. 348-353).

McKenna, S., Raja, Y., & Gong, S. (1999). Tracking colour objects using adaptive mixture models. *Image and Vision Computing,* (n.d), 225–231. doi:10.1016/S0262-8856(98)00104-8

Musa, Z., & Watada, J. (2008). Video tracking system: A survey. *An International Journal of Research and Surveys, 2,* 65–72.

Paragios, N., & Deriche, R. (2005). Geodesic active regions and level set methods for motion estimation and tracking. *Computer Vision and Image Understanding, 97,* 259–282. doi:10.1016/j.cviu.2003.04.001

Park, M., Makhmalbaf, A., & Brilakis, I. (2011). Comparative study of vision tracking methods for tracking of construction site resources. *Automation in Construction, 20,* 905–915. doi:10.1016/j.autcon.2011.03.007

Reddy, V. (2012). Object tracking based on pattern matching. *International Journal of Advanced Research in Computer Science and Software Engineering, 2.*

Rittscher, J., Kato, J., Joga, S., & Blake, A. (2000). A probabilistic background model for tracking. *European Conference on Computer Vision,* Vol. 2, (pp. 336-350).

Roh, H., Kang, S., & Lee, S. (2000). Multiple people tracking using an appearance model based on temporal color. *International Conference on Pattern Recognition,* (pp. 643-646).

Senior, A. (2002). Tracking with probabilistic appearance models. *ECCV Workshop on Performance Evaluation of Tracking and Surveillance Systems*, (pp. 48-55).

Senior, A., Hampapur, A., Tian, Y., Brown, L., Pankanti, S., & Bolle, R. (2006). Appearance models for occlusion handling. *Image and Vision Computing, 24*(11). doi:10.1016/j.imavis.2005.06.007

Shan, Y., Sawhney, H., Matei, B., & Kumar, R. (2006). Shapeme histogram projection and matching for partial object recognition. *IEEE Transactions on Pattern Analysis and Machine Intelligence, 28*, 568–577. doi:10.1109/TPAMI.2006.83

Sheikh, Y., Javed, O., & Kanade, T. (2009). Background subtraction for freely moving cameras. *IEEE 12th International Conference on Computer Vision* (pp. 1219-1225).

Shi, J., & Tomasi, C. (1994). Good features to track. *IEEE Computer Vision and Pattern Recognition, 1*, 593–600.

Stauffer, C., & Grimson, W. (2000). Learning patterns of activity using real time tracking. *IEEE Transactions on Pattern Analysis and Machine Intelligence, 22*, 747–767. doi:10.1109/34.868677

Sun, S., Haynor, D., & Kim, Y. (2003). Semiautomatic video object segmentation using VSnakes. *IEEE Transactions on Circuits and Systems for Video Technology, 13*, 75–82. doi:10.1109/TCSVT.2002.808089

Wren, C., Azarbayejani, A., & Pentland, A. (1997). Pfinder: Real time tracking of the human body. *IEEE Transactions on Pattern Analysis and Machine Intelligence, 19*(7), 780–785. doi:10.1109/34.598236

Yilmaz, A., Li, X., & Shah, M. (2004). Contour-based object tracking with occlusion handling in video acquired using mobile cameras. *IEEE Transactions on Pattern Analysis and Machine Intelligence, 26*(11). doi:10.1109/TPAMI.2004.96

Zhang, T., & Freedman, D. (2005). Improving performance of distribution tracking through background mismatch. *IEEE Transactions on Pattern Analysis and Machine Intelligence, 27*(2), 282–287. doi:10.1109/TPAMI.2005.31

Zhou, H., Yuan, Y., & Shi, C. (2008). Object tracking using SIFT features and mean shift. *International Journal of Computer Vision, 113*(3).

ADDITIONAL READING

Ali, A., & Aggarwal, J. (2001). Segmentation and recognition of continuous human activity. *IEEE Workshop on Detection and Recognition of Events in Video*, (pp. 28–35).

Avidan, S. (2001). Support vector tracking. *IEEE Conference on Computer Vision and Pattern Recognition*, (pp. 184–191).

Black, M., & Jepson, A. (1998). Eigentracking: Robust matching and tracking of articulated objects using a view-based representation. *International Journal of Computer Vision, 26*(1), 63–84. doi:10.1023/A:1007939232436

Bradski, G. (1998). Real-time face and object tracking as a component of a perceptual user interface. *Proceedings of the 4th IEEE Workshop on Applications of Computer Vision*, (pp. 214-219).

Caselles, V., Kimmel, R., & Sapiro, G. (1995). Geodesic active contours. *IEEE International Conference on Computer Vision*, (pp. 694–699).

Comaniciu, D., & Ramesh, V. (2000). Mean shift and optimal prediction for efficient object tracking. *International Conference on Image Processing*, (pp. 70-73).

Edwards, G., Taylor, C., & Cootes, T. (1998). Interpreting face images using active appearance models. *International Conference on Face and Gesture Recognition*, (pp. 300–305).

Fiegth, P., & Terzopoulos, D. (1997). Color-based tracking of heads and other mobile objects at video frame rates. *IEEE Conference on Computer Vision and Pattern Recognition*, (pp. 21–27).

Isard, M., & Maccormick, J. (2001). Bramble: A Bayesian multiple-blob tracker. *IEEE International Conference on Computer Vision*, (pp. 34–41).

Jepson, A., Fleet, D., & Elmaraghi, T. (2003). Robust online appearance models for visual tracking. *IEEE Transactions on Pattern Analysis and Machine Intelligence*, *25*(10), 1296–1311. doi:10.1109/TPAMI.2003.1233903

Lienhart, R., & Maydt, J. (2002). An extended set of Haar-like features for rapid object detection. *IEEE International Conference on Image Processing*, Vol. 1, (pp. 900-903).

Lowe, D. (2004). Distinctive image features from scale-invariant keypoints. *International Journal of Computer Vision*, *60*(2), 91–110. doi:10.1023/B:VISI.0000029664.99615.94

Mikolajczyk, K., & Schmid, C. (2002). An affine invariant interest point detector. *European Conference on Computer Vision*, Vol. 1, (pp. 128–142).

Monnet, A., Mittal, A., Paragios, N., & Ramesh, V. (2003). Background modeling and subtraction of dynamic scenes. *IEEE International Conference on Computer Vision*, (pp. 1305–1312).

Papageorgiou, C., Oren, M., & Poggio, T. (1998). A general framework for object detection. *IEEE International Conference on Computer Vision*, (pp. 555–562).

Paragios, N., & Deriche, R. (2002). Geodesic active regions and level set methods for supervised texture segmentation. *International Journal of Computer Vision*, *46*(3), 223–247. doi:10.1023/A:1014080923068

Piccardi, M. (2004). Background subtraction techniques: A review. *IEEE International Conference on Systems, Man, and Cybernetics*, Vol. 4, (pp. 3099-3104).

Radke, R., Andra, S., Al-Kofahi, O., & Roysam, B. (2005). Image change detection algorithms: A systematic survey. *IEEE Transactions on Image Processing*, *14*, 294–307. doi:10.1109/TIP.2004.838698

Roller, D., Daniilidis, K., & Nagel, H.-H. (1993). Model- based object tracking in monocular image sequences of road traffic scenes. *International Journal of Computer Vision*, *10*(3), 257–281. doi:10.1007/BF01539538

Rowley, H., Baluja, S., & Kanade, T. (1998). Neural network-based face detection. *IEEE Transactions on Pattern Analysis and Machine Intelligence*, *20*(1), 23–38. doi:10.1109/34.655647

Rui, Y., & Chen, Y. (2001). Better proposal distributions: Object tracking using unscented particle filter. *IEEE Conference on Computer Vision and Pattern Recognition*, (pp. 786-793).

Serby, D., Koller-Meier, S., & Gool, L. V. (2004). Probabilistic object tracking using multiple features. *IEEE International Conference of Pattern Recognition*, (pp. 184–187).

Shi, J., & Malik, J. (2000). Normalized cuts and image segmentation. *IEEE Conference Computer Vision and Pattern Recognition*, *22*(8), 888–905.

Veenman, C., Reinders, M., & Backer, E. (2001). Resolving motion correspondence for densely moving points. *Transactions on Pattern Analysis and Machine Intelligence*, *23*(1), 54–72. doi:10.1109/34.899946

Viola, P., Jones, M., & Snow, D. (2003). Detecting pedestrians using patterns of motion and appearance. *IEEE International Conference on Computer Vision* (pp. 734–741).

Yang, C., Duraiswami, R., & Davis, L. (2005). Fast multiple object tracking via a hierarchical particle filter. *IEEE International Conference on Computer Vision*, (pp. 212–219).

Zivkovic, Z., & Kröse, B. (2004). An EM-like algorithm for color-histogram-based object tracking. *IEEE Conference on Computer Vision and Pattern Recognition*, Vol. 1, (pp. 798-803).

KEY TERMS AND DEFINITIONS

Background Subtraction: Finding the difference between the current frame and a statistical background image.

Color Histogram: Represents an image as the number of pixels in each color range.

Kalman Filter: A mathematical estimation of the state of an object in the future according to previous measurements.

KLT Tracker: A tracker algorithm that finds good features to identify and tracks them from one frame to another.

Motion Histogram: Built by using the magnitude and direction of the KLT features where the peak number refers to the tracked object number.

Object Tracking: The process of locating an object in every frame of the video frames and using its locations to generate a trajectory.

Occlusion: Occurs when a part or full tracked object is occluded by other object.

Chapter 7
A Robust Color Watershed Transformation and Image Segmentation Defined on RGB Spherical Coordinates

Ramón Moreno
Universidad del País Vasco, Spain

Manuel Graña
Universidad del País Vasco, Spain

Kurosh Madani
University PARIS-EST Creteil, Senart-FB Institute of Technology, France

ABSTRACT

The representation of the RGB color space points in spherical coordinates allows to retain the chromatic components of image pixel colors, pulling apart easily the intensity component. This representation allows the definition of a chromatic distance and a hybrid gradient with good properties of perceptual color constancy. In this chapter, the authors present a watershed based image segmentation method using this hybrid gradient. Oversegmentation is solved by applying a region merging strategy based on the chromatic distance defined on the spherical coordinate representation. The chapter shows the robustness and performance of the approach on well known test images and the Berkeley benchmarking image database and on images taken with a NAO robot.

INTRODUCTION

Image pre-processing and image segmentation are key steps on robotic vision. On the one hand, humanoid robots are in continuous movement, changing of sceneries, changing of point of view, and in all cases illumination conditions could be unstable and different, e.g. a robot can go by walk from a corridor with natural illumination to a room with tungsten illumination. In this case, a robust image pre-processing is key in order to normalize the image respect to the illumination. On the other

DOI: 10.4018/978-1-4666-2672-0.ch007

hand, when the image is already normalized, the following step, is the information extraction. The most used is the segmentation process. This step divides the image in some regions such that these regions can be identified as objects in the scene. For this work, we have found robustness respect to the illumination through spherical coordinates, and for segmentation we will use a watershed transform with a region merging directed by the image chromaticity.

Color images have additional information over gray scale images that may allow the development of robust segmentation processes. There have been works using alternative color spaces with better separation of the chromatic components like HSI, HSL, HSV, Lab (Angulo & Serra, 2007; Hanbury & Serra, 2001) to obtain perceptually correct image segmentation. However, chromaticity's illumination can blur and distort color patterns. Color constancy is the perceptual mechanism which provides humans with color vision which is relatively independent of the spectral content of the illumination of a scene. It is the ability of a vision system to diminish or, in the ideal case, remove the effect of the illumination, and therefore "see" the true physical scene as the invariant to illumination changes. To obtain color constancy, one approach consists in the estimation of the illumination source chromaticity followed by the chromatic normalization of the image. There are several approaches in the literature to achieve this goal (Tan, 2003; Yoon, 2005; Toro, 2008) assuming a uniform chromaticity of the illumination all over the scene. Other approaches try to obtain segmentation procedures which are inherently robust to illumination effects (Mallick, 2005; Zickler, 2006). The segmentation method proposed in this paper has inherent color constancy due to the color representation chosen and the definition of the chromatic distance.

Color constancy is closely related to the response of the gradient operators (Geusebroek, 2003). Regions of constant color must have low gradient response, while color edges must have a strong gradient response. Image segmentation methods based on spatial gradients need a correct definition of the spatial color gradient and unambiguous contour definition. In fact, formulation of watershed segmentation methods in color images is still an open research issue (Aptoula, 2007). A straightforward but inexact approach is the independent application of the watershed segmentation on image channel (Tarabalka, 2010). This approach loses chromatic information, and has difficulties merging the subsequent independent segmentations into one.

In this work we will use the RGB spherical coordinates representation to achieve color constancy properties of our image segmentation approach (Moreno & Graña, 2010; Moreno & d'Anjou, 2010; Moreno & Zulueta, 2010). We define a chromatic distance on this representation. The robustness and color constancy of the approach is grounded in the dichromatic reflection model (DRM) (Shafer, 1984). We propose a chromatic gradient operator suitable for the definition of a watershed transformation on color images and a robust region merging for meaningful color image segmentation. The baseline chromatic gradient operator (R. Moreno & d'Anjou, 2010; Moreno, 2010) suffers from noise in the dark areas of the image. We propose in this work a hybrid gradient operator overcoming this problem. We use it to build a watershed transformation on color images. To achieve a natural segmentation, we perform region merging on the basis of our proposed chromatic distance over the chromatic characterization of the watershed regions. We give a general schema that combines watershed flooding with region merging in a single process. Finally, we specify our proposal as an instance of the aforementioned general schema.

This work is motivated for its application on robotic context; therefore it needs of a good speed-up (close to real time). The method is implemented in C# and the results are shown in the color image segmentation section where we will give more details. We have used a NAO robot. The images taken by the NAO robot are noisy due to the movements of the robot and to the quality limitations of the camera. In order to validate the method for general purposes, we test the method with the Berkeley image dataset too.

The chapter outline is as follows. The next section presents the RGB spherical interpretation, including the definition of the chromatic distance. The section following reviews the definition of spatial gradients, and gives our hybrid gradient operator proposal. After this, the chapter presents our image segmentation method, then shows the experimental results comparing our method and other approaches. Finally, the last section gives the conclusions of this work.

SPHERICAL COORDINATES IN THE RGB COLOR SPACE

There are several works that use a spherical/cylindrical representation of RGB color space points looking for photometric invariants (Mileva, 2007; van de Weijer & Gevers, 2004). We are interested in the correspondence between the angular parameters (ϑ, ϕ) of the spherical representation of a color point in the RGB space and its chromaticity Ψ

Chromaticity Interpretation of RGB Spherical Coordinates

An image pixel's color corresponds to a point in the RGB color space $c = \{R_c, G_c, B_c\}$. Chromaticity Ψ_c of a RGB color point c is given by two of its normalized coordinates $r_c = \dfrac{R_c}{R_c + G_c + B_c}$,

$g_c = \dfrac{G_c}{R_c + G_c + B_c}$, $b_c = \dfrac{B_c}{R_c + G_c + B_c}$, that

fulfill the condition $r_c + g_c + b_c = 1$ that is, chromaticity coordinates $\Psi_c = \{r, g, b\}$ correspond to the projection of c on the chromatic plane Π_Ψ which is defined by the collection of vertices $\{(1,0,0),(0,1,0),(0,0,1)\}$ of the RGB cube which define the Maxwell's chromatic triangle (Maxwell,

1885), along the chromatic line defined as $L_c = \{y = k.\Psi_c; k \in \mathbb{R}\}$. In other words, all the points in line L_c have the same chromaticity Ψ_c. Chromaticity is a robust color characterization, which is independent of illumination intensity and preserves the geometry of the objects in the scene (Shen et al., 2008; Tan et al., 2004; Zickler et al., 2006).

On the other hand, the vector going from the RGB space origin up to the point c, as shown in Figure 1, can be represented using spherical coordinates $c = \{\theta_c + \phi_c + l_c\}$, where θ_c is zenithal angle, ϕ_c is the azimuthal angle, and l_c is the vector's magnitude. The pair of angle values (θ_c, ϕ_c) define a line which is the same as the chromatic line L_c. Notice that $L_c = \|c\| = k\|\Psi_c\|$, with $k = R_c + G_c + B_c$. We have $\theta_c = cos^{-1}\left(\dfrac{R_c}{\|c\|}\right) = cos^{-1}\left(\dfrac{r_c}{\|\Psi_c\|}\right)$, and $\phi_c = tan^{-1}\left(\dfrac{G_c}{\|c\|}\right) = tan^{-1}\left(\dfrac{g_c}{\|\Psi_c\|}\right)$. Therefore, the point in the chromatic plane determined by the normalized RGB coordinates and the angular values of the spherical coordinates provide the same chromatic information.

Denoting a color image in RGB space as $I = \{I(x); x \in \mathbb{N}^2\} = \{(R, G, B)_x; x \in \mathbb{N}^2\}$, where x refers to the pixel coordinates in the image grid domain, we denote the corresponding spherical representation as $P = \{P(x); x \in \mathbb{N}^2\} = \{(\theta_c, \phi_c, l_c)_x; x \in \mathbb{N}^2\}$, so we will use (θ_c, ϕ_c) as the chromaticity representation of the pixel's color. Sometimes we use the notation $x = (i, j)$.

The Chromatic Distance in the RGB Space

First, we convert the RGB Cartesian coordinates of each pixel to polar coordinates, with the black

Figure 1. A vector in the RGB space

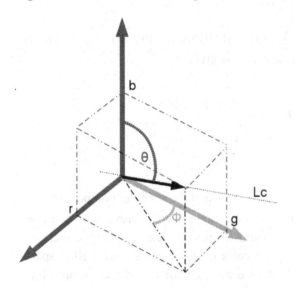

color as the RGB space origin. Let us denote the Cartesian coordinate image as $I(x) = I_x = (r, g, b)_x; x \in \mathbb{N}^2$ and the spherical coordinate as $P(x) = P_x; x \in \mathbb{N}^2 = (\theta, \phi, l)_x; x \in \mathbb{N}^2$, where p denotes the pixel position. For the remaining of the paper we discard the magnitude l_x because it does not contain chromatic information, therefore $P(x) = P_x; x \in \mathbb{N}^2 = (\theta, \phi, l)_x; x \in \mathbb{N}^2$. For a pair of image pixels x and y, the color distance between them is defined as:

$$\angle\left(P_x, P_y\right) = \sqrt{\left(\theta_x - \theta_y\right)^2 + \left(\phi_x - \phi_x\right)^2} \qquad (1)$$

That is, the color distance corresponds to the Euclidean distance of the Azimuth and Zenith angles of the pixel's RGB color polar representation. This distance does not take into account the intensity component and, thus, will be robust against specular surface reflections. It posses color constancy, because pairs of pixels in the same reflectance surface will have the same distance regardless of their intensity.

Color Constancy of the Spherical Coordinates Based on the DRM

The Dichromatic Reflection Model (DRM)(Shafer, 1984) explains the perceived color intensity $I(x) \in \mathbb{R}^3$ of each pixel in the image as the addition of two components, one diffuse $\mathcal{D} \in \mathbb{R}^3$ and another specular component $\mathcal{S} \in \mathbb{R}^3$. The diffuse component refers to the chromatic properties of the observed surface, while the specular component refers to the illumination color. Image bright spots are pixels with a high specular component.

The mathematical expression of the model, when we have only one surface color in the scene, is as follows:

$$I\left(x\right) = m_d\left(x\right)\mathcal{D} + m_s(x)\mathcal{S} \qquad (2)$$

where, m_d and m_s are weighting values for the diffuse and specular components and $\mathcal{D}$ and $\mathcal{S}$, are points in the three-dimensional space. In Figure 2 the stripped region represents a trapezoidal convex region of the plane $\prod_{dc}$ in the RGB space that contains all the possible colors expressed by the DRM Equation 2. This equation expressed by Cartesian coordinates, can be rewritten in spherical coordinates, where Λ is the diffuse chromaticity and Γ the specular chromaticity. Differently from the Cartesian expression, Λ and Γ are lines whereas $\mathcal{D}$ and $\mathcal{S}$ are point. A consequence of this formulation is that $\mathcal{S} \in \Lambda$ and $\mathcal{D} \in \Gamma$. The Equation 2 is rewritten in spherical coordinates as:

$$P\left(x\right) = \left(\theta_D, \phi_D, l_D(x)\right) + \left(\theta_S, \phi_S, l_S(x)\right)$$

where the diffuse chromaticity is defined as $\Gamma = (\theta_D, \phi_D)$, the magnitude of the pixel's color diffuse component is given by:

$$l_D\left(x\right) = \sqrt{\begin{array}{l}\left(m_d(x)\mathcal{D}_R\right)^2 + \\ \left(m_d(x)\mathcal{D}_G\right)^2 + \left(m_d(x)\mathcal{D}_B\right)^2\end{array}}$$

The specular chromaticity is defined as $\Gamma = (\theta_S, \phi_S)$ and the magnitude of the illuminant color is given by:

$$l_S\left(x\right) = \sqrt{\begin{array}{l}\left(m_d(x)\mathcal{S}_R\right)^2 + \left(m_d(x)\mathcal{S}_G\right)^2 \\ + \left(m_d(x)\mathcal{S}_B\right)^2\end{array}}$$

When the scene contains several surface colors, the DRM equation must assume that the diffuse component may vary spatially, while the specular component is constant across the image domain:

$$I\left(x\right) = m_d\left(x\right)\mathcal{D} + m_s(x)\mathcal{S} \qquad (3)$$

this in spherical coordinates is expressed as:

$$P\left(x\right) = \left(\theta_D(x), \phi_D(x), l_D(x)\right) + \left(\theta_S, \phi_S, l_S(x)\right)$$

where

$\Lambda(\mathrm{x}) = (\theta_D(x), \phi_D(x))$, the diffuse component magnitude is given by:

$$l_D\left(x\right) = \sqrt{\begin{array}{l}\left(m_d\left(x\right)\mathcal{D}_R(x)\right)^2 + \left(m_d\left(x\right)\mathcal{D}_G(x)\right)^2 + \\ \left(m_d\left(x\right)\mathcal{D}_B(x)\right)^2\end{array}}$$

$\Gamma = (\theta_S, \phi_S)$, and the specular component magnitude is the same that the previous because we are supposing uniform chromatic illumination.

The color image is plotted in the RGB space. Authors draw the diffuse and the specular lines where $\mathcal{D}$ belong to the diffuse line, and $\mathcal{S}$ to the specular line (the achromatic line)

The DRM expresses the pixel color as a vectorial sum, the sum of the diffuse component and the specular component. Pure diffuse pixels haven't got specular component, therefore in DRM they are expressed as $P\left(x\right) = \left(\theta_D(x), \phi_D(x), l_D(x)\right)$. For diffuse pixels inside a homogeneous surface color region the

Figure 2. Distribution of pixels in the RGB space

(a)

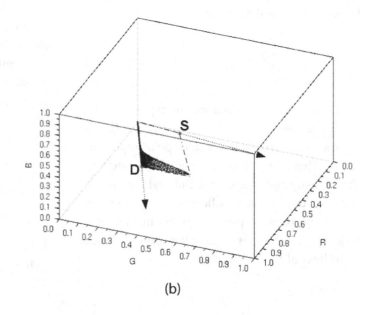

(b)

zenith ϕ and azimuthal θ angles are almost constant, because they have the same chromaticity and their magnitude depends on the geometry of the object (Lambert, 1760; Zickler, 2006). These angles have big changes among specular pixels because the specular component is not null, hence the vectorial sum gives different values in a neigborhood. This difference is bigger between diffuse pixels belonging to different color regions. Therefore, the chromatic distance between the vectors representing two neighboring pixels $P\left(x_p\right)$ and $P\left(x_q\right)$, $\angle(P_p, P_q)$ reflects the chromatic variation among them. For two diffuse pixels in the same chromatic regions, their vectorial representation will be along the same chromatic line and this chromatic distance must be $\angle\left(P_p, P_q\right) = 0$. This distance has color constancy properties, because changing the illumination source chromaticity corresponds to introducing a linear term in the Equation 3 which disappears when we consider the distance between pixels for segmentation purposes.

GRADIENTS

Let us denote $x = (i, j)$ the pixel coordinates. We recall the definition of the image spatial gradient

$$\nabla I\left(i, j\right) = \begin{bmatrix} G_i(i, j) \\ G_j(i, j) \end{bmatrix} = \begin{bmatrix} \dfrac{\partial}{\partial i} I(i, j) \\ \dfrac{\partial}{\partial j} I(i, j) \end{bmatrix} \tag{4}$$

where $I(i, j)$ is the image function at pixel (i, j). For edge detection, the usual convention is to examine the gradient magnitude:

$$G\left(i, j\right) = \left|G_i(i, j)\right| + \left|G_j(i, j)\right| \tag{5}$$

For color images, a simplistic approach to perform edge detection is to drop all color information, and convolve the intensity image with a pair of high-pass convolution kernels to obtain the gradient components and gradient magnitude. The most popular edge detectors are the Sobel and the Prewitt detectors, illustrated in Figure 3 because we will build our own spatial chromatic gradient operators following their pattern. To take into account color information, the straightforward approach is to apply the gradient operators to each color band image and to combine the results afterwards: $\nabla I = \left[\nabla I_r + \nabla I_g + \nabla I_b\right] / 3$. Figure 3 illustrates these approaches. It can be appreciated that the gradient magnitude amplifies noise when we combine the color band gradient magnitudes, and that the color edge goes undetected by the gradient operator applied to the intensity image, because the two color regions have quite near intensity values. The edge magnitude computed by these straightforward approaches is also misled by the specular surface reflections, which are highlighted as can be appreciated in Figure 3(d).

Chromatic Gradient Operator

We will formulate a pair of Prewitt-like gradient pseudo-convolution operations on the basis of the above distance. Note that the $\angle(P_p, P_q)(P_p, P_q)$ distance is always positive. Note also that the process is non linear, so we cannot express it by convolution kernels. The row pseudo-convolution is defined as

$$CR_R\left(P\left(i, j\right)\right) = \sum_{r=-1}^{1} \angle(P_{i-r, j+1}, P_{i-r, j-1})$$

and the column pseudo-convolution is defined as

$$CR_C\left(P\left(i,j\right)\right) = \sum_{c=-1}^{1} \angle\left(P_{i+1,j-c}, P_{i-1,j-c}\right)$$

so that the color distance between pixels substitutes the intensity subtraction of the Prewitt linear operator. The color gradient image is computed as

$$CG\left(x\right) = CR_R\left(x\right) + CR_C\left(x\right) \tag{6}$$

A Hybrid Gradient

Empirical experiments show that the aforegoing gradient is very susceptible to image noise. The angular distance of Equation 1, is more sensitive to noise for pixel colors lying close to the origin in RGB space. This is due to the fact that the angular distance between two points at a given Euclidean distance grows as the points are closer to the origin. Small perturbations as measured by the Euclidean distance are mapped into big angular differences. The background noise which has little effect in lighted regions is amplified in the dark regions.

Inspired in the human vision, we propose a hybrid gradient which is an intensity gradient when the illumination is poor, and a chromaticity gradient in better illuminated image regions. For intensity values below a threshold a it is an intensity gradient, for values above another threshold b it is a chromatic gradient, and for values between

both it is a mixture of the two kinds of gradients whose mixing coefficient is sinusoidal function of the image intensity. This idea is expressed mathematically as a convex combination of the two gradient operators:

$$HG\left(x\right) = \alpha\left(x\right)G\left(x\right) + \bar{\alpha}\left(x\right)CG(x) \tag{7}$$

where x is the pixel location, $G(x)$ is the intensity gradient magnitude of Equation 5, $CG(x)$ is the chromatic gradient and $\bar{\alpha}\left(x\right) = 1 - \alpha\left(x\right)$, hence $\bar{\alpha}\left(x\right) + \alpha\left(x\right) = 1$, $\alpha\left(x\right)$ is normalized to the range [0,1] and its expression is as follows:

$$\alpha\left(x\right) = \begin{cases} 1 & I\left(x\right) < a \\ .5 + \dfrac{\cos\left(\dfrac{x-a}{b-a}\pi\right)}{2} & a \le I\left(x\right) < b \\ 0 & b \le I\left(x\right) \end{cases} \tag{8}$$

where $I(x)$ is the pixel intensity. This factor is plotted in Figure 4.

This hybrid gradient does not suffer from noise sensitivity in dark regions of the image, the effect of bright spots is reduced because it is chromatically consistent in bright image regions, and it detects chromatic edges.

Figure 3. Effect of the illumination on gradients: (a) original synthetic RGB image, (b) intensity image, (c) gradient magnitude computed on the intensity image, (d) gradient magnitude combining the gradient magnitudes of each color band

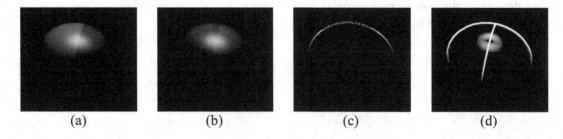

(a)   (b)   (c)   (d)

In the Figure 5 we show the response of different gradient operators on the same test image. Figure 5(a) presents the original image. Figure 5(b) contains the response of the intensity gradient, it shows false border detection in some diffuse regions, i.e. the yellow ball, green thorax. It also shows false edge detection in bright spots. Figure 5(c) shows the response of the chromatic gradient operator. It does not give false edge detection inside diffuse regions. It does not give false edge detection in bright spot areas. However, it is very sensitive to noise in the dark regions, showing false edge detections due to small random variations. Figure 5(d) presents the response of the hybrid gradient which has the good detection properties of the chromatic gradient operator and it is not sensitive to noise in the dark image regions.

COLOR IMAGE SEGMENTATION

In this section we will first recall some background on watershed transformations, then we present the general scheme of the watershed segmentation algorithm. The precise specification of our and competing approaches is achieved stating the concrete parameters of this general algorithm.

Watershed

Watershed transformation is a powerful mathematical morphology technique for image segmentation. It was introduced in image analysis by Beucher and Lantuejoul (1979), and subsequently a lot of algorithm variations and applications have been proposed (Dagher & Tom, 2008; Elwaseif & Slater, 2010; Tarabalka,2010).

The watershed transform considers a bidimensional image as a topographic relief map. The value of a pixel is interpreted as its elevation. The watershed lines divide the image into catchment basins, so that each basin is associated with one local minimum in the topographic relief map. The watershed transformation works on the spatial gradient magnitude function of the image. The crest lines in the gradient magnitude image correspond to the edges of image objects. Therefore, the watershed transformation partitions the image into meaningful regions according to the gradient crest lines.

The baseline watershed transformation computation method is as follows. The image pixel sites with local minimum gradient are selected as the sources of their respective catchment basins. A flooding process fills each catchment basin from its respective source. When a catchment basin is full, the contour points which are in touch with a neighbor catchment basin are the dam points. The process is finished when all gradient surface is covered. The closed lines defined by the dam points give us the watershed transformation, and, implicitly, the image partition. Usually, this partition is very fine, therefore a further step of region merging is needed to obtain partitions closer to the natural segmentation of the image. Region merging needs the specification of when and why two neighboring catchment basins are merged into one region. In other words, region merging criterion defines which watershed lines are going to be removed. Watershed regions are image regions with homogeneous properties. We look for homogeneous chromatic regions on the basis of the aforegoing hybrid chromatic gradient.

Figure 4. Hybrid gradient convex combination factor as a function of the image intensity

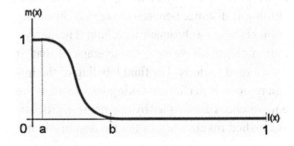

Figure 5. Response of different gradient operators: (a) original image, (b) intensity gradient, (c) chromatic gradient, (d) hybrid gradient

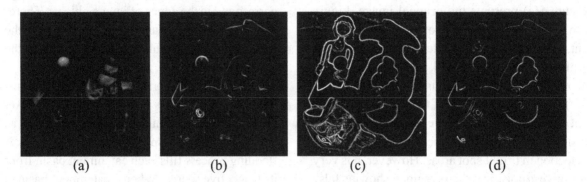

| (a) | (b) | (c) | (d) |

General Schema of a Watershed Method

The general schema of the watershed method perform a flooding process which performs a region growing based on the ordered examination of the level sets of the gradient image. In fact, an ordered succession of thresholds is applied to produce the progression of the flooding. The image is examined iteratively *n* times, each iteration step the threshold is raised and pixels of the gradient image below the new threshold are examined to be labeled with a corresponding region. Initially each region will contain the source of its catchment basin when the flooding level reaches it. Each flooded region is also characterized by a chromaticity value, which corresponds to the source pixel chromaticity. This chromaticity value is used to perform region merging simultaneously with the flooding process. A pixel whose neighboring pixels belong to different regions is a watershed pixel. When a watershed pixel is detected, the adjacent regions may be merged into one if the chromatic distance between the region chromatic values is below a chromatic threshold. The merged region chromatic value is the average of that of the merged regions. The final labeling of the image regions is performed taking into account the equivalences established by the merging process. Watershed pixels whose adjacent regions do not merge into one are labeled as region boundary pixels and retain their chromaticity.

The Proposed Approach

Our color image segmentation process proposal can be precisely specified by Algorithm 1 applied on the zenithal and azimuthal angles of the color representation $P(x)$ of Equation 3, the gradient magnitude image computed by the hybrid gradient $HG(x)$ of Equation 7, using the chromatic distance of Equation 1.

Algorithm 1: General scheme of watershed and region merge for color image segmentation.

Set number of iterations n and the chromatic distance threshold δ, initialize the pixel labels $\forall x; L_x = 0$, the region label counter $R = 0$. $\Omega(x)$ is the color image and $\Phi(x)$ is the gradient magnitude image.

1. Calculate $\Phi_{min} = \min_{x}\{\Phi(x)\}$ and $\Phi_{max} = \max_{x}\{\Phi(x)\}$

    a. Calculate the step at each interaction $s = (\Phi_{max} - \Phi_{min})/n$

    b. Initialize $t = \Phi_{min}$

2. Iterate n times, setting

a. Calculate threshold $t = t + s$;

b. Consider $X'(t) = \{x' \mid \Phi(x') < t\}$, for each $x \in X'(t)$ perform:

   i. If $L(x) = \varnothing$ the pixel is unprocessed, then one of the following cases apply

3. If $L_8(x) = \varnothing$

a. Assign new label $R \leftarrow R+1; L(x) = R$;

b. Assign the region chromatic value $\Psi_R = \Psi(\Omega(x))$

4. If $|L_8(x)| = 1$

a. $L(x) = L_8(x)$;

b. Update $\Psi_{L(x)}$ using $\Psi(\Omega(x))$

5. If $|L_8(x)| > 1$ there are at least two adjacent regions, x is a gradient watershed pixel. Consider all pairs of adjacent regions of labels r_1 and r_2

a. If $\Delta(\Psi_{r1}, \Psi_{r2}) < \delta$ then, we can merge both regions into one of label r^*

   i. Compute

$$\Psi_{r^*} = \left(\frac{|r_1|.\Psi_{r_1} + |r_2|.\Psi_{r_2}}{|r_1| + |r_2|} \right)$$ We keep record of the detected equivalence.

   ii. Update Ψ_{r^*} using $\Psi(\Omega(x))$.

   iii. $L(x) = r^*$

b. If " $(\Psi_{r1,}\Psi_{r2}) > \delta$ then, the pixel x is a region boundary pixel with a special label $L(x) = b$

6. From the recorded label equivalences compute the final region labels, and assign definitive labels.

7. Each region R has a corresponding chromatic value Ψ_R which can be used for visualization.

The Algorithm 1 gives the details of our method. In this algorithm $L(x)$ denotes the region label of pixel x, $L_8(x)$ denotes the set of labels of the 8 neighbors of pixel x, that can be expressed as $L(x) = \bigcup_{x' \in N_8(x)} L('x)$ where $N_8(x)$ is the 8-th neighborhood of pixel x. The algorithm may be applied to any color image $\Omega(x)$ and gradient magnitude image $\Phi(x)$. The algorithm needs the specification of a chromatic distance $\Delta(\Omega(x), \Omega(y))$ that gives a measure of the similarity between pixel colors $\Omega(x)$ and $\Omega(y)$. To label the regions we keep a counter R, and we build a map Ψ_R assigning to each region label a chromatic value. While the flooding process performs region growing, the region chromatic value is updated to the average chromaticity of the pixels in the region. Each region R has a corresponding chromatic value Ψ_R which can be used for visualization.

EXPERIMENTAL RESULTS

The watershed-merge algorithm is parametrized by:

- The number of iterations n, which determines the resolution of the flooding process going over the gradient magnitude image level sets.

The gradient operator used to compute the gradient magnitude image, which can be either the intensity gradient $G(x)$ of Equation 5 or the hybrid gradient $HG(x)$ of Equation 7.

The color representation of the image. Assuming the RGB space, it can be either the Cartesian representation $I(x)$ or the zenithal and azimuthal angles of the spherical representation $P(x)$. This selection determines the selection of the chromatic distance.

- The chromatic distance, which can be either the Euclidean distance in the RGB Cartesian space, or the chromatic distance of Equation 1.

The chromatic distance threshold δ, which determines the chromatic resolution of the region merging process.

This section reports results of three experiments. The first one compares our proposal of the hybrid gradient section with other instances of the algorithm, whereas in the second one we will to provide a more extensive qualitative validation our method using the well know Berkeley benchmark image collection (Martin, 2001) which provides hand-draw artistic shape boundaries. Finally we show some experimental results with images taken by a NAO robot.

Behavior

In this section we will use a well known benchmark image (Tan, 2004) to compare our proposed segmentation process with variations of the algorithm obtained with other parameter settings. The dark regions are critical to the perceptually correct gradient computation, while the bright spots may induce false edge detection. The algorithm does not compute any specular free image to remove this latter problem.

The operational parameter setting are $n = 100$ and $\delta = 0.1$. In Figure 6 we show the segmentation results on this image for all combinations of the remaining Algorithm 1 parameter settings. The column of images labeled "Gradient" has the gradient magnitude images. From top to bottom, Figures 6(a), 6(e), 6(i) show, respectively the result of the intensity gradient, the chromatic gradient of Equation (6), and the hybrid gradient of Equation (8).

The column of images labeled "Watershed" correspond to the image region partition performing only to the flooding process, without any

region merging, on the corresponding gradient magnitude images. It can be appreciated that the hybrid gradient watershed removes most of the dark microregions originated by the chromatic gradient. There are, however, some regions with different colors in this rough dark region which are not fully identified by the intensity gradient watershed of Figure 6(b) and are better detected by the hybrid gradient watershed in Figure 6(j).

The two image columns with the heading "segmentation" show the results of the region merging from the corresponding gradient watershed in the same row. The left column shows the results of using of the Euclidean distance on the RGB Cartesian coordinates. The right segmentation column show the results of the using the chromatic distance of Equation 1. If we want to ascertain the effect of the color representation and the chromatic distance we must compare the rightmost columns in Figure 6.

We find that the general effect is that the chromatic distance on polar coordinates is better identifying the subtle color regions in the darkest areas of the image, it detects better the shape of the objects, has better color constancy properties, and it is much less sensitive to bright spots or shining areas. Comparing the gradient operators attending to the final segmentation we observe that the hybrid gradient is better than the others in removing noise from the dark regions and maintain the object integrity. Overall the best result is obtained with our proposal as shown in Figure 6(l), where we can easily identify the subtle regions in the upper dark area, the shadow of the lowermost object, and we can clearly identify object with the same color unaffected by shading and bright spots.

Validation on the Berkeley Images

In the Figure 7 we can show the experimental results using the Berkeley DB (Martin, 2001). The first and fourth rows show the original im-

Figure 6. Image segmentation results with different parameterizations of algorithm 1

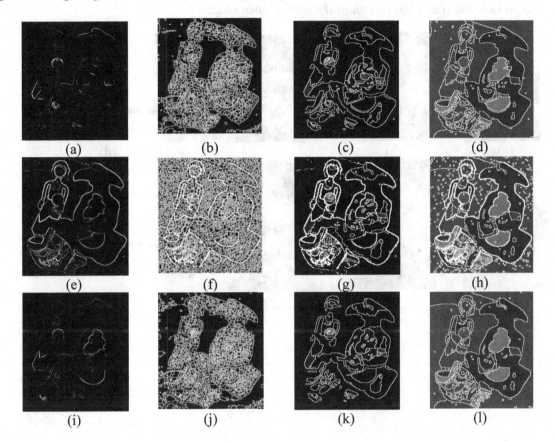

age, the second and fifth shows our respective outputs, whereas the third and sixth row shows the human segmentation reference. As we can see our method gives always homogeneous regions, and the segmentation output is close to the human segmentation.

Some facts that we find comparing our segmentation with the hand-drawn segmentation:

- Large chromatically smooth regions are well segmented by our approach despite variations in intensity, e.g. the face skin of the portraited man, the river, the road in the road race image.

- Some subtle chromaticity variations are detected and segmented, like the reflections in the water of the jungle river image.

- The algorithm does not use any spatial information to segment textured objects. However it can cope with some textured spatial intensity variations of chromatically constant regions, outlining the corresponding object, i.e. the clouds in the flying plane image, the yellow skirt in the jungle river image.

- The hand-draw contours obviate some regions of the image that the artist may have found irrelevant, i.e. the clouds in the sky in some images, the texture details of some bushes. Some of these regions cannot be segmented as a unit unless some spatial texture information is used, like the bushes in the jungle river image, or the skyscraper windows.

Figure 7. Segmentation results on some of the Berkeley images. Second and fourth rows show the results of our approach. Third and last row show the hand-drawn shapes.

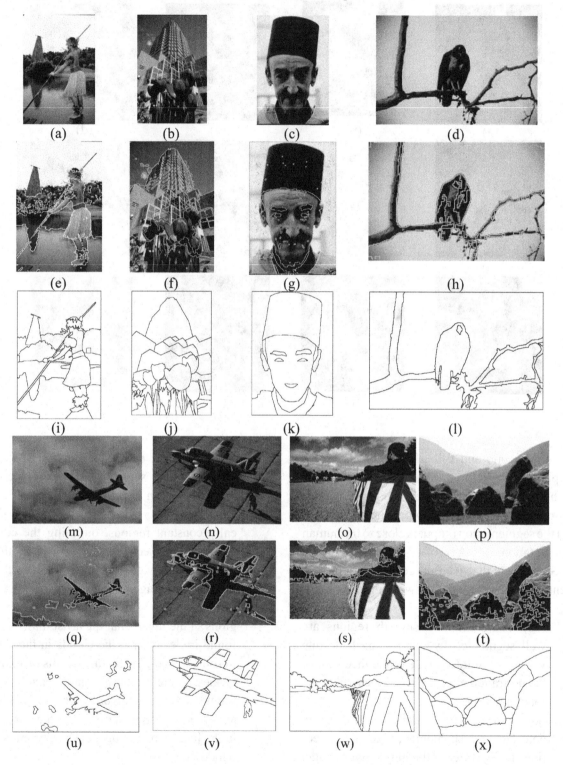

Figure 8. Validation over NAO images

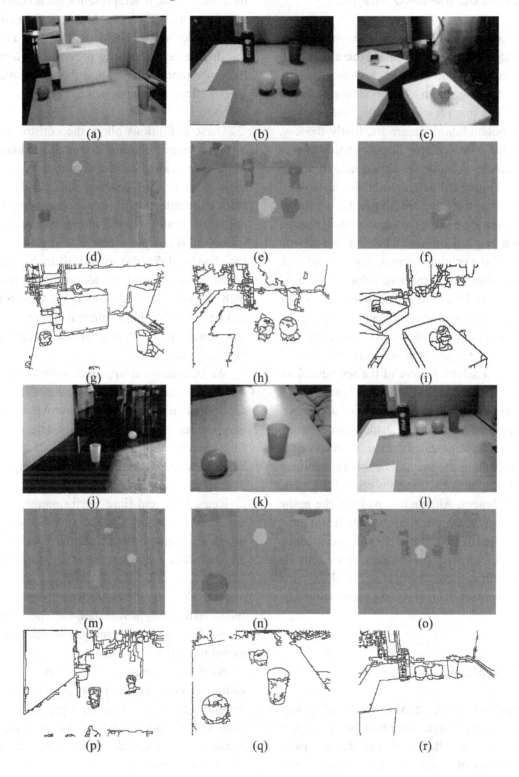

Validation on the NAO Images

In this section we show some experimental results with images obtained from the camera of a real robot Nao, Aldebaran Robotics, because we have developed the algorithm with robotic applications in mind. The main features of the images taken by the robot on-board camera are, firstly, the low signal to noise ratio, due to the poor quality of this cameras and, secondly, the appearance of many shines in the images due to illumination sources of the real environment where the robot is working.

In Figure 8 we show the segmentation results for some original images. The first and fourth rows show the original images. The second and fifth rows show the respective color region segmentations, and the third and sixth rows show the watersheds where watersheds and color region segmentation not correspond with the same parameters of the segmentation algorithm.

The most salient feature of the segmentation results obtained on these images is that most of the shines on the floor are not identified as distinct image regions and the object shadows are ignored as well. All the segmented regions correspond to actual objects in the scene. For robotic applications, this robustness may be critical for task accomplishment. Although it was not the main goal of our work at this point, to obtain real time performance (of the order of 50 milliseconds per image). This experiment has been carried out in a laptop with a processor Intel Core i3 M330 with 4GB of memory. The code has been written in C#.

CONCLUSION

The paper introduces a watershed and region merging segmentation algorithm based on the zenithal and azimuthal angles of the spherical representation of colors in the RGB space. We have shown that this representation is equivalent to the chromaticity representation of the color.

Considering the DRM we find that most of the diffuse reflectance is preserved. Moreover the color representation and the chromatic distance defined on it possesses has color constancy properties.

These definitions allow the construction of a robust hybrid chromatic spatial gradient that we use to realize robust chromatic watershed segmentation. This gradient operator has good color edge detection in lightened areas and does not suffer from the noise in the dark areas. The watershed is complemented by a region merging based on the defined chromatic distance. We give a general schema of the algorithm performing both watershed and region merging. Out proposal can be stated by this algorithm fixing the color representation, gradient operator, and region merging distance.

We compare our approach with other algorithms obtained with different setting of the general schema, obtaining the best qualitative segmentation. The results on the Berkeley database images find excellent approximations to the provided hand-drawn segmentations, without using spatial or semantic information.

Regarding real-time performance, current experiments on the NAO robot off-loading the image segmentation to an auxiliary workstation give real-time responses for small image frames (20 frame/second). We are working on the port of the code to a GPU in order to obtain further speed-up that may provide real-time performance for medium and large images which could be useful for real life applications.

As further works, we will study the optimal values for the a,b parameters of the alpha function having in account colorimetric properties of the image. On the other hand, we will perform GPU implementation in order to accelerate more the presented algorithm.

REFERENCES

Angulo, J., & Serra, J. (2007, Apr). Modelling and segmentation of colour images in polar representations. *Image and Vision Computing, 25*(4), 475–495. doi:10.1016/j.imavis.2006.07.018

Aptoula, E. (2007). A comparative study on multivariate mathematical morphology. *Pattern Recognition, 40*(11), 2914–2929. doi:10.1016/j.patcog.2007.02.004

Beucher, S., & Lantuejoul, C. (1979, September). *Use of watersheds in contour detection.* In International workshop on image processing: Real-time edge and motion detection/estimation, Rennes, France.

Dagher, I., & Tom, K. E. (2008, Jul). WaterBalloons: A hybrid watershed balloon snake segmentation. *Image and Vision Computing, 26*(7), 905–912. doi:10.1016/j.imavis.2007.10.010

Elwaseif, M., & Slater, L. (2010, Jul). Quantifying MB geometries in resistivity images using watershed algorithms. *Journal of Archaeological Science, 37*(7), 1424–1436. doi:10.1016/j.jas.2010.01.002

Geusebroek, J., van den Boomgaard, R., Smeulders, A. W. M., & Gevers, T. (2003, July). Color constancy from physical principles. *Pattern Recognition Letters, 24*(11), 1653–1662. doi:10.1016/S0167-8655(02)00322-7

Hanbury, A., & Serra, J. (2001). Mathematical morphology in the HLS colour space. *Proceedings of the 12th British Machine Vision Conference.*

James Clerk Maxwell, B. (1885). Experiments on colour, as perceived by the eye, with remarks on colour blindness. *Transactions of the Royal Society of Edinburgh, 21*(2), 274–299.

Lambert, J. (1760). *Photometria sive de mensure de gratibus luminis. Colorum umbrae.* Eberhard Klett.

Mallick, S., Zickler, T., Kriegman, D., & Belhumeur, P. (2005). Beyond Lambert: Reconstructing specular surfaces using color. In *IEEE Computer Society Conference on Computer Vision and Pattern Recognition, CVPR 2005* (Vol. 2, pp. 619-626).

Martin, D., Fowlkes, C., Tal, D., & Malik, J. (2001, July). A database of human segmented natural images and its application to evaluating segmentation algorithms and measuring ecological statistics. In *Proceedings of the 8th International Conference on Computer Vision* (Vol. 2, pp. 416–423).

Mileva, Y., Bruhn, A., & Weickert, J. (2007). Illumination-robust variational optical flow with photometric invariants. In *Pattern Recognition* (pp. 152–162).

Moreno, R. Graña, & d'Anjou, A. (2010, July). An image color gradient preserving color constancy. *In Fuzz-IEEE 2010* (p. 710-714).

Moreno, R., Graña, M., & d'Anjou, A. (2010). A color transformation for robust detection of color landmarks in robotic contexts. *In Trends in Practical Applications of Agents and Multiagent Systems* (pp. 665–672).

Moreno, R., Graña, M., & Zulueta, E. (2010, June). RGB colour gradient following colour constancy preservation. *Electronics Letters, 46*(13), 908–910. doi:10.1049/el.2010.0553

Moreno, R., Lopez-Guede, J., & d'Anjou, A. (2010). Hybrid color space transformation to visualize color constancy. *In Hybrid Artificial Intelligence Systems* (pp. 241–247).

Shafer, S. A. (1984, April). Using color to separate reflection components. *Color Research and Application, 10*, 43–51.

Shen, H., Zhang, H., Shao, S., & Xin, J. H. (2008, August). Chromaticity-based separation of reflection components in a single image. *Pattern Recognition, 41*, 2461–2469. doi:10.1016/j.patcog.2008.01.026

Tan, R. T., Nishino, K., & Ikeuchi, K. (2004, Oct). Separating reflection components based on chromaticity and noise analysis. *IEEE Transactions on Pattern Analysis and Machine Intelligence, 26*(10), 1373–1379. doi:10.1109/TPAMI.2004.90

Tan, T., Nishino, K., & Ikeuchi, K. (2003, 18-20 June). Illumination chromaticity estimation using inverse-intensity chromaticity space. In *Proceedings 2003 IEEE Computer Society Conference on Computer Vision and Pattern Recognition,* (Vol. 1, pp. 673-680).

Tarabalka, Y., Chanussot, J., & Benediktsson, J. A. (2010). Segmentation and classification of hyperspectral images using watershed transformation. *Pattern Recognition, 43*(7), 2367–2379. doi:10.1016/j.patcog.2010.01.016

Toro, J. (2008). Dichromatic illumination estimation without pre-segmentation. *Pattern Recognition Letters, 29,* 871–877. doi:10.1016/j.patrec.2008.01.004

van de Weijer, J., & Gevers, T. (2004). Robust optical flow from photometric invariants In *ICIP: 2004 International Conference on Image Processing,* Vols. 1- 5 (pp. 1835-1838).

Yoon, K. J., Chofi, Y. J., & Kweon, I. S. (2005, 11-14 Sept.). Dichromatic-based color constancy using dichromatic slope and dichromatic line space. In *IEEE International Conference on Image Processing, ICIP 2005* (Vol. 3, pp. 960-3).

Zickler, T., Mallick, S., Kriegman, D., & Belhumeur, P. (2006). Color subspaces as photometric invariants. In *2006 IEEE Computer Society Conference on Computer Vision and Pattern Recognition,* (Vol. 2, pp. 2000–2010).

ENDNOTES

[1] The source of this algorithm and others more are shared as free software on http://www.ehu.es/ccwintco/index.php/Robotic_Vision:_Technologies_for_Machine_Learning_and_Vision_Applications Also in this site we have a light application which helps to test the parameters a, b and the threshold δ on any image.

Chapter 8
Computer Vision Applications of Self-Organizing Neural Networks

José García-Rodríguez
University of Alicante, Spain

José Antonio Serra-Pérez
University of Alicante, Spain

Juan Manuel García-Chamizo
University of Alicante, Spain

Anatassia Angelolopoulou
University of Westminster, UK

Sergio Orts-Escolano
University of Alicante, Spain

Alexandra Psarrou
University of Westminster, UK

Vicente Morell-Gimenez
University of Alicante, Spain

Miguel Cazorla
University of Alicante, Spain

Diego Viejo
University of Alicante, Spain

ABSTRACT

This chapter aims to address the ability of self-organizing neural network models to manage video and image processing in real-time. The Growing Neural Gas networks (GNG) with its attributes of growth, flexibility, rapid adaptation, and excellent quality representation of the input space makes it a suitable model for real time applications. A number of applications are presented, including: image compression, hand and medical image contours representation, surveillance systems, hand gesture recognition systems, and 3D data reconstruction.

INTRODUCTION

The growing computational power of current computer systems and the reduced costs of image acquisition devices allow a very large audience to have an easier access to image analysis issues.

Self-organizing models (SOM), with their massive parallelism, have shown considerable promise in a wide variety of application areas, not related to vision problems only, and have been particularly useful in solving problems for with traditional techniques have failed or proved

DOI: 10.4018/978-1-4666-2672-0.ch008

inefficient. Accordingly, even hard video and image processing applications like some robotic operation, visual inspection, remote sensing, autonomous vehicle driving, automated surveillance, and many others, have been approached using neural networks.

The investigation in SOM in this context has received a great attention, on one hand, as imaging tasks are computationally intensive and high performances potentially can be reached in real time, and on the other hand, thanks to the versatility of neural approaches. Versatility means that a set of interesting properties are shared by neural systems; apart from the inherent parallelism, they allow a distributed representation of the information, easy learning by examples, high generalization ability, and a certain fault tolerance. There is a vast literature on NN, it deals with both theoretical investigation of their inherent mechanisms, and solutions to real problems. SOM have been and have been used extensively for pattern recognition problems, namely classification, clustering, and feature selection.

Several works have used self-organizing models for the representation and tracking of objects. Fritzke (Fritzke, 1997) proposed a variation of the GNG (Fritzke, 1995) to map non–stationary distributions that (Frezza-Buet, 2008). applies to the representation and tracking of people. In [4] it is suggested the use of self-organized networks for human-machine interaction. In [5], amendments to self-organizing models for the characterization of the movement are proposed.

From the works cited, only (Frezza-Buet, 2008) represents both the local movement and the global movement, however there is no consideration of time constraints, and do not exploits the knowledge of previous frames for segmentation and prediction in subsequent frames. Neither uses the structure of the neural network to solve the problem of correspondence in the analysis of the movement.

The time constraints of the problem suggest the need for high availability systems ca-pable to obtain representations with an acceptable quality in a limited time. Besides, the large amount of data suggests definition of parallel solutions.

In this paper we present applications of a neural architecture based on GNG that is able to adapt the topology of the network of neurons to the shape of the entities that appear in the images and can represent and characterize objects of interest in the scenes with the ability to track the objects through a sequence of images in a robust and simple way.

It has been demonstrated that using architectures based on (GNG), can be assumed the temporary restrictions on problems such as tracking objects or the recognition of gestures, with the ability to processing sequences of images offering an acceptable quality of representation and refined very quickly depending on the time available.

With regard to the processing of image sequences, it is possible to accelerate the track-ing and allow the architecture to work at video frequency. In this proposal, the use of the GNG to the representation of objects in sequences solves the costly problem of matching of features over time, using the positions of neurons in the network. Furthermore, can be ensured that from the map obtained with the first image will only be required to represent a fast re-adaptation to locate and track objects in subsequent images, allowing the system to work at video frequency.

The data stored throughout the sequence in the structure of the neural network about characteristics of the entities represented as position, colour, and others, provide information on deformation, merge, paths followed by these entities and other events that may be analyzed and interpreted, giving the semantic description of the behaviors of these entities.

The remainder of the paper is organized as follows: section 2 introduces the capacities of topology learning and preservation of GNG. Section 3 presents a number of video and image processing applications of GNG followed by our major conclusions.

TOPOLOGY LEARNING

One way to obtain a reduced and compact representation of 2D shapes or 3D surfaces is to use a topographic mapping where a low dimensional map is fitted to the high dimensional manifold of the shape, whilst preserving the topographic structure of the data. A common way to achieve this is by using self-organising neural networks where input patterns are projected onto a network of neural units such that similar patterns are projected onto units adjacent in the network and vice versa.

The approach presented in this paper is based on self-organising networks trained using the Growing Neural Gas learning method (Fritzke, 1995), an incremental training algorithm. The links between the units in the network are established through competitive hebbian learning. As a result the algorithm can be used in cases where the topological structure of the input pattern is not known a priori and yields topology preserving maps of feature manifold.

Growing Neural Gas

From the Neural Gas model (Martinetz, Berkovich, & Schulten, 1993) and Growing Cell Structures (Fritzke, 1993), Fritzke developed the Growing Neural Gas model, with no predefined topology of a union between neurons. A growth process takes place from minimal network size and new units are inserted successively using a particular type of vector quantisation (Bauer, Hermann, & Villmann, 1999), (Martinetz, & Schulten, 1994). To determine where to insert new units, local error measures are gathered during the adaptation process and each new unit is inserted near the unit which has the highest accumulated error. At each adaptation step a connection between the winner and the second-nearest unit is created as dictated by the competitive hebbian learning algorithm. This is continued until an ending condition is fulfilled,

as for example evaluation of the optimal network topology based on some measure. Also the ending condition could it be the insertion of a predefined number of neurons or a temporal constrain. In addition, in GNG networks learning parameters are constant in time, in contrast to other methods whose learning is based on decaying parameters.

The network is specified as:

- A set N of nodes (neurons). Each neuron $c \in N$ has its associated reference vector $w_c \in R^d$. The reference vectors can be regarded as positions in the input space of their corresponding neurons.
- A set of edges (connections) between pairs of neurons. These connections are not weighted and its purpose is to define the topological structure. The edges are determined using the competitive hebbian learning algorithm. An edge aging scheme is used to remove connections that are invalid due to the activation of the neuron during the adaptation process.

The GNG learning algorithm is as follows:

1. Start with two neurons a and b at random positions w_a and w_b in R^d.
2. Generate a random input signal ξ according to a density function $P(\xi)$.
3. Find the nearest neuron (winner neuron) s_1 and the second nearest s_2.
4. Increase the age of all the edges emanating from s_1.
5. Add the squared distance between the input signal and the winner neuron to a counter error of s_1:

$$\Delta error(s_1) = \left\| w_{s_1} - \xi \right\|^2 \quad (1)$$

6. Move the winner neuron s_1 and its topological neighbours (neurons connected to s_1)

towards ξ by a learning step ε_w and ε_n, respectively, of the total distance:

$$\Delta w_{s_1} = \varepsilon_w (\xi - w_{s_1}) \quad (2)$$

$$\Delta w_{s_n} = \varepsilon_n (\xi - w_{s_n}) \quad (3)$$

7. If s_1 and s_2 are connected by an edge, set the age of this edge to 0. If it does not exist, create it.

8. Remove the edges larger than a_{max}. If this results in isolated neurons (without emanating edges), remove them as well.

9. Every certain number λ of input signals generated, insert a new neuron as follows:
    a. Determine the neuron q with the maximum accumulated error.
    b. Insert a new neuron r between q and its further neighbor f :

    $$w_r = 0.5(w_q + w_f) \quad (4)$$

    c. Insert new edges connecting the neuron r with neurons q and f, removing the old edge between q and f.
    d. Decrease the error variables of neurons q and f multiplying them with a constant α. Initialize the error variable of r with the new value of the error variable of q and f.

10. Decrease all error variables by multiplying them with a constant β.

11. If the stopping criterion is not yet achieved, go to step 2.

In summary, the adaptation of the network to the input space takes place in step 6. The insertion of connections (step 7) between the two closest neurons to the randomly generated input patterns establishes an induced Delaunay triangulation in the input space. The elimination of connections (step 8) eliminates the edges that no longer comprise the triangulation. This is made by eliminating the connections between neurons that no longer are next or that they have nearer neurons. Finally, the accumulated error (step 5) allows the identi-

fication of those zones in the input space where it is necessary to increase the number of neurons to improve the mapping.

Visual Data Representation with Growing Neural Gas

The ability of neural gases to preserve the topology will be employed in this work for the representation and tracking of objects. Identifying the points of the image that belong to objects allows the network to adapt its structure to this input subspace, obtaining an induced Delaunay triangulation of the object.

Let an object $O = \left[A_G, A_V \right]$ that is defined by a geometric appearance A_G and a visual appearance A_V. The geometric appearance A_G is given by a morphologic parameter (local deformations) and positional parameters (translation, rotation and scale):

$$A_G = \left[G_M, G_P \right] \quad (5)$$

The visual appearance A_V is set of object characteristics such as colour, texture or brightness, among others.

In particular, considering objects in two dimensions. Given a domain support $S \subseteq R^n$, an image intensity function $I(x_0,..,x_n) \in R$ such that $I : S \rightarrow \left[0, I_{max}\right]$, and an object O, its standard potential field $\psi_T \left(x_0,..,x_n\right) = f_T \left(I \left((x_0,..,x_n) \right) \right)$ is the transformation $\psi_T : S \rightarrow \left[0,1\right]$ which associates to each point $\left((x_0,..,x_n) \right) \in S$ the degree of compliance with the visual property T of the object O by its associated intensity $I((x_0,..,x_n))$.

Considering:

- The space of input signals as the set of points in the image:
  $V = S$

$$\xi = \left(\left(x_0, .., x_n \right) \right) \in S \quad (6)$$

- The probability density function according to the standard potential field obtained for each point of the image:

$$p\left(\xi\right) = p\left(\left(x_0, .., x_n\right)\right) = \quad (7)$$
$$\psi_T\left(\left(x_0, .., x_n\right)\right)$$

Learning takes place following the GNG algorithm described in previous section. So, doing this process, a representation based on the neural network structures is obtained which preserves the topology of the object O from a certain feature T. That is, from the visual appearance A_V of the object is obtained an approximation to its geometric appearance A_G.

Henceforth we call the Topology Preserving Graph $TPG = \langle A, C \rangle$ to the non directed graph, defined by a set of vertices (neurons) A and a set of edges C that connect them, preserving the topology of an object from the considered standard potential field.

Getting different
$$\psi_T\left(x_0, .., x_n\right) = f_T\left(I\left(\left(x_0, .., x_n\right)\right)\right)$$ can be obtained, for example, the representation of objects in two dimensions (Figure 1 left) or the silhouette of these (Figure 1 right), which cause different structures in the network for each of these standard potential field.

EXAMPLES

A number of applications to demonstrate the ability of GNG to manage Video and Image processing problems are presented that includes: image compression (García-Rodríguez, Flórez-Revuelta, & García-Chamizo, 2007), hand (Angelopoulou, Psarrou, & García-Rodríguez, 2007) and medical image contours automatic landmark (Angelopoulou, Psarrou, García-Rodríguez, & Revett, 2005), surveillance systems (Cao, & Suganthan, 2003),

(Frezza-Buet, 2008), (Garcia-Rodriguez, Angelopoulou, Garcia-Chamizo, & Psarrou, 2010) and hand gesture recognition systems (Flórez, García, García, & Hernández, 2002). or 3D reconstruction (Holdstein, & Fischer, 2008).

Automatic Landmark

Shape training sets usually come from manually annotated boundaries. A landmark point is a point of correspondence on each shape of the training set; it identifies a salient feature such as high curvature and is present on any shape. We introduce a new and computationally inexpensive method for the automatic selection of landmarks along the contours of 2D hand shapes. The novelty in using the Growing Neural Gas method for unsupervised learning of 2D hand shapes is that we can automatically construct statistical shape models independently of closed or open shapes in

Figure 1. Different image features representation with GNG

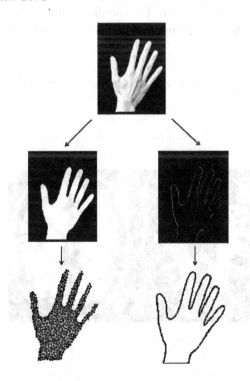

contrast to most methods which can be defined only for closed contours. Furthermore, the incremental neural network, the growing neural gas (GNG) is used to automatically annotate the training set without using a priori knowledge of the structure of the input patterns. Unlike other methods, the incremental character of the model avoids the necessity to previously specify a reference shape. GNG allows the extraction, in an autonomous way, of the contour of any object as a set of edges that belong to a single polygon and form a topology preserving graph. The method is based on the assumption that correspondences are the nodes (the cluster centers in a high-dimensional vector space) of a network. The automatic extraction and correspondence is performed with the GNG (Figure 2).

Image Compression and Hand Gesture Recognition

The main goal of image compression methods is to minimize the amount of data stored for the correct representation of the images with the smaller loss of information. The compression algorithms usually try to eliminate redundancy in the data, but

making possible the reconstruction of the image. The usual image data compression-decompression sequence consists on a characterization process that reduces data redundancy obtaining the most relevant image information, codifying these data and transmitting them. With these data, a decoding and reconstruction or synthesis of the images it is made.

In the compression applications, we study the capacities of characterization and synthesis of objects by using a self-organizing neural model, the Growing Neural Gas. These networks, by means of their competitive learning try to preserve the topology of an input space. This feature is being used for the representation of objects and their movement with topology preserving networks. We characterize the object to be represented by means of the obtained maps and kept information solely on the coordinates and the pixel color of the neurons. With this information it is made the synthesis of the original images, applying mathematical morphology and simple filters using the available information.

In the case of gesture recognition, GNG structure is able to characterize hand posture, as well as its movement. Topology of the self-organizing

Figure 2. Automatic landmarking with GNG

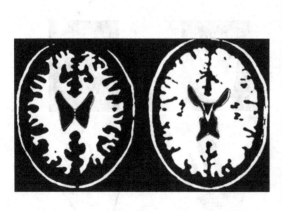

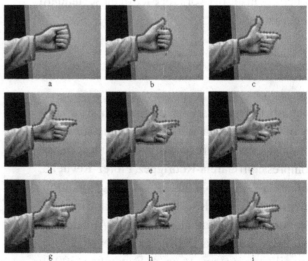

neural network determines the posture, whereas its adaptation dynamics throughout the time determines the gesture. This adaptive character of the network avoids the correspondence problem that other methods have, so that the gestures are modeled by the movement of the neurons. The reduced structure (neurons and edges) obtained for any image in the sequence can be used as a fixed markers (Figure 3) to follow along the sequence if we do not add or delete neurons after first frame, only redistribute neurons. That is, we take previous map obtained from previous image as a start point for next image representation. Trajectory of neurons reference vectors can be use to define the gesture described.

Surveillance Systems

It is proposed the design of a modular system capable of capturing images from a camera, target areas of interest and represent the morphology of entities in the scene. As well as analyzing the evolution of these entities in time and obtain semantic knowledge about the actions that occur at the scene. We propose the representation of these entities through a flexible model able to characterize their morphological and positional changes along the image sequence. The representation

should identify entities over time, establishing a correspondence of these during the various observations. This feature allows the description of the behavior of the entities through the interpretation of the dynamics of the model representation. The time constraints of the problem suggest the need for highly available systems capable to obtain representations with an acceptable quality in a limited time. Besides, the large amount of data suggests definition of parallel solutions.

To solve the problem, the representation of the objects and their motion is done with a modified self-growing model. We propose a neural architecture able to adapt the topology of the network of neurons to the shape of the entities that appear in the images, representing and characterizing objects of interest in the scenes. The model has also the ability to track the objects through a sequence of images in a robust and simple way.

With regard to the processing of image sequences, we have introduced several improvements to the network to accelerate the tracking and allow the architecture to work at video frequency. In this paper, the use of the GNG to represent objects in image sequences solves the costly problem of matching features over time by using the positions of neurons in the network. Likewise, the use of simple prediction facilitates

Figure 3. Image compression and hand gesture recognition applications

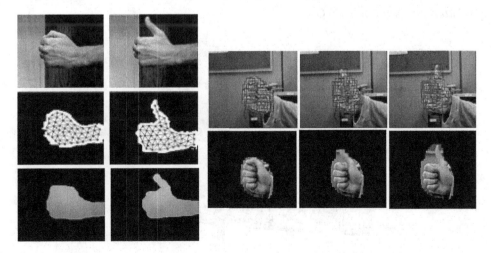

Figure 4. GNG-based surveillance system

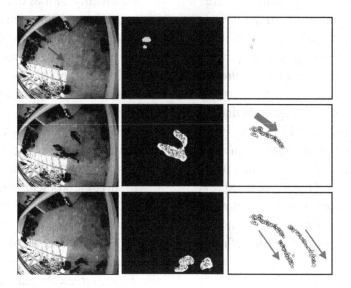

the monitoring of neurons and reduces the time to readapt the network between frames without damaging the quality and speed of system response.

The data stored throughout the sequence in the structure of the neural network about characteristics of the entities represented as: position, color, texture, labels or any interesting feature, provide information on deformation, merge, paths followed by these entities and other events. This information may be analyzed and interpreted, giving the semantic description of the behaviors of these entities (Figure 3).

3D Data Reconstruction

The ability of GNG to represent the topology of the input space (3D in this case) permits to obtain, based on the structure of the net (neurons and edges), an induced Delaunay Triangulation that can be used to obtain several features and reduced graphs. Although, the compact and reduced representation obtain permits to deal with computationally expensive algorithms.

Several recent works deal with 3D data. Data come from any kind of sensor (RGB depth, time of

Figure 5. 3D data representation with GNG

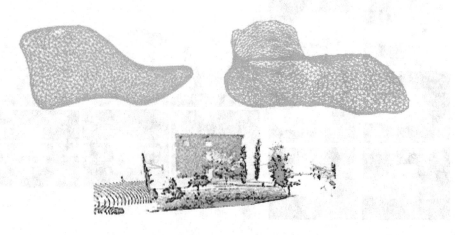

flight or stereo cameras and 3D lasers) providing a huge amount of unorganized 3D data. GNG is an efficient method to build complete 3D models with. Due to the facility of neural models to deal with noise, imprecision, uncertainty or partial data, GNG provides better results than other approaches Figure represents 3D data obtain from scanner for shoe lasts and foots in the upper images and outdoor images obtained with a 3D laser in lower one (Figure 5).

FUTURE RESEARCH DIRECTIONS

Since most of computer vision application have temporal constrain, further work will include the implementation of the GNG onto Graphic Processor Units to accelerate its calculation. Automatic segmentation and sophisticated methods of prediction like Kalman or particle filters should be included in future implementations to support real time video processing.

CONCLUSION

In this paper, we have demonstrated the capacity of Growing Neural Gas to solve some computer vision an image processing tasks. Demonstrating their capacity to segment, extract and represent 2D/3D objects in images. Establishing a suitable transformation function, the model is able to adapt its topology to images of the environment, to compress data, represent the movement and provide reduced and compact information about the structure of the entities in the images that permit real time processing video.

ACKNOWLEDGMENT

This work was partially supported by the University of Alicante project GRE09-16 and Valencia Government project GV/2011/034.

REFERENCES

Angelopoulou, A., Psarrou, A., & García-Rodríguez, J. (2007). *Robust modeling and 43 tracking of non-rigid objects using Active-GNG* (pp. 1–7). ICCV.

Angelopoulou, A., Psarrou, A., García-Rodríguez, J., & Revett, K. (2005). Automatic landmarking of 2D medical shapes using the growing neural gas network. *ICCV 2005 Workshop CVBIA* (pp. 210-219).

Bauer, H.-U., Hermann, M., & Villmann, T. (1999). Neural maps and topographic vector quantization. *Neural Networks*, *12*(4–5), 659–676. doi:10.1016/S0893-6080(99)00027-1

Cao, X., & Suganthan, P. N. (2003). Video shot motion characterization based on hierachical overlapped growing neural gas networks. *Multimedia Systems*, *9*, 378–385. doi:10.1007/s00530-003-0107-2

Flórez, F., García, J. M., García, J., & Hernández, A. (2002). hand gesture recognition following the dynamics of a topology-preserving network. In *Proceedings of the 5th IEEE International Conference on Automatic Face and Gesture Recognition*, (pp. 318-323). Washington, DC: IEEE.

Frezza-Buet, H. (2008). Following non-stationary distributions by controlling the vector quatization accuracy of a growing neural gas network. *Neurocomputing*, *71*, 1191–1202. doi:10.1016/j.neucom.2007.12.024

Fritzke, B. (1993). *Growing cell structures – A self-organising network for unsupervised and supervised learning*. Technical Report TR-93-026, International Computer Science Institute, Berkeley, California.

Fritzke, B. (1995). In Tesauro, G., Touretzky, D. S., & Leen, T. K. (Eds.). Advances in Neural Information Processing Systems: *Vol. 7. A growing neural gas network learns topologies.* Cambridge, MA: MIT Press.

Fritzke, B. (1997). A self-organizing network that can follow non-stationary distributions. *Proceedings of the International Conference on Artificial Neural Networks '97*, (pp. 613-618). Springer.

Garcia-Rodriguez, J., Angelopoulou, A., Garcia-Chamizo, J. M., & Psarrou, A. (2010). GNG based surveillance system. In International Joint Conference on Neural Networks (pp. 1-8).

García-Rodríguez, J., Flórez-Revuelta, F., & García-Chamizo, J. M. (2007). Image compression using growing neural gas. *In Proceedings of International Joint Conference on Artificial Neural Networks*, (pp. 366-370).

Holdstein, Y., & Fischer, A. (2008). Three-dimensional surface reconstruction using meshing growing neural gas (MGNG). *The Visual Computer*, *24*, 295–302. doi:10.1007/s00371-007-0202-z

Martinetz, T., Berkovich, S. G., & Schulten, K. J. (1993). Neural-gas network for vector quantization and its application to time-series prediction. *IEEE Transactions on Neural Networks*, *4*(4), 558–569. doi:10.1109/72.238311

Martinetz, T., & Schulten, K. (1994). Topology representing networks. *Neural Networks*, *7*(3), 507–522. doi:10.1016/0893-6080(94)90109-0

KEY TERMS AND DEFINITIONS

Growing Neural Gas: A self-organizing neural model where the number of units is increased during the self-organization process using a competitive Hebbian learning for the topology generation.

Hebbian Learning: A time-dependent, local, highly interactive mechanism that increases synaptic efficacy as a function of pre- and post-synaptic activity.

Non-Rigid Objects: A class of objects that suffer deformations changing its appearance along the time.

Object Representation: Is the construction of a formal description of the object using features based on its shape, contour or specific region.

Object Tracking: Is a task within the field of computer vision that consists on the extraction of the motion of an object from a sequence of images estimating its trajectory.

Self-Organising Neural Networks: A class of artificial neural networks that are able to self-organize themselves to recognize patterns automatically without previous training preserving neighbourhood relations.

Topology Preserving Graph: Is a graph that represents and preserves the neighbourhood relations of an input space.

Section 3
3D Computer Vision and Robotics

Chapter 9
A Review of Registration Methods on Mobile Robots

Vicente Morell-Gimenez
University of Alicante, Spain

José García-Rodríguez
University of Alicante, Spain

Sergio Orts-Escolano
University of Alicante, Spain

Miguel Cazorla
University of Alicante, Spain

Diego Viejo
University of Alicante, Spain

ABSTRACT

The task of registering three dimensional data sets with rigid motions is a fundamental problem in many areas as computer vision, medical images, mobile robotic, arising whenever two or more 3D data sets must be aligned in a common coordinate system. In this chapter, the authors review registration methods. Focusing on mobile robots area, this chapter reviews the main registration methods in the literature. A possible classification could be distance-based and feature-based methods. The distance based methods, from which the classical Iterative Closest Point (ICP) is the most representative, have a lot of variations which obtain better results in situations where noise, time, or accuracy conditions are present. Feature based methods try to reduce the great number or points given by the current sensors using a combination of feature detector and descriptor which can be used to compute the final transformation with a method like RANSAC or Genetic Algorithms.

INTRODUCTION

The task of registering three dimensional data sets with rigid motions is a fundamental problem in many areas as computer vision, medical images, mobile robotic, arising whenever two or more 3D data sets must be aligned in a common coordinate system. The registration problem is comprised of two related sub-problems: correspondence selection and motion estimation. In the former, candidate correspondences between data sets are chosen, while in the latter, rigid motions minimizing the distances between corresponding points are estimated.

In the robotic field, the registration problem can be used for a great variety of tasks: find the relative pose between two or more sensors, object reconstruction, object tracking or estimate the

DOI: 10.4018/978-1-4666-2672-0.ch009

movement (translation and rotation) of a mobile robot.

Our main goal is to use registration in order to help mobile robotics to solve the Simultaneous Localization And Mapping (SLAM) (Dissanayake et al., 2001) obtaining the egomotion at each step.

The remainder of this chapter is organized as follows; in the first section, we present the registration problem. Then, we describe the robot platforms and sensors we use. Next section reviews distance based registration methods, most of them based on the classical ICP. Then, some feature based registration methods are introduced, and finally, some experiments are done finishing by our main conclusions.

BACKGROUND

Registration problem is the process of transforming different sets of data into one coordinate system. On the field of mobile robotics, the registration problem is defined as the task of finding the transformation needed to fit one set called scene S (the set which its coordinate systems is unknown) to another set called model M (the set with the coordinate system known). Formally,

$$T = \arg \min \sum_{s \in S} \sum_{m \in M} w_{ms} \left| m - T(s) \right| \qquad (1)$$

where $s \in S$ are points of the scene, $m \in M$ are points of the model, and w_{ms} if the probability that the point m matches with the point s. The problem can be simplified when the correspondence pairs between scene and model are known,

$$T = \arg \min \sum_{i=1}^{N} \left\| m_i - T\left(s_i\right) \right\|^2 \qquad (2)$$

where N is the number of correspondence pairs, and m_i is the point of the model set which has a correspondence with the scene point s_i. In the field

of mobile robotic, the transformation T usually is assumed to be an affine transformation, trying to solve registration on mostly static scenes and assuming the object are rigid bodies.

2D/3D DATA ACQUISITION

We have used several robot platforms, depending on the perception system used. In Figure 1, two of these platforms are shown. The left one is a Magellan Pro from iRobot used for indoor experiments. For outdoors we have used a PowerBot from ActiveMedia. Furthermore, PowerBot can carry heavy loads like the 3D sweeping laser unit. Both come with an onboard computer.

In our research, we manage 3D data that can come from different sensor devices. For outdoor environments we use a 3D sweeping laser unit, a LMS-200 Sick laser mounted on a sweeping unit. Its range is 80 meters with an error of 1mm per meter. The main disadvantage of this unit is the data capturing time: it takes about one minute to get a complete frame. For indoor environments we use another two sensors. The first one is a SR4000 camera from Mesa Imaging, which is a time-of-flight camera, based on infrared light. Its range is limited to 5 or 10 meters, providing gray level from the infrared spectrum. Finally, a Kinect sensor has been included. This sensor provides 3D data together with RGB data, with a maximum range of 10 meters.

Distance Based Methods

Some of the 2D/3D registration methods use the distance information between the matched points to calculate the global transformation which better explains the change of the position of two data sets. Most of these methods use an Expectation-Maximization model (EM) to solve the two problems that have the registration task. These two problems are finding the correspondence (or matching) and the calculation estimation of the movement/transformation. The most used of these methods is the so called Iterative Closest Point (ICP) which was introduced in Chen and Medioni (1991) and Besl and McKay (1992). The structure of the ICP method can be shown in figure 2.

One of the data set it is named as a Model and the other one is called as a Scene. The ICP starts with a given initial transformation, and then continue to iterate. First, scene points are transformed using the current transformation. After that, the correspondence pairs are calculated using these scene and model points, and at the same time, the error of the current transformation is calculated usually using the distance between pairs. At the end of each iteration and using the correspondence pair information, the current transformation is calculated. Iterating over these three steps, the error of the current transformation will decrease and therefore, the accuracy of the transformation will be higher. Base scheme of the ICP normally gets a local optimum solution for the registration, depending of the initial transformation given to the ICP method. This is the basic structure of the ICP, from which a lot of variations have emerged which seek to change or improve any of the steps of the classical ICP. The different versions can be classified based on the following criteria:

- **Selection of points of the data sets:** The basic ICP algorithm does not include any policy of selection of points from data sets. The default case uses all of them (Besl & McKay, 1992). This approach leads to a poor performance and present problems as the high influence of the outliers (noise or bad located points) in the resulting transformation. Other papers, as in Turk and Levoy (1994), use an uniform sub-sambpling method to reduce the amount of points of the data sets. Another one, is the random selection of points (Masuda, Sakaue & Yokoya, 1996), which reduces the number of points quickly, at risking to lose some parts of the data sets structure. Another paper (Weik, 1997) uses additional information to do the selection of points like the color or variations of intensity of the points. Other papers try to use these techniques alternating them between the model and scene data sets, using different criteria to each set (Godin, Rioux & Baribeau, 1994). There are other works that make a more elaborated process on the data to select those points that meet certain more advanced features, similar to the feature detectors we will see in the next section, such as corner points, edges, distribution of the normals (Rusinkiewicz & Levoy, 2001), angles, etc.

Figure 1. Mobile robots used for experiments. From left to right: Magellan Pro unit used for indoors; PowerBot used for outdoors. SR4000 camera used with both robots; Kinect sensor used in indoors.

Figure 2. Basic scheme of the iterative closest point algorithm

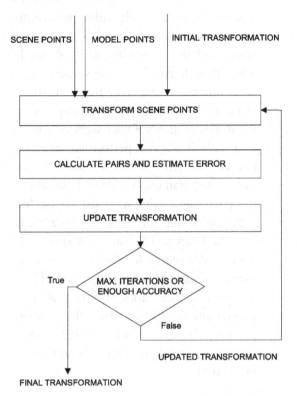

SCENE POINTS   MODEL POINTS   INITIAL TRASNFORMATION

TRANSFORM SCENE POINTS

CALCULATE PAIRS AND ESTIMATE ERROR

UPDATE TRANSFORMATION

MAX. ITERATIONS OR ENOUGH ACCURACY

True   False

UPDATED TRANSFORMATION

FINAL TRANSFORMATION

- **Matching points:** The matching point method is the task of finding the point correspondences between both data sets. Initially, it was based on matching those points with less Euclidean distance (Besl & McKay, 1992). This computation can be greatly accelerated using a kd-tree or a closest-point cache system (Simon, 1996). Another of the distances used to do the matching is the distance of the point from one data set (usually a scene point) to the plane formed by the normal of another point(usually a model point). In other words, given a point and its neighborhood, we compute the plane formed by them (this computation is similar to calculate the normal of the point). This method needs more calculations but it seems that it provides better correspondeces than other methods. Other of the distances that has been used is called the "reverse-calibration", and it is based on the projection of the source point onto the destination mesh, from the point of view of the destination mesh's range camera (Blais & Levine, 1995, Neugebauer, 1997). Other matching methods are based on projecting the source point onto the destination mesh and then performing a search of correspondence point in their image range. The search criteria can be the closest point (Benjemaa & Schmitt, 1999), point-to-ray distance (Dorai, Wang, Jain & Mercer, 1998), a similar intensity value (Weik, 1997) or similar color (Pulli & Shapiro, 1997).Other authors also propose to add extra conditions to the above methods (not only the projection criterion), conditions like similar color values (Godin, Rioux & Baribeau, 1994) or similar point normals (Pulli, 1999). The paper by Rusinkiewicz and Levoy (2001) states that projection based methods work best for complex data sets, but fail to converge in simpler situations and so, closest point methods are more robust in different situations.

- **Setting weigths to Pairs:** In the classical ICP, all point pair correspondences are weighted with the same value, and therefore all points have the same relevance in the error rate and the final transformation. Other papers like Godin et al. (1994) weight the pairs in function of their distance, giving less weight to pairs with more distance or also depending on the difference between the normal of the points. Another criteria for weighting pairs step is to assign different weights in function of the noise model ofthe sensors, for example, if it is known that the camera produces more noise to more distant points, give them less weight than the closest points to the camera.

- **Rejecting pairs:** Once the matching and weighting steps are done, most of the registration methods do a special selection of the better point pair to calculate a better tranformation, trying to exclude those pairs that do meet certain criteria. In the classical ICP there is no policy to reject pairs, therefore all of them are used. Thereafter, many authors reject pairs that are considered bad matching pairs or when its use causes erroneous transformations. It is very common to reject a pair with a distance greater than a given threshold or some precomputed value. Pulli (1999) proposes rejecting a percentage of the worst matches, according to some criterion, usually distance from the points. Other variants, like Masuda, Sakaue and Yokoya (1996), reject the pairs whose distance between its points is greater than a multiple of the standard deviation of the distances. Other more elaborated methods aim to reject those pairs that are not consistent with the rest, that is, assuming that objects are rigid, neighboring pairs should have a similar distance (Dorai et al., 1998) or the angle formed between them does not differ too much. Greg Turk and Marc Levoy (1994) reject pairs whose points are on the data set boundaries, since at most cases these points should not have a correspondence pair with the other data set and their use lead to a greater error in the registration result.

- **Computing transformation and estimating error:** There are many methods to achieve a good alignment, the classic ICP (Chen & Medioni, 1991) uses the current transformation to align sets of points, minimizing error and computing the next transformation. In Besl and McKay (1992) the transformation space is extrapoled to accelerate the convergence of the iteration minimization. Simon (1996) proposes starting the iterative algorithm performing various perturbations in the initial conditions, and then selecting the best result. This aproach avoids the convergence to a local minima in the error function, which is normally seen in the point-to-point error metric methods. Other works (Masuda et al. 1996) propose performing many iterative minimization computations using various randomly-selected subsets of the points and using a robust metric like the least median of squares. Other methods like Blais and Levine (1995) perform a stochastic search for the best transform, using a Simulated annealing method. These last methods need a lot of time to achieve a good solution, and that is why they are not tipically used in areas which a quick answer is needed, as in the case of the mobile robotics.

Many other methods have been presented more recently. Most of them propose a combination of the techniques described above and other methods proposed to replace certain parts of the iterative minimization algorithm. In Zinsser, Schmidt and Niemann (2001), an ICP algorithm, which is called Picky ICP, is introduced as a collection of variations of the classical ICP in order to make it more robust and efficient. The Picky ICP uses a hierarchical point sampling method, using few points at first's iterations of ICP and using more and more points in the following iterations. Picky ICP uses an optimized K-Dtree nearest neighbor algorithm to accelerate the correspondence computation and reject the erroneous pairs that are not bijective or one of their points has more than one correspondence and rejects pairs with a great distance. The experiments show that the picky ICP convergence is faster and more robust to outliers than classical ICP. Zhang, Choi and Park (2011) use a similar method to reject pairs. This method called Biunique Correspondence ICP (BC-ICP) searches multiple nearest points, then discard the correspondences with larger distance than a calculated threshold and the correspondence is biunique only if the point of the model has only one correspondence with the scene points. With this technique, the experiments show how this BC-ICP can manage large transformations and non-overlapping areas. In addition, in the experimentation results, they show how the BC-ICP can be successfully used on a 6DOF SLAM system.

Another version of the classical ICP that uses a hierarchical point sampling method is presented in Li, Xue, Du and Zheng (2010). In this ICP, the multi-resolution sub-sample operation is implemented by Gauss pyramid, and it is shown on experimentation that it achieves a better convergence rate and less execution time than the classical ICP.

In Carcassoni and Hancock (2002), the spectral correspondence approach is introduced, and it attempts to characterize the points of a given point set in terms of their relationship with other points. A point p in one data set would be considered to be in correspondence with some point q in another data set, if both p and q have the same relationship with other points in their respective data sets. In this and other previous works (Cross & Hancock, 1997, Carcassoni & Hancock, 2000) Hancock et al. presents an Expectation-Maximization (EM) algorithm to do the registration of the two data sets. At each iteration, they first calculate a likelihood function and then estimate the transformation matrix parameters and, finally, update correspondence matches using a hierarchical spectral graph analysis for a good global correspondence match. The main contribution of this paper is to handle two data sets that do not have the same point number. The key idea is that they consider each eigenvector of the proximity matrix as a cluster and then iteratively align the cluster centers using dual step EM algorithm.

In Fitzgibbon (2003), a general-purpose non-linear optimization (the Levenberg-Marquardt algorithm) is used to minimize the registration error. Extra functions evaluations are introduced to do a better robust estimation increasing the cost of each iteration of the ICP. However, the experimentation results show that a fewer number of iterations is needed to achieve a given accuracy.

In Chetverikov, Stepanov and Krsek (2005), a simple, robustified extension of the classic Iterative Closest Point (ICP) algorithm is presented, called Trimmed ICP (TrICP). This ICP algorithm is based on the consistent use of the Least Trimmed Squares approach in all phases of the operation. They use a simple boxing structure that partitions the space and use the box corresponding to the current point and the adjacent boxes to get the correspondence point. The paper shows that TrICP is fast, applicable to overlaps of 50%, robust to erroneous and incomplete measurements, and has easy-to-set parameters.

Another variation of the classical ICP is presented in Almhdie, Léger, Deriche and Lédée (2007). They called this ICP implementation "Comprehensive ICP" (CICP). The CICP main goal is to use a novel look up distance matrix for finding the best correspondence pairs. M is the size of the model data set, and S is the size of the scene data set. The CICP proposes the use of this MxS matrix to achieve unique and bijective correspondences between data sets. In the experimentation results, they show an improvement of the speed of convergence and a better resilience to an additive Gaussian noise and outlier than the classical ICP and the Picky ICP.

One more variation of the ICP is proposed in Du, Zheng, Meng and Yuan (2008). It works like the classical ICP but using a novel affine transformation model trying to achieve better results on affine registration than the classical ICP. Moreover, it uses an Independent Component Analysis (ICA) method to estimate the initial parameters of the ICP. In the experimentation, it is shown that this ICP can manage affine transformation when the classical ICP fails and this affine method presents better results than other affine methods like the presented in Ho, Ming-Hsuan, Rangarajan and Vemuri (2007).

In Sandhu, Dambreville and Tannenbaum (2008) a particle filtering approach is presented in order to solve the registration of two point sets that differ by a rigid body transformation. This method use a particle filtering scheme to drive the point set registration process that uses some iterations of the standard ICP. The robustness of the registration is increased by the use of a dynamic model of uncertainty for the transformation parameters. Some experiments are shown to demonstrate the robustness of the algorithm to initialization, noise, missing structures or differing point densities in both sets.

Another method to reject bad point pairs is presented in Xin and Pu (2010). In this paper, a combination of different pair rejection methods is made following next rules: if one point has more than one match, only the shorter match is accepted; the match with a distance greater than a given threshold is rejected; and it uses the pair formed by the centroids of each point cloud to compare this reference pair with all the pairs, and reject the pairs which differs in distance and/or angle. Some improvement of time and accuracy over the classic ICP is shown at the experimentation results.

The problem of scale transformation is dealt in Du, Zheng, Xiong, Ying and Xue (2010). This variation of the ICP introduces the scale as an extra parameter to fix in the registration problem. Given the current transformation, it tries different scales trying to find the scale with less error. Scale boundaries are needed to prevent wrong values (points of a set normally converge to a small subset of the other).

Feature Based Methods

In contrast to the methods based on calculation of distances between pairs of points, the feature based methods try to reduce the amount of points of both sets (model and scene) using a special detection and description method to represent the data of the points. The steps of this registration methods are feature detection/description, feature matching and transform model estimation. These steps can be showed at figure 3.

The feature detection step try to detect salient and distinctive objects (shapes, closed regions, contours, lines, line intersections, etc) in the data sets using a feature detection method, and then represent the point as a set of information normally called feature descriptor. The feature detection steps solve the problem wich is called Points of Interest (POI) or key points in the literature. The detection of key points step, calculate the choosen feature to all point and then takes those point that pass some given parameters tests. After this step we have the POI set that usually it is a much smaller subset of the initial data. That points can be represented in different ways depending on the method wich is used on the next step, the feature matching step. In this step, the correspondence between the features detected on both data sets is established using the information of the feature descriptors and/or the similarity measures along with spatial relationships among the features. Similar to methods based on ICP, these correspondence pairs can be selected using the information of the features of each point of the pair and information of the rest of pairs. Once the correspondences has been stablished and selected, a transformation is calculated using the pairs setwith similar strategies like the mentioned on ICP based systems.

There are a lot of features detectors and descriptors that work with 2D data, solutions such as the Canny algorithm (Canny, 1986), SUSAN (Smith & Brady, 1995) or Nitzberg-Harris are used to extract contourfeatures. The K-Means algorithm that divides the imagein k clusters (where k is the number of distinct colorsin the image, selected by the user) and JSEG (Deng & Manjunath, 2001), which uses a method of unsupervised region

Figure 3. Basic scheme of the feature registration model

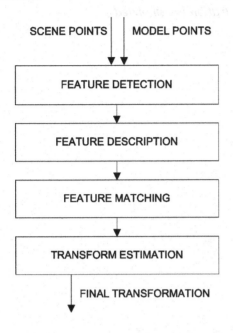

growing are two region segmentation algorithms examples. The SIFT (scale Invariant Features Transforms)(D. G. Lowe, 2004) and SURF (Speeded Up Robust Features) (Bay, Tuytelaars, & Gool, 2008) descriptors extract features invariant to scale, rotation and small changes of illumination known as interest points or invariant points. The latter features have been widely accepted and used by the robotics community in recent years. Besides those two feature extractors, we want to use other similar feature extractor, like MSER or Harris-affine, in order to detect which of them are better for recognition tasks.

Recently with the fast increase in the usage of 3D range scanners in many applications, a lot of 2D/3D features detectors systems has been presented. Using device sensors like Microsoft Kinect, range images can be obtained with the information in a 3D coordinate system and the color of each of these points. 2D feature detector/descriptor techniques can be used to detect the points of interest and then assign the 3D info to these descriptors. Only few general purpose pure 3D feature detectors/descriptors has been pre-

sented. In Przemysław (2009), some extensions of the famous 2D Harris detector are presented. In Rusu, Blodow, and Beetz (2009) a pure 3D descriptor is presented. It is called Fast Point Feature Histograms (FPFH) and is based on a histogram of the differences of angles between the normals of the neighbor points of the source point.

In Viejo and Cazorla (2007 & 2008) a feature extraction process is applied to the raw 3D data in order to obtain a complexity reduction. Those features are planar patches which are models representing surfaces from the 3D data. The basic idea is to take advantage of the extra knowledge that can be found in 3D models such as surfaces and its orientations. This information is introduced in a modified version of an ICP-like algorithm in order to reduce the outliers incidence in the results. In the mobile robotics area, the initial transformation usually comes from odometry data.

Nevertheless, this approach does not need an initial approximate transformation like ICP based methods do. It uses the global model structure to recover the correct transformation. This feature is useful for those situations where no odometry is available, or it is not accurate enough, such as legged robots. In this case, it exploits both the information given by the normal vector of the planar patches and its geometric position. Whereas original ICP computes both orientation and position at each iteration of the algorithm, this method takes the advantage on the knowledge about planar patches orientation for decoupling the computation of rotation and translation. At first, the method registers the orientation of planar patches sets and when the two planar patches sets are aligned, it address the translation registration. Figure 4 shows the steps performed for computing the alignment between two sets of planar patches. The top image shows a zenithal view of two planar patches sets computed from two consecutive 3D scenes obtained by a robot during its trajectory. The middle image shows the result of rotation registration. Finally, bottom image shows the result after the translation between planar patches sets is computed.

Figure 4. Planar patches matching example. For all the three images patches from the model are painted in dark grey whereas scene paths are represented in light grey. Top, initial situation. Middle, after rotation registration. Bottom, final result after translation registration is completed.

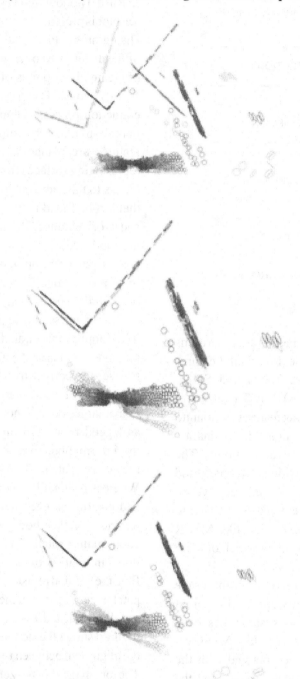

Figure 5. Example of a colored 3D dataset used in experimentation

Feature matching methods are commonly based on the distances between the feature descriptors. The most used distance is the euclidean distance, like the basic ICP. The feature matching methods usually use the euclidean distance to find the nearest/closest features in the other feature set.Once the feature descriptors and the correspondence pairs are calculated, the selection of pairs and the transform computation are the last step.

The most used selection of pairs/transform calculation method is based on the RANSAC algorithm (Fischler & Bolles, 1981). It is an iterative method that estimates the parameters of a mathematical model from a set of observed data which contains outliers. In our case, we look for a 3D transformation (our model) which best explain the data (matches between 3D features). At each iteration of the algorithm, a subset of data elements (matches) is randomly selected. These elements are considered as inliers; a model (3D transformation) is fitted to those elements; all other data are then tested against the fitted model and included as inliers if its error is below a threshold; if the estimated model is reasonably good (its error is low enough and it has enough matches),

Figure 6. RMS error on three registration implementations at different angle steps

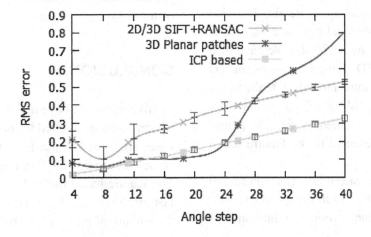

Table 1. Mean execution time for the different methods

	2D/3D SIFT	3D Planar Patches	ICP Based
Time (milliseconds)	1026.87	2154.25	20698.1833

it is considered as a good solution. This process is repeated a number of times and then, the best solution is returned.

Other registration methods are based on Genetic Algorithms (GA) like in Brunnstrom and Stoddart (1996). Using these strategies, the problem of registration is dealt as a search/optimization problem. With this method, the final transformation is generated using a genetic algorithm, gettingsome info of pairs to estimate the best tranformation of all the transformation generated by de GA.

EXPERIMENTATION

We have implemented some of the registration methods introduced above and used to register one static room captured using a kinnect mounted in the PowerBot. An example of the 3D colored data can be show in the figure 5.

We have implemented a ICP algorithm that incorporates the outlier rejection approach proposed in Dalley and Flynn (2002). KD-tree and point subsampling techniques are also included for reducing computational time as described in Rusinkiewicz and Levoy (2001). We have also implementeda 2D/3D method that uses the 2D SIFT detector/descriptor and calculate the transformation using a RANSAC algorithm. The other method implemented is the 3D planar patches features presented in the feature based methods.

We used the three methods over a data set (90 captures to cover 360°, 240k points aprox. each capture) to reconstruct a room, so the major of

movement between the different captures is a rotation over one axis.The measures are done between different angle steps, taking steps of 4 degrees from 4 to 40 degrees. The Root Means Square error on each step is shown in figure 6.

Figure 6 shows that the ICP is the method that has less error in the differentangle steps, followed by the 3D planar patches method,which has less error in when the angle is low. But the execution time is quite different between the different method. Table1 shows the execution times of the different experiments.

We can conclude from experiments that the ICP based method is the one which get better accuraccy but it is slower. 3D planar patches method has similar accuracy results than the ICP method at lesser angles, but it is 10 times faster. 2D/3D SIFT is the fastest method but it has more error than the other methods.

FUTURE RESEARCH DIRECTIONS

Future work includes the design of a hybrid method to compute the registration problems, using several methods or a combination of some of them to solve the registration. The idea is to get better local transformation to increase the speed and the accuracy of an SLAM system. One possible combination can be detecting when its better use one or another method depending on scene special situations, like big angle estimation, time restrictions or a fewer number of features founded.

CONCLUSION

In this chapter, we have reviewed different registration methods applied to mobile robotics. ICP based are a good choice for smaller data sets because they are fast and they can be time bounded. The feature based methods are good to achieve fast results on big data sets because they reduce the amount of points of the data sets to calculate the correspondence. The selection of a registration method depends on the scenes to register, the characteristic of sensors, accuracy required, time constraints, the type of transformation (scale problems), the amount of movement and type of objects (rigid, non-rigid, etc.), etc.

ACKNOWLEDGMENT

This work was partially supported by the University of Alicante project GRE09-16, Valencia Government project GV/2011/034 and by grant DPI2009-07144 from Ministerio de Ciencia e Innovación of the Spanish Government

REFERENCES

Almhdie, A., Léger, C., Deriche, M., & Lédée, R. (2007). 3D registration using a new implementation of the ICP algorithm based on a comprehensive lookup matrix: Application to medical imaging. *Pattern Recognition Letters, 28*(12), 1523–1533. doi:10.1016/j.patrec.2007.03.005

Bay, H., Tuytelaars, T., & Gool, L. V. (2008). Surf: Speeded up robust features. *Computer Vision and Image Understanding, 110*(3), 346–359. doi:10.1016/j.cviu.2007.09.014

Belongie, S., Malik, J., & Puzicha, J. (2001). Matching shapes. *Proceedings Eighth IEEE International Conference on Computer Vision, ICCV 2001,* Vol.1, (pp. 454-461).

Benjemaa, R., & Schmitt, F. (1999). Fast global registration of 3D sampled surfaces using a multi-z-buffer technique. *Image and Vision Computing, 17*(2), 113–123. doi:10.1016/S0262-8856(98)00115-2

Besl, P. J., & McKay, N. D. (1992). A method for registration of 3-D shapes. *IEEE Transactions on Pattern Analysis and Machine Intelligence, 14,* 239–256. doi:10.1109/34.121791

Blais, G., & Levine, M. D. (1995). Registering multiview range data to create 3D computer objects. *IEEE Transactions on Pattern Analysis and Machine Intelligence, 17*(8), 820–824. doi:10.1109/34.400574

Brunnstrom, K., & Stoddart, A. J. (1996), Genetic algorithms for free-form surface matching. *Proceedings of the 13th International Conference on Pattern Recognition,* Vol. 4, (pp. 689-693).

Canny, J. (1986). A computational approach to edge detection. *IEEE Transactions on Pattern Analysis and Machine Intelligence,* (n.d), 679–698. doi:10.1109/TPAMI.1986.4767851

Carcassoni, M., & Hancock, E. R. (2000). *Point pattern matching with robust spectral correspondence* (pp. 1649–1655). Computer Vision and Pattern Recognition. doi:10.1109/CVPR.2000.855881

Carcassoni, M., & Hancock, E. R. (2002). Alignment using spectral clusters. *Proceedings of the 13th British Machine Vision Conference* (pp. 213-222).

Chen, Y., & Medioni, G. (1991). Object modeling by registration of multiple range images. *Proceedings of the IEEE Conference on Robotics and Automation.*

Chetverikov, D., Stepanov, D., & Krsek, P. (2005). Robust Euclidean alignment of 3D point sets: The trimmed iterative closest point algorithm. *Image and Vision Computing, 23*(5), 299–309. doi:10.1016/j.imavis.2004.05.007

Cross, A. D. J., & Hancock, E. R. (1997). *Recovering perspective pose with a dual step EM algorithm.* Eighteenth Annual Conference on Neural Information Processing Systems.

Dalley, G., & Flynn, P. (2002). Pair-wise range image registration: A study in outlier classification. *Computer Vision and Image Understanding, 87*(1-3), 104–115. doi:10.1006/cviu.2002.0986

Deng, Y., & Manjunath, B. S. (2001). Unsupervised segmentation of color-texture regions in images and video. *IEEE Transactions on Pattern Analysis and Machine Intelligence*, (n.d), 800–810. doi:10.1109/34.946985

Dissanayake, M., Dissanayake, M., Newman, P., Clark, S., Durrant-Whyte, H., & Csorba, M. (2001). A solution to the simultaneous localization and map building (SLAM) problem. *IEEE Transactions on Robotics and Automation, 17*(3), 229–241. doi:10.1109/70.938381

Dorai, C., Wang, G., Jain, A. K., & Mercer, C. (1998). Registration and integration of multiple object views for 3D model construction. *IEEE Transactions on Pattern Analysis and Machine Intelligence, 20*(1), 83–89. doi:10.1109/34.655652

Du, S., Zheng, N., Meng, G., & Yuan, Z. (2008). Affine registration of point sets using ICP and ICA. *Signal Processing Letters, 15*, 689–692. doi:10.1109/LSP.2008.2001823

Du, S., Zheng, N., Xiong, L., Ying, S., & Xue, J. (2010). Scaling iterative closest point algorithm for registration of m–D point sets. *Journal of Visual Communication and Image Representation, 21*(5-6), 442–452. doi:10.1016/j.jvcir.2010.02.005

Fischler, M. A., & Bolles, R. C. (1981). Random sample consensus: A paradigm for model fitting with applications to image analysis and automated cartography. *Communications of the ACM, 24*, 381–395. doi:10.1145/358669.358692

Głomb, P. (2009). *Detection of interest points on 3D data: Extending the Harris operator. Computer Recognition Systems 3* (pp. 103–111). Advances in Intelligent and Soft Computing.

Godin, G., Rioux, M., & Baribeau, R. (1994). Three-dimensional registration using range and intensity information. *Proceedings of SPIE, 2350*, (p. 279).

Ho, J. Ming-Hsuan, Yang, Rangarajan, A., & Vemuri, B. (2007). *A new affine registration algorithm for matching 2D point sets*. IEEE Workshop on Applications of Computer Vision, WACV '07.

Li, C., Xue, J., Du, S., & Zheng, N. (2010). A fast multi-resolution iterative closest point algorithm. *Chinese Conference on Pattern Recognition (CCPR)*, (pp. 1-5, 21-23).

Lowe, D. G. (2004). Distinctive image features from scale-invariant keypoints. *International Journal of Computer Vision, 60*(2), 91–110. doi:10.1023/B:VISI.0000029664.99615.94

Masuda, T., Sakaue, K., & Yokoya, N. (1996). Registration and integration of multiple range images for 3-D model construction. *Proceedings of the 13th International Conference on Pattern Recognition*, Vol. 1, (pp. 879-883).

Neugebauer, P. J. (1997), Geometrical cloning of 3D objects via simultaneous registration of multiple range images. *Proceedings 1997 International Conference on Shape Modeling and Applications*, (pp. 130-139).

Pulli, K. (1999). Multiview registration for large data sets. *Proceedings Second International Conference on 3-D Digital Imaging and Modeling*, (pp. 160-168).

Pulli, K., & Shapiro, L. G. (1997). Surface reconstruction and display from range and color data. *Graphical Models, 62*(3), 165–201. doi:10.1006/gmod.1999.0519

Rusinkiewicz, S., & Levoy, M. (2001). Efficient variants of the ICP algorithm. In *Proceedings of the Third International Conference on 3D Digital Imaging and Modeling*, (pp. 145-152).

Rusu, R. B., Blodow, N., & Beetz, M. (2009). Fast point feature histograms (FPFH) for 3D registration. *IEEE International Conference on Robotics and Automation, ICRA '09*, (pp. 3212-3217).

Sandhu, R., Dambreville, S., & Tannenbaum, A. (2008). Particle filtering for registration of 2D and 3D point sets with stochastic dynamics. *IEEE Conference on Computer Vision and Pattern Recognition, CVPR 2008*, (pp. 1-8, 23-28).

Simon, D. (1996). *Fast and accurate shape-based registration*. Ph.D. Dissertation, tech. Report CMU-RI-TR-96-45, Robotics Institute, Carnegie Mellon University.

Smith, S. M., & Brady, J. M. (1997). SUSAN - A new approach to low level image processing. *International Journal of Computer Vision, 23*(1), 45–78. doi:10.1023/A:1007963824710

Turk, G., & Levoy, M. (1994). Zippered polygon meshes from range images. In *Proceedings of the 21st annual conference on Computer graphics and interactive techniques* (SIGGRAPH '94) (pp. 311-318). New York, NY: ACM.

Viejo, D., & Cazorla, M. (2007). 3D plane-based egomotion for slam on semi-structured environment. *IEEE/RSJ International Conference on Intelligent Robots and Systems, IROS 2007*, (pp. 2761-2766).

Viejo, D., & Cazorla, M. (2008). *3D model based map building*. International Symposium on Robotics, ISR 2008, Seoul, Korea.

Weik, S. (1997). Registration of 3-D partial surface models using luminance and depth information. *Proceedings of the International Conference on Recent Advances in 3-D Digital Imaging and Modeling*, (pp. 93-100).

Xin, W., & Pu, J. (2010). An improved ICP algorithm for point cloud registration. *2010 International Conference on Computational and Information Sciences (ICCIS)*, (pp. 565-568).

Zhang, L., Choi, S.-I., & Park, S.-Y. (2011). Robust ICP registration using biunique correspondence. *2011 International Conference on 3D Imaging, Modeling, Processing, Visualization and Transmission (3DIMPVT)*, (pp. 80-85).

Zinsser, T., Schmidt, J., & Niemann, H. (2003). A refined ICP algorithm for robust 3-D correspondence estimation. *Proceedings 2003 International Conference on Image Processing, ICIP 2003*, Vol. 2, (pp. 695-8).

KEY TERMS AND DEFINITIONS

Iterative Closest Points (ICP): An iterative algorithm to compute the transformation between two clouds of points that have minimum distance.

RANdom SAmple Consensus (RANSAC): An iterative method to estimate parameters of a mathematical problem, in this case, to estimate transformation parameters.

Scale-Invariant Feature Transform (SIFT): A method to locate and describe local features in 2D images that try to be robust to changes in illumination, noise, and minor changes in viewpoint.

Simultaneous Localization and Mapping (SLAM): A technique used in mobile robotics to build/update a map within an unknown environment at same time the location on this current location in this map.

Chapter 10
Methodologies for Evaluating Disparity Estimation Algorithms

Ivan Cabezas
Universidad del Valle, Colombia

Maria Trujillo
Universidad del Valle, Colombia

ABSTRACT

The use of disparity estimation algorithms is required in the 3D recovery process from stereo images. These algorithms tackle the correspondence problem by computing a disparity map. The accuracy assessment of a disparity estimation process has multiple applications such as comparing among different algorithms' performance, tuning algorithm's parameters within a particular context, and determining the impact of components, among others. Disparity estimation algorithms can be assessed by following an evaluation methodology. This chapter is dedicated to present and discuss methodologies for evaluating disparity estimation algorithms. The discussion begins with a review of the state-of-the-art. The constitutive components of a methodology are analysed. Finally, advantages and drawbacks of existing methodologies are presented.

INTRODUCTION

A 3D scene can be described as a set of points in space. It can be captured, from different viewpoints, by a stereo camera system generating a stereo image pair. The conjugate projections of a common scene point are captured into image planes in different image coordinates. These projections are corresponding points, and the shift or displacement between them is termed disparity (Hartley & Zisserman, 2003). Disparity is, in

essence, a vector related corresponding points. Disparity values are used to recover the 3D information of the captured scene. If the disparity of a point is known, its original 3D position can be recovered by a triangulation process. In fact, the 3D information recovery process from a stereo image is an inverse and ill-posed problem, due to depth ambiguity, data instability and lack of information. A stereo image pair can be modified in such a way that the displacement between corresponding points becomes horizontal (Fusiello

DOI: 10.4018/978-1-4666-2672-0.ch010

et al., 2000). Such modified stereo image pair is the input to disparity estimation algorithms, which produce as output an estimated disparity map. Nevertheless, even in horizontal parallax, there is still a lack of information about the correspondences between points. Such lack of information gives rise to the stereo correspondence problem. Moreover, the stereo correspondence problem involves two inherent problems: ambiguity and occlusion. The projections into image planes are ambiguous by nature, since different objects, at different depths, can generate equal projections into image planes. Moreover, multiple objects captured into a stereo image pair may have the same appearance. On the other hand, a point may lack of a corresponding point – be occluded – in the conjugate image plane. In fact, occlusion phenomena arise naturally in stereo images, but it is not known beforehand.

The assessment of disparity estimation is important since a small inaccuracy in an estimated disparity may produce a large error in the recovered 3D. The performance of disparity estimation algorithms can be assessed by following an evaluation methodology. This assessment has several applications such as comparing algorithms (Szeliski, 1999; Szeliski & Zabih, 2000; Scharstein & Szeliski, 2002; Tombari et al., 2010; Cabezas & Trujillo, 2011), comparing methods and procedures (Tombari et al. 2008; Hirschmuller & Scharstein, 2009; Bleyer & Chambon, 2010), tuning algorithm's parameters within a particular context (Hoyos et al., 2011), and identifying algorithm's advantages or drawbacks (Kostlivá et al., 2008), among others.

There are two main ways for assessing disparity estimation algorithms: qualitatively and quantitatively. A qualitative way is based on collecting and processing opinions from human observers by means of statistical techniques (Benoit et al., 2008; Lü et al., 2011). However, human observations may be biased due to factors such as the observation environment (i.e. illumina-

tion, experimental setup, display devices, among others), and/or observer characteristics (i.e. age, health conditions, mood, background knowledge, among others). Moreover, a qualitative evaluation may be subjective, time consuming and difficult to be routinely performed in many scenarios. Consequently, qualitative evaluation approaches have drawbacks that make them not suited to properly evaluate or compare disparity estimation algorithms. On the other hand, a quantitative way is based on objective, deterministic, and automatic methods capable of asserting signal fidelity (Wang & Bovik, 2009). Thus, quantitative approaches are robust to the above mentioned factors. A quantitative evaluation approach can be conducted based on ground-truth data (Scharstein & Szeliski, 2003; Strecha et al., 2008), or additional views from the original 3D scene (Szeliski, 1999; Leclerc et al., 2000). Evaluation approaches have to follow an evaluation methodology (Scharstein & Szeliski, 2000; Cabezas & Trujillo 2011). A quantitative evaluation methodology may include components such as: an imagery test-bed on which the evaluation is conducted, a set of evaluation criteria related to the aspects under analysis, a set of error measures that produces error scores, an evaluation model that operates on scores and produces a set of accuracy indicators, and a method for analysing indicators in order to obtain evaluation results.

This chapter presents each component of a quantitative evaluation methodology, doing emphasis on error measures and evaluation models. The purpose of the chapter is to provide a guideline for designing and conducting a proper evaluation of disparity estimation algorithms, along with being a basis for future research on the field. The structure of the chapter is as follows. Firstly, the state-of-the-art on evaluation methodologies is presented. Secondly, the components of an evaluation methodology are analysed taking into account the impact on evaluation results. Error measures commonly used, as well as two evaluation models are discussed in detail. Thirdly, the advantages and drawbacks of existing methodologies, as well as a set of principles for a proper evaluation process, are presented. Finally, concluding remarks are stated.

STATE-OF-THE-ART

An evaluation may be addressed as an inter-technique or an intra-technique according to the goal of the evaluation process. The intra-technique evaluation allows the comparison of different algorithms, whilst the inter-technique evaluation allows the comparison of different modules or parameters of a same algorithm. However, the main distinction among evaluation approaches is based on what it is used to compare against. Although the state-of-the-art on quantitative evaluation methodologies is commonly classified into ground-truth-based and prediction-error based approaches (Szeliski & Zabih, 1999), this classification may not suite all approaches. In this chapter, a distinction between evaluation in the presence of ground-truth data and evaluation in the absence of ground-truth data is presented. Evaluation methodologies in the presence of ground-truth data rely on measuring error by comparing an estimated disparity map against a ground-truth disparity map generated using an active stereo technique such as structured light or time of flight measurements, among others (Scharstein & Szeliski, 2003). However, the generation of disparity ground-truth may be too difficult or laborious and even impossible to achieve in some circumstances due to the limitations of active stereo techniques to be used in indoor or controlled environments (Strecha et al., 2008, Morales & Klette, 2010). In contrast, most of the evaluation methodologies, in the absence of ground-truth, data rely on using additional views. Nevertheless, the generation of additional views is a requirement that should be taken into account during the capturing process of the stereo image (Morales & Klette, 2009). In this case, a more complex camera setup or a more expensive system is required in order to fulfil this requirement. Methodologies in the presence of ground-truth data, and methodologies in the absence of ground-truth data are discussed in more detail in the following subsections.

Evaluation Methodologies in the Absence of Disparity Ground-Truth Data

The prediction-error approach is proposed for the evaluation of both, motion and disparity estimation algorithms in (Szeliski, 1999). A predicted view can be rendered based on a reference image and its associated estimated disparity map. Then, the rendered view is compared against an additional image (i.e. an image that was not used to compute the disparity map), captured from a known camera position with respect to the input stereo image. Forward and/or inverse predictions are the two alternatives to generate a rendered view. However, error scores reflect not only the accuracy of the disparity estimation algorithm, but also the accuracy of the selected rendering algorithm. This is not a trivial matter, since the rendering process of the predicted view has to deal with interpolation or extrapolation issues. In fact, this evaluation approach is best suited for applications domains where the output is a rendered view and human observers are final users. In these domains, the capability of bringing a visual comfort sensations to users could be more important that the accuracy of the estimation (Mittal et al., 2011). Szeliski (1999) considers an inter-technique evaluation scenario. The experimental evaluation conducted in this work focuses more in motion estimation, than in disparity estimation. The Root Mean Square (RMS) was used as error measure.

The self-consistency is presented as a property of the Human Visual System (HVS) in (Leclerc et al., 2000): perceptual inferences made by a HVS, from different viewpoints, are, most times, consistent among them. Thus, the self-consistency property can be used for assessing the performance of disparity estimation algorithms in intra and inter-technique comparisons. It measures the distance among triangulated 3D world coordinates of a set of corresponding points from multiple views. However, the assessment of disparity estimation algorithms by the self-consistency property requires reliable information about projection matrices, in a common coordinate system. In fact, this information can be considered as having ground-truth data. On the other hand, the self-consistency is a necessary but not a sufficient condition for a disparity estimation algorithm to be correct. Consequently, an algorithm can be self-consistent over several scenes, but it may produce severely biased or entirely wrong disparity estimations. Although an evaluation based on the self-consistency property is quantitative, it can be considered as heuristic.

Evaluation Methodologies in the Presence of Disparity Ground-Truth Data

An experimental comparison of disparity estimation algorithms is proposed in (Szeliski & Zabih, 2000). This proposal applies, in a separated way, a comparison against disparity ground-truth data, and the prediction-error approach of (Szeliski, 1999). The concept of error criteria is introduced in this proposal. The use of error criteria allows a detailed analysis of algorithms performance in relation to different image phenomena, such as specular surfaces, low texture regions, depth discontinuities, and occluded pixels, among others. In addition, an error function is also introduced. It defines an error as an estimation disagreeing from the ground-truth disparity value in more than a threshold. The error function is gathered according to error criteria. The Tsukuba and the Map stereo images were used as test-bed images. The ground-truth of the Tsukuba image was generated manually, which make it prune to errors, whilst the Map image is, in essence, an artificial image.

On the other hand, the prediction error approach is used considering how well the reference image and its estimated disparity map can be used to predict other views using a forward or inverse warping. As conclusions, consistent results were obtained, whilst each approach detects better particular kinds of errors: the prediction-error based approach does emphasis on errors over highly textured regions, and the ground-truth based approach does emphasis on errors over low texture regions.

The Middlebury's methodology was introduced in (Scharstein & Szeliski, 2002). It extends concepts previously introduced in (Szeliski & Zabih, 2000). This methodology can be used in both intra and inter-techniques evaluation. In addition, a taxonomy based on the modules of disparity estimation algorithm is introduced. These modules are: matching cost computation, cost aggregation, disparity optimization and disparity refinement. Thus, evaluation processes conducted following the introduced methodology allows comparing performance achieved by different algorithms in the above mentioned modules. The imagery test-bed includes four images, and their respective ground-truth data: Tsukuba, Map, Venus, and Sawtooth. The error function used in (Szeliski & Zabih, 2000) is properly formulated and termed as Bad Matched Pixels (BMP). In addition to the BMP, the RMS is also used for comparing estimated maps against ground truth data. Different error criteria are associated to image segments resulting in the following criteria:

- **All:** The entire image,
- **Nonocc:** Areas that are not-occluded,
- **Disc:** Areas near depth discontinuities, and
- **Textureless:** Areas of low texture.

In the evaluation model of this methodology, error scores are sorted, in descendent way, by error criterion and stereo image pair. A final rank is computed as the average of all ranks. In this way, the evaluation model of Middlebury's methodology can be seen as a linear combination of ranks, where a real value is associated to the accuracy of an algorithm. Different experiments were conducted and obtained results were plotted for analysis. As conclusions, the experimentation conducted produces a better understanding on shortcomings of some algorithms in regard to their particular composition in terms of building blocks, as well on the sensitivity of algorithms to setup of key parameters.

An online rank based on the Middlebury's methodology is available and keeps updated at (Scharstein & Szeliski, 2012). The imagery test-bed used is composed by Tsukuba, Venus, Cones and Teddy stereo image pairs (Scharstein & Szeliski, 2003). The BMP and the all, the nonocc and the disc are used as error measure and error criteria, respectively. The evaluation is focused on an inter-technique comparison, and algorithm's parameters have to be fixed for the entire test-bed. This online benchmarking has been widely used by the community. It contains a repository of disparity maps generated by, approximately, 130 algorithms (May, 2012). Nevertheless, it compares the entire set of algorithms reported by the community regardless their differences in requirements (e.g. used hardware, or execution time, among others). Thus, such comparison may be unfair.

The evaluation methodology proposed in (Kostlivá et al., 2007) is focused on parameter settings (i.e. it addresses an inter-technique evaluation scenario prior to an intra-technique evaluation). It considers both the accuracy and the density of the estimated disparity map in regard to parameters settings. Two errors are defined based on the accuracy and the density of estimated disparity maps: the error rate and the sparsity rate. The error rate is defined as the percentage of incorrectly estimated disparities, without considering a missing disparity as an error (i.e. mismatches and false positives). The sparsity rate is defined as the percentage of all missing disparity estimations which are not ruled out by any other incorrect estimation (i.e. false negatives). These errors are based on four

principles: orthogonality, symmetry, completeness and algorithm independence (Kostlivá et al., 2003). These principles can be outlined as follows.

- **Orthogonality:** Errors have to be mutually independent.
- **Symmetry:** Errors have to be invariant to the direction of search (i.e. from the left to the right view, or from the right to the left view) during the disparity estimation process.
- **Completeness:** Error definitions have to be valid in any scene of arbitrary geometry.
- **Algorithm independence:** An evaluation process has to be possible, disregarding the density, or semi-density, of estimated disparity maps.

A Receiver Operating Characteristics (ROC) analysis is adopted upon the error rate and the sparsity rate. In addition, an is better relation is defined based on the ROC curve. A particular parameter setting is better than another if it produces more accurate and denser results. However, the ROC curve can be computed on just one set of stereo images. Thus, the evaluation turns probabilistic when the imagery test-bed includes several stereo image pairs. Additionally, this evaluation methodology requires weight assigns in relation to the importance of each stereo image pair included in the test-bed. On the other hand, this methodology assumes that there exist different evaluation models depending on the density or semi-density of estimated disparity maps. Two stereo images, of artificial scenes with varying texture, were used in the experimental evaluation in addition to the Middlebury's imagery test-bed. As conclusions, disparity estimation algorithms with different occlusion models should not be compared, the selection of test-bed images is still an open debate, and algorithms execution time should be also considered during an evaluation process.

A cluster ranking intra-technique evaluation method is proposed in (Neilson & Yang, 2008).

It consists on using a statistical inference technique (ANOVA) to rank the accuracy of disparity estimation algorithms over a single stereo image pair, and the posterior combination of ranks from multiple stereo pairs, into a final rank. Thus, the same ranks are assigned to algorithms producing statistically similar results. This proposal is focused on comparing matching costs using a hierarchical belief propagation algorithm (Felzenszwalb & Huttenlocher, 2004). However, a different significance test (Friedman) has to be applied when the test-bed includes more than one stereo image pair. Moreover, a greedy clustering algorithm, which requires a threshold related to a confidence level, is used. The clustering algorithm used, computes iteratively the final ranks as the average of several ranks in a partition. Thus, assigned rank may be a real number which lacks of a concise interpretation. This conducted evaluation includes 90 synthetic images, with three different levels of noise, generated by a ray tracing method, and 18 images from the Middlebury's image repository, some of them captured with three different illuminatios and times exposure (Hirschmüller & Scharstein, 2009). The BMP measure is used, only, according to the nonocc error criterion. As conclusions, the selection of a matching cost has a large impact on the accuracy of estimated disparity maps, and there is not a single parameters setting working well for every matching cost metric or even every stereo image pair. Moreover, a particular setting, working fine in one case, may worked very poorly in other case. On the other hand, this class of study requires huge computational resources, which are not available for all developers or researchers on disparity estimation algorithms. Thus, the conducted study is difficult to be repeated. Consequently, this methodology does not take into account the capabilities or requirements of final users.

An inter-technique evaluation, involving estimation accuracy and computational efficiency, is proposed in (Tombari et al., 2010a). This proposal is focused on disparity estimation algorithms suitable to be used in application domains requiring near real-time performance (i.e. more than 5 frame per second), real-time performance (i.e. more than 25 frame per second) and/or to be executed on hardware platforms with limited resources (i.e. with a low-power architecture and limited memory) such as the offered by embedded devices. The imagery test-bed used includes the Middlebury's data set (with the addition of Gaussian noise) as well as a dataset related to common working conditions (i.e. uncontrolled illumination, photometric distortions, small defocus, and non-perfect rectification). This dataset, termed Lab, was acquired in an uncontrolled environment using an off-the-shelf stereo camera. It is composed by 6 stereo images containing different objects on a table (Tombari et al., 2010b). The complement of the BMP measure was used to gathering errors according to the nonocc and the disc criteria, whilst the computational efficiency was compared based on the quantity of disparities computed per second. This proposal uses the evaluation model of Middlebury's methodology. Nevertheless, accuracy and efficiency are, by nature, conflicting goals for a disparity estimation algorithm. As conclusions, in top performer algorithms, a small increment in estimation accuracy implies a large additional computational effort. In addition, the Lab dataset may be much more challenging than the Middlebury dataset.

An evaluation methodology, termed as A^* is proposed in (Cabezas & Trujillo, 2011). The methodology can be used for both: intra and inter-techniques evaluation. In this proposal, the evaluation of disparity maps is addressed as a multiobjective optimisation problem. The evaluation model of the A^* methodology is based on the Pareto Dominance relation (Van Velduizen et al., 2003). It computes a proper subset A^* from the set of stereo algorithms under evaluation. Moreover, the interpretation of results is based on the cardinality of the A^* set, which is composed by disparity estimation algorithms of comparable performance among them, and at the same time, of superior performance to the rest of algorithms under evaluation. As an advantage, the introduced evaluation model avoids a subjective interpretation of quantitative evaluation results. Additionally, it avoids the operation of factors that are incommensurable among them, such as the error score obtained for different evaluation criteria or different stereo image. Nevertheless, this work fails in considering an evaluation scenario on which a user is interested in an exhaustive evaluation of the entire set of algorithms, and not only in determining which ones are the most accurate. The Middlebury's imagery test-bed and the BMP error measure are used for evaluating disparity estimation algorithms. Different evaluation scenarios were considered by the combination of different error criteria and the selection of different sets of disparity estimation algorithms. As conclusions, obtained evaluation results for a particular evaluation scenario cannot be considered of general character or expected to be repeated under all possible evaluation scenarios.

The A^* Groups evaluation model is proposed in (Cabezas et al., 2012a). It extends the evaluation model of the A^* methodology, introducing the capability of performing an exhaustive analysis of algorithms under evaluation. The introduced capability is based on an iterative algorithm for computing partitions of the set of disparity estimations algorithms under evaluation –multiple groups, or instances of the A^* set. Each A^* set is identified with an ordinal label associated to the performance of algorithms composing the partition, and is composed by algorithms of comparable performance: not better neither worst. In particular, the A_1^* set is of special interest since it is composed by the top performer disparity estimation algorithms. The conducted evaluation is focused on showing the capabilities of the proposal, using the same test-bed imagery, the same error criteria and 112 disparity estimation

algorithms reported at (Scharstein & Szeliski, 2012). As another innovative aspect, this methodology uses the Sigma-Z-Error measure (SZE) (Cabezas et al., 2011). The SZE measure is based on the inverse relation between depth and disparity, considering the disparity error magnitude. As obtained results, seven different groups were computed. The first group –the A_1^* set– was composed by nine algorithms. Among these algorithms, seven use global optimisation techniques, whilst two use local techniques. The GC+occ, the MultiCamCG (Kolmogorov & Zabih; 2001, 2002) and the MultiResGC (Papadakis & Caselles, 2010) are among the approaches based on Graph Cuts (GP). All of them consider occlusion explicitly as a term of used energy functions. In addition, two unpublished algorithms are also based on GP. The DoubleBP algorithm (Yang et al., 2008) is based on Belief Propagation (BP). In particular, a hierarchical BP is applied twice, were occlusions and low-texture areas are first identified and filled using neighbouring values. The Segm+Visib algorithm (Bleyer & Gelautz, 2004) is based on colour segmentation. Segments are grouped into planar layers. Areas with low texture and depth discontinuities are handled by segmentation, whilst occlusions are detected by layers assignment. In regard to the algorithms using local optimisation techniques, the DistinctSM algorithm (Yoon & Kweon, 2007) uses a similarity measure based on the distinctiveness of image points, and the similarity between them, whilst the PatchMatch algorithm (Bleyer et al., 2011) finds a slanted support plane at each image point.

On the other hand, there are works related to a quantitative comparisons of algorithmic modules, such as (Hirschmuller & Scharstein, 2009) and (Bleyer & Chambon, 2010), among others. The first work compares the robustness of different matching cost against radiometric differences in a stereo image pair, whilst the second work matching costs using different colour spaces. Nevertheless, these works cannot be considered as evaluation methodologies, since they are, in fact, particular evaluation scenarios following the Middlebury's evaluation methodology.

COMPONENTS OF AN EVALUATION METHODOLOGY

A quantitative evaluation methodology can be addressed by following the steps illustrated in Figure 1 (Cabezas & Trujillo, 2011). Among the components that an evaluation methodology may involve, the following are clearly identified: test-bed images, error criteria and error measures, an evaluation model, and a method for interpreting results. Each component is discussed in this section.

Test-Bed Images

There are several issues in regard to the selection of an imagery test-bed in an evaluation process (Haeusler & Klette, 2010). A fundamental one is image content. The content of stereo images selected as test-bed should be related to the application domain where the disparity estimation algorithm is going to be used, since, in practice, does not exist a single disparity estimation algorithm performing well with all possible image contents. In fact, an application domain can be related to a type, or even several types of image content, which should be assessed independently. On the other hand, synthetic images (i.e. generated by a ray tracing method) do not reflect realistic capturing or imaging conditions (Sandino et al., 2011). Consequently, observations made upon them are improbable to be repeated on real imagery, even in the cases that synthetic images are deliberately contaminated with noise. In addition, if a limited number of images are considered during the evaluation process, obtained results may be biased (Neilson & Yang, 2008). In general terms, the evaluation results obtained using a particular test-bed set, cannot be extrapolated, neither expected, on others test-bed.

Error Criteria

Error criteria are commonly understood as a set of regions, or image segments, defined over a reference image. This segmentation provides a link among images content and error scores. Error criteria provide a basis for analysing algorithms performance, since they allow an evaluation on specific, problematic or challenging image regions on which error scores are gathered. The most commonly used error criteria are: regions near depth discontinuities –disc–, non-occluded regions, –nonocc–, and the entire image –all (Scharstein & Szeliski, 2012). However, alternative regions can be defined according to an application domain, such as foreground and background in unmanned automotive vehicles (Van der Mark & Gavrila, 2006), among others.

Error Measure

The goal of an error measure can be defined as comparing two signals/images in order to describe the degree of similarity, or the level of distortion between them, by a quantitative score. An error measure should fulfil the following properties: non-negativity, identity, symmetry, and triangular inequality (Zhou & Bovik, 2009). In this context, one of the compared signals is reliable (i.e. the disparity ground-truth data) whilst the other is contaminated by errors (i.e. the estimated disparity map). Thus, ground-truth based error functions can be considered (or analysed) as full reference metrics (Rehman & Wang, 2010). Different error functions have been proposed for evaluating disparity map. Most of them are discussed below.

Two indices for measuring smoothness of a noised disparity map –the disparity gradient and the disparity acceleration– are proposed in (Zhang et al., 2009). Three ground-truth disparity maps

Figure 1. Steps of an evaluation methodology

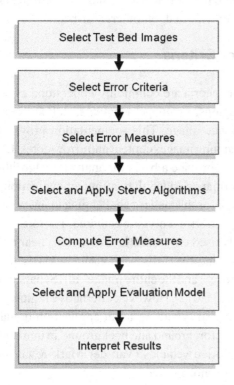

from the Middlebury repository (Scharstein & Szeliski, 2003) were artificially corrupted with noise. Then, the proposed indices were applied to the noisy images, and obtained results compared against the level of introduced noise. The indices require a threshold in order to consider an estimated disparity as inaccurate, or related to a depth discontinuity. However, no information is provided about how required thresholds can be fixed, neither about considered noise nor its relation to artefacts produced by a disparity estimation algorithm. Moreover, this work ignores the fact that a disparity map may vary smoothly, but being totally wrong. Consequently, the disparity gradient and the disparity acceleration indices may be not suited for properly evaluating an estimated disparity map.

A modification of the Multi-scale Structural Similarity index (MS-SSIM) measure is introduced in (Malpica & Bovik, 2009). The proposed measure, termed R-SSIM, is capable of handling

missing data in both, a disparity map under evaluation and used ground-truth data. As a conclusion, obtained results by the R-SSIM measure and obtained results by the BMP measure are statistically correlated. Nevertheless, the final ranking assigned to disparity estimation algorithms, using the evaluation model of the Middlebury methodology, varies considerably when the R-SSIM measure is used. In addition, the discussion about the analogy among the components of the MS-SSIM measure and the properties of a disparity map is not properly addressed. Consequently, the argumentation of why it is convenient to use the R-SSIM measure for evaluating disparity maps turns weak.

The BMP error measure, formulated in (Scharstein & Szeliski, 2000), is widely adopted for evaluating estimated disparity maps. It can be seen as binary function using a threshold ´. A value of 1 pixel is commonly used. Nevertheless, a score of zero does not necessarily imply that a disparity map is free of errors (Cabezas et al,. 2012b). It is, in essence, a measure of the quantity of errors in a disparity map, regardless their magnitude. Consequently, the BMP measure may conceal estimation errors of a large magnitude, and at the same time, it may penalise errors of low impact on the final 3D reconstruction. The BMP measure is formulated as follows:

$$\text{BMP} = \frac{1}{N} \sum\nolimits_{(x,y)} \mu(x,y) \qquad (1)$$

where δ. is the amount of compared pixels, and $\mu(x,y)$ is a binary function:

$$\mu(x,y) = \begin{cases} 1 & \text{if} |D_{\text{true}}(x,y) - D_{\text{estimated}}(x,y)| > \delta \\ 0 & \text{if} |D_{\text{true}}(x,y) - D_{\text{estimated}}(x,y)| \leq \delta \end{cases}$$

$$(2)$$

where δ is the error threshold, D_{true} is the ground-truth disparity map, and $D_{\text{estimated}}$ is the estimated disparity map.

The Mean Square Error (MSE) and the Mean Relative Error (MRE) have been used for comparing estimated disparity maps against disparity ground-truth data (Van der Mark & Gavrila, 2006; Cabezas et al,. 2012b). However, the mean is sensitive to outliers and may introduce bias in the evaluation. Moreover, the MSE ignores the inverse relation between depth and disparity, and penalise in the same way all estimation errors. Thus, different disparity maps, with very different types and levels of distortion may obtain the same MSE scores. In contrast, the MRE considers the error magnitude, as well as their impact in relation to the ground-truth value. The MSE and the MRE are formulated in Equations (3) and (4), respectively.

$$MSE = \frac{1}{N} \sum_{(x,y)} \left(D_{true}(x,y) - D_{estimated}(x,y) \right)^2$$

$$(3)$$

$$MRE = \frac{1}{N} \sum_{(x,y)} \left| \frac{D_{true}(x,y) - D_{estimated}(x,y)}{D_{true}(x,y)} \right|$$

$$(4)$$

The Sigma-Z-Error (SZE) measure is introduced in (Cabezas et al., 2011). It is based on the inverse relation between disparity and depth, using the magnitude of the disparity estimation error. It fulfils the properties of a metric, but it is unbounded. It aims to measure the impact on the 3D reconstruction of a disparity estimation error. The SZE does not use thresholds, but it requires information about the stereo camera system: the focal length and the system baseline (i.e. the distance between optical centres). In practice, information about the stereo camera system should be available to allow the 3D reconstruction process. The SZE measure is formulated as follows:

$$SZE = \sum_{(X,Y)} \left| \frac{f*B}{D_{true}(x,y) + \frac{1}{4}} - \frac{f*B}{D_{estimated}(x,y) + \frac{1}{4}} \right|$$

$$(5)$$

where f is the focal length, B is the baseline and $\frac{1}{4}$ is a small constant to avoid the instability caused by missing estimations.

Evaluation results may vary considerably according to the used error measures (Cabezas et al., 2011). Thus, the selection of an error measure is not a trivial issue. A characterisation of error measures, in order to provide the bases to select a measure during the evaluation process is presented in (Cabezas et al., 2012b). The characterisation introduces five attributes: automatic, reliable, meaningful, unbiased, and consistent. They are outlined at follows.

- **Automatic:** The error measure should be computed without human intervention.
- **Reliable:** The error measure should operate without being influences by external factors and in a deterministic way.
- **Meaningful:** The error measure should be intended for a particular purpose, has a concise interpretation and does not lead to ambiguous results.
- **Unbiased:** The error measure should allow performing impartially comparisons.
- **Consistent:** The score produced by an error measure should be compatible (i.e. in agreement of observations) with the scores produced by any other error measure with a common particular purpose.

In addition, a measure for quantifying the consistency attribute was presented. As conclusions, the MRE and the SZE have the highest consistency using the Middlebury's, and the A^* Groups evaluation models, respectively.

Evaluation Model

Among different evaluation models available in the literature, three of them will be analysed in detail in this subsection. These are the evaluation model of Middlebury's methodology (Scharstein

& Szeliski, 2000), the evaluation model of the A^* methodology, as well as the extension introduced to it by the A^* Groups methodology (Cabezas et al., 2012a). Required background is presented below.

Let A be a non-empty set of disparity estimation algorithms, as follows:

$$A = \{a \in A | a{:}(I_{stereo}) \rightarrow D_{esimated_{(a)}}\}, \quad (6)$$

where I_{stereo} is a non-empty set of stereo images (i.e. the imagery test-bed), and $D_{estimated(a)}$ is the set of estimated disparity maps by a particular disparity estimation algorithm.

Let $D_{estimated}$ be the set of disparity maps to be compared, defined as:

$$D_{estimated} = \{D_{estimated_{(a)}} \in D_{estimated} | \forall a \in A{:}\exists D_{estimated_{(a)}}\}. \quad (7)$$

Let D_{true} be the set of disparity ground-truth data associated to the I_{stereo} set. Let $R_{criteria}$ be the set of error criteria. Let $E_{measures}$ be the set of selected error measures. The comparison of the set $D_{estimated}$ against the D_{true} set is a fundamental step in the quantitative evaluation approach. This comparison can be formalised as follows. Let g be a function such that:

$$g{:}(D_{estimated_{(a)}} \times D_{true} \times R_{criteria} \times E_{measures}) \rightarrow V_a, \quad (8)$$

where $V_a \in \Re^k$ is a vector, and the magnitude of k is determined by the cardinality of the sets D_{true}, $R_{criteria}$, and $E_{measures}$.

Let V be the set obtained by applying the function g to the set $D_{estimated}$:

$$V = \{V_a \in V | \; \forall D_{estimated(a)} \in D_{estimated}{:}\exists V_a\} \quad (9)$$

The evaluation model of the Middlebury's methodology assigns a rank to each algorithm under evaluation, based on the error scores of estimated disparity maps. This rank is a real value. This evaluation model can be formalised as:

$$\forall V_a \in V{:}\exists r | \; r{:}(V_a) \rightarrow \Re \quad (10)$$

On the other hand, the evaluation model of the A^* methodology is based on the formulation of the evaluation process as a Multiobjective Optimisation Problem (MOP). In general, a MOP involves two different spaces: a decision space and an objective space. The nature of these spaces may depend on the nature of the particular MOP. In this context, the decision space is defined as the set of disparity estimation algorithms under evaluation –the A set–, and the objective space is defined as the set composed by vectors of errors scores –the V set. In particular, let p and q be elements of the decision space: $p,q \in A \wedge p \neq q$. Let V_p and V_q be pair of vectors belonging to the objective space, as follows:

$$V_p = g{:}(D_{estimated_{(p)}} \times D_{true} \times R_{criteria} \times E_{measures}), \quad (11)$$

$$V_q = g{:}(D_{estimated_{(q)}} \times D_{true} \times R_{criteria} \times E_{measures}), \quad (12)$$

Then, without loss of generalisation, the following relations between V_p and V_q can be considered:

$$V_p = V_q \text{ iff } \forall i \in \{1,2,...k\}{:}V_{pi} = V_{qi}, \quad (13)$$

$$V_p \leq V_q \text{ iff } \forall i \in \{1, 2, ...k\}: V_{pi} \leq V_{qi}, \quad (14)$$

$$V_p < V_q \text{ iff } \forall i \in \{1, 2, ...k\}: V_{pi} < V_{qi} \land V_{pi} \neq V_{qi} \quad (15)$$

Thus, for any two elements in the decision space, three possible relations do exist:

$$p \prec q \text{ (p dominates q) iff } V_p < V_q, \quad (16)$$

$$p \preceq q \text{ (p weakly dominates q)} \\ \text{iff } V_p \leq V_q, \quad (17)$$

$$p \sim q \text{ (p is comparable to q)} \\ \text{iff } V_p \not\leq V_q \land V_q \not\leq V_p. \quad (18)$$

The three relations above are the base for the evaluation model of the A^* methodology, which operates by defining a partition over the A set. It can be formalised as follows. Let d be a function such as:

$$d:(A) \rightarrow \\ A' \cup A^* | A' \cup A^* \subseteq A \land A' \cap A^* = \emptyset, \quad (19)$$

subject to:

$$\nexists A'_{(a)} \in A' | A'_{(a)} \prec A^*_{(a)} \in A^*, \quad (20)$$

where " $\prec$ " denotes the Pareto Dominance relation.

The A^* Groups methodology extends the evaluation model by iteratively computing a partition of the A set, assigning to each computed A^* set an ordinal label related to the partition established in a current iteration. Moreover, all algorithms in a specific partition are dominated by at least on algorithm in a partition with

a greater label. The evaluation model of the A^* Groups can be formalised as follows.

$$d:(A) \rightarrow \\ A'_n \cup A^*_n | A'_n \cup A^*_n \subseteq A_n \land A'_n \cap A^*_n = \emptyset, \quad (21)$$

subject to:

$$\nexists A'_{(a)} \in A'_n | A'_{(a)} \prec A^*_{(a)} \in A^*_n, \quad (22)$$

$$\forall n \ A^*_n \neq \emptyset, \quad (23)$$

$$\forall n \neq m \ A^*_n \cap A^*_m = \emptyset, \quad (24)$$

$$\bigcup_n A^*_n = A, \quad (25)$$

$$\forall q \in A^*_m \exists p \in A^*_n | p \prec q \land n < m. \quad (26)$$

Interpretation of Results

The evaluation model and the interpretation of results are inherently related. The evaluation model of the Middlebury's methodology computes a rank for each algorithm under comparison. In this way, the rank position of an algorithm is an indicator of performance, where a lower rank position means a higher performance on test-bed imagery. However, the specific rank assigned to an algorithm may be arbitrary, since the amount of algorithm's parameter tuning cannot be controlled (Scharstein & Szeliski, 2012). Thus, if two algorithms have very similar ranks, it is not clear which one performs better. Additionally, it is not possible to unambiguously determine which set of algorithms is the top-performer. Nevertheless, in this regard, may be reasonable to use a heuristic such as an amount expressed as a percentage (i.e. the 25%), instead of considering a fixed amount.

On the other hand, the A^* methodology, and consequently, the A^* Groups methodology as well, define an interpretation of results based on the cardinality of the A^* set, which, without loss of generalisation, is stated as follows:

$$\begin{cases} \text{if } \left|A_1^*\right| = 1 \\ \text{then } superior\ performance \\ \text{if } \left|A_1^*\right| > 1 \\ \text{then } comparable\ performance \end{cases}, \qquad (27)$$

where *superior performance* means that there exists a unique disparity estimation algorithm with a superior performance to the rest of algorithms under comparison, and *comparable performance* means that there exists a set of algorithms with a comparable performance among them, (i.e. not better, neither worst) and simultaneously, of superior performance to algorithms in a group with a different label. Moreover, the interpretation of results, in both of the above cases, is done in regard to the imagery test-bed considered. Thus, by definition, it cannot be extrapolated to other images neither be generalised to all possible capturing conditions.

Figure 2 illustrates the evaluation process discussed in this section using selected disparity estimation algorithms, the BMP measure, and the Middlebury's and the A^* evaluation methodologies.

DRAWBACKS AND ADVANTAGES OF EXISTING EVALUATION METHODOLOGIES

The evaluation model introduced in (Kostlivá et al., 2007) has a solid theoretical support. However, the evaluation of more than one stereo image pair is difficult due to the requirements of weights according to scene's relevance. In practice, the assignation of such weights may become a really hard task (i.e. subjective). Moreover, weights are used to compute a unique real value by a linear combination of evaluated factors, summarizing evaluation results into a unique value which may hide interesting facts about algorithms performance.

The work of (Neilson & Yang, 2008) points out that the conclusions made upon evaluation results by considering only a few stereo images, may lack of statistical validity. Consequently, a comparison of algorithms should involve a large quantity of stereo image pairs. However, the required computational effort for processing such amount of data makes this methodology not suited for a final user. In addition, several stereo image pairs used in this study were generated synthetically. Some of them may even challenge the depth perception of human observer. This arise a concern about if such evaluation scenario may be considered unrealistic.

Although the Middlebury's methodology (Scharstein & Szeliski, 2000) is widely used, it has some shortcomings: the use of the BMP measure along with the evaluation model. The BMP is a binary function using a threshold that may impact the evaluation results. Moreover, the BMP measure ignores the inverse relation between depth and disparity. Consequently, the BMP measure may not be suited to properly evaluate the accuracy of a disparity map. In regard to the evaluation model, it is possible that two or more algorithms have equal error scores in a same error criterion. In this case, the rank assigned to these algorithms became arbitrary. Moreover, different algorithms may have the same average rank. In addition, the cardinality of a set of top ranked is a free parameter. This fact may lead to discrepancies or controversy among researchers about the state-of-the-art in the field. Thus, the above shortcomings may introduce bias to evaluation results, and distort the interpretation of the state-of-the-art of stereo correspondence algorithms by the research community.

The A^* Groups methodology (Cabezas et al., 2012a) extends the A^* methodology, introducing the capability of exhaustively analysing an entire set of algorithms, by computing disjoint groups associated to different levels of performance. However, it may lack of a method for supporting an algorithm selection from a set of comparable algorithms.

On the other hand, the Middlebury's online ranking (Scharstein & Szeliski, 2012) is a point of information sharing, on which is possible to find the latest contributions on disparity estimation algorithms. It makes public a large ground-truth imagery test-bed of scenes with complex geometry and varying capturing conditions.

The work of (Tombari et al., 2010a) considers both estimation accuracy and time performance as evaluation criteria. However, the evaluation model of the Middlebury's methodology is used. Consequently, the drawbacks mentioned above about rank issues may apply to this work.

The A^* methodology, introduced in (Cabezas and Trujillo, 2011) has a proper theoretical support. It is able to determine, without ambiguity, which algorithms show a superior performance. Moreover, its nature makes it suited to include several factors (i.e. such as several error measures, density ratio of an estimated map, or false positives and false negatives of an specific task performed over estimated maps, among others). However, this methodology fails in considering an evaluation scenario on which a user is interested in an exhaustive evaluation of the entire set of algorithms.

Recommendations

Based on the drawbacks and advantages identified on the discussed methodologies, it is possible to define a set of principles that a user should keep in mind when he/she addresses an evaluation process. These principles are: planned scenario, domain relevance, fairness, and objectivity.

Planned scenario: Goals and scope of the evaluation should be clearly identified a-priori to the evaluation process. This principle will assist the selection of error measures, error criteria, algorithms and evaluation model.

Domain relevance: The image content should be relevant to the domain on which the algorithm is going to be used. This principle should be taken into account for the selection of imagery test-bed.

Fairness: The comparison of disparity estimation algorithms should involve not only the estimated disparity maps, but also resources (e.g. time, hardware) used in estimation process. This principle should be taken into account for selecting algorithms under evaluation.

Objectivity: A subjective description or interpretation of quantitative evaluation results should be avoided. This principle should be taken into account for selecting an evaluation model, as well as in the interpretation of results.

FINAL REMARKS

The quantitative evaluation of disparity estimation algorithms is a fundamental issue in a research process, since it brings arguments to choose and/or developing a particular approach to a specific problem. Although, there are a plethora of disparity estimation algorithms, contributions on quantitative evaluation of estimated disparity maps are not so many. Among these works, only few contributions have consolidated so far a large impact on the research community. In fact, most of contributions in the literature, in regard to quantitative evaluation, are focused on ground-truth based approaches. Although the prediction-error approach arose first as an option for evaluating disparity maps, it is not widely used by the community. Perhaps innovative works in this subject may change that trend, considering that the generation of domain specific disparity ground-truth data is a bottleneck in the evaluation process. Works addressing the parameters setting problem are not few, but, works showing how, or to what extent, an achieved setup for a particular imagery could be useful in a different imagery set, are still required. Moreover, it could be of interest to the research community, a development of works incorporating into stereo systems the

Figure 2. Illustration of the evaluation process using the: (a) Middlebury's methodology, (b) A* methodology

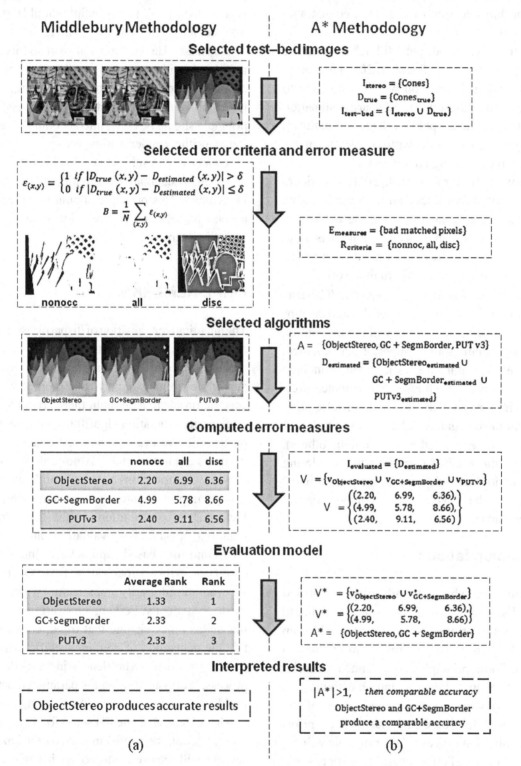

capability of determining, in an automatic way, when a change of parameters is required and how to achieve it in a real scenario (i.e. in ground-truth absence). Finally, a unique evaluation methodology properly handling all evaluation requirements of users may not exist. This may be due do to in existing methodologies, evaluation scenarios are fixed beforehand. However, evaluation requirements are changing according to particularities of conducted research. Thus, it may be a lack of evaluation environments where users are allowed to take decisions about the setup of evaluation scenarios.

REFERENCES

Benoit, A., Le Callet, P., Campisi, P., & Cousseau, R. (2008). Using disparity for quality assessment of stereoscopic images. In *Proceedings International Conference on Image Processing,* (pp. 389-392). IEEE Computer Society.

Bleyer, M., & Chambon, S. (2010). *Does color really help in dense stereo matching?* Paper presented at International Symposium 3D Data Processing, Visualization and Transmission.

Bleyer, M., & Gelautz, M. (2004). A layered stereo algorithm using image segmentation and global visibility constraints. In *Proceedings International Conference Image Processing,* (pp. 2997-3000). IEEE Computer Society.

Bleyer, M., Rhemann, C., & Rother, C. (2011). PatchMatch stereo- stereo matching with slanted support. In *Proceedings of the British Machine Vision Conference* (pp. 1-11).

Cabezas, I., Padilla, V., & Trujillo, M. (2011). A measure for accuracy disparity maps evaluation. In *Proceedings of the Iberoamerican Congress on Pattern Recognition. Lecture Notes in Computer Science, 7042,* 223–231. doi:10.1007/978-3-642-25085-9_26

Cabezas, I., Padilla, V., Trujillo, M., & Florian, M. (2012b). A non-linear quantitative evaluation approach for disparity estimation. In *Proceedings of the World Congress on Automation,* in press.

Cabezas, I., & Trujillo, M. (2011). A non-linear quantitative evaluation approach for disparity estimation. In *Proceedings of the International Conference on Computer Vision. Theory and Applications* (pp. 704–709).

Cabezas, I., Trujillo, M., & Florian, M. (2012a). A non-linear quantitative evaluation approach for disparity estimation. In *Proceedings of the International Conference on Computer Vision, Theory and Applications* (pp. 154–163).

Felzenszwalb, P., & Huttenlocher, D. (2004). Efficient belief propagation for early vision. In *Proceedings of Computer Vision and Pattern Recognition* (pp. 261–268). IEEE Computer Society.

Fusiello, A., Trucco, E., & Verri, A. (2000). A compact algorithm for rectification of stereo pairs. *Machine Vision and Applications, 12*(1), 16–22. doi:10.1007/s001380050120

Gallup, D., Frahm, J., Mordohai, P., & Pollefeys, M. (2008). Variable baseline/resolution stereo. In *Proceedings of Computer Vision and Pattern Recognition* (pp. 1–8). IEEE Computer Society.

Haeusler, R., & Klette, R. (2010). Benchmarking stereo data (not the matching algorithms). In M. Goesele, S. Roth, A. Kuijper, B. Schiele, & K. Schindler (Eds.), *Proceedings of the 32nd DAGM Conference on Pattern Recognition,* (pp. 383-392). Springer-Verlag.

Hartley, R., & Zisserman, A. (2003). *Multiple view geometry in computer vision.* Cambridge, UK: Cambridge University Press.

Hirschmüller, H., & Scharstein, D. (2009). Evaluation of stereo matching costs on images with radiometric differences. *IEEE Transactions on Pattern Analysis and Machine Intelligence, 31*(9), 1582–1599. doi:10.1109/TPAMI.2008.221

Hoyos, A., Congote, J., Barandiaran, I., Acosta, D., & Ruiz, O. (2011). Statistical tuning of adaptive-weight depth map algorithm. In A. Berciano, D. Diaz-Pernil, H. Molina-Abril, P. Real, & W. Kropatsch (Eds.), *Proceedings of the 14th International Conference on Computer Analysis of Images and Patterns: Part II* (pp. 563-572). Springer-Verlag.

Kelly, P., O'Connor, N., & Smeaton, A. (2007). A framework for evaluating stereo-based pedestrian detection techniques. *IEEE Transactions on Circuits and Systems for Video Technology, 18*(8), 1163–1167. doi:10.1109/TCSVT.2008.928228

Kolmogorov, V., & Zabih, R. (2001). Computing visual correspondence with occlusion using graph cuts. In *Proceedings of Eight International Conference on Computer Vision*, (pp. 508-515). IEEE Computer Society.

Kolmogorov, V., & Zabih, R. (2001). Computing visual correspondence with occlusion using graph cuts. In *Proceedings of European Conference on Computer Vision*, (pp. 82-96). Springer-Verlag.

Kostlivá, J., Cech, J., & Sara, R. (2003). Dense stereo matching algorithm performance for view prediction and structure reconstruction. In J. Bigun & T. Gustavsson (Eds.), *Proceedings of the 13th Scandinavian Conference on Image Analysis,* (pp. 101-107). Springer-Verlag.

Kostlivá, J., Cech, J., & Sara, R. (2007). *Feasibility boundary in dense and semi-dense stereo matching. Computer Vision and Pattern Recognition* (pp. 1–8). IEEE Computer Society.

Leclerc, Y. G., Luong, Q., & Fua, P. (2000). Measuring the self-consistency of stereo algorithms. In *European Conference on Computer Vision,* (pp. 282–298). Springer-Verlag.

Lü, C., Huang, J., & Shen, Y. (2011), Subjective assessment of noised stereo images. *International Conference on Multimedia Technology,* (pp.783-785).

Malpica, W. S., & Bovik, A. C. (2009). Range image quality assessment by structural similarity. *IEEE International Conference on Acoustics, Speech and Signal Processing,* (pp. 1149-1152).

Mittal, A., Moorthy, A. K., Ghosh, J., & Bovik, A. C. (2011). *Algorithmic assessment of 3D quality of experience for images and videos. Digital Signal Processing Workshop and IEEE Signal Processing Education Workshop (DSP/SPE)* (pp. 338–343). IEEE Computer Society.

Morales, S., & Klette, R. (2009). A third eye for performance evaluation in stereo sequence analysis. In X. Jiang & N. Petkov (Eds.), *Proceedings of the 13th International Conference on Computer Analysis of Images and Patterns* (pp. 1078-1086). Springer-Verlag.

Morales, S., & Klette, R. (2010). Ground truth evaluation of stereo algorithms for real world applications. In R. Koch & F. Huang (Eds.), *Proceedings of the 2010 International Conference on Computer Vision: Part II,* (pp. 152-162). Springer-Verlag.

Neilson, D., & Yang, Y. (2008). Evaluation of constructable match cost measures for stereo correspondence using cluster ranking. In *Computer Vision and Pattern Recognition* (pp. 1–8). IEEE Computer Society. doi:10.1109/CVPR.2008.4587692

Papadakis, N., & Caselles, V. (2010). Multi-label depth estimation for graph cuts stereo problems. *Journal of Mathematical Imaging and Vision, 38*(1), 70–82. doi:10.1007/s10851-010-0212-8

Rehman, A., & Wang, Z. (2010). Reduced-reference SSIM estimation. *IEEE International Conference on Image Processing (ICIP)* (pp. 289-292).

Scharstein, D., & Szeliski, R. (2002). A taxonomy and evaluation of dense two-frame stereo correspondence algorithms. *International Journal of Computer Vision*, *47*, 7–42. doi:10.1023/A:1014573219977

Scharstein, D., & Szeliski, R. (2003). High-accuracy stereo depth maps using structured light. In *Computer Vision and Pattern Recognition* (pp. I–195–I–202). IEEE Computer Society. doi:10.1109/CVPR.2003.1211354

Scharstein, D., & Szeliski, R. (2012). *Middlebury stereo evaluation - Version 2*. Retrieved April 30th, 2012, from http://vision.middlebury.edu/stereo/eval/

Strecha, C., von Hansen, W., Van Gool, L., Fua, P., & Thoennessen, U. (2008). On benchmarking camera calibration and multi-view stereo for high resolution imagery. In *Proceedings Conference on Computer Vision and Pattern Recognition*, (pp. 1-8). IEEE Computer Society.

Szeliski, R. (1999). Prediction error as a quality metric for motion and stereo. In *International Conference on Computer Vision*, Vol. 2, (pp. 781–788). IEEE Computer Society.

Szeliski, R., & Zabih, R. (2000). An experimental comparison of stereo algorithms. In *Proceedings of the International Workshop on Vision Algorithms*, (pp. 1–19). Springer-Verlag.

Tombari, F., Di Stefano, L., Mattoccia, S., & Mainetti, A. (2010b). A 3D reconstruction system based on improved spacetime stereo. In *Proceedings of International Conference on Control, Automation, Robotics and Vision*, (pp. 1886-1896). IEEE Computer Society.

Tombari, F., Mattoccia, S., & Di Stefano, L. (2010a). Stereo for robots: Quantitative evaluation of efficient and low-memory dense stereo algorithms. In *Proceedings of International Conference Control Automation Robotics and Vision*, (pp. 1231–1238). IEEE Computer Society.

Tombari, F., Mattoccia, S., Di Stefano, L., & Addimanda, E. (2008). Classification and evaluation of cost aggregation methods for stereo correspondence. In *Proceedings Conference on Computer Vision and Pattern Recognition*, (pp. 1-8). IEEE Computer Society.

Van der Mark, W., & Gavrila, D. (2006). Real-time dense stereo for intelligent vehicles. *IEEE Transactions on Intelligent Transportation Systems*, *7*(1), 38–50. doi:10.1109/TITS.2006.869625

Wang, Z., & Bovik, A. C. (2009). Mean squared error: Love it or leave it? A new look at signal fidelity measures. *Signal Processing Magazine*, *26*(1), 98–117. doi:10.1109/MSP.2008.930649

Yang, Q., Wang, L., Yang, R., Stewénius, H., & Nister, D. (2008). Stereo matching with color-weigthed correlation, hierarchical belief propagation and occlusion handling. In *Transactions on Pattern Analysis and machine. Intelligence*, *31*(3), 492–504.

Yoon, K., & Kweon, I. (2007). Stereo matching with the distinctive similarity measure. *Proceedings of International Conference on Computer Vision*, (pp. 1-7). IEEE Computer Society.

Zhang, Z., Hou, C., Shen, L., & Yang, J. (2009). An objective evaluation for disparity map based on the disparity gradient and disparity acceleration. In *Proceedings of International Conference on Information Technology and Computer Science* (pp. 452-455).

KEY TERMS AND DEFINITIONS

All: Evaluation criterion related to the whole image.

Corresponding Points: Projections on a stereo image of a common 3D point in space.

Disc: Evaluation criterion related to points nearby to depth discontinuities areas.

Disparity: Vector related corresponding points.

Disparity Map: Matrix of disparity magnitudes from the reference view, to the target view, in a stereo image pair.

Error Criteria: Set of challenging image regions for stereo algorithms on which the error measures are gathered.

Error Measure: Function that compares two images and provides a score of similarity between them.

Evaluation Model: Set of functions that operate on error scores to produce an indicator of algorithms performance.

Nonocc: Evaluation criterion related to points laying in non-occluded areas.

Test-Bed Imagery: Image data set on which the evaluation is conducted.

Chapter 11
Real–Time Structure Estimation in Dynamic Scenes Using a Single Camera

Ashwin P. Dani
University of Florida, USA

Nic Fischer
University of Florida, USA

Zhen Kan
University of Florida, USA

Warren E. Dixon
University of Florida, USA

ABSTRACT

In this chapter, an online method is developed for estimating 3D structure (with proper scale) of moving objects seen by a moving camera. In contrast to traditionally developed batch solutions for this problem, a nonlinear unknown input observer strategy is used where the object's velocity is considered as an unknown input to the perspective dynamical system. The estimator is exponentially stable, and hence, provides robustness against modeling uncertainties and measurement noise from the camera. The developed method provides first causal, observer based structure estimation algorithm for a moving camera viewing a moving object with unknown time-varying object velocities.

INTRODUCTION

Recovering the structure (i.e., the 3D Euclidean coordinates of feature points) of a static or dynamic scene using a monocular camera can be highly beneficial in robotic applications. Structure and motion information can be useful for numerous tasks such as simultaneous localization and mapping in an unstructured/unknown environment, obstacle avoidance, multi-robot coordination and control, etc. In this chapter, borrowing ideas

from the nonlinear control and estimation theory, a real-time (causal) algorithm is developed to estimate the structure of independently moving objects viewed by a camera undergoing 6 degree of freedom (DoF) motion.

BACKGROUND

Recovering the structure of a stationary object viewed by a moving camera is called structure

DOI: 10.4018/978-1-4666-2672-0.ch011

from motion (SfM). A number of solutions to the SfM problem exist in the form of batch or offline methods (Kahl & Hartley, 2008; Oliensis, 2000; Strum & Triggs, 1996) and causal or online methods (Dahl, Nyberg, & Heyden, 2007; Dani, Fischer, Dixon, 2012; Dani, Kan, Fischer, Dixon, 2012; Dani & Dixon, 2010; Dani, Rifai, & Dixon, 2010; Dixon, Fang, Dawson, & Flynn, 2003; Jankovic, Ghosh, 1995). The fundamental concept behind SfM algorithms is triangulation. Since the object being observed is stationary, two rays projected onto consecutive images and the camera baseline form a triangle. If the object is not stationary, then the projections of the object will be from different locations in the fixed inertial frame; hence, triangulation is not feasible and standard SfM techniques cannot be used to recover the structure of a moving object using a moving camera (Avidan & Shashua, 2000).

This chapter presents a result for the case when the viewed object is in motion. We term this problem (SaMfM) estimation. In the pioneering work by Avidan and Shashua (2000) developed an offline method, termed trajectory triangulation, to recover the structure of a moving object using a moving camera. In Avidan and Shashua (2000), a batch algorithm is applied for points moving in straight lines or conic trajectories given five or nine views, respectively. In Kaminski and Teicher (2004), a batch algorithm is presented for object motions represented by more general curves. In Han and Kanade (2004), a factorization-based batch algorithm is proposed where objects are assumed to be moving with constant speed in a straight line, observed by a weak perspective camera. An algebraic geometry approach is presented in Vidal, Ma, Soatto, and Sastry (2006) to estimate the motion of objects up to a scale given a minimum number of point correspondences. Yuan and Medioni (2006) developed an approximate batch algorithm to estimate the structure and motion of objects by assuming that one of the feature points of the moving object lies on the static background. In Park, Shiratori, Matthews, and Sheikh (2010), a batch algorithm is designed which requires approximation of the trajectories of a moving object using a linear combination of discrete cosine transform (DCT) basis vectors.

Batch algorithms use an algebraic relationship between 3D coordinates of points in the camera coordinate frame and corresponding 2D projections on the image frame collected over multiple images to estimate the structure; hence, batch algorithms are not useful in real-time control algorithms. For visual servo control or video-based surveillance tasks, online structure estimation algorithms are beneficial. The objective of an unknown input observer (UIO) for SaMfM is to estimate the structure of moving objects from a stream of images described using a continuous dynamical model, instead of algebraic relationships and geometric constraints. The use of a dynamical model enables the design of an online/causal algorithm which uses data from images up to the current time step. Avidan and Shashua (2000) point out that structure estimation of a moving object using a moving camera can only be obtained if some assumptions are made on the trajectories of the moving object. Recently, a causal algorithm is presented in Dani, Kan, Fischer, and Dixon (2010) to estimate the structure and motion of objects moving with constant linear velocities observed by a moving camera with known camera motions. In this chapter, a UIO provides an online algorithm which relaxes the constant object linear velocity assumption and does not require the time-varying object velocities to be approximated by a basis. A preliminary version of the method is presented in Dani, Kan, Fischer, and Dixon (2011).

Several UIO algorithms are present in literature for estimating the state when an exogenous time-varying unknown input is present in the system. For linear systems UIOs can be found in (Bhattacharyya, 1978; Darouach, Zasadzinski, & Xu, 1994; Guan & Said, 1991; Hou & Muller, 1992; Tsui, 1996). Linear UIO algorithms are extended to nonlinear systems in (Moreno, 2000; Koenig & Mammar, 2001; Pertew, Marquez, & Zhao, 2005; Chen & Saif, 2006a, 2006b; Liu, Farza, & M'Saad, 2006; Fridman, Shtessel, Edwards, & Yan, 2007; Moreno & Dochain, 2008; Hammouri & Tmar, 2010). In Moreno (2000), an UIO is designed for single-input single-output (SISO) nonlinear systems. In Pertew, Marquez, and Zhao (2005), a nonlinear UIO is presented based on H_∞ optimization. The observer is called a dynamic UIO which provides an extra degree of design freedom but increases the order of the system. In Chen and Saif (2006b), a nonlinear UIO is presented for a class of nonlinear systems based on an linear matrix inequality (LMI) approach but no necessary and sufficient observer existence conditions are developed. In Liu, Farza, and M'Saad (2006), a high gain observer for a class of nonlinear systems is presented for state and unknown input estimation but is achieved only up to a small bound which can be reduced by increasing the observer gains, (i.e., a uniformly ultimately bounded (UUB) result). A higher order sliding mode UIO is presented for nonlinear systems in Fridman, Shtessel, Edwards, and Yan (2007) which requires the original nonlinear system to satisfy geometric conditions for transforming the system into the Brunovsky canonical form. Based on a detectability notion, a sufficient condition for the existence of an UIO is derived by Moreno & Dochain (2008) for state-affine systems up to an output injection. Based on the geometric approach of Bhattacharyya (1978), necessary and sufficient conditions are derived in Hammouri and Tmar (2010) for the existence of an UIO for state affine systems up to a nonlinear unknown input dynamics; hence, the UIO can be used for a larger class of unknown inputs. UIOs are used extensively in fault detection and isolation for various classes of systems such as linear systems (Patton & Chen, 1993), control affine systems (De Persis & Isidori, 2001), bilinear systems (Hammouri, Kabore, & Kinnaert,

2001), and nonlinear systems (Koenig & Mammar, 2001; Hammouri, Kinnaert, & El Yaagoubi, 1999). In Koenig & Mammar (2001), an unknown input observer for a class of nonlinear systems is presented for fault diagnosis. The observer design relies on a coordinate transformation which decouples the nonlinear system into a system independent of unknown inputs and a system with the states that can be expressed as linear combinations of the outputs and the states of the first subsystem. The observer gain is obtained by solving a parametric Lyapunov equation which can be challenging to compute (Chen & Saif, 2006).

The contribution of this chapter is to provide a causal algorithm for estimating the structure of a moving object using a moving camera with relaxed assumptions on the object's motion. In the relative rigid body motion dynamics, the moving object's linear velocity can be viewed as an exogenous time-varying disturbance and is considered as an unknown input. Given the unique UIO method, no assumptions are made on the minimum number of tracked feature points or minimum number of views required to estimate the structure. In fact, the developed approach only requires single tracked feature point and camera velocity information.

STRUCTURE ESTIMATION OF A MOVING OBJECT

In this section, a UIO algorithm for SaMfM is presented. Specifically, objectives are defined, a pinhole camera model is introduced and a differential equation formulation of the problem is provided. A UIO is then designed to estimate the unknown structure given measurements from camera images and camera velocities. A procedure for selecting observer gains is developed using a linear matrix inequality (LMI) formulation. Finally, numerical simulations demonstrate the effectiveness of the algorithm.

UNKNOWN INPUT OBSERVER FOR STRUCTURE ESTIMATION

Structure and Motion from Motion (SaMfM) Objective

The objective of SaMfM is to recover the structure (i.e., Euclidean coordinates with a scaling factor) and motion (i.e., velocities) of moving objects observed by a moving camera, when all camera velocities are assumed to be known. In this section, an observer is presented which estimates the structure of a moving object with respect to an intrinsically calibrated moving camera. At least one feature point on the object is assumed to be tracked in each image frame, and camera velocities are recorded using sensors such as an Inertial Measurement Unit (IMU).

Euclidean Space to Image Space Mapping

Consider a scenario depicted in Figure 1 where a moving camera views moving point objects (such as feature points on a rigid object). In Figure 1, an inertial reference frame is denoted by $\mathcal{F}^*$. The reference frame $\mathcal{F}^*$ can be attached to the camera at the location corresponding to an initial point in time t_0 where the object is in the camera field of view (FOV). i.e., $\mathcal{F}^*$ is identical to $\mathcal{F}_c(t_0)$. After the initial time, a reference frame $\mathcal{F}_c$ attached to a pinhole camera undergoes some rotation $\bar{R}(t) \in SO(3)$ and translation $\bar{x}_f(t) \in \mathbb{R}^3$ away from $\mathcal{F}^*$.

The Euclidean coordinates $\bar{m}(t) \in \mathbb{R}^3$ of a point observed by a camera expressed in the camera frame $\mathcal{F}_c$ and the respective normalized Euclidean coordinates $m(t) \in \mathbb{R}^3$ are defined as

$$\bar{m}(t) \triangleq \left[X(t), \quad Y(t), \quad Z(t) \right]^T, \tag{1}$$

$$m(t) \triangleq \left[\frac{X(t)}{Z(t)}, \quad \frac{Y(t)}{Z(t)}, \quad 1 \right]^T. \tag{2}$$

Consider a closed and bounded set $Y \subset \mathbb{R}^3$. To facilitate the subsequent development, the state vector $x(t) = [x_1(t), \ x_2(t), \ x_3(t)]^T \in Y$ is constructed from (2) as

$$x = \left[\frac{X}{Z}, \quad \frac{Y}{Z}, \quad \frac{1}{Z} \right]^T. \tag{3}$$

Using projective geometry, the normalized Euclidean coordinates $m(t)$ can be related to the pixel coordinates in the image space as

$$q = A_c m \tag{4}$$

where $q(t) = \begin{bmatrix} \bar{u}(t) & \bar{v}(t) & 1 \end{bmatrix}^T$ is a vector of the image-space feature point coordinates $\bar{u}(t)$, $\bar{v}(t) \in \mathbb{R}$ defined on the closed and bounded set $I \subset \mathbb{R}^3$, and $A_c \in \mathbb{R}^{3\times3}$ is a constant, known, invertible intrinsic camera calibration matrix (Ma, Soatto, Kosecka, & Sastry, 2004). Since A_c is known, the expression in (4) can be used to re-

Figure 1. Moving camera looking at a moving object

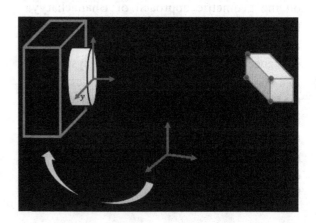

cover $m(t)$, which can be used to partially reconstruct the state $x(t)$ so that the first two components of $x(t)$ can be determined. The states defined in (3) contain unknown structure information about the object. The object structure can be determined by estimating the Euclidean coordinates of multiple feature points on the object; however, without loss of generality the following UIO development is applied to a single feature point, with Euclidean coordinates that can be recovered from the state $x(t)$ in (3).

Assumption 1: The relative Euclidean distance $Z(t)$ between the camera and the feature points observed on the object is upper and lower bounded by some known positive constants (i.e., the object remains some finite distance away from the camera). Therefore, the definition in (3) can be used to assume that

$$\bar{x}_3 \geq x_3(t) \geq \underline{x}_3 \tag{5}$$

where $\bar{x}_3, \underline{x}_3 \in \mathbb{R}$ denote known positive bounding constants. Likewise, since the image coordinates are constrained (i.e., the object is assumed to remain within the camera field of view (FOV)), the relationships in (2)-(4) along with the fact that A_c is invertible can be used to conclude that

$$\bar{x}_1 \geq |x_1(t)| \geq \underline{x}_1, \qquad \bar{x}_2 \geq |x_2(t)| \geq \underline{x}_2$$

where $\bar{x}_1, \bar{x}_2, \underline{x}_1, \underline{x}_2 \in \mathbb{R}$ denote known positive bounding constants.

Camera Motion Model and State Space Dynamics

Consider a moving camera viewing a moving point q. As shown in Figure 1, the point q can be expressed in the coordinate system $\mathcal{F}_c$ as

$$\bar{m} = \bar{x}_f + \bar{R}x_{oq} \tag{6}$$

where x_{oq} is a vector from the origin of coordinate system $\mathcal{F}^*$ to the point q expressed in the coordinate system $\mathcal{F}^*$. Differentiating (6), the relative motion of q as observed in the camera coordinate system can be expressed by the following kinematics (Ma, et.al., 2004)

$$\dot{\bar{m}} = [\omega]_\times \bar{m} + v_r \tag{7}$$

where $\bar{m}(t)$ is defined in (1), $[\omega]_\times \in \mathbb{R}^{3\times3}$ denotes a skew symmetric matrix formed from the angular velocity vector of the camera $\omega(t) = [\omega_1 \quad \omega_2 \quad \omega_3]^T \in W$, and $v_r(t)$ represents the relative velocity of the camera with respect to the moving point, defined as

$$v_r \triangleq v_c - \bar{R}\bar{v}_p. \tag{8}$$

In (8), $v_c(t) = [v_{cx} \quad v_{cy} \quad v_{cz}]^T \in V_c \subset \mathbb{R}^3$ denotes the camera velocity expressed in camera reference frame $\mathcal{F}_c$, $\bar{R}\bar{v}_p(t) \triangleq v_p(t) = [v_{px} \quad v_{py} \quad v_{pz}]^T \in V_p \subset \mathbb{R}^3$ denotes the velocity of the moving point q expressed in camera reference frame $\mathcal{F}_c$, and $\bar{v}_p(t) \triangleq [\bar{v}_{px} \quad \bar{v}_{py} \quad \bar{v}_{pz}]^T \in \bar{V}_p \subset \mathbb{R}^3$ denotes the velocity of the moving point in the inertial reference frame $\mathcal{F}^*$. The sets W, V_c and V_p are closed and bounded sets such that $W \subset \mathbb{R}^3, V_c \subset \mathbb{R}^3$ and $V_p \subset \mathbb{R}^3$. Let the linear and angular camera velocities be denoted by $u = [v_c \quad \omega]^T$.

The states defined in (3) contain unknown structure information of the object. To recover the 3D structure, the state $x(t)$ should be estimated. Using (3) and (7), the dynamics of the state vector $x(t)$ are expressed as

$$\dot{x}_1 = \Omega_1 + (v_{cx} - x_1 v_{cz})x_3 - v_{px}x_3 + x_1 v_{pz}x_3,$$

$$\dot{x}_2 = \Omega_2 + (v_{cy} - x_2 v_{cz})x_3 - v_{py}x_3 + x_2 v_{pz}x_3,$$

$$\dot{x}_3 = -v_{cz}x_3^2 - (x_2\omega_1 - x_1\omega_2)x_3 + v_{pz}x_3^2,$$

$$y = [x_1 \quad x_2]^T. \tag{9}$$

where $\Omega_1(u,y), \Omega_2(u,y) \in \mathbb{R}$ are defined as

$$\Omega_1(u,y) \triangleq -x_1 x_2 \omega_1 + \omega_2 + x_1^2 \omega_2 - x_2 \omega_3,$$

$$\Omega_2(u,y) \triangleq -\omega_1 - x_2^2 \omega_1 + x_1 x_2 \omega_2 + x_1 \omega_3,$$

Assumption 2: The camera velocities $\omega(t)$, and $v_c(t)$, the object velocity $v_p(t)$, and the feature points $y(t)$ are assumed to be upper bounded by constants.

Remark 1: The dynamics in (9) are not observable if: a) the camera is stationary, i.e., $v_c(t) = 0, \omega(t) = 0$, or b) the camera moves along the ray projected by the feature point on the image, i.e., $\left(v_{cx} - v_{px} - x_1\left(v_{cz} - v_{pz}\right)\right) = \left(v_{cy} - v_{py} - x_2\left(v_{cz} - v_{pz}\right)\right) = 0.$

Nonlinear Unknown Input Observer

In this section a UIO is developed for a class of nonlinear systems. Additionally, conditions on the moving object's velocity are stated so that the model in (9) satisfies the requirements of the UIO. The system in (9) can be represented in the form

$$\dot{x} = f(x,u) + g(y,u) + Dd$$

$$y = Cx \tag{10}$$

where $x(t) \in \mathbb{R}^3$ is a state of the system, $u(t) \in \mathbb{R}^6$ is a measurable control input, $d(t) \in \mathbb{R}$ is an unmeasurable disturbance input, $y(t) \in \mathbb{R}^2$ is output of the system, the functions $f : \mathbb{R}^3 \times \mathbb{R}^6 \to \mathbb{R}^3$ and $g : \mathbb{R}^2 \times \mathbb{R}^6 \to \mathbb{R}^3$ are given by

$$f(x,u) = \begin{bmatrix} (v_{cx} - x_1 v_{cz})x_3 \\ (v_{cy} - x_2 v_{cz})x_3 \\ -x_3^2 v_{cz} - x_2 x_3 \omega_1 + x_1 x_3 \omega_2 \end{bmatrix},$$

$$g(y,u) = \begin{bmatrix} -x_1 x_2 \omega_1 + (1 + x_1^2)\omega_2 - x_2 \omega_3 \\ -(1 + x_2^2)\omega_1 + x_1 x_2 \omega_2 + x_1 \omega_3 \\ 0 \end{bmatrix},$$

$$C = \begin{bmatrix} 1 & 0 & 0 \\ 0 & 1 & 0 \end{bmatrix}$$

is full row rank and $D \in \mathbb{R}^{3\times 1}$ is full column rank. The function $f(x,u)$ satisfies following Lipschitz conditions (Yaz & Azemi, 1993; Xie & Khargonekar, 2010)

$$\left\| f(x,u) - f(\hat{x},u) \right\| \leq \gamma_1 \left\| x - \hat{x} \right\| \tag{11}$$

$$\left\| f(x,u) - f(\hat{x},u) - A\left(x - \hat{x}\right) \right\| \leq \left(\gamma_1 + \gamma_2\right)\left\| x - \hat{x} \right\| \tag{12}$$

where $\gamma_1, \gamma_2 \in \mathbb{R}^+$, $A \in \mathbb{R}^{3\times 3}$, and $\hat{x}(t) \in \mathbb{R}^3$ is an estimate of the unknown state $x(t)$. Larger γ_1 would mean that the UIO is stable even in the presence of fast moving nonlinear dynamics in $f(x,u)$, i.e., for larger camera velocities.

The UIO objective is to design an exponentially converging state observer to estimate $x(t)$

in the presence of an unknown input $d(t)$. To quantify this objective an estimation error is defined as

$$e(t) \triangleq \hat{x}(t) - x(t). \tag{13}$$

Based on (13) and the subsequent stability analysis, the UIO for the system in (10) is designed as

$$\dot{z} = Nz + Ly + M\overline{f}(\hat{x}, u) + Mg(y, u)$$

$$\hat{x} = z - Ey \tag{14}$$

where $\overline{f}(x, u) = f(x, u) - Ax$, $z(t) \in \mathbb{R}^3$ is an auxiliary signal, the matrices $N \in \mathbb{R}^{3\times3}$, $L \in \mathbb{R}^{3\times2}$, $E \in \mathbb{R}^{3\times2}$, $M \in \mathbb{R}^{3\times3}$ are designed as

$$M = I_3 + EC$$

$$N = MA - KC$$

$$L = K(I_2 + CE) - MAE \tag{15}$$

where E is subsequently designed and $K \in \mathbb{R}^{3\times2}$ is a gain matrix which satisfies the inequality

$$Q = N^T P + PN + \left(\gamma_1 + \gamma_2\right)^2 \\ PMM^T P + 2I_3 < 0_{3\times3} \tag{16}$$

where $P \in \mathbb{R}^{3\times3}$ is a positive definite, symmetric matrix. Using (15), the equality

$$NM + LC - MA = 0_{3\times3} \tag{17}$$

is satisfied, where $0_{3\times3}$ denotes a zero matrix of the dimensions 3×3. E is selected as

$$E = F + YG \tag{18}$$

where $Y \in \mathbb{R}^{3\times2}$ can be chosen arbitrarily, and F and G are given by

$$F = -D(CD)^\dagger, \, G = \left(I_2 - (CD)(CD)^\dagger\right).$$

From the definitions of E, F, and G, $ECD = -D$ and the following equality is satisfied:

$$MD = \left(I_3 + EC\right)D = 0_{3\times1}. \tag{19}$$

Since $rank(CD) = 1$, the generalized pseudo inverse of the matrix CD exists and is given by

$$\left(CD\right)^\dagger = \left[\left(CD\right)^T \left(CD\right)\right]^{-1} \left(CD\right)^T.$$

Substituting (14) into (13), taking the time derivative of the result, and using (10) and (14) yields

$$\dot{e} = Nz + Ly + M\overline{f}(\hat{x}, u) - (I_3 + EC)Ax - \\ (I_3 + EC)\overline{f}(x, u) - (I_3 + EC)Dd. \tag{20}$$

Using (13) and (15), and the conditions in (17) and (19) the error system in (20) can be written as

$$\dot{e} = Ne + M\left(\overline{f}(\hat{x}, u) - \overline{f}(x, u)\right). \tag{21}$$

Theorem 1: The nonlinear unknown input observer in (14) is exponentially stable if the assumptions 1 and 2, and condition in (16) is satisfied.

Proof: Consider a positive definite (PD), radialy unbounded Lyapunov candidate function $V(e): \mathbb{R}^n \to \mathbb{R}$ defined as

$$V = e^T P e. \qquad (22)$$

which satisfies

$$\lambda_{min}\left(P\right)\parallel e \parallel^2 \leq V \leq \lambda_{max}\left(P\right)\parallel e \parallel^2$$
(23)

where $\lambda_{min}\left(\cdot\right)$ and $\lambda_{max}\left(\cdot\right)$ are the minimum and maximum eigenvalues of the matrix P. Taking the time derivative of (22) along the trajectories of (21) yields

$$\dot{V} = e^T\left(N^T P + PN\right)e + 2e^T PM\left(\overline{f}(\hat{x},u) - \overline{f}(x,u)\right)$$

$$\dot{V} \leq e^T\left(N^T P + PN\right)e + 2\parallel e^T PM \parallel$$
$$\left\| f(\hat{x},u) - f(x,u) - A\left(\hat{x} - x\right)\right\|$$

$$\dot{V} \leq e^T(N^T P + PN)e + 2\parallel e^T PM$$
$$\parallel \gamma_1 \parallel e \parallel + 2 \parallel e^T PM \parallel \gamma_2 \parallel e \parallel.$$

Using the norm inequality

$$2\gamma_i \parallel e^T PM \parallel \parallel e \parallel \leq \gamma_i^2 \parallel e^T PM$$
$$\parallel^2 + \parallel e \parallel^2, \quad \forall i = \{1,2\},$$

the upper bound on $\dot{V}$ is given by

$$\dot{V} \leq e^T\left(N^T P + PN\right)e + \left(\gamma_1^2 + \gamma_2^2\right)e^T PMM^T Pe + 2e^T e$$

$$\dot{V} \leq e^T\begin{pmatrix} N^T P + PN + \\ \left(\gamma_1^2 + \gamma_2^2\right)PMM^T P + 2I_3 \end{pmatrix}e$$

$$\dot{V} \leq e^T Q e. \qquad (24)$$

If the condition in (16) is satisfied, then $\dot{V} < 0$. Using (22)-(24) the upper bounds for $V(e)$ and $e(t)$ can be developed as

$$V\left(e\right) \leq V\left(e(0)\right)exp\left(-\xi t\right)$$

where $\xi = \dfrac{\lambda_{max}\left(Q\right)}{\lambda_{min}\left(P\right)}$ and

$$\parallel e(t) \parallel \leq \zeta \parallel e(0) \parallel exp\left(-\xi t\right)$$

where $\zeta = \dfrac{\lambda_{max}\left(P\right)}{\lambda_{min}\left(P\right)}$.

Remark 2: Model uncertainties can be represented by an additive disturbance term $d_1(t) \in \mathbb{R}^3$ in (10). Also, if the number of outputs, p, is less than the number of unknown disturbance inputs, q, then $q_1 < p$ number of unknown inputs can be represented by $d(t)$ and remaining $q - q_1$ can be represented by an additive disturbance $D_1 d_1(t) \in \mathbb{R}^{q-q_1}$. In each of the above mentioned cases, the estimation error will be uniformly ultimately bounded.

SaMfM UIO Conditions

In this section, conditions are developed for which an exact estimation of the 3D position of the feature point can be found. Conditions for choosing matrix A in (15) are stated and conditions on the object's trajectories are developed so that an exact estimation can be obtained. The UIO can be used even if the conditions on the object's velocity are not satisfied. In this special case, the state $x(t)$ can only be estimated up to a bounded estimation error, which can be reduced using an optimization procedure.

Conditions on Matrix A

Next, conditions for selecting A are provided based on the observer existence conditions. Sufficient conditions for the existence of the UIO in (14) can be summarized as: $\left(MA, C\right)$ is observable, $rank(CD) = rank(D) = q$, $MA - KC$ is

Hurwitz, and (16) holds. Since $rank(CD) = rank(D) = 1$, subsequent development indicates that observability of the pair (MA, C) is equivalent to the following rank condition (Darouach, Zasadzinski, & Xu, 1994; Hautus, 1983)

$$rank \begin{bmatrix} \lambda I_3 - A & D \\ C & 0_{2 \times 1} \end{bmatrix} = 4, \quad \forall \lambda \in C. \quad (25)$$

Thus, A in (15) should be selected so that (25) is satisfied and (MA, C) is observable. The condition in (25) facilitates the selection of A based on the system matrices and hence circumvents the computation of M for checking the observability of (MA, C). Another criteria on the selection of A is to minimize the Lipschitz constant in (11).

Conditions on Object Trajectories

Some assumptions on the motion of the camera and the viewed object are required to satisfy the constraints on disturbance inputs in (10). In this section, two specific scenarios are discussed.

Example 1: The camera is undergoing arbitrary purely translational motion, i.e., the angular velocities of the camera are zero and the object is moving along a straight line with time-varying unknown velocities. For this case, choices of $\bar{R}(t)$ and $\bar{v}_p(t)$ in (8) become

$$\bar{R} = I_3, \quad \text{and} \quad \bar{v}_p(t) = \begin{bmatrix} d_1(t) & 0 & 0 \end{bmatrix}^T, \quad \text{or}$$

$$\bar{v}_p(t) = \begin{bmatrix} 0 & d_1(t) & 0 \end{bmatrix}^T, \quad \text{or}$$

$$\bar{v}_p(t) = \begin{bmatrix} 0 & 0 & d_1(t) \end{bmatrix}^T, \quad \text{or}$$

$$\bar{v}_p(t) = \begin{bmatrix} d_1(t) & d_1(t) & 0 \end{bmatrix}^T, \quad \text{etc., where}$$

$d_1(t) \in \mathbb{R}$ is the unknown time-varying

object velocity. Physically, these object velocities would mean the object is moving along straight lines or moving along a circle.

Example 2: A downward looking camera is undergoing arbitrary translational motion along with the angular velocity along the Z-direction (an axis pointing downwards). The object is moving on a ground plane (i.e., X-Y plane) with arbitrary time-varying velocities. In this case, the rotation matrix $\bar{R}(t)$ is given by

$$\bar{R} = \begin{bmatrix} cos(\theta(t)) & sin(\theta(t)) & 0 \\ -sin(\theta(t)) & cos(\theta(t)) & 0 \\ 0 & 0 & 1 \end{bmatrix} = \begin{bmatrix} \bar{R}_s & 0_{2 \times 1} \\ 0_{1 \times 2} & 1 \end{bmatrix}$$

where $\theta(t) \in [-2\pi, 2\pi)$ is the rotation angle between the current camera coordinate frame and inertial coordinate frame, and $\bar{R}_s(t) \in \mathbb{R}^{2 \times 2}$ represents the upper left 2×2 block of the $\bar{R}(t)$. The object velocity in the inertial frame is represented as $\bar{v}_p(t) = \begin{bmatrix} \bar{d}_1(t) & \bar{d}_2(t) & 0 \end{bmatrix}^T$, where $\bar{d}_1(t), \bar{d}_2(t) \in \mathbb{R}$. The camera angular velocity is such that

$$\bar{R}\bar{v}_p = v_p \quad (26)$$

where $v_{p1}(t) = \begin{bmatrix} \bar{d}_3(t) & 0 & 0 \end{bmatrix}^T$ or $v_{p2}(t) = \begin{bmatrix} 0 & \bar{d}_4(t) & 0 \end{bmatrix}^T$, and $\bar{d}_3(t), \bar{d}_4(t) \in \mathbb{R}$ are unknown time-varying quantities. The equality in (26) can be achieved if $\bar{R}_s \begin{bmatrix} \bar{d}_1(t) & \bar{d}_2(t) \end{bmatrix}^T = v_{p3}$, where $v_{p3}(t) = \begin{bmatrix} \bar{d}_3(t) & 0 \end{bmatrix}^T$ or $v_{p3}(t) = \begin{bmatrix} 0 & \bar{d}_4(t) \end{bmatrix}^T$. For example, consider an object undergoing a circular motion with unknown and possibly time-varying radius in the X-Y plane with unknown time-varying velocities observed by a camera undergoing an arbitrary linear motion in the X-Y-Z plane and circular motion along the downward looking Z-direction.

Figure 2. Comparison of the actual and estimated x, y and z positions of a moving object with respect to a moving camera expressed in the camera frame for simulation case 1

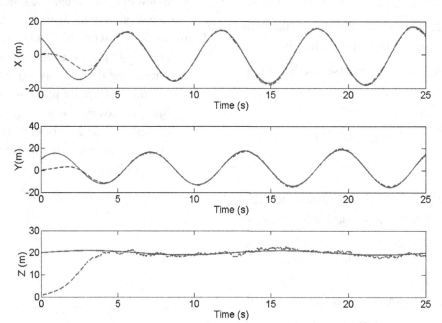

LMI Formulation for Observer Gain Selection

In this section, the condition in (16) is reformulated as an LMI feasibility problem. The matrices P, K and Y should be selected such that the sufficient condition for the observer error stability in (16) is satisfied. Substituting N and M from (15) into (16) yields

$$
\left(MA - KC\right)^{T} P + P\left(MA - KC\right) + 2I_{3} +
$$
$$
\left(\gamma_{1}^{2} + \gamma_{2}^{2}\right) P\left(I_{3} + EC\right)\left(I_{3} + EC\right)^{T} P < 0_{3\times3}.
$$
(27)

After using (18), the inequality in (27) can be expressed as

$$
A^{T}\left(I_{3} + FC\right)^{T} P + P\left(I_{3} + FC\right)A +
$$
$$
A^{T}C^{T}G^{T}P_{Y}^{T} + P_{Y}GCA - C^{T}P_{K}^{T} - P_{K}C
$$

$$
+2I_{3} + \left(\gamma_{1}^{2} + \gamma_{2}^{2}\right)\left(P + PFC + P_{Y}GC\right)
$$
$$
\left(P + PFC + P_{Y}GC\right)^{T} < 0_{3\times3}
$$
(28)

where $P_{Y} = PY$ and $P_{K} = PK$. For the observer synthesis, the matrices Y, K and $P > 0$ should be computed such that the matrix inequality in (16) is satisfied. Since $P > 0$, P^{-1} exists, and Y and K can be computed using $Y = P^{-1}P_{Y}$, and $K = P^{-1}P_{K}$. Using Schur's complement, the inequality in (28) can be transformed into the matrix inequality

$$
\begin{bmatrix} P_{1} & \beta R \\ \beta R^{T} & -I_{3} \end{bmatrix} < 0_{6\times6}
$$
(29)

where

$$
P_{1} = A^{T}\left(I_{3} + FC\right)^{T} P + P\left(I_{3} + FC\right)A +
$$
$$
A^{T}C^{T}G^{T}P_{Y}^{T} + P_{Y}GCA - C^{T}P_{K}^{T} - P_{K}C + 2I_{3},
$$

$$R = P + PFC + P_Y GC,$$

$$\beta = \sqrt{\gamma_1^2 + \gamma_2^2}.$$

The matrix inequality in (29) is an LMI in variables P, P_Y, and P_K. The LMI feasibility problem can be solved using standard LMI algorithms (Gahinet, Nemirovskii, Laub, & Chilali, 1994) and is a problem of finding P, P_Y and P_K such that β is maximized. Maximizing β is equivalent to maximizing γ_1 which means the observer can be designed for nonlinear functions with a larger Lipschitz constant. If the LMI in (29) is feasible, then a solution to (16) exists. Hence, the LMI feasibility problem is a sufficient condition for the stability of the observer.

Remark 3: In (14), the gain K is multiplied by the measurement $y(t)$. To reduce the effects of noise, the elements of gain K should not be large. Since $K = P^{-1}P_K$, the minimum eigenvalue of P should be maximized and the maximum eigenvalue of P_K should be minimized. This can be achieved via a multi-objective optimization problem with the objective function as $maximize \; \delta\lambda_{min}\left(P\right) - (1-\delta)\lambda_{max}\left(P_K\right)$ subject to the LMI in (29), where $\delta \in \left[0,1\right]$ (Boyd & Vandenberghe, 2004). In the presence of model uncertainties or disturbance inputs as stated in Remark 2, another optimization objective can be to choose gain K such that the $\mathcal{L}_2$ gain from the disturbance input to the estimation error is minimized.

Figure 3. Camera trajectory (blue), object trajectory (red), and comparison of the object trajectory with the estimated x, y and z positions of a moving object in the inertial frame for simulation case 2

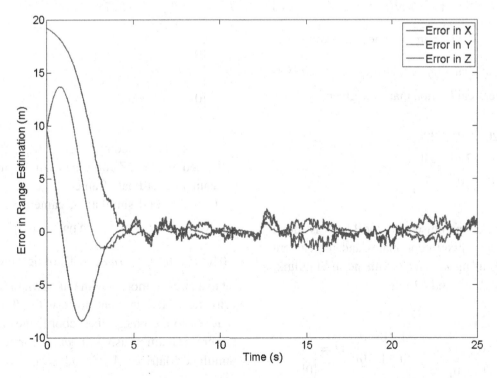

Figure 4. Error in the relative range estimation of the moving object in the camera reference frame for simulation case 2

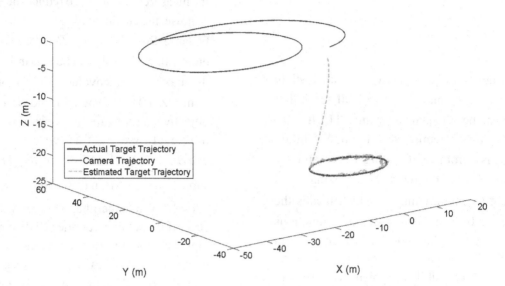

NUMERICAL SIMULATION

Two simulations are performed for a moving camera observing an object moving in a plane. For first simulation, camera velocities are given by $v_c(t) = \begin{bmatrix} 2 & 1 & 0.5cos(t/2) \end{bmatrix}^T m/s$, and $\omega(t) = \begin{bmatrix} 0 & 0 & 1 \end{bmatrix}^T rad/s$. The object velocity is selected such that $v_p(t) = \begin{bmatrix} 0.5sin(t) & 0 & 0 \end{bmatrix}^T m/s$. The camera calibration matrix is given by

$$A_c = \begin{bmatrix} 720 & 0 & 320 \\ 0 & 720 & 240 \\ 0 & 0 & 1 \end{bmatrix}.$$

Measurement noise with 20dB SNR is added to the image pixel coordinates and the camera velocities using $awgn()$ command in Simulink. Matrices A, C and D are

$$A = \begin{bmatrix} 0 & -1 & 2 \\ 1 & 0 & 1 \\ 0 & 0 & 0 \end{bmatrix}, C = \begin{bmatrix} 1 & 0 & 0 \\ 0 & 1 & 0 \end{bmatrix}, D = \begin{bmatrix} 1 \\ 0 \\ 0 \end{bmatrix}.$$

The matrix Y and the gain matrix K are computed using the LMI feasibility command '$feasp$' in Matlab and are given by

$$K = \begin{bmatrix} 0.8278 & 0 \\ 0 & 0.8278 \\ -1.5374 & 0 \end{bmatrix},$$

$$Y = \begin{bmatrix} 0 & 0 \\ 0 & -1 \\ 0 & -1.5374 \end{bmatrix}.$$

Figure 2 shows comparison of the actual and estimated X, Y and Z coordinates of the object in the camera coordinate frame.

In the second simulation, camera velocities are given by $v_c(t) = \begin{bmatrix} 2 & 1 + 0.02t & -0.01 \end{bmatrix}^T m/s$, $\omega(t) = \begin{bmatrix} 0 & 0 & 0.1 \end{bmatrix}^T rad/s$. The object motion is set to a circular motion with unknown radius and velocities. Measurement noise with 20dB SNR is added to the image pixel coordinates and the camera velocities using $awgn()$ command in Simulink. Matrices A, C and D are:

$$A = \begin{bmatrix} 0 & -0.1 & 2 \\ 0.1 & 0 & 1 \\ 0 & 0 & 0 \end{bmatrix},$$

$$C = \begin{bmatrix} 1 & 0 & 0 \\ 0 & 1 & 0 \end{bmatrix}, \; D = \begin{bmatrix} 0 \\ 1 \\ 0 \end{bmatrix}.$$

The matrix Y and the gain matrix K are computed using the LMI feasibility command $'feasp'$ in Matlab and are given by

$$K = \begin{bmatrix} 1.0667 & 0 \\ 0 & 1.0667 \\ 0 & 0.056 \end{bmatrix},$$

$$Y = \begin{bmatrix} -1 & 0 \\ 0 & 0 \\ -0.5598 & 0 \end{bmatrix}.$$

Figure 3 shows the comparison of actual and estimated X, Y, and Z coordinates of the object

in the Inertial coordinate frame in the presence of measurement noise. Figure 4 shows the structure estimation error. From the two results in two simulations, it can be seen that the observer is insensitive to the disturbance input.

EXPERIMENTS

Experiments are conducted on PUMA 560 serial manipulator to test the performance of the proposed estimator. Results of one of the test cases are presented in this section. A mvBlueFox-120a USB camera is rigidly attached to the end-effector of PUMA. A two-link robot mounted on the same platform as the PUMA is used as a moving object. The PUMA is controlled using a workstation to generate camera motion, and the camera linear and angular velocities are computed using the joint encoders and forward velocity kinematics of the PUMA. The data from the camera is recorded on the PUMA workstation computer at 25 frames per second for further processing and is time synchronized with the camera images.

Figure 5. Camera angular velocities

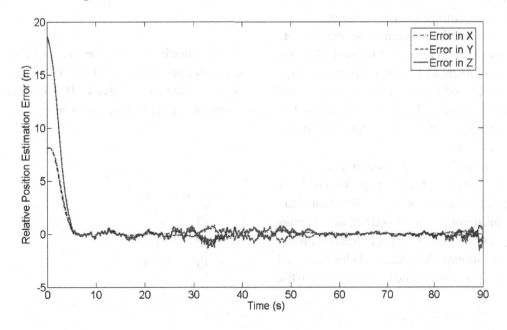

Figure 6. Camera linear velocities

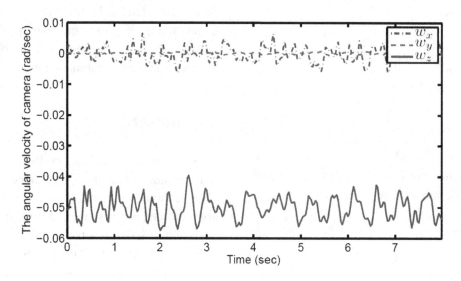

Two-link robot was controlled using another workstation which was time-synchronized with the PUMA workstation for ground truth verification. The video data recorded from the camera is post-processed to track a feature point on second link of the two-link robot using KLT feature point tracker. The image processing code is implemented in Visual C++ 2008 using OpenCV library. Since the PUMA and two-link robot was mounted on the same table, a precise location of the base of both the robots is computed and forward kinematics is used to compute the position of the camera and the feature point on a link of the two-link robot. Hence, a 3D distance between camera mounted on the PUMA and feature point on the two-link robot can be computed precisely and is used as a ground truth for testing the performance of the proposed estimator.

A downward looking camera attached to PUMA is moved with randomly chosen linear and angular velocities in X-Y-Z directions, and the two-link robot is commanded certain velocities in X-Y plane. The observer in (14)-(15) is implemented to estimate the relative 3D location of a moving feature point on the two-link robot in the camera reference frame. The observer parameters are selected as

$$A = \begin{bmatrix} 0 & -0.05 & 0 \\ 0.05 & 0 & -0.3 \\ 0 & 0 & 0 \end{bmatrix},$$

$$C = \begin{bmatrix} 1 & 0 & 0 \\ 0 & 1 & 0 \end{bmatrix}, \quad D = \begin{bmatrix} 1 \\ 0 \\ 0 \end{bmatrix}. \tag{30}$$

The matrix Y and the gain matrix K are computed using the LMI feasibility command *feasp* in Matlab to solve LMI in (29) using the matrices in (30) and are given by

$$K = \begin{bmatrix} 1.2204 & 0 \\ 0 & 1.2204 \\ 0.3731 & 0.056 \end{bmatrix},$$

$$Y = \begin{bmatrix} 0 & 0 \\ 0 & -1 \\ 0 & 7.4618 \end{bmatrix}.$$

Figure 5 and figure 6 shows the camera linear and angular velocities. The comparison of the estimated relative 3D position of the moving point in the camera reference frame and ground truth 3D position is shown in Figure 7.

In this section, some preliminary studies of the structure (3D location) of a moving object using a moving camera with known camera velocities is presented. More testing and experimental verification of the proposed algorithm is required to verify the performance of the algorithm in various real-life situations.

FUTURE RESEARCH DIRECTIONS

In this chapter, a method to estimate the relative structure of the feature points on a moving object with respect to a moving camera is presented. Special cases are presented for which the structure estimation can be exact but for general motions of the camera and the object, structure (i.e., the relative 3D coordinates of the feature points) estimation can be obtained within a certain ball

of the true structure. Future work will focus on improving the current method using multiple feature points to perform joint estimation so that the constraints on object and camera motions can be relaxed. Also, the development of camera-in-the-loop control algorithms for vision-based tracking of ground vehicle using unmanned aerial vehicles (UAVs) should be explored. In some applications, motion estimation along with the structure is also desirable; the current algorithm can be extended to infer the motion (velocities) of the moving object also. The structure estimates can be used to compute the motion of the objects using estimator, such as the one presented in Chitrakaran, Dawson, Dixon, and Chen (2005).

A UIO design with time-varying or state dependent D matrix so that a larger class of object motions can be considered. Existence conditions for a time-varying D matrix for state-affine systems have been developed recently in Hammouri and Tmar (2010). Extension of the result in Hammouri and Tmar (2010) to nonlinear systems would generalize the developed result for larger set of time varying object velocities. Using the

Figure 7. Comparison of the estimated and ground truth 3D position of the moving object measured in the moving camera frame

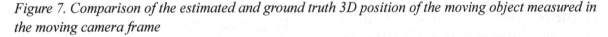

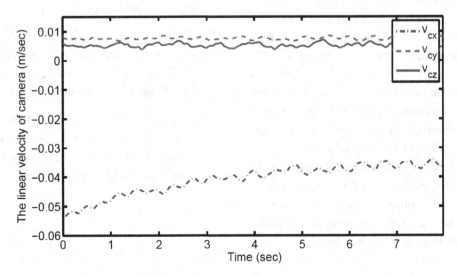

proposed estimator for motion-planning of robots and position-based visual servo control tasks such as landing on a moving platform should also be pursued.

CONCLUSION

A nonlinear observer is developed to estimate the structure of a moving object using a single moving monocular camera with less restrictive assumptions on the object motion than existing methods. The algorithm improves on our previous work in Dani, Kan, Fischer, and Dixon (2010) by relaxing the constant object velocity assumption to arbitrary object motion in a straight line or in a plane. The observer-based approach is causal and does not assume a minimum number of views or feature points. Cases are provided to illustrate that the structure estimation is insensitive to object motion. The estimation algorithm has a good potential in applications where real-time estimation is required. The estimator is tested in simulations and in experiments with a single feature point. More experimental studies on multiple feature points in real-time scenarios and in different environments are required. Also, various experiments should be conducted to test the estimator performance in different computer vision applications such as, structure estimation of deformable objects, structure and motion estimation of cars, etc.

For some computer vision applications estimation of the entire shape of the object, i.e., the 3D locations of the dense set of feature points on the object are required to be estimated. In some applications the entire object can be tracked as a single feature point and one would be interested in estimating the 3D location of a point object, e.g., tracking of a car from an aerial vehicle. In this chapter, the estimation of 3D locations of a point on a moving object is presented. Same estimator can be used for estimating 3D locations of multiple feature point on an object.

Since the camera image data from only current image frame is used the estimator is computationally very efficient, but is more susceptible to the noise because there is no optimization/bundle adjustment present in the algorithm. Future efforts will focus on adding optimization over small number of keyframes along with the estimation algorithm to improve the accuracy of the estimation.

REFERENCES

Avidan, S., & Shashua, A. (2000). Trajectory triangulation: 3D reconstruction of moving points from a monocular image sequence. *IEEE Transactions on Pattern Analysis and Machine Intelligence, 22*(4), 348–357. doi:10.1109/34.845377

Bhattacharyya, S. (1978). Observer design for linear systems with unknown inputs. *IEEE Transactions on Automatic Control, 23*(3), 483–484. doi:10.1109/TAC.1978.1101758

Boyd, S., & Vandenberghe, L. (2004). *Convex optimization*. New York, NY: Cambridge University Press.

Chen, W., & Saif, M. (2006a). Fault detection and isolation based on novel unknown input observer design. *Proceedings of the American Control Conference*, Minneapolis, MN.

Chen, W., & Saif, M. (2006b). Unknown input observer design for a class of nonlinear systems: An LMI approach. *Proceedings of the American Control Conference*, Minneapolis, MN.

Chitrakaran, V., Dawson, D. M., Dixon, W. E., & Chen, J. (2005). Identification of a moving object's velocity with a fixed camera. *Automatica, 41*(3), 553–562. doi:10.1016/j.automatica.2004.11.020

Dahl, O., Nyberg, F., & Heyden, A. (2007). Nonlinear and adaptive observers for perspective dynamic systems. *Proceedings of the American Control Conference*, (pp. 966–971). New York City, NY.

Dani, A. P., & Dixon, W. E. (2010). Single Camera Structure and Motion Estimation. In Chesi, G., & Hashimoto, K. (Eds.), *Visual servoing via advanced numerical methods* (pp. 209–229). Springer Lecture Notes in Control and Information Sciences. doi:10.1007/978-1-84996-089-2_12

Dani, A. P., Fischer, N., & Dixon, W. E. (2012). Single camera structure and motion. *IEEE Transactions on Automatic Control*, *57*(1), 241–246. doi:10.1109/TAC.2011.2162890

Dani, A. P., Kan, Z., Fischer, N., & Dixon, W. E. (2010). Structure and motion estimation of a moving object using a moving camera. *Proceedings of the American Controls Conference*, (pp. 6962-6967). Baltimore, MD.

Dani, A. P., Kan, Z., Fischer, N., & Dixon, W. E. (2011). Estimating structure of a moving object using a moving camera: An unknown input observer approach. *IEEE Conference on Decision and Controls*, (pp. 5005-5010), Orlando, FL.

Dani, A. P., Kan, Z., Fischer, N., & Dixon, W. E. (2012). Globally exponentially convergent robust observer for vision-based range estimation. *IFAC Mechatronics. Special Issue on Visual Servoing*, *22*(4), 381–389.

Dani, A. P., Rifai, K., & Dixon, W. E. (2010). *Globally exponentially convergent observer for vision-based range estimation* (pp. 801–806). Yokohama, Japan: IEEE Multi-Conference on System and Control. doi:10.1109/ISIC.2010.5612878

Darouach, M., Zasadzinski, M., & Xu, S. (1994). Full-order observers for linear systems with unknown inputs. *IEEE Transactions on Automatic Control*, *39*(3), 606–609. doi:10.1109/9.280770

De Persis, C., & Isidori, A. (2001). A geometric approach to nonlinear fault detection and isolation. *IEEE Transactions on Automatic Control*, *46*, 853–865. doi:10.1109/9.928586

Dixon, W. E., Fang, Y., Dawson, D. M., & Flynn, T. J. (2003). Range identification for perspective vision systems. *IEEE Transactions on Automatic Control*, *48*(12), 2232–2238. doi:10.1109/TAC.2003.820151

Fridman, L., Shtessel, Y., Edwards, C., & Yan, G. (2007). Higher-order sliding-mode observer for state estimation and input reconstruction in nonlinear systems. *International Journal of Robust and Nonlinear Control*, *18*, 399–412. doi:10.1002/rnc.1198

Gahinet, P., Nemirovskii, A., Laub, A., & Chilali, M. (1994). The LMI control toolbox. *IEEE Conference on Decision and Control*, (pp. 2038–2041).

Guan, Y., & Saif, M. (1991). A novel approach to the design of unknown input observers. *IEEE Transactions on Automatic Control*, *36*(5), 632–635. doi:10.1109/9.76372

Hammouri, H., Kabore, P., & Kinnaert, M. (2001). Geometric approach to fault detection and isolation for bilinear systems. *IEEE Transactions on Automatic Control*, *46*(9), 1451–1455. doi:10.1109/9.948476

Hammouri, H., Kinnaert, M., & El Yaagoubi, E. H. (1999). Observer based approach to fault detection and isolation for nonlinear systems. *IEEE Transactions on Automatic Control*, *44*, 1879–1884. doi:10.1109/9.793728

Hammouri, H., & Tmar, Z. (2010). Unknown input observer for state affine systems: A necessary and sufficient condition. *Automatica*, *46*, 271–278. doi:10.1016/j.automatica.2009.11.004

Han, M., & Kanade, T. (2004). Reconstruction of a scene with multiple linearly moving objects. *International Journal of Computer Vision*, *59*(3), 285–300. doi:10.1023/B:VISI.0000025801.70038.c7

Hautus, M. (1983). Strong detectability and observers. *Linear Algebra and Its Applications, 50,* 353–368. doi:10.1016/0024-3795(83)90061-7

Hou, M., & Muller, P. (1992). Design of observers for linear systems with unknown inputs. *IEEE Transactions on Automatic Control, 37*(6), 871–875. doi:10.1109/9.256351

Jankovic, M., & Ghosh, B. (1995). Visually guided ranging from observations points, lines and curves via an identifier based nonlinear observer. *Systems & Control Letters, 25*(1), 63–73. doi:10.1016/0167-6911(94)00053-X

Kahl, F., & Hartley, R. (2008). Multiple-view geometry under the L1-norm. *IEEE Transactions on Pattern Analysis and Machine Intelligence, 30*(9), 1603–1617. doi:10.1109/TPAMI.2007.70824

Kaminski, J., & Teicher, M. (2004). A general framework for trajectory triangulation. *Journal of Mathematical Imaging and Visualization, 21*(1), 27–41. doi:10.1023/B:JMIV.0000026555.79056.b8

Koenig, D., & Mammar, S. (2001). Design of a class of reduced order unknown inputs nonlinear observer for fault diagnosis. *Proceedings of the American Control Conference,* (pp. 2143–2147). Arlington, VA.

Liu, F., Farza, M., & M'Saad, M. (2006). *Unknown input observers design for a class of nonlinear systems - application to biochemical process.* IFAC Symposium on Robust Control Design.

Ma, Y., Soatto, S., Kosecka, J., & Sastry, S. (2004). *An invitation to 3-D vision.* Springer.

Moreno, J. A. (2000). Unknown input observers for SISO nonlinear systems. *IEEE Conference on Decision and Control,* (pp. 790–801). Sydney, NSW Australia.

Moreno, J. A., & Dochain, D. (2008). Global observability and detectability analysis of uncertain reaction systems and observer design. *International Journal of Control, 81,* 1062–1070. doi:10.1080/00207170701636534

Oliensis, J. (2000). A critique of structure-from-motion algorithms. *Computer Vision and Image Understanding, 80,* 172–214. doi:10.1006/cviu.2000.0869

Park, H., Shiratori, T., Matthews, I., & Sheikh, Y. (2010). 3D reconstruction of a moving point from a series of 2D projections. *European Conference on Computer Vision, Vol. 6313,* (pp. 158–171).

Patton, R. J., & Chen, J. (1993). Optimal selection of unknown input distribution matrix in the design of robust observers for fault diagnosis. *Automatica, 29,* 837–841. doi:10.1016/0005-1098(93)90089-C

Pertew, A., Marquez, H., & Zhao, Q. (2005). Design of unknown input observers for Lipschitz nonlinear systems. *Proceedings of American Control Conference,* (pp. 4198–4203).

Rajamani, R. (1998). Observers for Lipschitz nonlinear systems. *IEEE Transactions on Automatic Control, 43*(3), 397–401. doi:10.1109/9.661604

Soatto, S., & Perona, P. (1998). Reducing "structure from motion": A general framework for dynamic vision, part 1: Modeling. *IEEE Transactions on Pattern Analysis and Machine Intelligence, 20*(9), 933–942. doi:10.1109/34.713360

Sturm, P., & Triggs, B. (1996). A factorization based algorithm for multi-image projective structure and motion. *Lecture Notes in Computer Science, 1065,* 709–720. doi:10.1007/3-540-61123-1_183

Tsui, C.-C. (1996). A new design approach to unknown input observers. *IEEE Transactions on Automatic Control, 41*(3), 464–468. doi:10.1109/9.486653

Vidal, R., Ma, Y., Soatto, S., & Sastry, S. (2006). Two-view multibody structure from motion. *International Journal of Computer Vision, 68*(1), 7–25. doi:10.1007/s11263-005-4839-7

Xie, L., & Khargonekar, P. P. (2010). Lyapunov-based adaptive state estimation for a class of nonlinear stochastic systems. *Proceedings of the American Controls Conference*, (pp. 6071–6076). Baltimore, MD.

Yaz, E., & Azemi, A. (1993). Observer design for discrete and continuous nonlinear stochastic systems. *International Journal of Systems Science, 24*(12), 2289–2302. doi:10.1080/00207729308949629

Yuan, C., & Medioni, G. (2006). 3D reconstruction of background and objects moving on ground plane viewed from a moving camera. *Computer Vision and Pattern Recognition, 2*, 2261–2268.

KEY TERMS AND DEFINITIONS

Exponential Stability: The estimator error will converge to zero with an exponential bound and all the signals of the estimator will ramain bounded for all time.

Linear Matrix Inequality: A matrix inequality which is linear in the unknown matrix variable. Often such matrix inequalities arise in systems and control theory and there are efficient numerical algorithms to determine the feasibility of the inequality.

Observer: A state estimator for deterministic linear/nonlinear systems.

Structure: Structure referes to the 3D shape of the object. Typically, an object is represented by a dense set of feature points.

Structure And Motion: In this work structure is refered to as 3D coordinates of feature points tracked on the object in the field of view. Motion refers to the angular and linear velocities of a moving camera.

Structure And Motion From Motion: The structure and motion from motion referes to estimation of the 3D shape of a moving object and motion of the moving object given the camera linear and angular velocities.

Unknown Input Observer: An observer which can estimate the state even if the input to the system is not measurable. Input to the system can be an external disturbance acting on the system.

Chapter 12

Intelligent Stereo Vision in Autonomous Robot Traversability Estimation

Lazaros Nalpantidis
Royal Institute of Technology – KTH, Sweden

Ioannis Kostavelis
Democritus University of Thrace, Greece

Antonios Gasteratos
Democritus University of Thrace, Greece

ABSTRACT

Traversability estimation is the process of assessing whether a robot is able to move across a specific area. Autonomous robots need to have such an ability to automatically detect and avoid non-traversable areas and, thus, stereo vision is commonly used towards this end constituting a reliable solution under a variety of circumstances. This chapter discusses two different intelligent approaches to assess the traversability of the terrain in front of a stereo vision-equipped robot. First, an approach based on a fuzzy inference system is examined and then another approach is considered, which extracts geometrical descriptions of the scene depth distribution and uses a trained support vector machine (SVM) to assess the traversability. The two methods are presented and discussed in detail.

INTRODUCTION

Autonomous robots need to be able to operate in unknown, harsh environments and stereo vision can provide the means to cope with them in an efficient way. A basic but essential aspect of robots' autonomous operation is the assessment of terrain traversability, i.e. whether the robot is able to move towards a specific area or whether there are obstacles in front of it. Obstacle detection and traversability evaluation are important, as they provide crucial information for the safe navigation of mobile robots either for use in outdoor operations (Baudoin et al., 2009), or for use

DOI: 10.4018/978-1-4666-2672-0.ch012

in space exploration, e.g. ESA's ExoMars rover (Kostavelis, Boukas, Nalpantidis, Gasteratos, & Aviles Rodrigalvarez, 2011), shown in Figure 1. However, the actual methodology to extract this kind of information in a resource-efficient manner is still an active and interesting research topic.

There is a plethora of non-vision sensors that are used by autonomous robots. Yet, stereo vision tends to be a favorable solution due to its versatile nature and its inherent biomimetic origin that have been proven to be of great reliability under a variety of circumstances. As a result, autonomous robots often use stereo vision as their primary source of information about the structure of the world around them. The requirements placed by robotic applications are very strict and, in general, different from the requirements usually adopted by the computer vision community. As such, special care has to be attributed to the characteristics of the vision system. However, solid stereo perception constitutes only the first step towards more advanced operations. When it comes for autonomous robots, the next big issue is terrain traversability estimation, and as a consequence obstacle avoidance. In this chapter two alternative approaches are discussed, both suitable for stereo vision equipped autonomous robots. More specifically, the first approach (Nalpantidis & Gasteratos, 2011) detects obstacles in the scene and avoids them based on a fuzzy inference system (FIS). 3D vision information is used by the algorithm to analyze depth maps and deduce the most appropriate direction for the robot to avoid any

existing obstacles. On the other hand, the second approach considered (Kostavelis, Nalpantidis, & Gasteratos, 2011) uses a machine learning method based on a support vector machine (SVM) for the traversability classification of the terrain. A v-disparity image calculation and processing step extracts suitable features about the scene's characteristics. The resulting data are used as input for the training of the SVM. This approach is able to classify the scenes either as traversable or non-traversable with high success rates. The two approaches are different in essence and each one mainly focuses on different aspects, as it will be discussed.

BACKGROUND

Depth Estimation Using Stereo Vision

Stereo vision is often used in vision-based robotics, instead of monocular sensors, due to the simpler calculations involved in the depth estimation. A correspondence search between the two stereo images can provide dense information about the depth of the depicted scene. In the case of stereo vision-based navigation, the accuracy and the refresh rate of the computed disparity maps are the cornerstones of its success (Schreer, 1998). Dense local stereo correspondence methods calculate depth for almost every pixel of the scene, taking into consideration only a small neighbor-

Figure 1. Robots equipped with potential stereo vision traversability estimation abilities

hood of pixels each time (Scharstein & Szeliski, 2002). On the other hand, global methods are significantly more accurate but at the same time more computationally demanding, as they account for the whole image (Torra & Criminisi, 2004). However, since the most important constraint in autonomous robotics is the real-time operation, such applications usually utilize local algorithms. Muhlmann et al. in (Muhlmann, Maier, Hesser, & Manner, 2002) describe a local method that uses the sum of absolute differences (SAD) correlation measure for RGB color images. Applying a left to right consistency check, the uniqueness constraint and a median filter, it can achieve 20 fps, for 160x120 pixel images. Another fast local AD based algorithm is presented in (Di Stefano, Marchionni, & Mattoccia, 2004). It is based on the uniqueness constraint and rejects previous matches as soon as better ones are detected. It achieves 39.59 fps for 320x240 pixel images with 16 disparity levels and the root mean square error for the standard Tsukuba pair is 5.77. The algorithm reported in (S. Yoon, Park, Kang, & Kwak, 2005) achieves almost real-time performance. It is once more based on SAD but the correlation window size is adaptively chosen for each region of the picture. Apart from that, a left to right consistency check and a median filter are utilized. The algorithm is able to compute 7 fps for 320x240 pixel images with 32 disparity levels. Another possibility in order to obtain accurate results in real-time is to utilize programmable graphic processing units (GPU). In (Zach, Karner, & Bischof, 2004) a hierarchical disparity estimation algorithm is presented. This method can process either rectified or non-calibrated image pairs using a local SAD based bidirectional process. This algorithm is implemented on an ATI Radeon 9700 Pro GPU and can achieve up to 50 fps for 256x256 pixel input images. On the other hand, an interesting but very computationally demanding local method is presented in (K.-J. Yoon & Kweon, 2006). It uses varying weights

for the pixels in a given support window, based on their color similarity and geometric proximity. However, the execution speed of the algorithm is far from being real-time. The running time for the Tsukuba image pair with a 35x35 pixels support window is about one minute. The error ratio is only 1.29%, 0.97%, 0.99%, and 1.13% for the Tsukuba, Sawtooth, Venus and Map image sets respectively. A detailed taxonomy and presentation of dense stereo correspondence algorithms can be found in (Scharstein & Szeliski, 2002). Additionally, more recent advances in the field as well as the aspect of hardware implementable stereo algorithms are covered in (Nalpantidis, Sirakoulis, & Gasteratos, 2008).

Obstacle Avoidance

As far as obstacle avoidance techniques are concerned, a wide range of sensors and various methods have been proposed in the relevant literature. The use of ultrasonic, laser and infrared sensors is well-studied and the depth measurements can be accurate and easily available. However, such sensors suffer either from achieving only low refresh rates (Vandorpe, Van Brussel, & Xu, 1996) or being extremely expensive. On the other hand, vision sensors either monocular, stereo or multicamera ones, can combine high frame rates and appealing prices. Interesting details about developed sensor systems and proposed detection and avoidance algorithms can be found in (Borenstein & Koren, 1990) and (Ohya, Kosaka, & Kak, 1998). Movarec proposed the Certainty Grid method in (Moravec, 1987) and Borenstein (J.Borenstein & Koren, 1991) has proposed the Virtual Force Field method for robot obstacle avoidance. Then the Elastic Strips method was proposed in (Khatib, 1996, 1999) treating the trajectory of the robot as an elastic material to avoid obstacles. Moreover, (Kyung Hyun, Minh Ngoc, & M. Asif Ali, 2008) presented a modified Elastic Strip method for mobile robots operating in uncertain environments. Finally, the concept of using fuzzy logic for obstacle avoidance purposes was covered by (Reignier, 1994), but only up to a theoretical level. Reviews of popular obstacle avoidance algorithms covering them in more detail can be found in (Manz, Liscano, & Green, 1993), (Kunchev, Jain, Ivancevic, & Finn, 2006) and more recently in (Ortigosa, Morillas, & Peris-Fajarns, 2011).

Traversability Estimation

The estimation of terrain traversability remains of interest for the mobile robotics community during the last decades. In 1994 a statistics-based method for classifying field regions as traversable or not was proposed in (Langer, 1994). More recently, the development of autonomous planetary rovers and the DARPA Grand Challenge have triggered rapid advancements in the field (Howard et al., 2006; Singh et al., 2000; Thrun et al., 2006). Machine learning methodologies have often been employed (Shneier, Shackleford, Hong, & Chang, 2006) and stereo vision has been widely used as input for such systems (Happold, Ollis, & Johnson, 2006; Kim et al., 2006). One of the most popular methods for terrain traversability analysis is the initial estimation of the v-disparity image (Labayrade, Aubert, & Tarel, 2002; De Cubber, Doroftei, Nalpantidis, Sirakoulis, & Gasteratos, 2009). This method is able to confront the noise in low quality disparity images (Zhao, Katupitiya, & Ward, 2007; Soquet, Aubert, & Hautiere, 2007) and model the terrain as well as any existing obstacles.

Several researchers have proposed robot navigation methods based on terrain classification. These methods use features derived from remote sensor data such as color, image texture and surface geometry. Initially, color-based methods have been proposed for the classification of outdoor scenes using a mixture of Gaussians models (Manduchi, 2004). Additionally, in (C.S. Dima & Hebert, 1994) a terrain classification method based on texture features of the images was introduced. A more

sophisticated and computationally demanding method based on 3D point clouds that uses the statistical distribution in space, has been proposed in (Vandapel, Huber, Kapuria, & Hebert, 2004). Those navigation methods are applied both in traversable and non-traversable scenes consuming valuable resources and time. For efficient robot navigation, non-traversable scenes should not be examined in detail. Therefore, scenes should firstly be inspected and classified according to their overall traversability. Supervised machine learning techniques have the advantage that while the training phase is slow, the classification of new unseen instances is performed rapidly.

STEREO VISION FOR AUTONOMOUS ROBOTS

Stereo-equipped robots need to extract depth information, or equivalently disparity maps, from the image pairs provided by the stereo camera. As a result, the efficiency and accuracy of the utilized stereo algorithm is crucial for the performance of every subsequent task. The stereo algorithm considered in the rest of this work is a local stereo correspondence algorithm with certain design choices that improve the quality of the produced disparity maps. The main attribute that differentiates this algorithm from the majority of the other ones is that the matching cost aggregation step consists of a sophisticated Gaussian-weighted rather than a simple summation. Furthermore, the disparity selection step is a winner-takes-all (WTA) choice, as the absence of any iteratively updated selection process significantly reduces the computational payload of the overall algorithm. Finally, intelligent filtering techniques remove pixels with non-reliable disparity estimation.

Image Preprocessing

The initially captured images are processed in order to extract the edges in the depicted scene.

The utilized edge detector is the Laplacian of Gaussian (LoG), using a zero threshold. This choice produces the maximum possible edges. The LoG edge detection method smoothens the initial images with a Gaussian filter in order to suppress any possible noise. Afterwards, a Laplacian kernel is applied that marks regions of significant intensity change. Actually, the combined LoG filter, with standard deviation equal to 2, is applied at once and the zero crossings are found. The extracted edges are, afterwards, superimposed to the initial images, as shown in Figure 2. The outcome of this procedure is a new version of the original images having more striking features and textured surfaces, which facilitate the following stereo matching procedure.

Stereo Calculation

The matching cost function utilized is the truncated absolute differences (AD). AD is inherently the simplest metric of all, involving only summations and finding absolute values. The AD are truncated if they excess the 4% of the maximum intensity value to suppress the influence of noise in the final result. This is very important for stereo algorithms that are intended to be applied to outdoor scenes, which usually suffer from noise induced by a variety of reasons, e.g. lighting differences

and reflections. For every pixel of the reference (left) image, AD are calculated for each of its candidate matches in the other (right) image as in Equation 1.

$$AD(x,y,d) = \left| I_{left}(x,y) - I_{right}((x-d),y) \right|$$

(1)

where I_{left} and I_{right} denote the intensity values for the left and right image respectively, d is the value of the disparity under examination ranging for 0 to D-1 and x, y are the coordinates of the pixel.

The computed matching costs for every pixel and for all its potential disparity values comprise a 3D matrix, usually called as disparity space image (DSI). The DSI values for constant disparity value are aggregated inside fix-sized square windows. The dimensions of the chosen aggregation window play an important role in the quality of the final result. Generally, small dimensions preserve details but suffer from noise, whereas large dimensions might produce coarser results but significantly suppress noise. Both of these phenomena are owed to the averaging of the disparity values preformed during aggregation. The choice of a suitable window size depends on the targeted application. Not recovering the absolutely exact objects boundaries does not necessarily affect traversability estimation, as

Figure 2. Image preprocessing

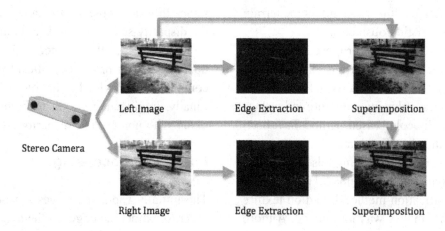

it might happen with other robotic tasks, e.g. manipulation or grasping. On the other, traversability estimation can be misled by the existence of numerous incorrect depth estimations, owed to noise. Moreover, outdoor robots are even more susceptible to spurious estimations due to difficult lighting (Nalpantidis & Gasteratos, 2011) or visibility conditions (Baudoin et al., 2009). Such considerations usually lead to choosing larger window sizes when it comes to outdoor robotic applications. The aggregation window's dimensions used in the discussed algorithm are 19x19 pixels, a choice that compromises real-time execution speed and noise cancelation. The AD aggregation step of the discussed algorithm is a weighted summation, i.e. each pixel is assigned a weight depending on its Euclidean distance from the central pixel. A 2D Gaussian function determines the weight's value for each pixel, as in Equation 2. The center of the function coincides with the central pixel. The standard deviation is equal to the one third of the distance from the central pixel to the nearest window-border. The applied weighting function can be calculated once and then be applied to all the aggregation windows without any further change. Thus, the computational load of this procedure is kept within reasonable limits.

$$DSI(x, y, d) = \sum_{i=-w}^{w} \sum_{j=-w}^{w} \begin{array}{l} gauss(i, j) \cdot \\ AD(x + i, y + j, d) \end{array}$$

(2)

where, the pixel ranges *[−w, w]* define the weighted aggregation window.

Finally, the optimum disparity value for each pixel *disp(x,y)*, i.e. the disparity map, is chosen by a simple, non-iterative WTA step, as in Equation 3. In the resulting disparity maps, smaller values indicate more distant objects, while bigger disparity values indicate objects lying closer.

$$disp(x, y) = \arg(\min(DSI(x, y, d)))$$

(3)

Disparity Filtering

The results of the per pixel optimum disparity values are filtered at two consequent steps. Firstly, the reliability of the selected disparity value is validated. That is, for every pixel of the disparity map a certainty measure is calculated indicating the likelihood of the pixel's selected disparity value to be the correct one. The certainty measure *cert* is calculated for each pixel *(x,y)* as in Equation 4.

$$cert(x, y) = \left| DSI(x, y, disp(x, y)) - \frac{\sum_{d=0}^{D-1} DSI(x, y, d)}{D} \right|$$

(4)

According to this, the certainty *cert* for a pixel *(x,y)* that the computed disparity value *disp(x,y)* is actually right is equal to the absolute value of the difference between the minimum matching cost value *DSI(x, y, disp(x,y))* and the average matching cost value for that pixel when considering all the *D* candidate disparity levels for that pixel. What the aforementioned measure evaluates is the amount of differentiation of the selected disparity value with regard to the rest candidate ones. The more the disparity value is differentiated, the most possible it is that the selected minimum is actually a real one and not owed to noise or other effects. A threshold is applied to this metric and only the pixels whose certainty to value ratio $cert(x, y) / DSI(x, y, disp(x, y))$ is equal to or more than 30% are counted as valid ones. The value of this threshold has been chosen after exhaustive experimentation so as to reject as many false matches as possible, while retaining the majority of the correct ones. Moreover, a bidirectional consistency check is applied. The selected disparity values are approved only if they are consistent, irrespectively to which image is the reference and which one is the target. This way, even more false matches are disregarded. The outcome of the considered stereo algorithm is a sparse disparity map, as shown in Figure 3(c), containing disparity values only for the most reliable pixels. The rest pixels, shown in black in Figure 3(c) are not considered at all.

FUZZY OBSTACLE ANALYSIS

The approach discussed in this section detects obstacles in the scene and avoids them using a FIS. The depth of objects, as described by the produced stereo disparity maps, is used by the algorithm to analyze the structure of the scene and deduce the most appropriate direction for the robot to avoid any existing obstacles.

This method processes each pair of stereoscopic images and indicates an obstacle-avoiding direction of movement for a robot, such as the one shown in Figure 4(a). First, the stereo image pair is given as input to a stereo vision algorithm and a depth map of the scene is obtained. This depth map is thereafter used as input of the fuzzy obstacle analysis and direction decision module. This fuzzy module indicates the proper direction of movement. The direction of movement ranges from −30° to +30°, considering 0° as the direction the robot is heading to. This angle range is dictated by considering a typical stereo camera with 60° horizontal field of view (HFoV). Furthermore,

in cases when the scene is full of obstacles or the depth map is too noisy to conclude safely a "move backwards" signal is foreseen. Figure 4(b) presents the mobile robot, shown as the "R" in the center, and the possible positions after the application of the discussed algorithm, shown by the bold red regions of the outer circle.

Theoretical Formulation

The calculated depth maps are used to extract useful information about the navigation of the robot. In contrast to many implementations that involve complex calculations upon the disparity maps, the discussed decision making algorithm is focused on computational efficiency. This is feasible due to the absence of significant noise in the produced disparity maps. The goal of the obstacle analysis module is to assess the traversability of three possible directions of movement, i.e. forward, left and right. In order to achieve that, the developed method divides each disparity map into three equal windows, as in Figure 5.

The division of the disparity map excludes the boundary regions, in this case a peripheral frame of 20 pixels width, because the disparity calculation in such regions is often problematic. In the each window w, the pixels p whose disparity value $disp(p)$ is greater than a defined threshold value *Thres* are enumerated. The enumeration results are normalized towards the widow's pixels

Figure 3. (a) Left image, (b) right image of a stereo image pair, and (c) the final disparity map

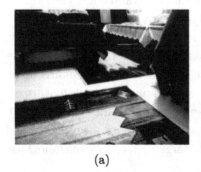 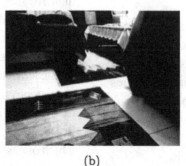 

(a) (b) (c)

Figure 4. (a) Stereo camera equipped mobile robotic platform and (b) floor plan of the robot's environment

(a)

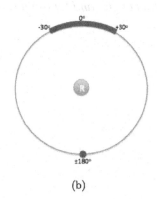

(b)

population and then examined. The more traversable the corresponding direction is the smaller the enumeration result should be. Thus, the traversability of the left, central and right window, respectively *TL*, *TC* and *TR*, is assessed.

The values of the parameter *Thres* play an important role to the algorithm's behavior. More specifically, small values of *Thres* favor hesitancy in moving forward, ensuring obstacle avoidance but at the same time being susceptible to false alarms due to noise. On the other hand, the opposite scenario is less susceptible to false alarms but may be proven risky for the robot. In this work the value *Thres*=120 was chosen after exhaustive experimentation as a fair compromise between the two extreme behaviors.

The results of the traversability estimation for the three windows, i.e. the left, central, and right one, are used as the three input values of a FIS that decides the proper direction of movement for the robot. The outputs of the FIS are the angle of the direction that the robot should follow and an indicator that the robot should move backwards. Figure 6 shows the membership functions (MF) for the three inputs (all having identical MF, which is shown in Figure 6(a)) and the two outputs (Figure 6(b) and 6(c)).

A direction angle of 0° indicates forward movement, negative angles indicate movement towards left and positive angles indicate movement towards right. The second output variable indicates that

the robot should move backwards in order to acquire a broader view of the scene. This should happen if all the possible directions within the robot's field of view are, or at least seem, non-traversable.

The set of fuzzy rules that were used for the FIS is:

- IF "*TL*" is "Large" AND "*TC*" is not "Large" AND "*TR*" is not "Large" THEN "*Angle*" is "Left".
- IF "*TL*" is not "Large" AND "*TC*" is "Large" AND "*TR*" is not "Large" THEN "*Angle*" is "Forward".

Figure 5. Depth map's division in three windows

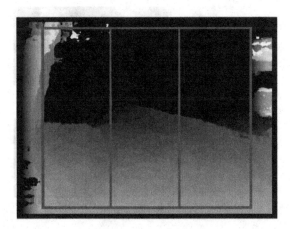

Figure 6. Fuzzy membership functions: (a) input MF: traversability of the left/central/right window, (b) output MF: direction angle, and (c) output MF: "move backwards" indicator

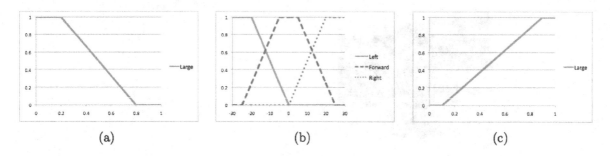

(a)   (b)   (c)

- IF "*TL*" is not "Large" AND "*TC*" is not "Large" AND "*TR*" is "Large" THEN "*Angle*" is "Right".
- IF "*TL*" is not "Large" AND "*TC*" is not "Large" AND "*TR*" is not "Large" THEN "move backwards" is "Large".

This small set of simple rules is enough to ensure that the robot will adapt its direction of movement so as to avoid any obstacles in its way. Experimental validation of the discussed system indicated that the option to move back should actually be adopted when the "move backwards" output of the FIS is larger than 0.65 (see Figure 6(c)). Of course, this value was determined em-

pirically and smaller or bigger values could be used instead, resulting in more "brave" or hesitant behaviors respectively.

Experimental Validation

The discussed method was tested on image pairs acquired by the robot shown in Figure 4(a). The image pairs were captured in real-life outdoor environments under natural lighting conditions. The experimental results covered various situations of different complexity that the robot had to deal with. For each case the reference image of the stereo pair and the computed disparity map are given. Figure 7(a), 7(b), 7(c), 7(d) show

Figure 7. Reference images and disparity maps for various decisions of the fuzzy obstacle analysis algorithm: (a) forward, (b) back, (c) left and (d) right

(a)   (b)

(c)   (d)

Figure 8. V-disparity images for a stereo image pair: (a) calculated v-disparity image and (b) v-disparity image with terrain modeled by the continuous line and the tolerance region shown between the two dashed lines

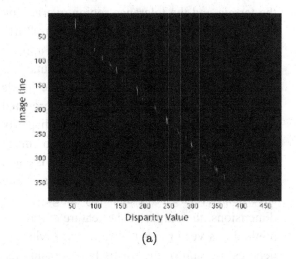

(a)

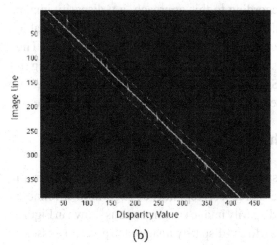

(b)

some cases where the robot has decided to move forward, backward, left and right, respectively.

The disparity maps obtained by stereo vision almost inevitably, even after filtering, contain some erroneously estimated pixels. Such inaccuracies are also present in the discussed method's resulting disparity maps. Many of the obstacle avoidance algorithms, including the ones found in (Moravec, 1987; J.Borenstein & Koren, 1991; Khatib, 1999; Kyung Hyun et al., 2008), rely on the availability of highly accurate and reliable depth estimations. As a consequence, a solely vision-based obstacle avoidance algorithm could not use any of the aforementioned stereo computation methods. On the contrary, the discussed algorithm can tolerate such problems as shown by the experimental results. Furthermore, the inherent ability of the algorithm to decide whether the robot should move backwards prevents entrapments into local minima, which often happens in methods such as the Virtual Force Field (J.Borenstein & Koren, 1991) and the Elastic Strips (Khatib, 1999).

The behavior of the method has been tested in real outdoor data sets of various scenes. The

algorithm has correctly chosen to move forward in 12 cases, turn left in 4 cases, turn right in 4 cases, and move backwards in 1 case. The detailed results of the experimental validation can be found in (Nalpantidis & Gasteratos, 2011). As shown by the experimental results the discussed algorithm has succeeded in deciding an obstacle free direction of movement. It manages to process even erroneous vision data without the need of input from additional sensors. The used FIS decides the exact value of the steering angle in an continuous manner. In this sense, the discussed method combines attributes of the methods presented in (Sabe et al., 2004) and (Reignier, 1994) while adding further capabilities and maintaining the computational load in very low levels.

V-DISPARITY FEATURE EXTRACTION AND SVM-BASED LEARNING

While the approach discussed in the previous section is very simple, more complex scenes often require more advanced methodologies. As a result, the following approach, which adds a small computational overhead but is able to deal with more complex scenes, is also of interest. According to this approach, a v-disparity image calculation and processing step extracts suitable features about the scene's characteristics. The resulting data are used as input for the training of a SVM. The final result is a classification of each scene either as traversable or non-traversable.

Theoretical Formulation

Using the sparse disparity map obtained from the stereo correspondence algorithm a reliable v-disparity image is computed, as shown in Figure 8(a). In a v-disparity image each pixel possesses a positive integer value, which denotes the number of pixels in the input image that lie on the same image line (ordinate) and have disparity value equal to its abscissa. The terrain in the v-disparity image is modeled by a linear equation, the parameters of which can be estimated using the Hough transform (De Cubber et al., 2009), condition to the fact that the majority of the input images' pixels belong to the terrain and not to obstacles. A tolerance region on both sides of the terrain's

linear segment is considered and any point outside this region can be safely considered as originating from an obstacle. The linear segments denoting the terrain and the tolerance region overlaid on the v-disparity image are shown in Figure 8(b).

A feature extraction procedure is then applied to the v-disparity image. For each horizontal line of a v-disparity image the values of the pixels lying outside the tolerance region are aggregated to produce a feature vector for each v-disparity image (or equivalently, for each stereo image pair). The dimension of this vector space is equal to the number of the image's scanlines. More specifically, for a given v-disparity image of MxN dimensions, the output of the feature extraction method is a vector $x = [x_1, x_2, ..., x_M]$ where x_i denotes the sum of the pixels lying outside the tolerance region for the row i and $i = 1,2,..M$ indicates the number of rows of the disparity map. As an example, the pixels of the v-disparity image whose values are aggregated so as to obtain the x_{50} component of the feature vector are the ones lying within the red rectangular regions of Figure 9. In this figure the red rectangular regions are exaggerated for the purpose of readability.

Figure 9. Features extraction for the 50th image line

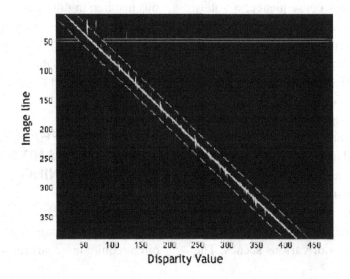

Figure 10. (a) Reference image and experimental results for a traversable scene tested: (b) sparse disparity map, (c) v-disparity image, (d) obstacles highlighted on the reference image, and (e) histogram of features

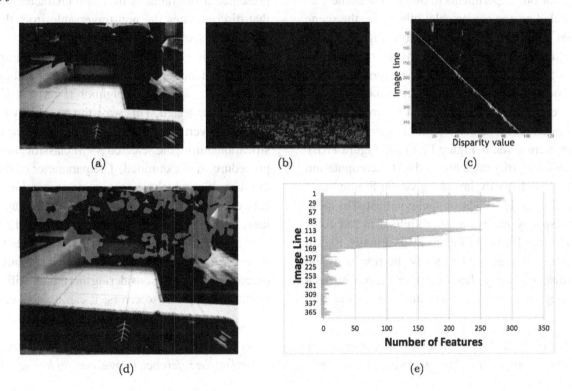

The feature vectors for all the stereo pairs of the used data set comprise a data matrix $D=[x_1,x_2,...,x_L]^T$, where each vector x_j corresponds to a stereo pair and $j = 1, 2, ...L$ denotes the number of samples. For the evaluation of this method a diminished data set constituted of 23 traversable and 10 non-traversable indoor scenes was used. One traversable and one non-traversable sample reference image are given in Figure 10(a) and Figure 11(a) respectively.

The next step of the discussed methodology deals with the training procedure for the aforementioned data set. This is an off-line procedure and, therefore, a non time-critical one. Thus a SVM classifier (Vapnik, 1995) has been chosen to be used. This approach constructs a binary classifier for each pair of classes by building a function that will be positive for one class (i.e. traversable) and negative for the other one (i.e.

non-traversable). Linear, polynomial and Gaussian kernels have been tested. The model regularization parameter C, which penalizes large errors, is chosen equal to 100, in order to optimize data separation. The optimal parameter γ used to control the width of the Gaussian distribution was set to 0.1. As for the polynomial kernel, the second order polynomial function was chosen for testing, whereas the third order polynomial did not offer any additional classification gain.

Experimental Validation

The methodology described above does no require a balanced data set for the SVM training phase. In the used input image pairs the traversable scenes were more than the non-traversable ones (i.e. 23 traversable scenes and 10 non-traversable ones). Each scene was manually labeled according to the distance of the nearest depicted obstacle. If that distance was less than a threshold value, set for our experiments to 50 cm, the scene was labeled as non-traversable; otherwise, the scene was considered to be traversable. Figure 10 and Figure 11 present the experimental result for an indicative traversable and non-traversable scene, respectively. The reference (left) images of each stereo image pair is shown in Figure 10(a) and Figure 11(a). The corresponding sparse disparity maps are given in Figure 10(b) and Figure 11(b). These disparity maps are used for the computation of the v-disparity images, given in Figure 10(c) and Figure 11(c). The obstacles indicated by these v-disparity images are highlighted in red color in Figure 10(d) and Figure 11(d). Finally, Figure 10(e) and Figure 11(e) show the results of the v-disparity image feature extraction process as a histogram of the detected features in each scanline.

Apparently, the existence or lack of outliers in the lower scanlines of the images, evident in the presented histograms, is the main characteristic that discriminates the non-traversable from the traversable scenes, respectively.

With a view to a better validation of the results, the performance of the proposed methodology employing a k-nearest neighbor (k-nn) classifier was also tested. This way, the possibility of obtaining overtraining results and polarized classification during the selected SVM classification procedure, was examined. The parameter of the k-nn classifier, which gave the greatest separability between the two classes, was also selected using a leave-one-out cross validation procedure and set to $k=5$ neighbors. The k-nn classifier manages to achieve a 74.3% classification rate, in leave-one-out cross validation. Considering that this classifier is inherently prone to errors, it can be deduced

Figure 11. (a) Reference image and experimental results for a non-traversable scene tested: (b) sparse disparity map, (c) v-disparity image, (d) obstacles highlighted on the reference image, and (e) histogram of features

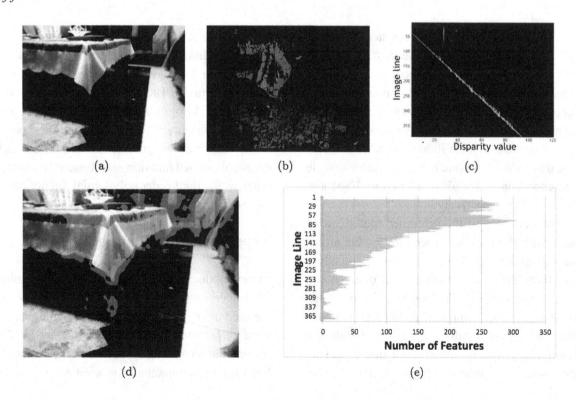

that all the preprocessing procedures are efficient and the feature extraction method indeed creates features that contain substantial information about the traversability of the scenes. Additionally, the SVM classifier succeeded a 91.2% classification rate using the second order polynomial kernel. This shows great separability between the two classes, taking into consideration that this classification rate corresponds to 30 correct out of 33 classified samples. The SVM classifier using a linear kernel achieved 87.88% and using a Gaussian kernel 81.83% classification rates, which places them in-between the other two classifiers. More detailed experimental results can be found in (Kostavelis, Nalpantidis, & Gasteratos, 2011). It should be noted that the aforementioned classification rates were obtained using a very limited set of input images, i.e. only 33 stereo image pairs. Moreover, the set of input images was not balanced, but the traversable scenes were more than twice the number of the non-traversable ones.

CONCLUSION AND DISCUSSION

Autonomous mobile robots need to ensure that they will avoid any collision during their operation. Towards this direction, stereo vision can provide the necessary information to achieve this kind of behavior. Given that robots have to execute a variety of processes in parallel, simple solutions, not involving too many calculations are favorable for traversability estimation and obstacle avoidance. This is in some sense analogous to the reflex actions of living organisms, which even if they are crucial for the very existence of the organisms, require a minimum of cognitive resources. In this chapter, stereo vision suitable for use by autonomous robots, combined with two different traversability estimation methods have been considered and discussed in detail.

It has been shown that stereo vision can provide accurate and reliable results if careful design choices are made. We have argued that for robot navigation purposes it is better to drop from dense to sparse disparity maps, through intelligent filtering techniques. Thus, even if information is lost for certain parts of the images, the remaining information is reliable and also enough for successful traversability estimation.

The traversability estimation approach described first is able to provide an analysis of the obstacles in a scene, using limited resources. The discussed method is based on stereo vision input processed by a simple but effective fuzzy obstacle analysis and direction decision module. The discussed algorithm has been shown to effectively detect and avoid obstacles. The behavior of the method has been validated by real outdoor data sets of various scenes. The algorithm exhibits robust behavior and is able to ensure collision-free autonomous mobility to robots. Moreover, the robot's direction changes within a continuous range of direction angles due to the inherent characteristics of the fuzzy system. On the other hand, the second approach is more elaborate but also able to handle more complex scenes. The disparity maps are first converted to v-disparity images. The v-disparity images are then coded, in a simple and intuitive way, so as to form the feature vectors of a two-class data set and a SVM classifier is trained to separate the traversable from the non-traversable scenes. The discussed approach has showed high classification ability, which is owed to the noise-tolerant histogram computation leading to the v-disparity images. This initial step is very important for the success of this method as the input feature vectors of the SVM classifier contain crucial and concise information for the traversability of the scene.

The methodologies covered in this chapter and the results discussed give raise to confidence that simple stereo vision methods can provide the necessary means for reliable robotic behaviors. As a result, we argue that future research directions in the field of robot traversability estimation should focus on the simplicity of the methods and the reliability of the results. Noise-tolerant computational tools, such as the ones used in the discussed two methodologies, i.e. computation of depth histograms and offline trained SVM, but also similar others can play a decisive role in the future. As a result, more elaborate methods for robotic applications can be developed employing or combining such tools.

To conclude, in tis chapter we have discussed about stereo vision-based traversability estimation. We have focused on efficient approaches that would fit the needs of autonomous robots. Such methods that provide lightweight "reflex"-like behaviors are essential for the successful operation of robots and for the safe development of more advanced autonomous behaviors.

REFERENCES

Baudoin, Y., Doroftei, D., De Cubber, G., Berrabah, S. A., Pinzon, C., Warlet, F., et al. (2009). View-finder: Robotics assistance to fire-fighting services and crisis management. In *IEEE International Workshop on Safety, Security, and Rescue Robotics* (pp. 1–6). Denver, Colorado, USA.

Borenstein, J., & Koren, Y. (1990). Real-time obstacle avoidance for fast mobile robots in cluttered environments. *IEEE Transactions on Systems, Man, and Cybernetics, 19*(5), 1179–1187. doi:10.1109/21.44033

Borenstein, J., & Koren, Y. (1991). The vector field histogram-fast obstacle avoidance for mobile robot. *IEEE Transactions on Robotics and Automation, 7*(3), 278–288. doi:10.1109/70.88137

De Cubber, G., Doroftei, D., Nalpantidis, L., Sirakoulis, G. C., & Gasteratos, A. (2009). *Stereo-based terrain traversability analysis for robot navigation*. In IARP/EURON Workshop on Robotics for Risky Interventions and Environmental Surveillance. Brussels, Belgium.

Di Stefano, L., Marchionni, M., & Mattoccia, S. (2004). A fast area-based stereo matching algorithm. *Image and Vision Computing, 22*(12), 983–1005. doi:10.1016/j.imavis.2004.03.009

Dima, C. S., & Hebert, M. (1994). Classifier fusion for outdoor obstacle detection. In *IEEE International Conference on Robotics and Automation* (pp. 665–671).

Happold, M., Ollis, M., & Johnson, N. (2006, August). *Enhancing supervised terrain classification with predictive unsupervised learning*. In Robotics: Science and systems. Philadelphia, USA.

Howard, A., Turmon, M., Matthies, L., Tang, B., Angelova, A., & Mjolsness, E. (2006). Towards learned traversability for robot navigation: From underfoot to the far field. *Journal of Field Robotics, 23*(11–12), 1005–1017. doi:10.1002/rob.20168

Khatib, O. (1996). Motion coordination and reactive control of autonomous multi-manipulator system. *Journal of Robotic Systems, 15*(4), 300–319.

Khatib, O. (1999). Robot in human environments: Basic autonomous capabilities. *The International Journal of Robotics Research, 18*(7), 684–696. doi:10.1177/02783649922066501

Kim, D., Sun, J., Min, S., James, O., Rehg, M., & Bobick, A. F. (2006). *Traversability classification using unsupervised on-line visual learning for outdoor robot navigation*. In IEEE International Conference on Robotics and Automation.

Kostavelis, I., Boukas, E., Nalpantidis, L., Gasteratos, A., & Aviles Rodrigalvarez, M. (2011, November). *SPARTAN system: Towards a low-cost and high-performance vision architecture for space exploratory rovers*. In 2nd International Workshop on Computer Vision in Vehicle Technology: From Earth to Mars, in Conjunction with ICCV 2011. Barcelona, Spain.

Kostavelis, I., Nalpantidis, L., & Gasteratos, A. (2011). Supervised traversability learning for robot navigation. In *12th Conference towards Autonomous Robotic Systems* (Vol. 6856, pp. 289–298). Sheffield, UK: Springer-Verlag.

Kunchev, V., Jain, L., Ivancevic, V., & Finn, A. (2006). Path planning and obstacle avoidance for autonomous mobile robots: A review. In *International Conference on Knowledge-Based and Intelligent Information and Engineering Systems* (Vol. 4252, pp. 537–544). Springer-Verlag.

Kyung Hyun, C., Minh Ngoc, N., & Asif Ali, R. (2008). A real time collision avoidance algorithm for mobile robot based on elastic force. *International Journal of Mechanical. Industrial and Aerospace Engineering, 2*(4), 230–233.

Labayrade, R., Aubert, D., & Tarel, J.-P. (2002). Real time obstacle detection in stereovision on non flat road geometry through "v-disparity" representation. In IEEE Intelligent Vehicle Symposium (Vol. 2, pp. 646–651). Versailles, France.

Langer, D. (1994). A behavior-based system for off-road navigation. *IEEE Transactions on Robotics and Automation, 10*(6), 776–783. doi:10.1109/70.338532

Manduchi, R. (2004). Learning outdoor color classification from just one training image. In *European Conference on Computer Vision* (Vol. 4, pp. 402–413).

Manz, A., Liscano, R., & Green, D. (1993). A comparison of realtime obstacle avoidance methods for mobile robots. In *Experimental Robotics II* (pp. 299–316). Springer-Verlag. doi:10.1007/BFb0036147

Moravec, P. (1987). Certainty grids for mobile robots. In *NASA/JPL Space Telerobotics Workshop* (Vol. 3, pp. 307–312).

Muhlmann, K., Maier, D., Hesser, J., & Manner, R. (2002). Calculating dense disparity maps from color stereo images, an efficient implementation. *International Journal of Computer Vision, 47*(1–3), 79–88. doi:10.1023/A:1014581421794

Nalpantidis, L., & Gasteratos, A. (2010). Stereo vision for robotic applications in the presence of non-ideal lighting conditions. *Image and Vision Computing, 28*(6), 940–951. doi:10.1016/j.imavis.2009.11.011

Nalpantidis, L., & Gasteratos, A. (2011). Stereovision-based fuzzy obstacle avoidance method. *International Journal of Humanoid Robotics, 8*(1), 169–183. doi:10.1142/S0219843611002381

Nalpantidis, L., Sirakoulis, G. C., & Gasteratos, A. (2008). Review of stereo vision algorithms: from software to hardware. *International Journal of Optomechatronics, 2*(4), 435–462. doi:10.1080/15599610802438680

Ohya, A., Kosaka, A., & Kak, A. (1998). Vision-based navigation of mobile robot with obstacle avoidance by single camera vision and ultrasonic sensing. *IEEE Transactions on Robotics and Automation, 14*(6), 969–978. doi:10.1109/70.736780

Ortigosa, N., Morillas, S., & Peris-Fajarns, G. (2011). Obstacle-free pathway detection by means of depth maps. *Journal of Intelligent & Robotic Systems, 63*, 115–129. doi:10.1007/s10846-010-9498-4

Reignier, P. (1994). Fuzzy logic techniques for mobile robot obstacle avoidance. *Robotics and Autonomous Systems, 12*(3-4), 143–153. doi:10.1016/0921-8890(94)90021-3

Sabe, K., Fukuchi, M., Gutmann, J.-S., Ohashi, T., Kawamoto, K., & Yoshigahara, T. (2004). Obstacle avoidance and path planning for humanoid robots using stereo vision. In *IEEE International Conference on Robotics and Automation* (Vol. 1, pp. 592–597).

Scharstein, D., & Szeliski, R. (2002). A taxonomy and evaluation of dense two-frame stereo correspondence algorithms. *International Journal of Computer Vision, 47*(1–3), 7–42. doi:10.1023/A:1014573219977

Schreer, O. (1998). Stereo vision-based navigation in unknown indoor environment. In *5th European Conference on Computer Vision* (Vol. 1, pp. 203–217).

Shneier, M. O., Shackleford, W. P., Hong, T. H., & Chang, T. Y. (2006). Performance evaluation of a terrain traversability learning algorithm in the darpa lagr program performance evaluation of a terrain traversability learning algorithm in the DARPA LAGR program. In *Performance Metrics for Intelligent Systems Workshop* (pp. 103–110). Gaithersburg, MD, USA.

Singh, S., Simmons, R., Smith, T., Stentz, A., Verma, I., Yahja, A., et al. (2000). Recent progress in local and global traversability for planetary rovers. In *IEEE International Conference on Robotics and Automation* (Vol. 2, pp. 1194–1200).

Soquet, N., Aubert, D., & Hautiere, N. (2007). Road segmentation supervised by an extended V-disparity algorithm for autonomous navigation. In *IEEE Intelligent Vehicles Symposium* (pp. 160–165). Istanbul, Turkey.

Thrun, S., Montemerlo, M., Dahlkamp, H., Stavens, D., Aron, A., & Diebel, J. (2006, September). Stanley: The robot that won the DARPA grand challenge: Research articles. *Journal of Robotic Systems, 23*, 661–692.

Torra, P. H. S., & Criminisi, A. (2004). Dense stereo using pivoted dynamic programming. *Image and Vision Computing, 22*(10), 795–806. doi:10.1016/j.imavis.2004.02.012

Vandapel, N., Huber, D., Kapuria, A., & Hebert, M. (2004). Natural terrain classification using 3-D ladar data. In *IEEE International Conference on Robotics and Automation* (Vol. 5, pp. 5117–5122).

Vandorpe, J., Van Brussel, H., & Xu, H. (1996). Exact dynamic map building for a mobile robot using geometrical primitives produced by a 2d range finder. In *IEEE International Conference on Robotics and Automation* (pp. 901–908). Minneapolis, USA.

Vapnik, V. N. (1995). *The nature of statistical learning theory*. New York, NY: Springer.

Yoon, K.-J., & Kweon, I. S. (2006). Adaptive support-weight approach for correspondence search. *IEEE Transactions on Pattern Analysis and Machine Intelligence, 28*(4), 650–656. doi:10.1109/TPAMI.2006.70

Yoon, S., Park, S.-K., Kang, S., & Kwak, Y. K. (2005). Fast correlation-based stereo matching with the reduction of systematic errors. *Pattern Recognition Letters, 26*(14), 2221–2231. doi:10.1016/j.patrec.2005.03.037

Zach, C., Karner, K., & Bischof, H. (2004). Hierarchical disparity estimation with programmable 3D hardware. In *International Conference in Central Europe on Computer Graphics, Visualization and Computer Vision* (pp. 275–282).

Zhao, J., Katupitiya, J., & Ward, J. (2007). Global correlation based ground plane estimation using V-disparity image. In *IEEE International Conference on Robotics and Automation* (pp. 529–534). Rome, Italy.

KEY TERMS AND DEFINITIONS

Disparity: The difference of an observed point's image coordinates when viewed under different viewpoints.

Disparity Map: An image constituted by the disparity values of each pixel, being thus equivalent to a depth map.

Fuzzy Inference System: A system that maps a set of inputs to output values using fuzzy logic.

Obstacle Avoidance: The procedure of avoiding obstacles during robot navigation.

Stereo Vision: The usage of two slightly differentiated images of the same scene in order to perceive the depth of the depicted objects.

Support Vector Machine: A supervised machine learning technique that performs non-probabilistic binary classification.

Traversability Estimation: The procedure of determining whether a robot can physically move across a region.

Section 4
Social Robotics

Chapter 13
Gesture Learning by Imitation Architecture for a Social Robot

J.P. Bandera
University of Málaga, Spain

L. Molina-Tanco
University of Málaga, Spain

J.A. Rodríguez
University of Málaga, Spain

A. Bandera
University of Málaga, Spain

ABSTRACT

Learning by imitation allows people to teach social robots new tasks using natural and intuitive inter-action channels. Vision is the main of these channels. This chapter describes a learning-by-imitation architecture that uses stereo vision to perceive, recognize, learn, and imitate social gestures.

This description is based on the identification of a set of generic components, which can be found in any learning by imitation architecture. It highlights the main contribution of the proposed architecture: the use of an inner human model to help perceiving, recognizing and learning human gestures. This allows different robots to share the same perceptual and knowledge modules. Experimental results show that the proposed architecture is able to meet the requirements of learning by imitation scenarios. It can also be integrated in complete software structures for social robots, which involve complex attention mechanisms and decision layers.

INTRODUCTION

Robots have been massively used in industrial environments for the last fifty years. Industrial robots are designed to perform repetitive, pre-dictable tasks, but are not able to easily adapt or learn new behaviours (Craig, 1986). In order to execute their programmed tasks, they have to sense only a constrained set of environmental parameters, thus perceptual systems mounted on industrial robots are simple, practical and task-oriented. On the other hand, they are designed to work in environments in which human presence is limited and controlled, if allowed. Thus, while their usefulness is evident, industrial robots are strongly limited. In order to remove these limita-tions, a new generation of robots began to appear more than thirty-five years ago (Inoue, Tachi,

DOI: 10.4018/978-1-4666-2672-0.ch013

Nakamura, Hirai, Ohyu, Hirai, Tanie Yokoi & Hirukawa, 2001). These robots were designed to cooperate with people in everyday activities, to adapt to uncontrolled environments and new tasks, and to become engaging companions for people to interact with. They usually benefit from sharing human perceptual and motor capabilities, and thus the term humanoid robot was used to name these agents. In the last decade, however, the difficulties of creating robots that resemble human beings have favoured the use of the more generic term social robot. Thus, today it is accepted that, although humanoid robots are certainly designed to be social, social robots do not need to be humanoid.

According to an early definition of social robot (Dautenhahn & Billard, 1999) social robots are agents designed to be part of an heterogeneous group. They should be able to recognize, explicitly communicate with and learn from other individuals in this group. They also possess history (i.e. they sense and interpret their environment in terms of their own experience). While this is a generic definition, in practice social robots are designed to work in human societies. Thus, later definitions of social robots present them as agents that have to interact with people (Breazeal, Brooks, Gray, Hancher, McBean, Stiehl & Strickon 2003). In this chapter the same ideas are followed, and social robots are understood as "robots that work in real social environments, and that are able to perceive, interact with and learn from other individuals, being these individuals people or other social agents" (Bandera, 2010, pp. 9).

Social robots have different options to achieve learning. Individual learning mechanisms (e.g. trial-and-error, imprinting, classical conditioning, etc.) are one of these options. However, their application to a social robot may lead it to learn incorrect, disturbing or even dangerous behaviours. Thus, they should be restricted to specific scenarios and tasks (e.g. games based on controlled stigmergy) (Breazeal et al., 2003; Bandera, 2010). Social learning mechanisms are a different option,

which allows the human teacher to supervise the learning process avoiding most issues of individual learning. Among different social learning strategies, learning by imitation appears as one of the most intuitive and powerful ones.

This chapter describes a RLbI architecture that provides a social robot with the ability to learn and imitate upper-body social gestures. This architecture, that is the main topic of the first author's Thesis (Bandera, 2010), uses an interface based on a pair of stereo cameras, and a model-based perception component to capture human movements from input image data. Perceived human motion is segmented into discrete gestures and represented using features. These features are subsequently employed to recognize and learn gestures. One of the main differences of this proposal with respect to previous approaches is that all these processes are executed in the human motion space, not in the robot motion space. This strategy avoids constraining the perceptual capabilities of the robot due to its physical limitations. It also eases sharing knowledge among different robots. Only if the social robot needs to perform physical imitation, a translation module is used that combines different strategies to produce valid robot motion from learned human gestures.

The rest of the chapter is organized as follows: A Background section firstly details state-of-the-art solutions and discusses their advantages and drawbacks. Then, the main corpus of the chapter starts introducing the set of components that can be identified in any RLbI architecture. These components are used to describe the RLbI architecture proposed in this chapter, which is deeply analyzed in the following sections. Thus, the Human Motion Perception section gives a description of the perceptual modules used to capture human gestures. In the Gesture Representation and Recognition section the methods used by the robot to encode and recognize gestures are detailed. The Learning section describes the learning system used to update the repertoire of the robot. As commented above, the differences between the bodies of the human and the robot suggest the use of a translation module to map the movements of the first to the later. The Motion Translation section describes the combined retargeting algorithm used to achieve this task.

The proposed RLbI architecture has been tested in daily life indoor environments, where different untrained users executed social gestures without using specific markers or color patches. The results of these experiments are detailed in this chapter, which concludes discussing the main conclusions and future research lines related to the proposed architecture.

BACKGROUND

As detailed above, social robots should be able to perceive, interact with and learn from people. The first step for a social robot to achieve these capabilities is to use natural and intuitive interaction and perception channels. Speech recognition (Breazeal et al., 2003) and tactile sensors (Asfour, Gyarfas, Azad & Dillmann, 2006) are important features for a social robot. But vision represents the sensory input that usually provides more information to people (Breazeal et al., 2003; Bandera, 2010). It is desirable for a social robot, then, to include a vision-based interface. Although recent alternatives to 3D visual perception, such as the combination of monocular images with light-based 3D scanners or TOF cameras, have begun to gain a growing popularity (Blanc, Oggier, Gruener, Weingarten, Codourey & Seitz, 2004; Kanekazi, Nakayama, Harada & Kuniyoshi, 2010), most social robots rely on stereo-vision systems for this interface. These systems can be easily mounted on the head of a robot, and they provide 3D information. Besides, they are more similar to human eyes than other solutions, and thus it may be easier for them to adapt to everyday environments, that are designed to be perceived by people (Azad,

Ude, Asfour & Dillmann, 2007). Stereo-vision systems, however, present drawbacks that have to be considered: they have limited resolution, field of view and frame rate, and they also have to deal with noisy images and depth ambiguity (Moeslund, Hilton & Krüger, 2006; Hecht, Azad & Dillmann, 2009).

Capturing images is just the first step in the perceptual process. It is necessary to extract only relevant information from the huge amount of visual input data if on-line response is required. Biological entities filter perceived information by attention (Bandura, 1969). Social robots focus also attention only in certain relevant features of the environment. Thus, Breazeal et al. (2003) considers controlled scenarios in which only certain defined objects are detected and tracked. Hecht et al. (2009) label different body parts and track them using particle filters. While their particular implementation may vary, attentional mechanisms are present in all perceptual components proposed for social robots (Schaal, 1999; Demiris & Hayes, 2002; Mühlig, Gienger, Hellbach, Steil & Goerick, 2009; Mohammad & Nishida, 2009).

According to the given definition, social robots are not only aware of their surroundings, but they are also able to learn from, recognize and communicate with other individuals. Robots that could learn from its observations and experiences, and from human teachers, would be able to adapt to new situations and perform new tasks, or improve already known ones. While other strategies are possible, robot learning by imitation (RLbI) represents a powerful, natural and intuitive mechanism to teach social robots new tasks (Schaal, 1999). In RLbI scenarios, a person can teach a robot by simply demonstrating the task that the robot has to perform. There are many issues that have to be addressed regarding RLbI. It is desirable to avoid invasiveness and controlled environments, the robot should provide on-line feedback for the human teacher, it may be required to research methods for sharing attention (Breazeal et al.,

2003), etc. One of the main of these issues is the translation from human to robot activities. This problem is more important as the differences from human to robot bodies grow (Nehaniv & Dautenhahn, 2002). Despite all these issues, the important advantages of RLbI systems over other learning methods (Schaal, 1999) have moved many researchers in the last decade to address the objective of providing social robots with RLbI architectures (Schaal, 1999; Demiris & Hayes, 2002; Breazeal, Brooks, Gray, Hoffmann, Kidd, Lee, Lieberman, Lockerd & Mulanda, 2004; Mühlig et al., 2009; Mohammad & Nishida, 2009). These architectures will allow social robots to perceive, recognize, learn and imitate behaviours exhibited by human companions.

MAIN FOCUS OF THE CHAPTER

System Overview

There have been many proposals of RLbI architectures in the last decade (Schaal, 1999; Demiris & Hayes, 2002; Breazeal et al., 2004; Domínguez, Zalama, García-Bermejo & Pulido, 2006; Mühlig et al., 2009; Bandera, 2010). These architectures are composed by a set of modules that address concrete tasks and that are connected between them. Some architectures divide these modules into two main parts: perceptual and motor (Schaal, 1999; Demiris & Hayes, 2002). However, other architectures may find difficulties in adapting to this structure, specially regarding the location of two key elements of any RLbI architecture: the knowledge database and the learning mechanism.

In this chapter, a recently proposed description of RLbI architectures is employed (Bandera, 2010). This description is based on the identification of a set of key components that are common to all RLbI architectures regardless their objectives, considered perceptual inputs, number of modules, levels of abstraction or structure. It is in the elements inside each of these components,

and the relations that are established between these elements, where the differences between RLbI architectures lie. A description of these components is given below:

- **Input:** This component includes all the sensory inputs that are available for the architecture. Visual input is the main of these inputs, as it is necessary to achieve imitation in RLbI scenarios (Schaal, 1999; Breazeal et al., 2003; Bandera, 2010). There are, however, additional sensory inputs that can help in the social learning process. Thus, many authors include audio sensors in the input component of RLbI architectures (Breazeal et al., 2004; Domínguez et al., 2006). Some other contributions also propose the use of tactile or proprioceptive -own state-perception (Breazeal et al., 2004; Lopes & Santos-Víctor, 2005; Calinon, 2007).

- **Perception:** The perception component contains all modules that are used to extract useful information from available perceptual channels. Raw data provided by sensors are useless for the robot. Thus, the audio information should be filtered to select only important messages. In order to offer on-line response for the artificial agent, a mechanism to focus attention should also be used, that is able to extract only relevant information from the huge amount of visual input data (Moeslund et al., 2006; Marfil, Bandera, Rodríguez & Sandoval, 2008). These attention mechanisms are also convenient to ease social interaction and provide the social robot with human-like features (e.g. people interacting with the robot appreciate that it focus on the interaction process, instead of distracting itself processing continuously every signal or object).

- **Knowledge:** People use their memories continuously in their daily life. For instance, facial expressions or social gestures are necessary to achieve social interaction. Thus, a social robot should know how to identify and execute them. The knowledge component represents the memory of the social robot. This component contains all elements that are used to store information units, both learned and preprogrammed. It also includes elements used to organize these data, reduce their dimensionality or transform them to a format that can be useful for other components.

- **Learning:** Social robots work in everyday environments where it is not possible to predict all possible situations they may face. Thus, social robots are provided with some learning mechanisms that allow them to adapt and learn from these new situations. The learning component is a key component of a RLbI architecture. It mainly affects the knowledge component, adding new items to the knowledge database, but also modifying already stored items or deleting old ones.

- **Action generation:** RLbI requires the social robot to be able to imitate behaviours. Imitation involves translating the perceived or learned motion to the robot, and generating a sequence of action commands. The action generation component contains all elements that are responsible of generating a sequence of robot actions.

- **Output:** Action commands are received by this component of the RLbI architecture, which uses the abilities of the social robot to physically execute them.

Figure 1 depicts the components of the RLbI architecture described in this chapter. This architecture captures, recognizes and learns upper-body human gestures. It is influenced by previous approaches, but it also incorporates new elements. Thus, as many previous RLbI architectures (Schaal, 1999; Demiris & Hayes, 2002; Mühlig et

al., 2009), it contains modules related to perception, and others related to action. It also includes a filtering process in the perceptual component. Finally, it relies on a database of known gestures to conform the core of the knowledge component (Schaal, 1999; Demiris & Hayes, 2002; Breazeal et al., 2003), as Figure 1 depicts.

The proposed architecture also presents important differences respect to most previous proposals. Thus, many of these proposals appeal for a unified representation of perception and action in the robot motion space (Schaal, 1999; Demiris & Hayes, 2002; Lopes & Santos-Víctor, 2005), following the idea that such a representation is present in biological entities (Meltzoff & Moore, 1989). But biological observers perform imitation and social learning from demonstrators of the same species (Bandura, 1969), or who are perceived as belonging to the same species. Small children may find difficulties in learning from imitation when the demonstrator is a machine (Meltzoff & Moore, 1989). It may be reasonable to suppose that the opposite situation will find similar difficulties. Using the robot motion space to represent perceived human motion constraints the perceptual capabilities of the robot due to its motor limitations, as the robot will not recognize gestures it is not able to imitate. In this chapter the idea of 'using a human model to perceive human movements' is proposed as an alternative to previous approaches. It has the additional advantage that translation -or retargeting- modules are executed only if imitation is required.

The explicit presence of a retargeting module is another important characteristic of the proposed architecture. It is considered here that human and robot may be very different, and thus translating motion from human to robot may become an important issue. This issue should not be addressed using simple one-to-one mapping techniques, but more complex strategies, that consider the physical differences in the bodies of the demonstrator and the imitator.

Finally, it is important to emphasize that the proposed RLbI architecture has been designed not as a theoretical framework, but as a system to be mounted in a real social robot. Thus, experiments have been performed in real RLbI scenarios, in which limited perception, uncontrolled environments and untrained users are present.

Human Motion Perception

In order to learn new gestures through imitation, the social robot needs to focus its attention in perceiving the motion of people that interact with it. This section describes the Human Motion Capture (HMC) system used by the proposed architecture. This HMC system relies on a model-based approach to meet the stringent requirements of RLbI scenarios. It includes the input and perception components of the RLbI architecture depicted in Figure 1.

Input

There are different options to achieve HMC using only visual information. Some proposals employ monocular images, but they have to use complex computation, such as probabilistic matching or optical flow, to extract human 3D pose (Kulic, Lee & Nakamura, 2009). Thus, they may find difficulties in meeting the time requirements of a system that has to be used in human interaction scenarios (Moeslund et al., 2006). Stereo vision, on the other hand, can access 3D data from the scene using disparity information. The RLbI architecture proposed in this chapter uses stereo vision to achieve visual perception. This system has a baseline close to the human one (about 10 centimeters) in order to be able to perceive human motion at typical social distances, between 1 and 2.5 meters. This solution is commonly used for visual perception systems that have to be mounted in social robots (Azad et al., 2007; Hecht et al., 2009).

Perception

It is necessary to extract useful information from the huge amount of data provided by a pair of stereo cameras. Vision-based HMC systems include very different approaches to achieve this, that can be divided into model-free and model-based approaches (Moeslund et al., 2006). Model-free methods directly map visual perception to pose space. This is a powerful tool to extract human pose from complex input data, such as cluttered natural scenes. However, these approaches are time consuming due to the involved search and matching process. Their results are also restricted to the poses demonstrated in the training phase, and extension to wider vocabularies of poses may introduce ambiguities in the mapping. Finally, they also have to deal with the problem of local minima or silhouette ambiguities (Agarwal & Triggs, 2006). Recent methods based on probabilistic assembly of body parts have achieved promising results in overcoming these limitations, but they are still bounded by training requirements (Demirdjian, Ko & Darrell, 2005), and are in general more suitable to surveillance or people detection applications, than to on-line HMC in uncontrolled scenarios (Moeslund et al., 2006).

Model-based approaches rely on the use of a human model to help in the human pose detection and tracking processes. Traditionally, model-based approaches required controlled environments and/or visual marks to achieve HMC (Moeslund et al., 2006). On the other hand, when compared to model-free approaches, they provide faster results, they are more robust against ambiguities and noisy input data, and they do not require executing complex training phases as the model itself stores information about human kinematics and dynamics (Agarwal & Triggs, 2006; Moeslund et al., 2006). In the last years different approaches have appeared that benefit from these advantages while they are able to be used in real dynamic environments (Kojo, Inamura, Okada

Figure 1. System overview

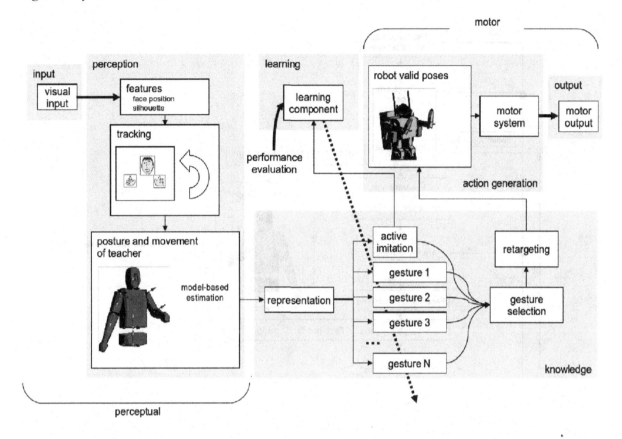

& Inaba, 2006; Azad et al., 2007; Hecht et al., 2009; Bandera, 2010). These approaches offer on-line response but, on the other hand, their use of a short stereo baseline provides only one perspective of the performed motion (Hecht et al., 2009). Thus, these systems are sensitive to occlusions. Besides, there are certain body parts, e.g. the elbows or shoulders, that are not going to be directly perceived by a markerless HMC system, but have to be estimated. These issues introduce errors in the HMC process that should be carefully considered.

The RLbI architecture proposed in this chapter uses a model-based HMC system to infer the upper-body pose of a human performer in daily life environments. This performer wears no markers, color patches nor specific garments. The system will be affected by the issues previously men-

tioned, but different strategies that are explained below have been used to minimize them. Figure 2 shows the flow diagram of the proposed HMC system. The different modules that compose it are described below.

The first step to capture the motion of a human is to locate him/her. This task is achieved by looking for a human face that is close enough to the cameras, and that looks to the face of the social robot (i.e. a frontal face perceived by the cameras of the robot). The 'Evaluation of salient regions' module depicted in Figure 2 performs this task. This module is designed to deal with scenes in which different regions, not only human faces, can drive the attention of the robot. Thus, it enables to integrate the RLbI architecture into complex attention mechanisms. An example of such integration has been implemented in the recent pro-

Figure 2. Flow diagram of the proposed HMC system

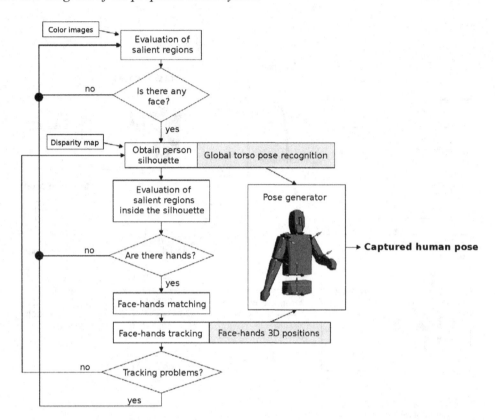

posal of Palomino, Marfil, Bandera, and Bandera (2011).

Once a human face has been located, its 3D position is extracted, using disparity information. The 3D position of the face is then employed to extract the silhouette of the human performer from the disparity map. This process is achieved by following these steps, that are deeply explained in Cruz, Bandera, and Sandoval (2009): (i) threshold the disparity map around the disparity value of the face; (ii) apply connected components to filter undesired objects; (iii) refine silhouette borders performing an and logical operation between the disparity silhouette and Canny borders of the color image, extracted in the k vicinity of the disparity silhouette; and (iv) perform a closing morphological operation to the resulting silhouette to reduce noise.

After having obtained a robust silhouette for the human performer, colour regions inside it

are evaluated, with the exception of the already labeled face region. The two biggest skin coloured regions found in this search process are labeled as the hands of the human. Once face and hands have been located using this procedure, they are tracked using a hierarchical algorithm (Marfil, Bandera, Rodríguez & Sandoval, 2004). This algorithm tracks desired regions on-line, and it implements an occlusion recovery method based on the use of stored region centroids and shapes. 2D position data obtained by this tracking algorithm for the face and the hands are used to obtain 3D positions of head and hands for each frame. As Figure 2 depicts, this process is only restarted if tracking problems (e.g. due to prolonged occlusions or face-hand overlapping) are detected (Bandera, 2010).

Disparity silhouettes and tracked 3D positions of face and hands are the features used to estimate upper-body human pose. As commented above, a human model is used to perform this task using an algorithm divided into two steps. Firstly, anthropometric relations are applied to the extracted silhouette to infer torso bending from the medium axis of the perceived torso. Torso rotation is also estimated from the comparison of the disparity values in the estimated positions of left and right shoulders. Once these parameters have been computed, they are applied to the torso of the human model to make it imitate the torso pose of the human, under the rigid torso assumption (Cruz et al., 2009). The second step of the proposed algorithm computes the pose of the arms from the 3D positions of the left and right hands with respect to the head. This second step of the pose generator uses an inverse kinematics algorithm that provides a fast candidate for the pose of the arms. Then, this candidate is evaluated in order to check if the resulting pose is valid or not (invalid poses can be produced by partial or noisy input data). If the pose is not valid (i.e. it produces a collision between different body segments or it makes a joint angle go beyond its limits) alternative poses are evaluated that put the hand of the model as close as possible to the detected human hand. A complete description of this algorithm can be found in a previous contribution of the authors (Bandera, Marfil, Molina-Tanco, Rodríguez, Bandera & Sandoval, 2007). When the definitive joint angles for the arms have been obtained, they are applied to the human model, that finally adopts a pose that imitates perceived upper-body pose.

Quantitative Evaluation

The HMC system conformed by the input and perception components was quantitatively tested in a set of experiments, in which a certain human motion was perceived using both the proposed vision-based HMC system and a Codamotion CX1 HMC system from Charnwood Dynamics Ltd., based on active markers. The latter was used to obtain a reliable ground-truth (Bandera, 2010). Stereo images for the proposed system were captured using a STH-DCSG-VAR-C stereo pair provided by Videre Design. Extensive tests were conducted, in which the positions of the real markers used by the Codamotion CX1 system were compared against the positions of virtual markers, located on the virtual human model used by the proposed vision-based HMC system. Table 1 depicts obtained errors (Bandera, 2010).

The benefits of using a human model to filter perceived data have also been evaluated qualitatively. The model guarantees that the poses in the inner representation the robot uses are valid. It filters outliers and noisy perceptual data (Bandera, 2010), provides a set of valid joint angles for each set of perceived 3D positions, and helps the system dealing with occlusions and tracker losses. For instance, shadows in the elbow cause frequent silhouette crops that eliminate the tracked arm. The model allows estimating arm poses for some frames using third order spline interpolation, which eases producing smooth and natural movements in the inner model.

Table 1. Tracking errors averaged over 5300 frames

Marker	Left Shoulder	Left Elbow	Left Hand	Abdomen
Mean error (cm)	5.74	12.53	11.51	7.76
Standard deviation (cm)	3.13	6.06	6.55	1.18
Marker	Right Shoulder	Right Elbow	Right Hand	Head
Mean error (cm)	6.72	12.41	11.47	6.77
Standard deviation (cm)	5.01	6.94	7.63	5.27

Gesture Representation and Recognition

As Figure 1 depicts, the knowledge component of the proposed RLbI architecture contains modules that firstly encode the gestures in an efficient representation, and then use these representations to compare the perceived gesture with the repertoire of the robot.

Gesture Representation

Before perceived motion can be compared against stored gestures, it has to be segmented into discrete gestures. It may be complex to find a general method to perform this task, as different authors propose different alternatives (Breazeal et al., 2003; Kojo et al., 2006). This chapter follows the criterion that a short pause in the motion signals the start and the end of a gesture (Calinon, 2007). Segmentation is based on dynamic time thresholds that detect these pauses. Once perceived motion has been segmented into discrete gestures, the proposed RLbI architecture represents them in an adequate format to be analyzed, recognized and learned. Different symbolic representations have been proposed to help encoding behaviours in an efficient way to allow generalization (Demiris & Hayes, 2002; Breazeal et al., 2004; Alissandrakis, Nehaniv & Dautenhahn, 2007). These symbolic representations ease interactive learning, but they rely on pre-determination of the observed cues and the efficiency of the segmentation process, thus restricting the possible movement repertoire. Besides, it may be difficult to find the optimal granularity level to represent generic gestures (Schaal, 1999).

Another approaches to encode perceived gestures represent them as continuous streams of data. While learning becomes a more complex issue for these solutions, they are more suitable to generalize and refine stored gestures (Calinon, 2007). The RLbI architecture proposed in this chapter perceives gestures using stereo vision, thus these streams of data will be composed by Cartesian 3D trajectories followed by different body parts. While it is possible to use these representations to achieve recognition and learning, increasing the sampling ratio, the number of tracked body parts or the gesture length can lead to excessively large descriptors. This problem can be tackled by dimensionality reduction methods, such as PCA or LWPR, although they can be overcome by methods specifically oriented to the proposed scenario (Bandera, Marfil, Molina-Tanco, Rodríguez, Bandera & Sandoval, 2009). Hidden Markov Models (HMMs) may also be used to encode gestures. They ease generalization, but are very sensitive to segmentation errors (Calinon, 2007). Other option is to define, and select for each gesture, a reduced set of features that describe the trajectories associated to different body parts. Previous works have addressed this problem of trajectory representation using global trajectory features, which are defined in relation to an external reference (Croitoru, Agouris & Stefanidis, 2005), or using local trajectory features, which are based on differential measures (Rodríguez, Last, Kandel & Bunke, 2004). The main advantage of the global features is their robustness to outliers and noise. On the contrary, they face major difficulties in capturing fine details of trajectories. Local features are superior in discriminating fine details, but they are usually highly sensitive to outliers and noise (Alajlan, Rube, Kamel & Freeman, 2007).

In this chapter both local and global features are used. Local matching provides discrimination ability, but local results are reinforced using global features, in order to increase the robustness of the results respect to outliers and noise. Thus, local features are computed as dominant points of perceived 3D trajectories. Dominant point extraction is based on computing the curvature function of each trajectory and selecting points in which curvature experiences high variations. Additional points may be added to properly encode trajectories in which the motion is smooth (Bandera et al., 2009). While the use of standard curvature estimations has been reported to originate segmentation problems in RLbI scenarios (Calinon, 2007), the proposed RLbI architecture uses an adaptive curvature function to alleviate these problems and provide accurate and robust curvature estimation. Global features, on the other hand, are used not to discriminate between gestures, but to reinforce local matching. Thus only some robust and simple global cues are extracted from perceived trajectories. These global features are measures of perceived absolute and relative amplitudes. A complete description of these local and global features, and an evaluation of their ability to efficiently encode gestures, can be addressed in a previous contribution of the authors (Bandera et al., 2009). As detailed there, the proposed local features outperform other feature extraction algorithms, while global features are robust and can be extracted from input trajectories using simple analytic relations.

Gesture Recognition

Once local and global features are extracted, they are used to recognize perceived gestures. While Hidden Markov Models (HMMs) represent the current state of the art in gesture recognition, their training requirements represent an issue for systems that have to learn incrementally in interactive, on-line scenarios (Rajko, Qian, Ingalls & James, 2007). As commented above, the limitation in the number of states that can be implemented also becomes a drawback for a HMM-based recognizer.

Dynamic Programming (DP) alignment techniques have been widely used for matching purposes, became a standard in speech recognition and are commonly used to match 3D trajectories (Croitoru et al., 2005). They are not bounded by training requirements as HMM, Support Vector Machines, PCA or LDA, and they can deal with time shifting, that commonly appears when matching different performed gestures. Thus, recent works have incorporated these techniques to gesture recognition systems (Calinon, 2007; Mühlig

et al., 2009). The RLbI architecture proposed in this chapter follows these contributions and uses the Dynamic Time Warping (DTW) algorithm to compare local features. Other DP techniques and traditional classification techniques have also been evaluated, but DTW overcome them (Bandera, 2010).

The resulting local distances are reinforced by a global similarity value, obtained by applying analytic relations to the global features of compared gestures. Thus, both local and global features are considered to obtain the final distances, expressed as confidence values (Bandera et al., 2009). These values indicate the degree of similarity between each gesture in the knowledge database and the perceived gesture. They will be further used in the learning component to decide whether the gesture is recognized or not.

As Figure 1 depicts, the whole representation and recognition processes are performed in the human motion space. Thus, even social robots which body is very different to the human body will be able to correctly perceive, understand and learn human gestures when using the proposed architecture.

Learning Component

The learning component of the proposed RLbI architecture uses the confidence values, obtained in the recognition stage, to add new gestures to the repertoire of the robot. The dataflow of this component is depicted in Figure 3. It can be seen that learning is based on a double threshold. The first threshold Ω allows to directly recognize gestures that are very similar to a stored one (i.e. the biggest obtained confidence value C_{iI} is over Ω). Recognized gestures do not modify the knowledge database. On the other hand, as Figure 3 shows, the second threshold ω is relative. Gestures that do not satisfy this second threshold are candidates to be included in the repertoire as new gestures. Human supervision is required when adding new gestures to the database. Besides, the first experiments involving this algorithm showed that the first steps in the learning process were critical. Thus, the amount of human supervision was incremented in this stage of the process (Figure 3), as suggested by previous works that emphasize the importance of supervision in robot learning processes (Breazeal et al., 2004; Kojo et al., 2006; Calinon, 2007).

Motion Translation

As commented above, one of the main issues related to RLbI is the translation of the perceived human motion to the robot motion space. This translation has been referred to by some computer animation researchers as retargeting (Gleicher, 1998). If the bodies of the imitator and the demonstrator are very similar, translation may be performed using a straightforward, one-to-one mapping. But if these bodies are different more complex strategies are required, that involve many-to-one, one-to-many and many-to-many mappings. Partial mappings

also appear as an useful approach, in which only some characteristics of the perceived motion are mapped to the imitator (Alissandrakis et al., 2007). On the other hand, some authors state that direct mapping should not be used even when the same embodiment is shared, as different factors that influence the behaviours of the demonstrator and the imitator have to be taken into account (e.g. close objects) (Nehaniv & Dautenhahn, 2002).

The RLbI architecture proposed in this chapter implements a retargeting system based on previous works in the field of 3D computer animation (Shin, Lee, Shin & Gleicher, 2001). Thus, it uses a fast retargeting system based on a weighted combination of two different strategies, which tend to preserve: (i) end-effector positions; or (ii) joint angles. The weights of the factors used in the final solution depend on the amplitude of the perceived gesture (Bandera, 2010). The combined retargeting algorithm will try to preserve the positions of the end-effectors for static, or slow, precise gestures. On the other hand, social gestures like

Figure 3. Flow diagram of the learning component

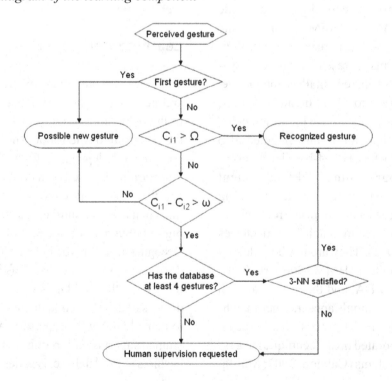

Table 2. Description of the social gestures performed to test the RLbI system

Gesture name	Gesture description
Shake hands	Shake hand gesture. The performer offers his/her right hand. Sometimes he/she shakes it.
Hi	The performer waves his/her right hand near his/her face.
Attention	This movement is similar to Hi, but the movement is made with both hands.
Left	The performer points left with his/her left hand.
Right	The performer points right with his/her right hand.
Shrug	Shoulders shrug and right and left hand moves up slightly.
Fish	Both hands move forward and perform a parallel little up-down movement. Different distances between hands ('fish sizes') were considered.
Me	The performer reaches his/her chest with his/her right hand.

waving hands to mean 'hello' or 'goodbye' do not require a very precise end-effector position, but they need to preserve the main characteristics of joint angle trajectories in order to be correctly imitated. Thus, the combined retargeting strategy will reinforce the weight of the second strategy for these gestures. This differentiation satisfies the classification established by Smyth and Pendleton (1990) for human movements, in which they are divided into location movements, that may be identified with the prior, and configured movements, that includes the later.

As Figure 1 depicts, the retargeting process does not need to be executed if the robot is not going to imitate the perceived motion.

The retargeted motion is not directly sent to the motors of the robot. The action generation component uses a virtual model of the robot before to check that the resulting poses are valid. The robot model adopts desired poses using the same algorithms employed in the perception component to pose the human model. Once valid poses have been obtained for the robot, they are sent to its motor system (i.e. the output component), to make the robot physically imitate perceived or recognized gestures.

EXPERIMENTAL RESULTS

Different experiments have been performed in order to analyze the different steps that lead to correctly imitate a set of perceived gestures. These gestures are recorded using a STH-DCSG-VAR-C stereo pair provided by Videre Design. The height of the cameras is 1.50 meters, and its distance to the performer varies from 1.30 to 1.80 meters. No specific markers nor garments were used, and the test were conducted in daily life, uncontrolled in-

Table 3. Gesture recognition using different local distance functions. Results after applying reinforcement with global distances are shown under the columns marked with R.

	DTW		EDR		ERP		LCSS		Euclidean dist.	
		R		R		R		R		R
$\Delta conf$	0.16	0.19	0.19	0.22	0.15	0.12	0.16	0.2	0.13	0.12
secs.	0.06	0.06	0.06	0.06	0.08	0.08	0.03	0.03	$\rightarrow 0$	$\rightarrow 0$
1-NN	93%	100%	72%	76%	21%	45%	69%	86%	24%	41%
3-NN	76%	79%	48%	52%	28%	41%	48%	62%	14%	21%

Figure 4. Upper-body social gestures used to test the proposed RLbI system. The trajectories of the left and right hands have been marked over the left frame.

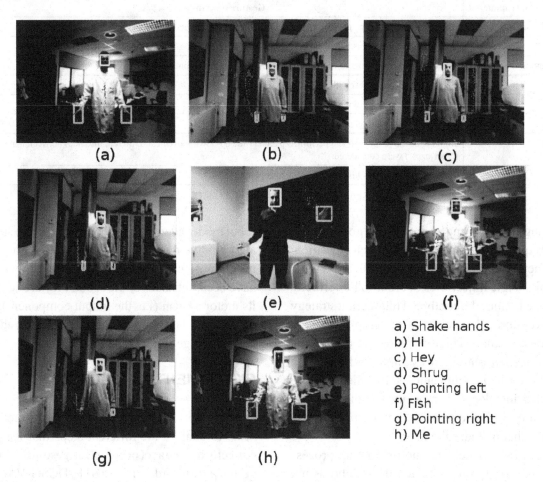

a) Shake hands
b) Hi
c) Hey
d) Shrug
e) Pointing left
f) Fish
g) Pointing right
h) Me

door environments, affected by light variations and partial occlusions. All experiments use a dataset consisting of 53 upper-body gestures performed by six different people. For each of these gestures, the motion of the left and right hands is recorded at an average sampling rate of 15 Hz. The average amount of samples per gesture is 103.5, thus each gesture, composed by two trajectories, is characterized by an average of 207 XYZ values.

The gestures in the dataset are different executions of 8 upper-body gestures, which are commonly found in social interaction scenarios. Table 2 gives their description, and indicates whether they are performed using one hand or both hands. Figure 4 shows the perceived trajectories for one

execution of each of these gestures. These trajectories are superimposed to one of the images taken by the left camera of the Videre pair during these tests. Bandera (2010) provides more images depicting additional performers and scenarios.

A social robot, named NOMADA, is currently being developed at ISIS Group, in the University of Málaga (Spain). While the construction of this robot is still not finished, the final tests of the system will retarget perceived motion to a virtual model of the NOMADA robot. Then, the motion of the right arm (that has been already implemented) will be translated from this model to the physical arm to check the performance of the system when real motion is involved.

Gesture Recognition

For these tests the system is provided with three demonstrations of each particular gesture. These demonstrations are correctly classified and labeled. Then, the remaining 29 gestures in the dataset are fed to the gesture recognition system. Table 3 show obtained results when Dynamic Time Warping (DTW) technique, used in the proposed system, is compared against Edit Distance on Real Sequence (EDR) (Chen, Tamer & Oria, 2005), Edit Distance with Real Penalty (ERP) (Chen et al., 2005), Longest Common Subsequences (LCSS) (Croitoru et al., 2005) and Euclidean distance. *k*-nearest neighbor (*k*-NN) algorithm is used to measure the performance of evaluated methods. Two different *k* values (1, 3) are used. Execution time of the distance evaluation is also measured. Δ*conf* is the distance from the first to the fourth nearest neighbour. For each tested method, results

before and after (**R**) applying global reinforcement are depicted.

ERP is too sensitive to time shifts and thus its results are not good in a RLbI scenario, in which different performers, and different executions of the same performer, will present important temporal differences. EDR and LCSS offer poor recognition rates even when reinforced with global features. Euclidean distance is the fastest method, but it is not robust against noise, outliers and time shifting. This leads to poor recognition rates. The results provided by DTW, on the other hand, are the best of the comparative study in terms of both recognition rates and robustness. DTW proves also to be able to adapt to the noisy trajectories used in this scenario, as its 1-NN recognition rates are nearly similar to those obtained when matching gestures perceived using an HMC system based on active markers (Bandera et al., 2009).

Figure 5. Right hand xyz trajectory, over which gestures starting and ending points have been marked using slashed and solid vertical bars, respectively

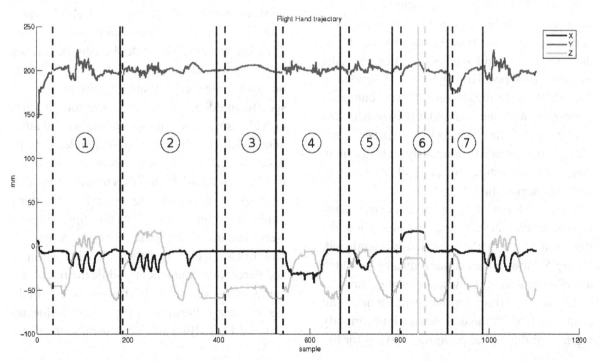

Table 4. Experiments and measures in frames per second

Measured module	Original RLbI	Nerve-RLbI (local)	Nerve-RLbI (distributed)
Human Motion Perception	8.41	10.16	14.98
Motion translation	26.80	27.50	40.23

Results depicted in Table 3 have been compared against results obtained using a commercial, model-free system based on active markers (Bandera et al., 2009; Bandera, 2010). While this second system provides more accurate 3D positions for the used markers (the average error is ten times lower than in the proposed perceptual system), obtained results are similar, due to the use of the human model to filter input data, and to the robustness and adaptability of the DTW algorithm (Bandera, 2010).

Gesture Segmentation and Learning

The last series of experiments tests the ability of the proposed RLbI architecture to learn and recognize gestures from different human demonstrators in a real interaction scenario. As commented above, these tests will involve physical imitation of the executed gestures. These gestures, described in Table 2, are performed by four different untrained performers and one additional person who had frequently used the system before. Each performer executes the gestures continuously, leaving a small pause between gestures in order to allow the system correctly segmenting the motion.

Perceived motion for these last experiments is segmented into discrete gestures using the threshold-based method previously mentioned. Figure 5 shows an example sequence in which a performer executes seven different gestures. It can be seen that results are adequate for nearly all gestures, although the sixth gesture is uncorrectly segmented due to the performer staying for too long in the same position during the execution. If the duration of the segment of the sixth gesture that produces the error is compared against the interval between consecutive gestures, it can be concluded that these errors are not going to be common as long as the performers are told to pause for a short time between gestures, and try not to stop for a long time while performing a certain gesture. In any case, these incorrect gestures will be detected as unrecognized executions, thus they can be discarded in the supervised learning process.

The trajectories followed by the inner human model are translated to the robot motion space using the combined retargeting strategy previously described. Bandera (2010) provides a deep evaluation of the position and angle errors that affect the final robot pose when position, angle, or combined retargeting strategies are used. Results discussed there (Bandera, 2010, pp. 170-171) show that the proposed combined retargeting algorithm approaches the optimal solution for both static and dynamic gestures.

The last performed tests also involved an evaluation of the performance of the overall system. This evaluation was conducted in three different implementations of the proposed architecture. The first of these implementations used a multi-threaded process to execute the complete RLbI architecture. The second and third implementations integrate the RLbI architecture in Nerve, a specific middleware for networked robotics that provides scalability and quality-of-service attributes (Martínez, Romero-Garcés, Bandera, Marfil & Bandera, 2012). While the second implementation divides the RLbI in different multi-threaded processes that run on the same PC, the third implementation fully exploits the capabilities of Nerve and distributes these processes in two different PCs. One of the PCs deals with the Input and Perception components, while the rest of the components are executed in the second PC. Table 4 summarizes the frames per second achieved in these tests. Both the implementation of the

RLbI architecture using Nerve and the complete experimental setup and results of these tests are deeply detailed in Martínez et al. (2012).

The results of these tests show that the RLbI architecture is able to capture the motion, retarget it to the NOMADA arm, recognize gestures and update the knowledge database on-line, using the supervised learning algorithm previously detailed. All users were satisfied with the system response, which was reported as adequate for human interaction rates.

Finally, the proposed RLbI architecture has been qualitatively compared against other RLbI architectures. Bandera (2010) includes a detailed description of the main RLbI architectures related to this proposal, and a comparative discussion about the characteristics, performance and limitations of the proposed system (Bandera, 2010, pp.210-213).

FUTURE RESEARCH DIRECTIONS

Limitations detected in the stereo vision system suggest that future work should mainly focus on increasing the perceptual capabilities of the social robot, using multimodal interfaces. More precisely, speech recognition is currently being incorporated to the robot as it is a key element in most social interactions. Other sensory inputs such as laser range finders or TOF cameras are also being considered. On the other hand, the inner representation the robot uses to store perceived data is limited to a virtual human model, thus interactions with objects of with other people cannot be perceived nor learned using this proposal. The inclusion of objects, multiple agents –people or other social robots-, environmental and inner conditions, or semantics data in the inner representation of the social robot are main topics in the current research of the authors. Other research work that is currently being conducted by the authors is related to the use of adaptable human models to increase the accuracy and functionality

of the HMC process. Finally, the complete RLbI architecture will be integrated in a more complex system, that will include higher level decision layers. A first example of this integration process has been recently published (Palomino et al., 2011).

CONCLUSION

The main contribution of this chapter is a RLbI architecture that can be integrated in a social robot. This architecture works at human interaction rates, it is non-invasive and can be used in uncontrolled environments, and by untrained users. Conducted tests show that upper-body gestures can be efficiently perceived, recognized and learned using the proposed architecture.[1]

REFERENCES

Agarwal, A., & Triggs, B. (2006). Recovering 3D human pose from monocular images. *IEEE Transactions on Pattern Analysis and Machine Intelligence, 28*, 44–58. doi:10.1109/TPAMI.2006.21

Alajlan, N., Rube, I. E., Kamel, M., & Freeman, G. (2007). Shape retrieval using triangle area representation and dynamic space warping. *Pattern Recognition, 40*, 1911–1920. doi:10.1016/j.patcog.2006.12.005

Alissandrakis, A., Nehaniv, C., & Dautenhahn, K. (2007). Correspondence mapping induced state and action metrics for robotic imitation. *IEEE Transactions on Systems, Man, and Cybernetics - Part B: Special Issue on Robot Learning by Observation. Demonstration and Imitation, 37*(2), 299–307.

Asfour, T., Gyarfas, F., Azad, P., & Dillmann, R. (2006). Imitation learning of dual-arm manipulation tasks in humanoid robots. In *6th IEEE RAS International Conference on Humanoid Robots* (pp. 40-47). Genoa, Italy: IEEE Press.

Azad, P., Ude, A., Asfour, T., & Dillmann, R. (2010). Stereo-based markerless human motion capture for humanoid robot systems. In *IEEE International Conference on Robotics and Automation (ICRA 2007)* (pp. 3951-3956). IEEE Press.

Bandera, J. P. (2010). *Vision-based gesture recognition in a robot learning by imitation framework*. Málaga, Spain: University of Málaga Press.

Bandera, J. P., Marfil, R., Bandera, A., Rodríguez, J. A., Molina-Tanco, L., & Sandoval, F. (2009). Fast gesture recognition based on a two-level representation. *Pattern Recognition Letters, 30*, 1181–1189. doi:10.1016/j.patrec.2009.05.017

Bandera, J. P., Marfil, R., Molina-Tanco, L., Rodríguez, J. A., Bandera, A., & Sandoval, F. (2007). Robot learning by active imitation. In Hackel, M. (Ed.), *Humanoid robots: Human-like machines* (pp. 519–544). Vienna, Austria: I-Tech Education and Publishing.

Bandura, A. (1969). Social learning theory of identificatory processes. In *Handbook of socialization theory and research* (pp. 213–262). Chicago, IL: Rand-McNally.

Blanc, N., Oggier, T., Gruener, G., Weingarten, J., Codourey, A., & Seitz, P. (2004). Miniaturized smart cameras for 3d-imaging in real-time. In *IEEE Sensors* (pp. 471–474). Vienna, Austria: IEEE Press. doi:10.1109/ICSENS.2004.1426202

Breazeal, C., Brooks, A., Gray, J., Hancher, M., McBean, J., Stiehl, D., & Strickon, J. (2003). Interactive robot theatre. *Communications of the ACM, 46*, 76–84. doi:10.1145/792704.792733

Breazeal, C., Brooks, A., Gray, J., Hoffman, G., Kidd, C., & Lee, H. (2004). Humanoid robots as cooperative partners for people. *International Journal of Humanoid Robots, 1*(2), 1–34.

Calinon, S. (2007). *Continuous extraction of task constraints in a robot programming by demonstration framework*. Unpublished doctoral dissertation, École Polytechnique Fédérale de Lausanne, Lausanne, EPFL.

Chen, L., Tamer, M., & Oria, V. (2005). Robust and fast similarity search for moving object trajectories. In *Special Interest Group on Management of Data (SIGMOD 2005)* (pp. 491-502). New York, USA.

Craig, J. (1986). *Introduction to robotics: Mechanics and control*. Boston, MA: Addison-Wesley.

Croitoru, A., Agouris, P., & Stefanidis, A. (2005). 3D trajectory matching by pose normalization. In *13th ACM International Symposium on Advances in Geographic Information Systems (ACM-GIS'05)* (pp. 153-162).

Cruz, A., Bandera, J. P., & Sandoval, F. (2009). Torso pose estimator for a robot imitation framework. In *12th International Conference on Climbing and Walking Robots and the Support Technologies for Mobile Machines (CLAWAR 2009)* (pp. 901-908). Istanbul, Turkey.

Dautenhahn, K., & Billard, A. (1999). Bringing up robots or -The psychology of socially intelligent robots: From theory to implementation. In *Third Annual Conference on Autonomous Agents* (pp. 366-367). Seattle, Washington, USA.

Demirdjian, D., Ko, T., & Darrell, T. (2005). Untethered gesture acquisition and recognition for virtual world manipulation. *Virtual Reality (Waltham Cross), 8*, 222–230. doi:10.1007/s10055-005-0155-3

Demiris, J., & Hayes, G. (2002). *Imitation as a dual-route process featuring predictive and learning components: A biologically plausible computational model*. Cambridge, MA: MIT Press.

Domínguez, S., Zalama, E., García-Bermejo, J., & Pulido, J. (2006). Robot learning in a social robot. In S. Nolfi, G. Baldassarre, R. Calabretta, J. Hallam, D. Marocco, J.A. Meyer, O. Miglino, & D. Parisi (Eds.), *From Animals to Animats 9, Vol. 4095 of LNCS* (pp. 691-702). Berlin, Germany: Springer.

Gleicher, M. (1998). Retargetting motion to new characters. In *25ᵗʰ Annual Conference on Computer Graphics and Interactive Techniques (SIGGRAPH '98)* (pp. 33-42). New York, NY: ACM.

Hecht, F., Azad, P., & Dillmann, R. (2009). Markerless human motion tracking with a flexible model and appearance learning. In *IEEE International Conference on Robotics and Automation (ICRA 2009)* (pp. 3173-3179). Kōbe, Japan: IEEE Press.

Inoue, H., Tachi, S., Nakamura, Y., Hirai, K., Ohyu, N., & Hirai, S. Tanie, K., Yokoi, K., & Hirukawa, H. (2001). Overview of humanoid robotics project of meti. In *32nd International Symposium on Robotics* (pp. 1478-1482).

Kanekazi, A., Nakayama, H., Harada, T., & Kuniyoshi, Y. (2010). High-speed 3D object recognition using additive features in a linear subspace. In *IEEE International Conference on Robotics and Automation (ICRA 2010)* (pp. 3128-3134). Anchorage, AK: IEEE Press.

Kojo, N., Inamura, T., Okada, K., & Inaba, M. (2006). Gesture recognition for humanoids using proto-symbol space. In *6th IEEE RAS International Conference on Humanoid Robots* (pp. 76-81). Genoa, Italy: IEEE Press.

Kulic, D., Lee, D., & Nakamura, Y. (2009). Whole body motion primitive segmentation from monocular video. In *IEEE International Conference on Robotics and Automation (ICRA 2009)* (pp. 3166-3172). Kōbe, Japan: IEEE Press.

Lopes, M., & Santos-Victor, J. (2005). Visual learning by imitation with motor representations. *IEEE Transactions on Systems, Man, and Cybernetics. Part B, Cybernetics, 35,* 438–449. doi:10.1109/TSMCB.2005.846654

Marfil, R., Bandera, A., Rodríguez, J. A., & Sandoval, F. (2004). Real-time template-based tracking of non-rigid objects using bounded irregular pyramids. In *IEEE/RSJ International Conference on Intelligent Robotics and Systems,* Vol. 1 (pp. 301-306). Sendai, Japan: IEEE Press.

Marfil, R., Bandera, A., Rodríguez, J.A., & Sandoval, F. (2008). A novel hierarchical framework for object-based visual attention. *Attention in Cognitive Systems, LNCS 5395.*

Martínez, J., Romero-Garcés, A., Bandera, J. P., Marfil, R., & Bandera, A. (2012). (in press). A DDS-based middleware for quality-of-service and high-performance networked robotics. *Concurrency and Computation.* doi:doi:10.1002/cpe.2816

Meltzoff, A., & Moore, M. (1989). Imitation in newborn infants: Exploring the range of gestures imitated and the underlying mechanisms. *Developmental Psychology, 25,* 954–962. doi:10.1037/0012-1649.25.6.954

Moeslund, T., Hilton, A., & Krüger, V. (2006). A survey of advances in vision-based human motion capture and analysis. *Vision and Image Understanding, 104,* 90–126. doi:10.1016/j.cviu.2006.08.002

Mohammad, Y., & Nishida, T. (2009). Interactive perception for amplification of intended behavior in complex noisy environments. *AI & Society, 23*(2), 167–186. doi:10.1007/s00146-007-0137-y

Mühlig, M., Gienger, M., Hellbach, S., Steil, J. J., & Goerick, C. (2009). Task-level imitation learning using variance-based movement optimization. In *IEEE International Conference on Robotics and Automation (ICRA 2009)* (pp. 1177-1184). Kōbe, Japan: IEEE Press.

Nehaniv, C. L., & Dautenhahn, K. (2002). The correspondence problem. In Dautenhahn, K., & Nehaniv, C. L. (Eds.), *Imitation in animals and artifacts* (pp. 41–61). Cambridge, MA: MIT Press.

Palomino, A., Marfil, R., Bandera, J. P., & Bandera, A. (2011). A novel biologically inspired attention mechanism for a social robot. *EURASIP Journal on Advances in Signal Processing*, (n.d), 2011.

Rajko, S., Qian, G., Ingalls, T., & James, J. (2007). Real-time gesture recognition with minimal training requirements and on-line learning. In *IEEE Conference on Computer Vision and Pattern Recognition (CVPR'07)* (pp. 1-8). IEEE Press.

Rodríguez, W., Last, M., Kandel, A., & Bunke, H. (2004). 3-dimensional curve similarity using string matching. *Robotics and Autonomous Systems*, *49*, 165–172. doi:10.1016/j.robot.2004.09.004

Schaal, S. (1999). Is imitation learning the route to humanoid robots? *Trends in Cognitive Sciences*, *3*(6), 233–242. doi:10.1016/S1364-6613(99)01327-3

Shin, H. J., Lee, J., Shin, S. Y., & Gleicher, M. (2001). Computer pupettry: An importance based approach. *ACM Transactions on Graphics*, *20*(2), 67–94. doi:10.1145/502122.502123

Smyth, M. M., & Pendleton, L. R. (1990). Space and movement in working memory. *The Quarterly Journal of Experimental Psychology Section A*, *42*, 291–304. doi:10.1080/14640749008401223

KEY TERMS AND DEFINITIONS

Autonomous Robots: Robots that can perform tasks in unstructured environments without requiring continuous human guidance.

Feature-Based Trajectory Representation: A method to reduce dimensionality of perceived trajectories, in which these trajectories are represented using features. These features can be local or global.

Human Motion Capture: The process of extracting the motion of a person from perceived input data.

Retargeting: Translation or mapping from the motion space of one agent to the motion space of another agent.

Robot Learning By Imitation: A specific social learning mechanism applied to social robots. The robot learns behaviours, tasks or motion by observing the demonstrations performed by a human teacher.

Social Robots: Robots that work in real social environments, and that are able to perceive, interact with and learn from other individuals, being these individuals people or other social agents.

Visual Perception: The process in which useful information is extracted from perceived images or sequences of images.

ENDNOTES

[1] This work has been partially granted by the Spanish Ministerio de Educación y Ciencia (MEC) and FEDER funds, Project n. TIN2008-06196 and by the Junta de Andalucía, Project n. P07-TIC-03106.

Chapter 14
Computer Vision for Learning to Interact Socially with Humans

Renato Ramos da Silva
Institute of Mathematics and Computer Science, University of Sao Paulo, Brazil

Roseli Aparecida Francelin Romero
Institute of Mathematics and Computer Science, University of Sao Paulo, Brazil

ABSTRACT

Computer vision is essential to develop a social robotic system capable to interact with humans. It is responsible to extract and represent the information around the robot. Furthermore, a learning mechanism, to select correctly an action to be executed in the environment, pro-active mechanism, to engage in an interaction, and voice mechanism, are indispensable to develop a social robot. All these mechanisms together provide a robot emulate some human behavior, like shared attention. Then, this chapter presents a robotic architecture that is composed with such mechanisms to make possible interactions between a robotic head with a caregiver, through of the shared attention learning with identification of some objects.

INTRODUCTION

Robots have been in human's minds for centuries, but robot technology was primarily developed in the mid and late 20th century (Goodrich and Schultz, 2007). At this first time, this technology was designed for a scientific proposes to solve, improve or increases industrial processes. Nowadays, their presence at home and general society become ever more common (Argall and Billard, 2010). Then, a study of robot behavior with a human is necessary to design, understanding, and evaluation of robotic system, that involve humans

and robots interacting through communication. For this, multidisciplinary fields of human-robot interaction (HRI) emerge in the mid 1990s and early years of 2000 (Goodrich and Schultz, 2007).

Fields that help scientists to construct social or sociable robots have contributions from linguistics, psychology, philosophy, engineering, mathematics, cognitive and computer science (Brezeal, 2002). From a computer view, several aspects have great influence on all development of the robots, such as autonomy, information exchange, learning, etc (Goodrich and Schultz, 2007). In order to create a computational system

DOI: 10.4018/978-1-4666-2672-0.ch014

that supplies a minimal requirement of social robot, some mechanisms are considered as fundamental and crucial: computer vision, learning, pro-active and voice mechanism. Because this, we will call them as basic elements. Moreover, others mechanisms, as emotion expression, are examples of others important features that help a robot to be more social and make the interaction easily and pleasant.

The basic elements are responsible to process external information, analyze it and performance actions that express acceptable human behavior.

We can note that in the beginning years of the child's life, some important behaviors and cognitive skills are learned and it is essential for her survival and develops a more complex behavior. One example of this is the shared attention ability. It is defined as redirecting attention to match another's focus of attention, based on the other's behavior (Deák and Triesch, 2006). It is important for a person transfer his/her intentions about the environment to another who are interacting.

The shared attention ability has high importance for people during the interaction and make necessary to develop a system that emulate such mechanism in robots. In spite of this, we present a simplest robotic architecture to provide a real robot head the capability of learning the shared attention ability by interacting with a caregiver. Furthermore, this robotic architecture was designed according to biological plausible reinforcement learning mechanism (Deák et al., 2001; Triesch et al., 2006; Triesch et al., 2007; Kim et al., 2008). In our case, a relational representation of data was used, because it is considered the most economical representation. All this facts have been made our architecture different than others and the results have been showed the capacity of learning shared attention ability. Therefore, the results show that the architecture is a potential tool to control sociable robots during interactions in a social environment.

The remaining of this chapter is organized as follows. First, a background section presents a brief overview of systems developed to provide a robot learn shared attention. After, we present our perspective on the issues, controversies, problems, etc., as they relate to theme and arguments supporting our position. Then, we discuss solutions and recommendations in dealing with the issues, controversies, or problems presented in the preceding section. In future research, we discuss future and emerging trends. Provide insight about the future of the book's theme from the perspective of the chapter focus. Finally, a discussion of the overall coverage of the chapter and concluding remarks.

BACKGROUND

Computer vision researches aim to extract information from images using many ways, such as video sequences, views from multiple cameras, or multi-dimensional data. It helps to solve some task, or "understand" the scene in either a broad or limited sense. Applications range from industrial machine vision systems which operate in a production line, medicine helping in disease medical diagnostic to artificial intelligence, helping to the computers or robots to comprehend better the world around them.

In this direction, the first step of all robots is to perceive the environment and to encode in a useful representation. This process is particularly complex, because of large number details of the environment, to filter what information is important and encode it without lose information. Breazeal classified this task as a hard problem to attempt for matching human performance using actual technology, both hardware (cameras, laser, etc) and software (Breazeal & Scassellati, 1999).

Robots have more difficulties than others to percept the environment. This difference is about the sensor used by robots, the type of the environment or the applications for which these robots will be use for. Robots can use a simple camera or a sophisticated laser to capture information from environment. All of the components surroundings, which are part of the environment can change randomly, named as dynamic environment, or without surrounding changes, in a deterministic way, named as static environment. Many requests must be considered when of the creating a robot aiming to attend the task for what the robot is being projected. For the robot navigation task, in most cases the robot need to know the target and the obstacle, but for interaction with a human it needs to percept more details from environment to correspond the expectative of plausible behavior.

In different levels of vision system the work can range from low, which is responsible for operations near the hardware, to high level, where computational operations are more required. The last option is used to create a basic system for a simple robot interaction we focused on the stage of visual attention, movement detection and face pose estimation.

At any moment, human eyes are faced with a huge load of visual stimuli. However, it is impossible to process all the information that reaches the eye at once (Tsotsos et. All, 1995). The visual attention is a very important feature for the survival, turning it able to interact with the environment, identifying areas of greatest interest or relevance in a scene, in a very fast way.

This process can occur when the agent automatically triggers the attention of a salient objet or event happens (Passive Attention) or when the agent is involved in an intentionally directed process to particular feature (Active Attention) (Kaplan & Hafner, 2004). There are two distinct components that drive the human visual attention: first the topdown attention, where voluntary visual control guides the attention for specific features or known objects in the image, and second, the bottom-up attention, where the attention is involuntarily guided by visual features, such as colors, deep, etc. (Desimone et al., 1995). The active attention has highly tolerant to the problems of sensor noise and incomplete input image information (Jones et all., 1997). However, it is necessary

prior knowledge or some indication to the robot which region of interest. Otherwise, it guided by primitive image features as color, intensity and orientation, the bottom-up model is more indicated to dynamic environment (Itti, 2000).

Other important function for a human to actively perceive in the environment is gaze property. Gaze behavior is a crucial element of social interactions and helps to establish triadic relations between self, other, and the world. In others words, this ability is very important for communication among humans because it helps a person expresses his or her intentions around external entities (Kaplan & Hafner, 2004; Schilbach et all., 2010).

The shared attention (SA) is one type of gaze behavior. Recently, SA has received an increasing interest, due to its importance in the development of interactions and communications. Some works have focused on modeling and understanding human developmental processes (Sumioka et. all, 2010), others only use it as a part of interaction process with sociable robot. Overall, we can find several systems in literature with both purposes (Brezeal, 2002; Triesch et al.,2006; Policastro,2009).

The observable behavior of an individual attending to an object or event that other person observes can be regarded as shared attention (Triesch et al.,2006). A formal definition can be made by a sequence constitutes by four steps (Kaplan & Hafner, 2004):

1. **Mutual Gaze:** This occurs when the establishment of eye contact between two people happens, that is there is an attention between them.
2. **Gaze Following:** This occurs when one person intentionally look at an object of interest or some event, by which another person is looking at.
3. **Imperative Pointing:** Pointing to an object or event that the partner changes the foci.

4. **Declarative Pointing:** This consist of some assumption about the scene in an appropriate sociable context.

In addition, there is the term "joint attention" that is found in the literature and it can cause some confusion with the shared attention concept. A process can be considered joint attention when it is comprised by a sequence of mutual gaze and gaze following. Many researchers have focused on joint attention or gaze following, because these processes are simpler then the complete shared attention process.

However, one of the first papers in the literature that developed a system to provide the ability of shared attention for an upper-torso humanoid robot was proposed by Scasselati (Scassellati, 1996; Scassellati, 1999). He proposed a nonlinguistic mechanism of shared attention (Scassellati, 1996), in which, he based on the three first steps mentioned.

Other similar work, it was proposed in (Mutlu et al., 2006), a model of human gaze to be used on a humanoid robot for creating a natural, human-like behavior for storytelling. In (Yamazaki et al., 2008), it was performed experiments with a guide robot designed using data from human experiments to turn its head towards the audience at important points during its presentation.

A study of the importance of eye-tracking for giving the ability of joint attention to the robot from human eye gaze is found in (Staudte & Crocker, 2009). Mutlu et al. (Mutlu et al., 2009) studied how a robot can establish the participant roles (address, bystander, and overhearer) of its conversational partners using gaze cues (Yu et al., 2010).

All of these works are focused on the mechanism of SA or part of SA process, to turning the interaction more natural. It is a step to a better interaction but it does not reflect how the SA emerges in humans, by learning in early stage of life. Then, computer scientists focus on researches

for modeling system to provide a robot the ability to learn the SA process.

Two robotic models using neural network with template matching extracting caregiver's face (edge features and motion information) was proposed by Yukie Nagai (Nagai et al., 2003; Nagai, 2005). For the first one, she proposed a developmental learning model with caregiver's evaluation with two modules, a task evaluation for caregiver and a neural network for robot. The first module uses a neural network to process information from object and robot's face to update the neural network's connections of other module that use the image from to select an action. The second module uses a visual attention with a three-layered back propagation neural network. These models are considered an important step in the construction of models of gaze following, especially since it learns to follow gaze without the need for a supervision signal telling it when it has achieved the goal and when not. However, both models had not considered the value of the mutual gaze.

Next, Shon et al. proposed a probabilistic algorithm to follow the gaze of a human and identify salient objects in a scene employing Bayesian inference (Shon, 2005). For this, they used gaze vectors in conjunction with the saliency maps of the visual scenes to produce maximum a posteriori (MAP) estimates of objects looked at by the caregiver. The resulting distribution forms the inter-modal representation of the common goal of robot and human. The relevant contribution of use the saliency maps in robots perception mechanism to demonstrate how it is helpful to learning mechanism in social-communicative interaction in mediating infants' gaze following of a robot (Meltzoff et al., 2009). However, there is a necessity of head's model to compute the probability.

Kim et al. implements a model proposed by Triesh with some improvements for a robot head (Kim et al., 2008). In their model, the computer vision was developed using harris corner detectors, k-means and hough transform was used in salient object process and the head pose processing is divided into three stages, detection, pose discrimination and depth estimation to distinguish one pose from the other. The result of this combination, called visual area, is a state vector u which comprises a body-centered saliency map s and the caregivers head pose map h. The main improvment was made on reinforcement learning module. Here, Kim used an actor-critic reinforcement learning scheme. Then, with this modification Kim et all. show the feasibility of their model in a real-world controlled environment experiment(Kim et al., 2008).

Social Robots Architectures

The robotic architectures to control social robots have been composed by mechanism and structures able to understand and reproduce all richness and complexity of natural human social behavior to interact with us. In this section, we present some of the main robotic architectures proposed to control sociable robots which served as the basis for this work, and some recently proposed in the literature.

The EGO (Emotionally GrOuded Architecture) (Arkin et all.,2001;Arkin et all.,2003) architecture was developed for behavior systems based on ethology and psychology theory. It used a bottom-up and top-down hierarchical approach, where the more abstract components are in high level of the hierarchy.

It is composed by a motivational module using an emotional and instinct model. The first model is based on Ekman studies (Ekman,1999) that emulate six basic emotion and the second model used other six variables to simulate a process of homeostasis. Furthermore, the system also uses a group of perceptual-motor structures relate to stimuli and responses and the activation component in each hierarchical level. The information flows from activation component by acquiring

information from the environment and after the system select an action, to perform a behavior, by computing the values from activation and motivational module. The behavior is selected from high level to lower level and then, send to a state machine, which have a specific sequence to translate action in behavior.

The robot Kismet has a social robot architecture composed by a perception, emotional, necessity, behavior and motor system. It is the first architecture able to express facial expression. The system captures external information by perception system which is influenced by behavior and emotional system to define what the robot look at. The information is processed by activator, composed by levels of arousal (A), valence (V) and stance (s), then, in a process of winner-takes-all the activator compete and the winner evokes a facial expression or/and a behavior.

The SARS (Social Robot Architecture) architecture (duffy,2005) is a hybrid architecture for social robots composed of a reactive layer, a deliberative layer, a mechanism of social reasoning and a layer that performs the interface between architecture and the physical components of the robot. The key components of this architecture are the hardware abstraction layer for applications in heterogeneous robotic platform, the integration between the reactive and deliberative layers, and a mechanism for beliefs-desires-intentions to support social interactions explicit. The physical layer provides portability of the architecture in relation to the physical structure of the robot. The reactive layer oversees the physical layer and is responsible for data acquisition, process sensory, reflex responses to unexpected or dangerous events, and more complex actions. The deliberative level of SARS follows a scheme of multi-agent organization, with several agents who oversee the different functional levels of the robot. In particular, the social level provides mechanisms for controlling interactions with other robots and humans, through a language of communication between agents (LCA) and mechanisms specified in the FIPA architecture (www.fipa.org). LCA provides the message handling that can trigger behavior rules which are sent to the deliberative level to be integrated into the current sequence of tasks running. These tasks are then passed to the reactive level that arbitrates the execution of them while no danger or unexpected event is detected.

The AD (Automatic-Deliberative) architecture (Salichs, 2006) is a hybrid architecture based on skills. A skill is the ability of the robot to perform a particular task. All actions and perceptions capabilities were pre-developed. This architecture is composed of two layers, a deliberative and a reactive. The path planner, the world modeling and task supervisor are some of the skills included in the deliberative level. At the reactive level, are the sensory-motor and perceptual skills. Additionally, an Emotional Control System (ECS) influence the deliberations of the architecture by controlling the robot's basic needs, determine goals and tendencies of actions for the robot, and generate the emotional state of the robot. Thus, when the level deliberative detects an event sent by the automatic level, your task sequencer selects new skills, considering the influence of the ECS. So an order of execution of this ability is passed to the reactive level that performs the requested task while monitoring the environment to detect unexpected events.

ROBOTIC ARCHITECTURE

Issues, Controversies, Problems

The fact of extract useful information from environment using cameras and represent it to a robotic architecture for a complex propose, as interact with a human, is a hard problem. Face detection mechanism to identify a person who the robot will interact, head pose estimation to identify some clues of a person about the environment and object detection to make and understand inferences about some properties of the environment are essential. These are the basic properties for a simple human robot interaction.

Nagai first attempt a neural network to process a camera image and after a salient feature detection. The problem of both methods was not defining any reinforcement for the human frontal face contact and object. Furthermore, the first method is so complex to reproduce and the second, the features are competitive, not associative (Nagai, 2004).

Shon and Kim use different forms of saliency maps to provide information to learning mechanism. However, the eye contact with a human not provide any reinforcement and the representation of the environment used by their system is unique for learning SA, it was not developed to a more complex interaction.

Nowadays, the researchers have still looking for an explanation about how and when the SA developed in infants. Furthermore, they already know that it emerge in early years of a person by interacting with his or her parents (Brooks & Meltzoff, 2005; Deák, & Triesch, 2006). A computer approach, reinforcement learning method, can be addressed to this biological explanation. Because this, the methods proposed by Nagai and Shon are not recommend.

Moreover, all systems cited here are developed only to solve SA. Then, they did not concerned in develop a robotic architecture to more sophisticated human-robot interaction.

Solutions and Recommendations

A basic robotic architecture developed to human-robot interaction need to be composed, at least, with three components: a computer vision module, learning module and a pro-active module. They are responsible to take important information from environment and use it to be executed with social acceptance. Then, our robotic architecture (Policastro et al., 2009; Silva & Romero, 2012) is constituted by mechanisms and structures inspired on Science of Behavior Analysis (Catania,

2006; Cooper, 2007; Pierce & Cheney, 2008) and is composed by the following modules:

- Stimulus Perception that acquires and codifies the state of the environment
- Consequence Control that simulates internal necessities for a pro-active interaction
- Response Emission that contains a learning mechanism for choice an action with input data from other modules

The general organization of the architecture and the interaction among these three main modules are illustrates in the Figure 1. In this figure, arrows indicate the flow of information in the three modules of the architecture. The circles indicate the methods and component structures of the modules (Policastro et al., 2009).

During an interaction, the Stimulus Perception module acquires and codifies the state of the environment and deploys this coded state for the Response Emission and the Consequence Control modules. After, the Consequence Control module checks the internal state of the robot and sets the active necessities, if there is one. Then, all infor-

mation is used by Response Emission to select an action. Afterward, the selected response is emitted by executing a motor script. Finally, the Stimulus Perception module acquires and encodes the new current environment state and the architecture control enters a loop that may be finished either at the end of an interaction or when the robot reaches its goal (Policastro et al., 2009).

In addition, the architecture also employs a working memory to exchange all information among the three main modules. This memory is used to keep information about stimuli (antecedents and consequents), the last emitted response and internal necessities. Each element inserted into the working memory has a counter that keeps the notion of time. When a new element is inserted in the working memory, its age counter is set to zero and it is incremented by 1 whenever new subsequent predicates are inserted.

So, elements persist by a number of time steps in the memory. This mechanism is employed to control the chronology of facts and events (Policastro et al., 2009).

A voice recognition system is based on the Nuance solution (Nuance, 2001) and it is able to

Figure 1. General organization of the architecture (Policastro et al., 2009)

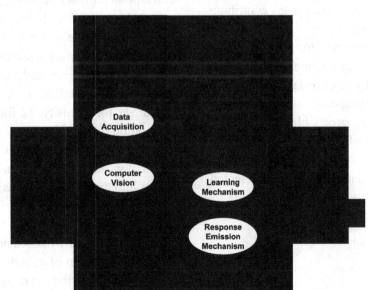

recognize naturally spoken Portuguese utterances. It contains a speech recognizer and grammatical knowledge base. The speech synthesis is performed by joining pre-recorded prompts in order to build complete phrases. This strategy enables short conversations with the robot (Policastro et al., 2009). This module is responsible by the ability of our architecture be able to do a declarative pointing.

We describe the main structures and modules of our architecture as it follows.

Stimulus Perception Module and Knowledge Representation

The stimulus perception module employs algorithms of vision system based on the work proposed by Breazeal and Scassellati (Breazeal and Scassellati, 1999). It is composed by two mechanisms: the visual attention mechanism (based on saliency (Itti, 2000)) and face recognition with head pose estimation (based on Adaptive Appearance Model (Morency et al., 2003)).

The visual attention mechanism is composed by several feature maps. Each one encodes an elementary property of the image, like color, intensity and orientation. Separately, the similar features try to be the most salient at the map, while different features among maps contribute to create a salient map.

Given an input image, the first processing step consists of decomposing this input into a set of distinct "channels," by using linear filters at eight spatial scales, tuned to specific stimulus dimension, such as color(red, green, blue and yellow), luminance (on and off) or local orientations $(0°, 45°, 90°, 135°)$.

The color channel and intensity are represented as (Itti, 2000):

$$R = r - \frac{(g + b)}{2} \text{ (Red)}$$

$$G = g - \frac{(r + b)}{2} \text{ (Green)}$$

$$B = b - \frac{(r + g)}{2} \text{ (Blue)}$$

$$Y = \frac{(r + g)}{2} - \|r - g\| - b \text{ (Yellow)}$$

$$I = \frac{(r + g + b)}{3} \text{ (Intensity)}$$

Local orientation information is obtained from I using Gabr pyramids $O(\sigma, \theta)$, where $\sigma \in [0..8]$ represents the scale and $\theta \in (0°, 45°, 90°, 135°)$ is the preferred orientation (Itti, 2000).

The second step consists of progressively low-pass filtering and sub-sampling, named as Gaussian pyramids. The result of this process, different spatial scales are created. Pyramids have a depth of 9 scales, providing horizontal and vertical image reduction factors ranging from $1 : 1$ (level 0; the original input image) to 1:256 (level 8) (Itti, 2000).

Three pyramids resulted enter in center-surround operations. The center of the receptive field correspond to pixel at level $c \in \{2, 3, 4\}$ in the pyramid, and the surround to the correspond pixel at level $s = c + \delta$, with $\delta \in \{3, 4\}$. Across-scale difference between two maps is obtained by interpolation to finer scale and point-by-point subtraction. In total, 42 feature maps are computed: six for intensity, 12 for color and 24 for orientation (Itti, 2000).

After normalization, feature maps for intensity, color, and orientation are summed across scales into three separate conspicuity maps. The purpose of the saliency map is to represent the conspicuity— or "saliency"—at every location in the visual field by a scalar quantity, and to guide the selection of attended locations, based on the spatial distribution of saliency. After, they

Figure 2. Experiment to detect objects in controlled environment with little noise

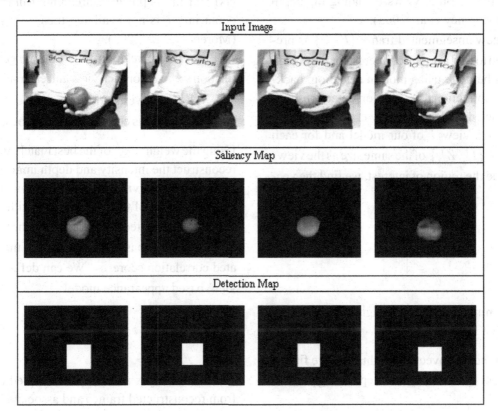

are normalized and summed into the final input saliency map (Itti, 2000).

The Figure 2 shows the previous experiment to detect objects in controlled environment with littler noise. It was done to validate the system and shows the efficiency of the process to focus on the object in the scene. All steps cited above was developed using LTI library (LTI, 2003).

Identify a person is an important aspect of human-robot interaction. In the literature, we found two models: Feature or Appearance based. The first aims to find invariant features of the face and the other, use learning and training with set of images. Then, the second strategy is more indicated. Moreover, the *eigenfaces,* which is based Karhunen–Loève transform (PCA), is motivated by the efficient of figure representation.

Other important point is related of estimating the pose of a rigid object or a person accurately

and robustly for a wide range of motion. It is a classic problem in computer vision and has many useful applications, mainly when the focus is with head pose tracking. A method used to estimate use intensity and depth view-based *eigenfaces* (Morency et al., 2003). It can be divided in two parts:

- Creating the model a priori given initial frame
- Calculation of pose changes using Appearance Based Method

The purpose of prior model reconstruction is to generate a set of views for use in the pose estimation module. Given a single example frame near one of the views in the model, we want to reconstruct all the other available views of our model. In our case, given one image, we can recreate the 27 other views including the depth images (Morency et al., 2003).

The new unsegmented frame $\{I_t, Z_t\}$ is pre-processed to find a region of interest for the object. This can be done using motion detection, background subtraction, flesh color detection or a simple face detector (Morency et al., 2003).

For each view i of our model and for each subregion $\{I'_t, Z'_t\}$ of the same size as the views in P inside the region of interest, we find the vector $\vec{w}_i$ that minimizes

$$E_i = \left| I'_t - \bar{I}_i - \vec{w}_i \times V_{I_i} \right|$$

The minimization is straightforward using linear least squares.

From the eigenvector weights $\vec{w}_i$, we first reconstruct the intensity and depth images

$$I_{R_i} = \bar{I}_i + \vec{w}_i \times V_{I_i}$$

$$Z_{R_i} = \bar{Z}_i + \vec{w}_i \times V_{Z_i}$$

The reprojection step is done for every view i. After the reconstruction of all intensity and depth images I_{R_i} and Z_{R_i}, we search for the best projection minimizing the correlation function:

$$\frac{\left(I'_t - \bar{I}'_t\right) \times (I_{R_i} - \bar{I}_{R_i})}{\left|I'_t - \bar{I}'_t\right|\left|I_{R_i} - \bar{I}_{R_i}\right|} +$$
$$\gamma \frac{\left(Z'_t - \bar{Z}'_t\right) \times (Z_{R_i} - \bar{Z}_{R_i})}{\left|Z'_t - \bar{Z}'_t\right|\left|Z_{R_i} - \bar{Z}_{R_i}\right|}$$

where is constant to compensate for the difference between intensity measurements(brightness lev-

els) and the depth measurements (mm). If the depth image is not available then [3] is set to 0 (Morency et al., 2003).

From the correlation function, we get a correlation score c_i for each view and each sub region $\{I'_t, Z'_t\}$. The lowest correlation c_i^* over all views and all sub regions corresponds to the best match.

Using the weights $\vec{w}^*$ of the best match, we again reconstruct the intensity and depth images of the object in all the views using Equations (3) and (4). The output of the matching algorithm is the set of reconstructed frames I_i^*, Z_i^*, the pose ε_i^* of the best view-based eigenspace and the associated correlation score c_i^*. We can define a prior view-based appearance model

$$M_p = \{\{I_{P_i}, Z_{P_i}, \varepsilon_{P_i}\}, \Delta_P\}$$

where the images and poses are copied directly from reconstructed frames and associated poses, and the covariance matrix Δ_P is initialized as the identity matrix times a small constant (Morency et al., 2003).

Appearance Based model also follow two steps:

- Calculation of relative pose between new frame and each pose model using AVAM algorithm.
- Integrating measured pose using Kalman filter to produce absolute pose.

When depth information is available in the new frame, we can register the frames using a hybrid error function which combines robustness of the ICP (Iterative Closest Point) algorithm and the precision of the normal flow constraint (NFC) (Morency et al., 2003).

The Figure 3 presents the previous experiment to estimate the head pose. Such as done with visual mechanism, some test we done to validate the method. The entire estimation pose was developed using WATSON library (Watson).

The perception module, by integration these two mechanisms cited above, is capable to identify the indication of other individual's head pointing and the object of other's focus. The high importance of the perception module to acquire this information from the environment and transfer to others module, show us how this module is fundamental.

Thus, this module detects the state from the environment and encodes this state using an appropriate representation. The knowledge representation adopted for the architecture is based on a relational representation (Driessen, 2004; Otterlo, 2005), enabling the representation of large spaces in an economical way. This representation is also known of first order logic and we adopt the same logic bases of Morrik (Morik et al., 1993).

The architecture encodes knowledge as stimuli, facts, responses, behavior rules and constraint rules, which are represented by associating each stimulus and fact with a binary number. This association is made by a function, $(f = 2^x)$, in which ranges from 0 to $n-1$, and n is the total number of stimuli and facts. This function allows us to represent the state of environment by a binary sum and we can separate it without lose any fact.

This simple technique is also known as binarization by conversion of a categorical attribute to asymmetric binary attributes (Tan et al., 2005).

Stimuli encode all signals received from the environment and they are represented as atoms or objects that may have properties like color, size, shape, position and pose (for faces). These atoms represent objects detected from the real world, relevant in the problem domain. Properties of the

Figure 3. Experiment to estimate the head pose

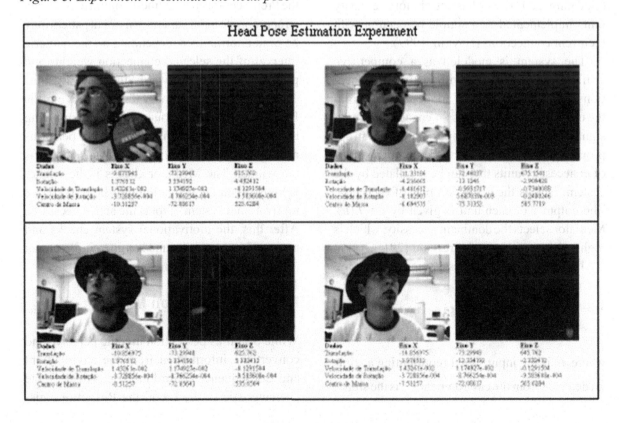

stimuli are set by the Stimulus Perception Module, in order to encode the current environment state. For example, one may define face as a stimulus of the environment to be detected and set a routine to encode skin color in its color property, its position and pose angles (pan and tilt). The environment state is encoded employing perception predicates: see(X), hear(Y), at(Z) and smell(W). The perception predicates relate all detected stimuli to build a representation from the current environment state (Policastro et al., 2009).

Consequence Control Module

An artificial motivational system may enable a robot to pro-actively interact with the environment, driving its behaviors to satiate its artificial internal necessities. The consequence control module is composed by a motivational system that simulates internal necessities of an individual. The motivational system is based on works presented by Breazeal (Brezeal, 2002) and Gadanho (Gadanho, 2001) and it has one or more necessity units implemented as a simple perceptron with recurrent connections (Haykin, 1999).

The system is modeled as a competitive artificial neural network with recurrent connections. Stimulus is a set of input stimulus from the environment. Preprocessor encodes the input signals received from the environment into an appropriate form. Units $i, j, ...m$ represent an array of m necessity units that can be simulated by the system. Bias is the activation bias of each unit. The output y of each unit is given by $y = f(u)$. Mediator selects the dominant necessity, which is higher than a predefined threshold [29].

The activation of a necessity unit is given by:

$$u = \left(\sum_{j=1}^{n} w_j \times i_j \right) + w_r \times i_r + b$$

where i_j is the input signal representing a stimuli detected from the environment, i_r is the signal

from the recurrent connection, w_j are connection weights of the input signal, w_r is connection weight of the recurrent signal, and b is the bias of the unit. All weights and bias are empirically defined according to the necessity being simulated. The output of a necessity unit is given by:

$$y = \frac{1}{1 + e^{-(u+')}}$$

where u is the activation value and δ is the sigmoid function inclination. A necessity unit simulates the internal necessities of an individual. Additionally, the motivational system has an output mediator that mediates activation among several necessity units, employing competition and an activation threshold (Policastro et al., 2009).

The motivational system works as it follows. Initially, the stimuli detected from the environment are sent to the consequence control module. Then, the Preprocessor encodes these stimuli to construct an appropriate input pattern. This input pattern may be or not normalized, depending on the numeric interval of the selected connection weights and problem domain. Afterwards, the necessity units calculate their activation values employing the Equation 1, and their output values employing the Equation 2. After this, the Mediator performs a competition among all unit outputs and selects the winner. The Mediator checks if the winner is higher than the activation threshold. If so, the motivational system outputs the active necessity. After this, the motivational system checks and reports if any necessity unit has got reinforcement (Policastro et al., 2009).

Response Emission Module

While Stimulus perception module perceive and convert the information from the environment and Consequence control module provide some characteristic to the robot, the Response emission module is responsible to act socially after processing all information from others modules.

The algorithm used by this module was proposed by us, the Economic TG (ETG). It is based on the works of Driessens (Driessens, 2004) where ETG is an enhancement of TG algorithm. Due to the characteristics of the architecture in relation to the learning mechanism, some changes were necessary to be done in ETG algorithm. Basically, these are related to the necessity value produced by consequence control module, in which is used in the process of example storage in the tree and by the fact that ETG do not use any property of TILDE system (Blockeel & De Raedt, 1998).

TG algorithm interacts with the environment until find final state. After, the examples are inserted in the tree using a specific metric to insert only good examples from final state until the first state. For this a look up table is used. In the other hand, the ETG algorithm, the examples are inserted in the tree at each interaction. This assumption was done to not use an auxiliary table to store the examples and metrics before insert them into the tree.

In short, the ETG algorithm learns a control policy for an agent as it moves through the environment and receives rewards for its actions. An agent perceives a state s_i, decides to take some action a_i, makes a transition from s_i to s_{i+1} and receives the reward r_i. The task of the agent is to maximize the total reward it gets while doing actions. Agents have to learn a policy which maps states into actions.

The Figure 2 shows the processing of ETG algorithm (Silva & Romero, 2012). It starts by initializing the *Q-function* and creates an empty regression tree.

The learning mechanism takes from the environment the current state, an agent necessity value, and then, using the current policy, it selects an action and performs it over the environment. This process changes the state and the agent receives its reward. The reward can be either positive (equals

to 10) or negative (equals to -1). After this occurred, the *qvalue* is computed by:

$$Q_i \leftarrow Q\left(s_i, a_i\right) + \alpha[r_{i+1} + \gamma \max\left(Q\left(s_{i+1}, a_{i+1}\right)\right) - Q\left(s_i, a_i\right)]$$

Then, the set of quadruples (state, action, *qvalue*, necessity) is presented to relational regression engine. This process is repeated until there are not more interactions to be executed. All processing can be found in Figure 4.

The relational regression engine (RRE) receives a set of (state, action, *qvalue*, necessity) and tests the internal nodes if the state already exists. If this condition is false, the state is inserted in the tree and the leaf receives the action with the *qvalue* and necessity values, forming a new branch. Otherwise, it updates the *qvalue* for respective action in the leaf node.

In a leaf node, more than one action can be considered. For an easier access to the most adequate action, these actions can be ordered in decreasing order according to their *qvalue* always that an example is inserted or updated. Each leaf also has a necessity associated with action and it refers to a necessity of the robot to choose this action in that state. Here, we use only the attention necessity.

A relational regression engine adopted in ETG algorithm is presented in Figure 5.

After an explanation about the algorithm, a resume of the process will presented. For this, the Figure 4, 5, 6, 7 and 8 help us to better understand providing an example of the process. For both figures we assume the presence of the robot placed in an environment with a caregiver.

The Figure 4 represents the beginning of all process. The robot perceives the environment, which is represented by "". After, the state "*X*" is encoded for a useful representation. This process is done by identify the characteristic of the environment, in the figure is represented as a binary

Figure 4. ETG algorithm (Silva & Romero, 2012)

Algorithm   ETG Algorithm

---

initialize the **Q-function** hypothesis $\hat{Q}_0$ and create a tree with a single leaf
$i \leftarrow 0$
repeat
  take state s_i
  take necessity n_i
  choose a_i for s_i using a policy derived from
  the current hypothesis $\hat{Q}_i$
  take action a_i, observe r_i and s_{i+1}
  Update $\hat{Q}_i$ using the Eq. 4
  Update relational regression algorithm using
  $x = (s_i, a_i, \hat{Q}_i, n_i)$ to produce $\hat{Q}_{i+1}$ {Use Algorithm 3}
  $i \leftarrow i + 1$
until forever

---

number and after, we done a binary sum. The value resulted of this process is used by other modules.

The consequence module takes this information, normalize and process it with a neural network. The neural network process it as explained above and as a result, if it has a necessity and the reinforcement value, positive (1) or negative (0). In the figure the necessity and the reinforcement as set with 1.

All this information is received by the response emission module to select an action to be executed. First, the algorithm verifies if there is a good action for the perceived state. The RRE is empty, when this or when an action was not selected for a specific state, the algorithm chooses an action randomly, in the figure is exemplified by letter a. The action is executed and the environment could change. Moreover, the state, action and reinforcement are stored in a working memory.

After the first interaction, the environment change and now it is represented by "", as represented in the Figure 5. Here and in the next figure, the same process occur in the perception and consequence module. The result of this process

is a state represented by 66, a necessity by 1 an reinforcement by 0.

Response emission module is different, in the second state some differences happen in the end. The algorithm tries again to take the best action from RRE and it still empty. Then, the algorithm chooses an action randomly, b. Until now there is no difference, but with the information about last turn the algorithm calculates the Q-value. So, with state, action, necessity and Q-value from last turn, the algorithm has all data to update the RRE. The process is simple; first the process tries to find the state over the tree by comparing each characteristic (binary number) in the intermediate nodes. If in a node the binary number is the same of the actual state, the algorithm goes to the next node on the left side; otherwise, it goes to the right side. When no node was found, the algorithm inserts a new node. In Figure 5, as the RRE was empty. The state value was inserted without test.

The Figure 6 show what happen when the RRE is not empty and, in this case and all subsequent turns, will follow this sequence. The response emission module takes all information and tries to find the best action consulting the RRE. Each state characteristic is compared to indentify a state. First, the node represented by value 4 is compared with current state; it is represented in the figure by red number 1. The positive response make the algorithm follow down over the tree by its left side. Over the path comparisons was done (red number 2, 3 and 4). All them the answer is positive, but in the red number 4 or leaf node, the necessary value is compared too, with an adverse result. In this case, an action was chosen randomly again. The action is executed in the environment and the Q-value is calculated. The RRE is updated with the information about the last turn. For this, the comparison with the first node (with the value 4) was done with negative response. Then, the algorithm follows by the right side and it not find a node. The new state is inserted in the right side of the first node of the structure.

The Figure 7 and 8 is inserted to show all possibilities and better reader's comprehension. For the Figure 7, the new state has some features of other state already inserted in the tree and it show what happen in a update procedure for a existing state with different action. In the first case, the algorithm has done a comparison in RRE structure. In the first node, with red number 1, the result is positive and the algorithm follow b the left side. But in the second node, red number 2, the result is negative and the algorithm follow to the right side. Here the algorithm does not find a node and calls the random function to chose an action, a. To update the RRE mechanism, the algorithm follows by the left side, because the state is the same. In the leaf node the action with its necessity and Q-value is inserted.

The last figure presents the case when an action is selected by RRE mechanism. The actual state is compared in each node and follow by left side, as show in the figure with the red numbers. In the leaf node the necessity is equal and the action is selected.

This example uses a simplified state for a better comprehension. Other point to remember, each action stored in the RRE mechanism has its Q-value.

In the end of this section, we will present an analysis of complexity of ETG algorithm and we star with memory requirements. The ETG algorithm stores all of information in a binary tree. As in a leaf node, more than one action can be considered, it is necessary a vector for saving all actions. However, this vector has a dimension limited by the number of actions, n, considered in a share attention task. So, considering that $h > 1$ denote the depth of the binary tree, we have a total of

$$\sum_{i=0}^{h-2} 2^i + 2^{h-1} \times n <$$

$$\frac{1}{2(h-1)}[(h-1).1] + n < n+1$$

nodes to be stored.

In terms of memory access for recovering information, the ETG algorithm made a treatment for an easier access to the most adequate action. The actions are stored in decreasing order according to their Q-value, the total time consumed is of order

$$O\left(\log_2 m\right)$$

since a binary tree structure is being used.

EXPERIMENTAL RESULTS

The main results from a set of experiments carried out to evaluate our architecture are presented here. We evaluate the performance of the robotic architecture controlling a robotic head using the learning method chosen.

Figure 5. Regression engine of ETG algorithm (Silva & Romero, 2012)

Algorithm ETG-Regression Engine
repeat
sort the state down the tree using the tests of the internal nodes until to reach leaf node or null
if the node is a leaf **then**
if action exists **then**
the *Q-value* is updated for the action in the leaf node according to the example {time indicates to update the Q value of rule}
else
the *Q-value* is inserted and the necessity for the action in the leaf node according to the example {time indicates the creation of a rule}
end if
else if the null is attained **then**
generate a node
end if
until the example in a branch
if necessary **then**
order actions in decreasing *Q-value*
end if

Figure 6. First round: first step the interaction of the robot in the environment

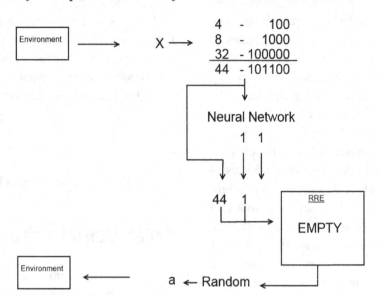

Before we start to talk about the experiment performed, we introduce the metric used for analysis the learning algorithm. The metric have used named as correct gaze index (CGI) measure is used in both experiment and it is based on measures proposed by Whalen (Whalen and Schreibman, 2003) defined as the frequency of gaze shifts from the human to the correct location where the human is looking at, given by:

$$CGI = \frac{\# \, shifts \; from \, the \; human \; to \; correct \; location}{\# \; shifts \; from \; the \; human \; to \; any \; location}$$

The purpose of this experiment was to evaluate the capabilities of the proposed architecture on exhibit appropriate sociable behavior and learning in a real and controlled environment. In the experimental scenario, a caregiver established eye contact with the robot and then presented three different objects (fruits represented by an apple, lemon and orange) in order to teach the robot for

following his gaze and after to do a declarative pointing of it.

For this, four stimuli have been declared: face, object, attention, and environment, in which attention was configured as reinforcer stimuli. Three facts have been declared to define that red, orange and green objects are fruits. Six facts have been declared in order to differentiate the human's head pose: in frontal pose, two poses of left profile, two poses of right profile and one pose of down profile. Additionally, two more facts have been declared to define when the robot is focusing the human or a fruit.

The response emission module was configured as it follows. The constant learning (alpha parameter) has been set with a value equal to 0.2. The discount factor (gamma parameter) was set to 0.9. The exploration factor (epsilon parameter) was set to 0.05. Seven responses have been defined so that the robot could look to the humans or search for a fruit in the five defined regions (down, left, right, left down and right down) by turning its head to the left or right side. This was done to divide the environment in areas of interest that would turn

Figure 7. Second round: inserting the first example in RRE

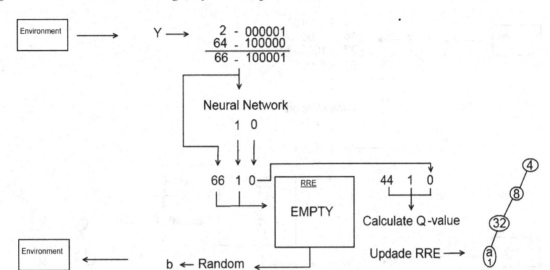

possible for the robot to learn following the gaze of a human being to correct places.

The motivational system was configured as it follows. The necessity unit was created: socialize. The activation threshold of the motivational system has been set to 0.50. The sigmoid function inclination (delta parameter) has been set with a value equal to 0.20. For the necessity unit socialize, the bias has been set with value equal to 1.00 and weight of its connection has been set with a value equal to 0.5. The weight of the recurrent connection has been set with a value equal to 1.00. The weights of the connections of the input units (hear (attention), see (frontal (face)), see (looking_frontal (face)), and see (looking_fruit (object)) have been set, respectively, with the values 1.50, 0.95, 0.50 and 1.50. All the constants have been chosen empirically.

The experiments were composed by learning phase of 100 time units or runs. During the learning

Figure 8. Third round: inserting other example in RRE

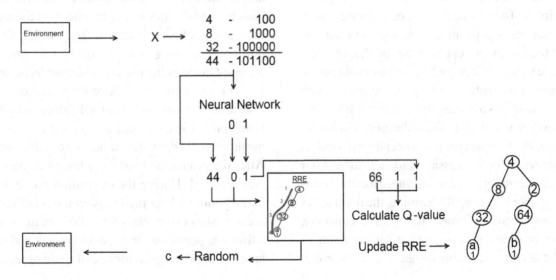

Figure 9. Fourth round: when example already exist in RRE

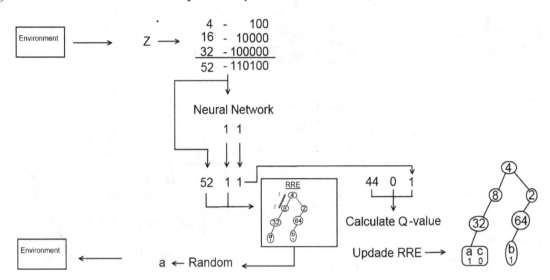

phase, the human being initially kept the focus on the robot until it establishes eye contact with him, characterized by looking each other. Then, one fruit were positioned in the environment and the caregiver turned his gaze for this fruit. Afterwards, the robot does a declarative pointing with safety or uncertainty about a particular fruit. Finally, the fruit is then removed from the environment and the human turns his gaze to the robot, keeping the eye contact with the robot again. This procedure is done in order to simulate an interaction where two agents are keeping eye contact and then one turns his gaze to an interesting event or object.

In the first 30 time units of learning phase, no objects are positioned in the environment and the caregiver kept his focus on the robot the whole time, so the robot has learned to obtain the caregiver's attention by keeping eye contact with him, satisfying its necessity of socialization. This procedure was done to shape the robot's behavior of looking for a caregiver and keeping eye contact whenever it feels necessity of socialization. After this, the learning phase was resumed using a fruit as stated above. During the learning, the robot looks for him whenever it wants to socialize. However, when an object is positioned in the environment and the caregiver turns his gaze to it, the robot

looses his attention and starts to seek anything in the environment that can satisfy its internal states of socialization. Additionally, if the robot looks for a fruit which the caregiver is keeping his gaze, the person returns his attention to the robot, in relation to the fruit. In this way, after a history of reinforcement, the robot will learn to follow the caregiver's gaze to receive his attention and to satisfy its needs of socializing.

The learning capabilities of the architecture was analyzed by observing the robot interacting with the caregiver and the environment, and computing a CGI measure. To quantify the learning capabilities of the architecture through the learning of gaze following, at specific points during the learning process, we the temporarily interrupt the learning phase to evaluate its behavior. This evaluation was done by 5 runs of 100 time units. For each run, the CGI value, given by Equation 1 was computed and after the 5 runs a mean and standard deviation were calculated. After the evaluation phase, the learning process was resumed. During the evaluation phase, the human initially kept the focus on the robot until it establishes eye contact with the human. Then a fruit was positioned in the environment and the human turned his gaze for one of these objects.

However, in the evaluation phase, the object to which the human should turn his gaze was place on a position given by pre-established sequence (to prevent non determinism in the results). Once the robot turns its head to any direction, the software in robotic head verifies if it is looking for the correct position in the environment or not, and update the CGI measure. Afterwards, the objects are then removed from the environment and the human turns his gaze to the robot, keeping eye contact with robot again.

To test the ability of the robotic head to do a declarative pointing step, it was taught for it how to recognize fruits, such as apple, lemon and orange. For the learning of the fruits, it was used a neural network ART2 (Haykin, 1999) to recognize different colors. The ART2 network is a good technique for clustering data. It was evaluated this neural network with 5 possibilities: Unknown, Correct guess, Incorrect guess, Error or Success. At first time, when the robot sees a new fruit, the caregiver tells its name for the robot. After, the robot tries to express the name of the fruit doing the association between name of fruit and its color. The presentation phase was repeated 5 times (5 executions) and we did not change the light conditions during the experiments. Then,

after 5 executions, the average value and standard deviation, for each measure in the 5 executions, have been calculated. The neural network ART2 was successful to do this task classifying correctly 73% of the patterns trained.

In Figure 9, it is presented the curve that demonstrates the progress of the learning during the experiments. This figure shows a chart that presents the value of the average of the CGI values, in specific points during the learning phase. The obtained results show that CGI values increases over the learning phase, demonstrating the learning capacities of the architecture with ETG algorithm.

A great increasing in a learning curve is noted in the beginning until the second running when the robotic head builds its first knowledge. After this, based on all new facts that happened in the environment, it adjusts its knowledge base. This process is shown in low increasing the curve. At the certain time, when the curve becomes fixed, or it has a little change, for a time period, the robot has the optimal reply for this problem.

In Table 1 is showed the average values and standard deviation of the 5 measures for the 5 executions carried out during the experiments (Silva et. al., 2008).

Figure 10. Fifth round: last example in RRE

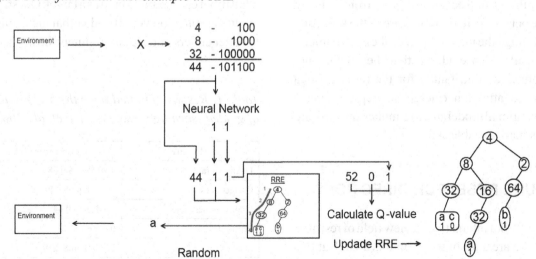

Figure 11. Learning evolution during the experiments (Silva & Romero, 2012)

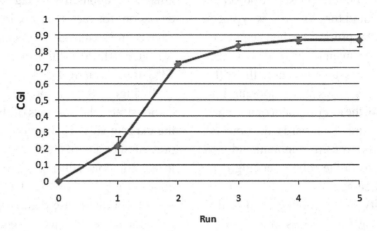

The robotic head shows be able to associate visual and auditory stimulus if we analyze the results of the declarative pointing step. With the good performance to identify a fruit, the architecture was able to learning with a caregiver's tutelage. The complete result of the declarative pointing step can be seen at (Silva et. al., 2008).

These experiments was carried out by the modeling of the behavior of looking for a human, followed by the behavior of follow the human's gaze and, finally, to perform the declarative pointing step. The results show that the architecture is able to exhibit appropriate behaviors during a real and controlled social interaction. Additionally, the results show that the proposed architecture is able to acquire basic sociable abilities from existing innate behaviors in the repertoire of the robot and also through the interaction with the environment. The results also evidence that the architecture constitutes a contribution for the research area of shared attention emergence, representing a computational model able to simulate the learning of this hard sociable skill.

FUTURE RESEARCH DIRECTIONS

Human-robot interaction is a new field of research and there are a lot of works in this area. But this chapter is focused only in computer science problems. Then, here we look forward on question of computer vision system, pro-active system, learning mechanism, auxiliary mechanism and applications, like shared attention.

Classic problems in computer vision as extract and represent information in real time still open. The filter used to extract useful information can change instantly, depending of robot necessities. Better filters that contain these changes need more concern.

Learning mechanism is applied on specifics mechanism as shared attention, face recognitions, to follow a person and others. Improvements in a general mechanism for learning like a baby to an adult robot can be considered as a target. Questions about learning non verbal and verbal mechanism to better understand the environment, how to learn

Table 1. Results obtained after the 5 executions of guided learning sessions (Silva et. al., 2008)

Measure	Average rate (%)
Unknown rate	7.23 $\pm$ 0.58
Correct guess rate	17.47 $\pm$ 1.28
Incorrect guess rate	1.8 $\pm$ 0.44
Error rate	0.6 $\pm$ 0.2
Success rate	72.89 $\pm$ 2.19

to do new functions (game, to do new work, etc) using old knowledge and how to choose the best action accord with the result of pro-active system.

Finally, improvements on architecture need to aggregate abilities like attention, shared attention, emotional cues and other to better interaction among robots and humans. Other important improvement is the implementation method, in this case is sequential, but a parallel implementation of the interaction is a step forward.

CONCLUSION

A robotic architecture was presented for a simple interaction between a caregiver and a robot face. This is possible because the stimulus perception module is capable to extract information from non-verbal information from a caregiver and convert the perception to a representation used by other modules. The voice recognition helps the system for a better basic human robot interaction.

REFERENCES

Argall, B. D., & Billard, A. G. (2010). A survey of tactile human-robot interactions. *Robotics and Autonomous Systems*, *58*(10), 1159–1176. doi:10.1016/j.robot.2010.07.002

Arkin, R. C., Fujita, M., Takagi, T., & Hasegawa, R. (2003). An ethological and emotional basis for human–robot interaction. *Robotics and Autonomous Systems*, *42*(3–4), 191–201. doi:10.1016/S0921-8890(02)00375-5

Blockeel, H., & De Raedt, L. (1998). Top-down induction of first-order logical decision trees. *Artificial Intelligence*, *101*, 285–297. doi:10.1016/S0004-3702(98)00034-4

Breazeal, C. (2002). *Designing sociable robots*. Cambridge, MA: MIT Press.

Breazeal, C., & Scassellati, B. (1999). A context-dependent attention system for a social robot. In T. Dean (Ed.), *Proceedings of the Sixteenth International Joint Conference on Artificial Intelligence (IJCAI '99)* (pp. 1146-1153). San Francisco, CA: Morgan Kaufmann Publishers Inc.

Brooks, R., & Meltzoff, A. N. (2005). The development of gaze following and its relation to language. *Developmental Science*, *8*(6), 535–543. doi:10.1111/j.1467-7687.2005.00445.x

Catania, A. C. (2006). *Learning* (4th ed.). Santa Rosa, CA: Sloan.

Cooper, J. O., Heron, T. E., & Heward, W. L. (2007). *Applied behavior analysis* (2nd ed.). Prentice Hall.

Deák, G. O., Fasel, I., & Movellan, J. (2001). The emergence of shared attention: Using robots to test developmental theories. In C. Balkenius, et al. (Eds.), *Proceedings 1st International Workshop on Epigenetic Robotics:* Vol. 85. (pp, 95-104). Lund University Cognitive Studies.

Deák, G. O., & Triesch, J. (2006). The emergence of attention-sharing skills in human infants. In Fujita, K., & Itakura, S. (Eds.), *Diversity of cognition*. University of Kyoto Press.

Desimone, R., & Duncan, J. (1995). Neural mechanisms of selective visual attention. *Annual Review of Neuroscience*, *18*, 193–222. doi:10.1146/annurev.ne.18.030195.001205

Driessens, K. (2004). *Relational reinforcement learning*. Unpublished doctoral dissertation, Katholieke Universiteit Leuven, Leuven.

Duffy, B. R., Dragone, M., & O'Hare, G. M. P. (2005). *Social robot architecture: A framework for explicit social interaction android science*. Towards Social Mechanisms, CogSci 2005 Workshop Stresa.

Ekman, P. (1999). Basic emotions. In Dagleish, T., & Power, M. (Eds.), *Handbook of cognition and emotion*. Sussex, UK: John Wiley & Sons.

Goodrich, M. A., & Schultz, A. C. (2007). Human-robot interaction: A survey. *Foundations and Trends in Human-Computer Interaction*, *1*(3), 203–275. doi:10.1561/1100000005

Haykin, S. (1999). *Neural networks—A comprehensive foundation*. Englewood Cliffs, NJ: Prentice Hall.

Itti, L. (2000). *Models of bottom-up and top-down visual attention*. Unpublished doctoral dissertation, California Institute of Technology, Pasadena, California.

Jones, M. J., Sinha, P., Vetter, T., & Poggio, T. (1997). Top–down learning of low-level vision tasks. *Current Biology*, *12*(7), 991–994. doi:10.1016/S0960-9822(06)00419-2

Kaplan, F., & Hafner, V. (2004). The challenges of joint attention. *Interaction Studies: Social Behaviour and Communication in Biological and Artificial Systems*, *7*(2), 67–74.

Kim, H., Jasso, H., Deak, G., & Triesch, J. (2008). A robotic model of the development of gaze following. *7th IEEE International Conference on Development and Learning* (ICDL 2008) (pp. 238-243).

Meltzoff, A. N., Brooks, R., Shon, A. P., & Rao, R. P. N. (2010). Social robots are psychological agents for infants: A test of gaze following. *Neural Networks*, *23*(8-9), 966–972. doi:10.1016/j.neunet.2010.09.005

Morency, L.-P., Sundberg, P., & Darrell, T. (2003). Pose estimation using 3D view-based Eigenspaces. ICCV Workshop on Analysis and Modeling of Face and Gesture, (pp. 45-52). Nice, France, October.

Morik, K., Wrobel, S., Kietz, J.-U., & Emde, W. (1993). *Knowledge acquisition and machine learning: Theory, methods, and applications*. San Francisco, CA: Academic.

Mutlu, B., Hodgins, J. K., & Forlizzi, J. (2006). A storytelling robot: modeling and evaluation of human-like gaze behavior. In *Proceedings of HUMANOIDS'06, 2006 IEEE-RAS International Conference on Humanoid Robots*, (pp. 518–523)

Mutlu, B., Shiwa, T., Kanda, T., Ishiguro, H., & Hagita, N. (2009). Footing in human-robot conversations: how robots might shape participant roles using gaze cues. In *HRI'09: Proceedings of the 4th ACM/IEEE International Conference on Human Robot Interaction* (pp. 61–68)

Nagai, Y. (2004). *Understanding the development of joint attention from a viewpoint of cognitive developmental robotic*. Unpublished doctoral dissertation, Osaka University, Japan.

Nagai, Y. (2005). The role of motion information in learning human-robot joint attention. In *Proceedings of the 2005 IEEE International Conference on Robotics and Automation* (ICRA), (pp. 2069–2074)

Nagai, Y., Hosoda, A., & Asada, M. (2003). A constructive model for the development of joint attention. *Connection Science*, *15*(4), 211–229. doi:10.1080/09540090310001655101

Nuance. (2001). *Nuance: Introduction to the Nuance System*. South Yarra, Australia: Nuance Communications Inc.

Otterlo, V. (2005). *A survey of reinforcement learning in relational domains*. CTIT Technical Report, TRCTIT-05-31.

Pierce, W. D., & Cheney, C. D. (2008). *Learning* (4th ed.). Hove, UK: Psychology Press.

Policastro, C. A., Romero, R. A. F., Zuliani, G., & Pizzolato, E. (2009). Learning of shared attention in sociable robotics. *Journal of Algorithms*, *64*(4), 139–151. doi:10.1016/j.jalgor.2009.04.005

Salichs, M. A., Barber, R., Khamis, A. M., Malfaz, M., Gorostiza, J. F., & Pacheco, R. … Garcia, D. (2006). Maggie: A robotic platform for human-robot social interaction. *IEEE Conference on Robotics, Automation and Mechatronics*, (pp. 1-7).

Scassellati, B. (1996). *Mechanisms of shared attention for a humanoid robot* (pp. 102–106).

Scassellati, B. (1999). *Imitation and mechanisms of joint attention: A developmental structure for building social skills on a humanoid robot* (pp. 176–195). New York, NY: Springer. doi:10.1007/3-540-48834-0_11

Schilbach, L., Wilms, M., Eickhoff, S. B., Romanzetti, S., Tepest, R., & Bente, G. (2010). Minds made for sharing: Initiating joint attention recruits reward-related neurocircuitry. *Journal of Cognitive Neuroscience*, *22*(12), 2702–2715. doi:10.1162/jocn.2009.21401

Shon, A. P., Grimes, D. B., Baker, C. L., Hoffman, M. W., Zhou, S., & Rao, R. P. N. (2005). Probabilistic gaze imitation and saliency learning in a robotic head. *Proceedings of the 2005 IEEE International Conference on Robotics and Automation (ICRA 2005)*, (pp. 2865-2870)

Silva, R. R., Policastro, C. A., Zuliani, G., Pizzolato, E., & Romero, R. A. F. (2008). Concept learning by human tutelage for social robots. *Learning and Nonlinear Models*, *6*, 44–67.

Silva, R. R., & Romero, R. A. F. (2012). Modelling shared attention through relational reinforcement learning. *Journal of Intelligent & Robotic Systems*, *66*(1), 167–182. doi:10.1007/s10846-011-9624-y

Staudte, M., & Crocker, M. W. (2009). Visual attention in spoken human-robot interaction. In *HRI '09: Proceedings of the 4th ACM/IEEE International Conference on Human Robot Interaction*, (pp. 77–84).

Sumioka, H., Yoshikawa, Y., & Asada, M. (2010). Reproducing interaction contingency toward open-ended development of social actions: Case study on joint attention. *IEEE Transactions in Autonomous Mental Development*, *2*(1), 40–50. doi:10.1109/TAMD.2010.2042167

Tan, P.-N., Steinbach, M., & Kumar, V. (2005). *Introduction to data mining*. Boston, MA: Addison-Wesley Longman.

Triesch, J., Teuscher, C., Deak, G. O., & Carlson, E. (2006). Gaze following: Why (not) learn it? *Developmental Science*, *9*(2), 125–147. doi:10.1111/j.1467-7687.2006.00470.x

Tsotsos, J. K., Culhane, S. M., Wai, W. Y. K., Lai, Y., Davis, N., & Nuflo, F. (1995). Modeling visual attention via selective tuning. *Artificial Intelligence*, *78*, 507–545. doi:10.1016/0004-3702(95)00025-9

Watson. (n.d.). *WATSON: A real-time head tracking and gesture recognition*. Retrieved from http://projects.ict.usc.edu/vision/watson/

Whalen, C., & Schreibman, L. (2003). Joint attention training for children with autism using behavior modification procedures. *Journal of Child Psychology and Psychiatry, and Allied Disciplines*, *44*(3). doi:10.1111/1469-7610.00135

Yamazaki, A., Yamazaki, K., Kuno, Y., Burdelski, M., Kawashima, M., & Kuzuoka, H. (2008). Precision timing in human-robot interaction: Coordination of head movement and utterance. In *CHI '08: Proceeding of the Twenty-Sixth Annual SIGCHI Conference on Human Factors in Computing Systems*, (pp. 131–140).

Yu, C., Scheutz, M., & Schermerhorn, P. (2010). Investigating multimodal real-time patterns of joint attention in an HRI word learning task. In *HRI '10: Proceeding of the 5th ACM/IEEE International Conference on Human–Robot interaction*, (pp. 309–316).

ADDITIONAL READING

Arkin, R. C., Fujita, M., Takagi, T., & Hasegawa, R. (2001). Ethological modeling and architecture for an entertainment robot. *Proceedings 2001 ICRA. IEEE International Conference on Robotics and Automation,* Vol. 1, (pp. 453- 458).

Asgharbeygi, N., Nejati, N., Langley, P., & Arai, S. (2005). Guiding inference through relational reinforcement learning. *Proceedings of the Fifteenth International Conference on Inductive Logic Programming, Vol. 3625,* (pp. 20-37). Springer.

Begum, M., & Karray, F. (2011). Visual attention for robotic cognition: A survey. *IEEE Transactions on Autonomous Mental Development, 3*(1), 92–105. doi:10.1109/TAMD.2010.2096505

Begum, M., Karray, F., Mann, G., & Gosine, R. (2009). A probabilistic approach for attention-based multi-modal human-robot interaction, *The 18th IEEE International Symposium on Robot and Human Interactive Communication, RO-MAN 2009,* (pp. 200-205).

Blockeel, H., & De Raedt, L. (1998). Top-down induction of first-order logical decision trees. *Artificial Intelligence, 101*(1-2), 285–297. doi:10.1016/S0004-3702(98)00034-4

Deák, G. O., & Triesch, J. (2006). Origins of shared attention in human infants. In Fujita, K., & Itakura, S. (Eds.), *Diversity of cognition* (pp. 331–363). University of Kyoto Press.

Dzeroski, S., de Raedt, L., & Driessens, K. (2001). Relational reinforcement learning. *Machine Learning, 43,* 7–52. doi:10.1023/A:1007694015589

Grossberg, S., & Vladusich, T. (2010). How do children learn to follow gaze, share joint attention, imitate their teachers, and use tools during social interactions? *Neural Networks, 23*(8-9), 940–965. doi:10.1016/j.neunet.2010.07.011

Hashimoto, M., Kond, H., & Tamats, Y. (2008). Effect of emotional expression to gaze guidance using a face robot. *The 17th IEEE International Symposium on Robot and Human Interactive Communication, RO-MAN 2008,* (pp. 95-100).

Klein, M., Kamp, H., Palm, G., & Doya, K. (2010). A computational neural model of goal-directed utterance selection. *Neural Networks, 23*(5), 592–606. doi:10.1016/j.neunet.2010.01.003

Kwisthout, J., Vogt, P., Haselager, P., & Dijkstra, T. (2008). Joint attention and language evolution. *Connection Science, 20*(2-3), 155–171. doi:10.1080/09540090802091958

Marin-Urias, L. F., Sisbot, E. A., Pandey, A. K., Tadakuma, R., & Alami, R. (2009). Towards shared attention through geometric reasoning for human robot interaction. *9th IEEE-RAS International Conference on Humanoid Robots, Humanoids 2009* (pp. 331-336).

Otterlo van. M., & Kersting, K. (2004). *Challenges for relational reinforcement learning.* In Workshop on Relational Reinforcement Learning of the International Conference on Machine Learning, ICML '04, 4-8 July 2004, Banff, Alberta, Canada.

Ravindra, P., De Silva, S., Tadano, K., Lambacher, S. G., Herath, S., & Higashi, M. (2009). Unsupervised approach to acquire robot joint attention, *4th International Conference on Autonomous Robots and Agent, ICARA 2009,* (pp. 601-606).

Ravindra, P., De Silva, S., Tadano, K., Saito, A., Lambacher, S. G., & Higashi, M. (2009). Therapeutic-assisted robot for children with autism. *IEEE/RSJ International Conference on Intelligent Robots and Systems, IROS 2009*, (pp. 3561-3567).

Robins, B., Dickerson, P., Stribling, P., & Dautenhahn, K. (2004). Robot-mediated joint attention in children with autism. A case study in robot-human interaction. *Interaction Studies: Social Behaviour and Communication in Biological and Artificial Systems, 5*(2), 161–198. doi:10.1075/is.5.2.02rob

Silva, R. R., Policastro, C. A., & Romero, R. A. F. (2008). An enhancement of relational reinforcement learning. In Proceedings of 2008 International Joint Conference on Neural Networks Hong Kong (pp. 2056-2061).

Silva, R. R., Policastro, C. A., & Romero, R. A. F. (2009). Relational reinforcement learning applied to shared attention. In Proceedings of 2009 International Joint Conference on Neural Networks, Atlanta (pp. 2943-2949).

Silva, R. R., & Romero, R. A. F. (2010). Using only aspects of interaction to solve shared attention. In *3rd International Workshop on Evolutionary and Reinforcement Learning for Autonomous Robot Systems (ERLARS 2010), 2010, The 19th European Conference on Artificial Intelligence* (ECAI 2010), Vol. 1 (pp. 43-51).

Silva, R. R., & Romero, R. A. F. (2011). Improvements on relational reinforcement learning to solve joint attention. In *Proceedings of the Third International Conference on Advanced Cognitive Technologies and Applications* (COGNITIVE 2011), Roma, (pp. 63-69).

Silva, R. R., & Romero, R. A. F. (2011). Relational reinforcement learning and recurrent neural network with state classification to solve joint attention. In Proceedings of the 2011 International Joint Conference on Neural Networks, San Jose, (pp. 1222-1229).

Staudte, M., & Crocker, M. W. (2011). Investigating joint attention mechanisms through spoken human-robot interaction. *Cognition, 120*, 268–291. doi:10.1016/j.cognition.2011.05.005

Tadepalli, P., Givan, R., & Driessens, K. (2004). Relational reinforcement learning: An overview. In *Proceedings of the ICML'04 Workshop on Relational Reinforcement Learning*, Vol. 4, (pp. 1-9).

Triesch, J., Jasso, H., & Deák, G. O. (2007). Emergence of mirror neurons in a model of gaze following. *Adaptive Behavior, 15*, 149–165. doi:10.1177/1059712307078654

Yucel, Z., Salah, A. A., Merigli, C., & Mericli, T. (2009). Joint visual attention modeling for naturally interacting robotic agents, *24th International Symposium on Computer and Information Sciences, 2009. ISCIS 2009* (pp. 242-247).

KEY TERMS AND DEFINITIONS

Computer Vision: Aim to extract information from images using many ways, such as video sequences, views from multiple cameras, or multi-dimensional data, help to solve some task, or "understand" the scene in either a broad or limited sense.

Human–Robot Interaction (HRI): Is a field of study dedicated to understanding, designing, and evaluating robotic systems for use by or with humans.

Joint Attention: Is part of shared attention process. It is composed by maintain eye contact and follow other person gaze by some object or event.

Relational Reinforcement Learning: Learning only from interacting with the environment by receiving positive and negative reward using relational representation.

Robotic Architecture: Software framework, organized with modules, each one with some properties, for controlling robots.

Shared Attention: Is a sequence process of maintain eye contact, follow other person gaze, point the object or event, and finish by made some assumption about the fact.

Social Robots: Robots capable to interact with human like humans interact with themselves.

Chapter 15
Learning Robot Vision for Assisted Living

Wenjie Yan
University of Hamburg, Germany

Nils Meins
University of Hamburg, Germany

Elena Torta
Eindhoven University of Technology,
The Netherlands

Cornelius Weber
University of Hamburg, Germany

Raymond H. Cuijpers
Eindhoven University of Technology,
The Netherlands

David van der Pol
Eindhoven University of Technology,
The Netherlands

Stefan Wermter
University of Hamburg, Germany

ABSTRACT

This chapter presents an overview of a typical scenario of Ambient Assisted Living (AAL) in which a robot navigates to a person for conveying information. Indoor robot navigation is a challenging task due to the complexity of real-home environments and the need of online learning abilities to adjust for dynamic conditions. A comparison between systems with different sensor typologies shows that vision-based systems promise to provide good performance and a wide scope of usage at reasonable cost. Moreover, vision-based systems can perform different tasks simultaneously by applying different algorithms to the input data stream thus enhancing the flexibility of the system. The authors introduce the state of the art of several computer vision methods for realizing indoor robotic navigation to a person and human-robot interaction. A case study has been conducted in which a robot, which is part of an AAL system, navigates to a person and interacts with her. The authors evaluate this test case and give an outlook on the potential of learning robot vision in ambient homes.

DOI: 10.4018/978-1-4666-2672-0.ch015

INTRODUCTION

The phenomenon of population ageing is becoming a serious problem of this century. According to the estimate of the U.S. Census Bureau, the American population aged over 65 will grow from 13% to 20% until 2030 (Hootman & Helmick, 2006). In Europe, more than 20% of the population will be beyond 60 by 2020 (Steg, Strese, Loroff, Hull, & Schmidt, 2006) and by 2050 this group will even exceed 37% (OECD, 2007). Ageing societies would benefit from the design of "intelligent" homes that provide assistance to the elderly (Steg et al., 2006). In this context the research field of robotics is focusing attention on AAL systems which refer to a set of technological solutions that permit the elderly population to maintain their independence at home for a longer time than would otherwise be possible (O'Grady, Muldoon, Dragone, Tynan, & O'Hare, 2010). Ambient homes will not only react passively, like turning on lights when the lighting condition changes, but they will also provide active help via home electronics, motorized actuators or - in the future - socially assistive robots. They can assist the elderly effectively in everyday tasks such as communication with the external world or the ambient system and can provide medicine and health check reminders in a proactive fashion.

A number of research topics are involved in the design of the functionalities of a socially assistive robot. Among them, robotic navigation and human-robot interaction are particularly relevant. Robotic navigation in ambient homes, in particular mutual positioning between the robot and a person, is an important task for a robot that strongly influences the quality of human-robot interaction. A robot should find a way to approach a target person after localization and go to the person without colliding with any obstacles, which is very challenging due to the complexity of real-home environments and the possible dynamical changes. A vision-based system is a potential way to tackle those challenges. Compared with other kinds of sensors, a vision system can provide far more information, good performance and a wide scope of usage at reasonable cost. A robot can perform different tasks and adapt its behavior by learning new features if equipped with sophisticated vision algorithms.

Human-robot interaction is a very broad research field. Therefore, in the context of this book chapter, we understand it as the study of how robots can communicate interactively with users. Computer vision algorithms are essentials for achieving this because they can be used to acquire feedback related to the user state during interaction. Unlike an industrial robot, that, in most cases, runs preprogrammed behaviors without being interactive, service robots should be able to adapt their behavior in real time for the purpose of achieving natural and easy interaction with the user. This requires the generation of appropriate verbal and non-verbal behaviors that allow the robot to participate effectively in communication. Vision algorithms can gather information about the user's attention, emotion and activity, and allow the robot to evaluate non-verbal communication cues of the user. The benefits of non-verbal communication cues become apparent when the conversation is embedded in a context, or when more than two persons are taking part in a conversation. Particularly, head gestures are important for a smooth conversation, because cues signaled by head gestures are used for turn taking. But head gestures serve many more purposes; they influence the likability of the observer, communicate the focus of attention or the subject of conversation, and they can influence the recollection of the content of conversation (Kleinke, 1986).

In this chapter we aim at introducing the reader to the computer vision techniques used in a robotics scenario for Ambient Assisted Living (AAL) in the context of the European project KSERA: Knowledgeable SErvice Robots for Aging. Our project develops the functionalities of a socially assistive robot that acts as an intelligent interface between the assisting environment and the elderly person. It combines different vision-based methods for simultaneous person and robot localization, robot navigation, human-robot interaction with online face recognition and head pose estimation, and adapt those techniques in ambient homes. The system is able to detect a person robustly by using different features, navigate a robot towards the person, establish eye contact and assess whether the person is giving attention to the robot.

This chapter is organized as follows: The section "Related Works" presents a brief review of the state of the art of algorithms and technology related to robotics and AAL. Section "Methods" provides insight in the computer vision algorithms developed and applied in order to increase the skills of a socially assistive robot in Ambient Homes. Section "Case Study" describes a detailed case study of AAL systems, which combines the different robotic vision techniques for allowing the humanoid robot Nao (Louloudi, Mosallam, Marturi, Janse, & Hernandez, 2010) to navigate towards a person in a cluttered and dynamically changing environment and to interact with the person. Section "Conclusion and Future Research Directions" summarizes the findings providing an outlook on the potential of learning robot vision in ambient home systems and outlines the possible developments of the algorithms and methods introduced in this chapter.

RELATED WORKS

This section provides an overview of the state-of-the-art vision technologies related to our work, which combines localization, navigation, face detection and head pose estimation algorithms for use in an assisted living home environment. We show that these vision algorithms are efficient to let a robot help a person in an AAL environment.

Simultaneous Person and Robot Localization

Person tracking based on vision is a very active research area. For instance, stereo vision systems (Muñoz-Salinas, Aguirre, & Garcá-Silvente,

2007; Bahadori, Iocchi, Leone, Nardi, & Scozzafava, 2007) use 3D information reconstructed by different cameras to easily distinguish a person from the background. Multiple ceiling-mounted cameras are combined (Salah, et al., 2008) to compensate for the narrow visual field of a single camera (Lanz & Brunelli, 2008), or to overcome shadowing and occlusion problems (Kemmotsu, Koketsua, & Iehara, 2008). While these multi-camera systems can detect and track multiple persons, they are expensive and complex. For example, the camera system has to be calibrated carefully to eliminate the distortion effect of the lenses and to determine the correlations between different cameras. A single ceiling-mounted camera is another possibility for person tracking. West, Newman and Greenhill (2005) have developed a ceiling-mounted camera model in a kitchen scenario to infer interaction of a person with kitchen devices. The single ceiling-mounted camera can be calibrated easily or can be used even without calibration. Moreover, with a wide-angle view lens, for example a fish-eye lens, the ceiling-mounted camera can observe the entire room. Occlusion is not a problem if the camera is mounted in the center of the ceiling and the person can be seen at any position within the room. The main disadvantage of the single ceiling-mounted camera setup is the limited raw information retrieved by the camera. Therefore, sophisticated algorithms are essential to track a person.

There are many person detection methods on computer vision area. The most common technique for detecting a moving person is background subtraction (Piccardi, 2004), which finds the person based on the difference between an input and a reference image. Furthermore, appearance-based models have been researched in recent years. For instance, principal component analysis (PCA) (Jolliffe, 2005) and independent component analysis (ICA) (Hyvärinen & Oja, 2000) represent the original data in a low dimensional space by keeping major information. Other methods like scale-invariant feature transformation (SIFT) (Lowe, 2004) or a speeded-up robust feature

(SURF) (Bay, Tuytelaars, & Van Gool, 2006) detect interest points, for example using Harris corner (Harris & Stephens, 1988) for object detection. These methods are scale- and rotation invariant and are able to detect similarities in different images. However, the computation complexity of these methods is high and they perform poorly with non-rigid objects. Person tracking based on body part analysis (Frintrop, Königs, Hoeller, & Schulz, 2010; Hecht, Azad, & Dillmann, 2009; Ramanan, Forsyth, & Zisserman, 2007) can work accurately, but requires a very clear body shape captured from a front view. A multiple camera system has to be installed in a room environment to keep obtaining the body shape. The color obtained from the clothes and skin can be a reliable tracking feature (Comaniciu, Ramesh, & Meer, 2000; Muñoz-Salinas, Aguirre, & Garcá-Silvente, 2007; Zivkovic & Krose, 2004), but this has to be adapted quickly when the clothes or the light condition changes. The Tracking-Learning-Detection algorithm developed by Kalal, Matas and Mikolajczyk (2010) works for an arbitrary object, however, it requires an initial pattern to be selected manually, which is not possible in a real AAL setting.

Navigation as Part of HRI

People tend to attribute human-like characteristics to robots and in particular, to socially assistive robots (Siino & Hinds, 2004), (Kahn et al., 2012). Therefore, when the robot behavior does not match prior expectations, humans tend to experience breakdowns in human-robot interaction (e.g., Mutlu and Forlizzi, 2008). As a consequence, mobile robots that share the same space with humans need to follow societal norms in establishing their positions with respect to humans (Kirby, 2010). By doing so, they will likely produce a match between their mobile behavior and people's expectations (Syrdal, Lee Koay, & Walters, 2007). Models of societal norms for robot navigation are mostly derived from the observation of human-human interaction scenarios. For instance, Nakauchi and

Simmons (2002) developed a control algorithm that allows a robot to stand in line using a model of personal space (PS) derived from observation of people standing in line. Similarly, Pacchierotti and Christensen (2007) define the robot's behavior in an avoidance scenario based on human-human proxemic distances derived from the work of Hall (1966) and Lambert (2004).

Navigation in dynamic and cluttered environments, such as ambient homes, in the presence of a human is still a challenge because the robot needs to cope with dynamic conditions while taking into account the presence of its human companion. Traditionally, two distinct approaches have been proposed to tackle this issue. The first approach includes societal norms in the navigation algorithm at the level of global path planning. As an example, Kirby (2010) described social conventions like personal space and tending to the right, as mathematical cost functions that were used by an optimal path planner to produce avoidance behaviors for a robot that were accepted by users. On the same line, Sisbot et al. (2010) introduce a human aware motion planning for a domestic robot that, besides guaranteeing the generation of safe trajectories, allows the robot to reason about the accessibility, the vision field and the preferences of its human partner when approaching him/her. The second approach to robotic navigation in the presence of a human includes societal norms at the level of reactive behaviors. As an example, Brooks and Arkin (2006) proposed a behavior-based control architecture for humanoid robot navigation that takes into account the user's personal space by introducing a proportional factor for mapping human-human interpersonal distances to human-robot distances. Along the same line, Torta et al. (2011) present a behavior-based navigation architecture that defines the robot's target positions through the solutions of a Bayesian filtering problem which takes into account the user's personal space. On the contrary of Brooks and Arkin, the model of the personal space was derived by means of a psychophysical experiment.

Face Detection and Head Pose Estimation

One of the most well known face detection methods for real-time applications is the one introduced by Viola and Jones (2002), which uses Haar-like features for image classification. Their work inspired subsequent research that extended the original method improving its accuracy. Some of them extended the spatial structure of the Haar-like features. For example Lienhart and Maydt (2002) have further developed the expression of the integral image that allows calculating Haar-like features which are rotated with respect to the original. Their method can cope better with diagonal structures. A different way to extend the original Haar-like features is presented by Mita, Kaneko and Hori (2005). In their work they join multiple threshold classifiers to one new feature, that they call "Joint Haar-like feature". By using this co-occurrence their method needs less Haar-like features for reaching the same accuracy. Beside the Viola and Jones method, other technologies and algorithms have been developed to solve the face detection problem. For instance, (Osuna, Freund, & Girosi, 1997) propose the use of support vector machines which yields a higher accuracy but is more computational expensive. Principal component analysis (Belhumeur, Hespanha, & Kriegman, 1997) was also used as well as convolution neural networks (Lawrence, Giles, Tsoi, & Back, 1997). The latter has the ability to derive and extract problem specific features.

Head gestures are specifically interesting for measuring the attention or engagement of the user related to what the robot is telling. Mutlu, Hodgins and Forlizzi (2006) have shown that the amount of gaze of the robot to the listener in a story telling context relates to the amount of information subjects remembered about the story. An advanced application of head pose estimation is joint attention. Joint attention entails reciprocal eye contact, but it can also be used to signal the subject of speech. Both interlocutors (the user and the robot) in this case focus on the same subject.

This serves a communicative purpose (Kaplan & Hafner, 2006); by using the estimation of gaze direction, the robot is able to infer the object that the user refers to. For example, the user might ask the robot to "pick up that cup", while looking at the cup on the table. The system can now control the robot to pick up the cup on the table, and not the one on the mantelpiece. Yücel and Salah (2009) proposed a method for establishing joint attention between a human and a robot. A more advanced application that requires the robot to establish and maintain a conversation is turn taking during a conversation with its human partner (Kendon, 1967). To give the floor to the conversation partner, people often look at the partner just before finishing their turn. This can also be implemented in a robot by allowing the robot to estimate the gaze direction of the user. This allows the robot to use these cues to time its turn correctly. As robots are joining humans in everyday tasks, head pose estimation based on computer vision has seen a revival of interest. Different vision-based head pose estimation systems have been compared and summarized in Murphy-Chutorian and Trivedi (2009). Voit, Nickel and Stiefelhagen (2007) developed a neural network-based model for head pose estimation which has been further developed by van der Pol, Cuijpers and Juola (2011) and will be discussed in the "Head Pose Estimation" section.

METHODS

Here we present methods for realizing robotic navigation in an AAL environment as well as human-robot interaction. We first describe a hybrid neural probabilistic model for localizing a robot and a person using a ceiling-mounted camera. Then, we present a behavior-based model for robotic navigation which integrates the information provided by the localization algorithm with the robot's own sensor readings. A real-time face detection model is employed using the robot's camera to allow the robot to make eye contact

Figure 1. The person and robot localization algorithm using input data from the ceiling mounted camera. The weights of particles are assigned with a polynomial combination of visual cues.

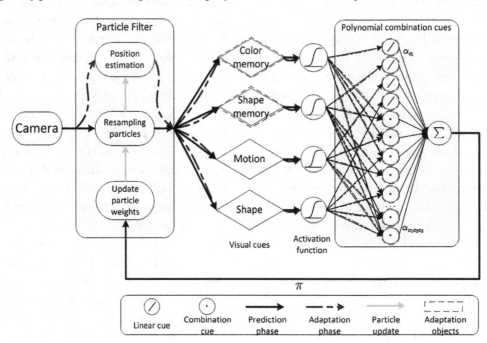

with a person. The robot determines whether the user is paying attention to it with the head pose estimation method. The details of each model will be described in the following sections.

Simultaneous Person and Robot Localization

Inspired by a model of combining different information for face tracking (Triesch & Malsburg, 2001), we combine different visual information to detect and localize a person's position reliably. The system can track a person with or without motion information, and it is robust against environmental noise such as moving furniture, changing lighting conditions and interaction with other people. The work flow (Figure 1) can be split into two parts: prediction and adaptation.

In the prediction phase, each particle segments a small image patch and evaluates this patch using pre-defined visual cues. Four cues are used: 1) color memory cue based on a color histogram, 2)

shape memory cue based on SURF features, 3) motion cue based on background subtraction and 4) fixed shape cue based on a neural network. When a person is detected in the image patch by for example the motion cue, the value of this cue will increase until it reaches a maximum value. The higher the visual cues' values are, the more likely is the target person present inside the image patch. We generate some extra polynomial combination cues using a Sigma-Pi network architecture (Weber & Wermter, 2007) to increase the weights when multiple cues are active at the same time. The output of the evaluation will be set to the particle filters, which provides robust object tracking based on the history of previous observations. Two particle filters are employed to estimate the position of a person and a robot. Particle filters are an approximation method that represents a probability distribution of an agent's state s_t with a set of particles $\{i\}$ and weight values $\pi^{(i)}$, which is usually integrated in partially observable Markov decision processes (POMDPs) (Kaelbling,

Littman & Cassandra, 1998). A POMDP model consists of unobserved states of an agent s, in our case the x, y position of the observed person based on the image frame, and observations of the agent z. A transition model $P(s_t \mid s_{t-1}, a_{t-1})$ describes the probability that the state changes from s_{t-1} to s_t according to the executed action a_{t-1}.

If the agent plans to execute the action a_{t-1} in state s_{t-1}, the probability of the next state can be predicted by the transitions model $P(s_t \mid s_{t-1}, a_{t-1})$ and validated by the observation $P(z_t \mid s_t)$. Hence, the agent's state can be estimated then as:

$$P(s_t \mid z_{0:t}) = \eta P(z_t \mid s_t) \int P(s_{t-1} \mid z_{0:t-1}) P(s_t \mid s_{t-1}, a_{t-1}) ds_{t-1}, \quad (1)$$

where η is a normalization constant, $P(z_t \mid s_t)$ is an observation model and $P(s_t \mid z_{0:t})$ is the belief of the state based on all previous observations. This distribution can then be approximated with weighted particles as:

$$P(s_t \mid z_{0:t}) \approx \sum_i \pi_{t-1}^{(i)} \delta(s_t - s_{t-1}^{(i)}), \quad (2)$$

where π denotes the weight factor of each particle with $\sum \pi = 1$ and δ denotes the Dirac impulse function. The higher the weight value, the more important this particle is in the whole distribution. The mean value of the distribution can be computed as $\sum_i \pi_{t-1}^{(i)} s_t$ and may be used to estimate the state of the agent if the distribution is unimodal.

In the person tracking system, the person's position is represented by the x- and y- coordinates in the image, i.e. $s = \{x, y\}$. The direction of a person's motion is hard to predict, because for example, an arm movement during rest could be wrongly perceived as a body movement into the corresponding direction. Hence, we do not use direction of movement information, but describe the transition model $P(s_t \mid s_{t-1}^{(i)}, a_{t-1})$ of the person with a Gaussian distribution:

Figure 2. Vision-based localization. Left: in the initial state particles are uniformly distributed in the image space. Blue circles represent particles describing the robot's pose while yellow circles represent particles for describing the person's location. Right: after initialization particles converge to the true person and robot locations

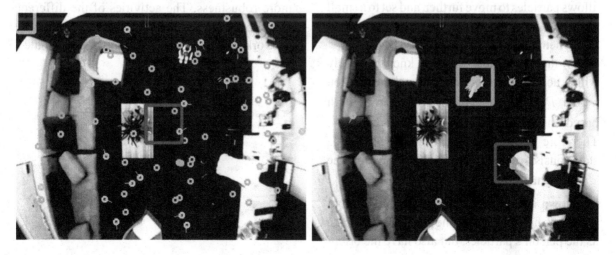

Figure 3. The equations of navigation rely on the visually obtained information from the localization module and on the information of the robot's proximity sensors. Left: relevant angles for the navigation algorithm. Right: view from the ceiling mounted camera.

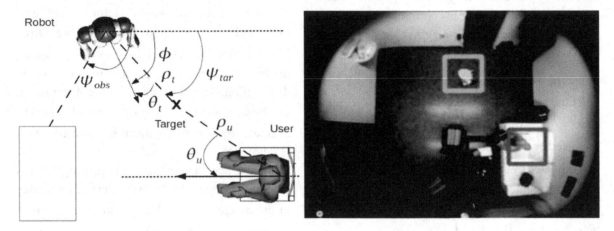

$$P(s_t \mid s_{t-1}^{(i)}, a_{t-1}) = \frac{1}{\sqrt{2\pi\sigma(a_{t-1})^2}} e^{-\frac{(s_{t-1}^{(i)} - s_t^{(i)})^2}{2\sigma(a_{t-1})^2}},$$

(3)

where $\sigma(a_{t-1})^2$ is the variance of the Gaussian related to the action, a_{t-1}, $s_{t-1}^{(i)}$ are the previous states and $s_t^{(i)}$ is the current states. In case of a moving person, the action a_{t-1} is a "binary" variable containing only information whether the person is moving or not. The variance $\sigma(a_{t-1})^2$ will be set larger when motion is detected, which allows particles to move further, and set to a small value to "stick" the particles on the current position when no motion is detected.

On the other hand, since we know precisely which actions the robot executes, the transition model $P(s_t \mid s_{t-1}^{(i)}, a_{t-1})$ of the robot can be built based on the robot's physical motion behavior. Therefore for the robot localization, the robot's state consists of three coordinate information: the x, y position and the orientation ϕ, i.e. $s = \{x, y, \phi\}$.

As shown in Figure 2, the particles of the robot in cyan do not only encode the position coordinate as the person's particles, but also have the orien-

tation information visualized by a short line. We can calculate the expected position of the robot x', y' and o' based on the designed feed-forward model and add Gaussian noise, as described by Equation (3).

When the person's and the robot's position are estimated, the particles will be resampled and their position will be updated (green arrows in Figure 1). After that, in the adaptation phase, the weight factor $\pi^{(i)}$ of particle i will be computed with a weighted polynomial combination of visual cues, inspired by a Sigma-Pi network (Weber & Wermter, 2007). The combination increases the effective number of cues and thereby leads to more robustness. The activities of the different visual cues are set as the input of the Sigma-Pi network and the particle weights are calculated as:

$$\pi^{(i)} = \sum_{j=1}^{4} \alpha_{c_j}^{l}(t) A_{c_j}(s_{t-1}^{(i)}) + $$

$$\sum_{\substack{j,k=1 \\ j>k}}^{4} \alpha_{c_j c_k}^{q}(t) A_{c_j}(s_{t-1}^{(i)}) A_{c_k}(s_{t-1}^{(i)}) + $$

(4)

$$\sum_{\substack{j,k,l=1 \\ j>k>1}}^{4} \alpha_{c_j c_k c_l}^{c}(t) A_{c_j}(s_{t-1}^{(i)}) A_{c_k}(s_{t-1}^{(i)}) A_{c_l}(s_{t-1}^{(i)}),$$

where $A_c(s_{t-1}^{(i)}) \in [0,1]$ is the activation function that signals activity of cue c at the state s_{t-1} (i.e. the position) of particle i. The activities of visual cues are generated via activation functions and scaled by their reliabilities α_c. We use a sigmoid activation function:

$$A(x) = \frac{1}{1 + e^{-(g \cdot x)}}, \qquad (5)$$

Here, x is the function input and g is a constant scale factor. Through the polynomial combination of cues represented by a Sigma-Pi network, the weights of particles are computed. The coefficient of the polynomial cues, i.e. the network weights $\alpha_{c_j}^l(t)$ denote the linear reliabilities, $\alpha_{c_j c_k}^q(t)$ the quadratic combination reliabilities and $\alpha_{c_j c_k c_l}^c(t)$ are the cubic. Compared with traditional multi-layer networks, the Sigma-Pi network contains not only the linear input but also the second-order correlation information between the input values. The reliability of some cues, like motion, is non-adaptive, while others, like color, need to be adapted on a short time scale. This requires a mixed adaptive framework, as inspired by models of combining different information (Bernardin, Gehrig, & Stiefelhagen, 2008; Weber & Wermter, 2007). An issue is that an adaptive cue will be initially unreliable, but when adapted may have a high quality in predicting the person's position. To balance the changing qualities between the different cues, the reliabilities will be evaluated with the following equation:

$$\alpha_c(t) = (1 - \varepsilon)\alpha_c(t-1) + \varepsilon(f(s_t') + \beta), \qquad (6)$$

where ε is a constant learning rate and β is a constant value. $f(s_t')$ denotes an evaluation function and is computed by the combination of visual cues' activities:

$$f_c(s_t') = \sum_{i \neq c}^{n} A_i(s_t')A_c(s_t'), \qquad (7)$$

where s_t' is the estimated position and n is the number of the reliabilities. In this model n is 14 and contains 4 linear, 6 quadratic and 4 cubic combination reliabilities. We use a Hebbian-like learning rule to adapt the reliabilities. When the cue c is active together with several others, the function $f_c(s_t')$ is large, which leads to an increase of the cue's reliability α_c. For details of each visual cue please refer to (Yan, Weber, & Wermter, 2011).

Behavior-Based Robot Navigation

The general view of behavior-based robotics states that complex behaviors can be generated by the coordination of simpler ones. In the case of mobile robot's navigation each simple behavior solves a navigational subtask without the need of high level world representation (Arkin, 1998; Althaus, Ishiguro, Kanda, Miyashita, & Christensen,

Figure 4. Misalignments of the robot with respect to the person's position

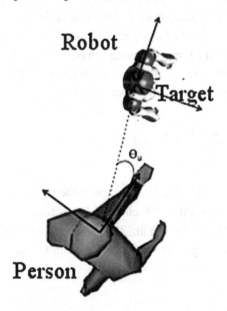

Figure 5. Left: integral image; right: haar-like feature

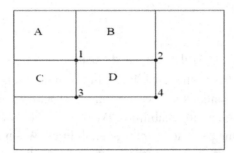

2004). Behavior-based robotics provides real-time adaptation to dynamically changing environments and can be adopted by robots with very limited sensory capabilities, such as humanoid robots (Bicho, 1999). These characteristics make this navigation framework convenient for applications in ambient homes.

There are several frameworks for behavior-based robotic navigation but here we focus on the dynamical system approach to mobile robot navigation (Schöner, Dose, & Engels, 1995; Bicho, 1999). This choice is due to the fact that the equations of the navigation algorithm do not rely on a complete world representation but on the visually obtained estimates of the user and robot positions, which can be obtained from the localization model. A behavior can be described by means of a behavioral variable that, in our work, is chosen to be the robot's heading direction $\phi(t)$, and by the temporal evolution of it. The evolution is controlled by a non-linear dynamical equation that can be generally expressed as:

$$\omega = \frac{d\phi(t)}{dt} = F(\phi(t)), \qquad (8)$$

where $F(t)$ defines how the value of the behavioral variable $\phi(t)$ changes over time (Bicho, 1999), (Althaus et al., 2004). Multiple behaviors are aggregated by means of a weighted sum:

$$\frac{d\phi(t)}{dt} = \sum_{i=1}^{m} w_i f_i(\phi(t)) + d, \qquad (9)$$

where m represents the number of behaviors that are needed for the robot to accomplish its task. The term $f_i(\phi(t))$ represents the force produced by the ith behavior and w_i represents the weight associated to the ith behavior. The term d represents a stochastic term that is added to guarantee escape from repellers generated by bifurcation in the vector field (Monteiro & Bicho, 2010). Attractor and repulsive functions, $f_i(\phi(t))$, are modeled with opposite signs. We can identify two basic behaviors whose coordination brings the robot from a generic location in the environment to a target location. The process of reaching a target point is represented by an attractor dynamic whose expression is:

$$f_1(t) = -\sin(\phi(t) - \psi_{tar}(t)), \qquad (10)$$

where the term $(\phi(t) - \psi_{tar}(t))$ accounts for the angular location of the target with respect to the robot at time t and can be obtained by the visual estimation reported in the Section "Simultaneous Person and Robot Localization". The attractor dynamic acts to decrease the difference between $\phi(t)$ and ψ_{tar}; a graphical representation of those angles is visible in Figure 3.

The ability to obtain collision-free trajectories is encoded by a repulsive dynamic whose mathematical expression is given by:

$$f_2 = e^{\left(-\frac{1}{\beta_2}(d_{obs}(t) - R)\right)} \underbrace{}_{\text{term } A}$$

$$\underbrace{(\phi(t) - \psi_{obs}(t))e^{\left(-\frac{(\phi(t) - \psi_{obs}(t))^2}{2\sigma_{obs}^2}\right)}}_{\text{term } B}. \qquad (11)$$

It generates a force which decays exponentially with the detected distance between the robot and the obstacle through the term A and with the angular separation between obstacle and robot thorough the term B. The detected distance between the robot and the obstacle at time t is represented by the term $d_{obs}(t)$, while the direction of the obstacle with respect to the robot at time t is encoded in the term $(\phi(t) - \psi_{obs}(t))$. The location of the obstacle with respect to the robot and the distance from the obstacles can be obtained from the robot's proximity sensors. The coefficients β_2 and σ_{obs} determine the range at which the repulsion strength acts. The repulsion force acts to increase the terms $(\phi(t) - \psi_{obs}(t))$ and $d_{obs}(t)$. A graphical visualization of $\psi_{obs}(t)$ is visible in Figure 3.

Referring to the stage of behaviors coordination reported in Equation(9), it is possible to obtain collision-free trajectories, if the weight w_2 associated with the repulsion force is greater than the weight w_1 associated with the attractor force. The aforementioned equations allow the robot to move towards a person and to avoid collisions on the way based on the localization estimate of the person and the robot and on the information of the robot's proximity sensors. We suppose to fix the target's position with respect to the user reference frame at a point in front of him located in the user's personal space described by:

$$X = (\rho_u \cos(\theta_u), \rho_u \sin(\theta_u)), \qquad (12)$$

where ρ_u and θ_u represent the target point expressed in polar coordinates. While the coordinates of the target point are fixed with respect to the user reference frame, their expression with respect to the robot reference frame $(\rho_u \cos(\theta_u), \rho_u \sin(\theta_u))$ changes as the user moves. Therefore, knowing the position of the user with respect to the robot, allows us to derive the position of the target point with respect to the robot and from there to derive the motion equation that brings the robot in a position for facing the user.

Figure 6. Face detection ensemble

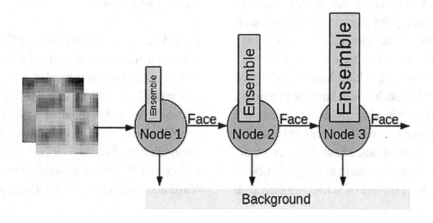

Algorithm 1. AdaBoost Algorithm

Initialize weights $w_{1,i} = \frac{1}{2m}, \frac{1}{2l}$ for $y_i = 0, 1$

for $t = 1, \dots, T$ **do**

1. Normalize the weights, $w_{t,i} \leftarrow \frac{w_{t,i}}{\sum_{j=1}^{n} w_{t,j}}$ so that w_t is a probability distribution.

2. For each feature, j, train a classifier h_j which is restricted to using a single feature. The error is

   evaluated with respect to w_i, $\epsilon_j = \sum_i |h_j(x_i) - y_i|$.

3. Choose the classifier, h_t with the lowest error ϵ_j.

4. Update the weithgs: $w_{t+1,i} = w_{t,i}\beta_t^{1-e_i}$ where $e_i = 0$ if example x_i is calssified correctly, $e_i = 1$

   otherwise, and $\beta_t = \frac{\epsilon_t}{1-\epsilon_t}$.

end for

The final strong classifier is:

$$h(x) = \begin{cases} 1 & \sum_{t=1}^{T} \alpha_t h_t(x) \le \frac{1}{2}\sum_{t=1}^{T} \alpha_t \\ 0 & \text{otherwise} \end{cases} \quad \text{where } \alpha_t = \log \frac{1}{\beta_t}$$

Once the robot has reached its target point, its orientation might not be suitable to start interacting. For example it could not be facing the user since the equations we described before do not control the robot's final orientation (see Figure 4). For this reason, once the robot stops, it looks for the face of the person and adjusts its body orientation according to the detected location of the face in its visual field. This passage requires face detection techniques then the user interaction can start. This is described in the next section.

Human Robot Interaction

A fundamental prerequisite for smooth human-robot interaction is joint attention. Joint attention means that both the person and the robot jointly focus their attention on a single object. The same applies to eye contact, where human and robot mutually attend to each other. For this to happen the robot must first be able to localize a person's face and then estimate where a person is paying attention to. The feedback about the user's estimated head pose can be used to modify the robot's behavior with the purpose of achieving effective communication. In particular, if the robot wants to deliver a message to a person, and this person is not paying attention, the robot should attract attention using verbal or non-verbal cues. As soon as the person pays attention, the robot can deliver the message while monitoring whether the person is still paying attention to the robot. We focus on describing two typical tasks of non-verbal interaction between robot and user: building up eye-contact through face detection and estimating user's attention using head pose estimation. We apply computer vision methods to the robot's head camera to realize these functions.

Face Detection

Face detection can serve many different purposes in human-robot interaction and one of them refers to the correction of the robot's alignment. With this model, the robot is able to align its orientation with respect to the user's face position in the robot's visual field. Different computer vision methods have been developed for face detection and one of the most well-known was proposed by Viola and Jones (2002). This algorithm is based on

Haar-like features, threshold classifiers, AdaBoost and a cascade structure.

Haar-like features (Figure 5 right) are digital image features whose shapes are similar to Haar-Wavelets. A simple rectangle Haar-like feature can be described as the difference of the sum of the pixels within two rectangle areas (Viola & Jones, 2002). In order to speed up the computation, an intermediate representation of an image is processed which is called integral image (Figure 5 left). The transformation of an integral image can be performed with following equations:

$$s(x,y) = s(x,y-1) + i(x,y)$$
$$ii(x,y) = ii(x-1,y) + s(x,y), \quad (13)$$

where (x,y) indicates the pixel position of an image, $s(x,y)$ is the cumulative row sum with $s(x,-1) = 0$, $i(x,y)$ represents the value of the pixel at location (x,y) in the initial image, $ii(x,y)$ represents the value of the integral image at location (x,y) with $ii(-1,y) = 0$.

The calculation of the Haar-like features can be simplified by computing four array references. As shown in Figure 5 left, the value of an integral image at location 1 is the sum of the pixels in rectangle A and the value at location 2 is the sum of pixels in rectangle $(A+B)$. Similarly, the sum within area D equals then 4+1-(2+3) (Viola & Jones, 2002). Since the set of rectangle Haar-like features is overcomplete (Viola & Jones, 2002), an AdaBoost algorithm (Freund & Schapire, 1995) is employed to select a small number of significant features. A set of threshold classifiers are built using these Haar-like features. Each classifier has only a low detection rate therefore they are also called weak classifiers. To improve the classification results, the classifiers are combined with a cascade structure which rejects most of the background within few classification steps. The detailed AdaBoost algorithm is shown in Algorithm 1(Viola & Jones, 2002).

Once a face is detected in the robot's camera, we use a closed-loop control mechanism for centering the user's face in its visual field. Our humanoid robot's head has two degrees of freedom namely yaw and pitch. We control the yaw and the pitch angle to minimize the distance in the vertical and horizontal direction between the detected face and a relevant point in the image (X_t, Y_t). The information we gather from the face detection module is denoted as (X_0, Y_0) and is

Figure 7. Face tracking is based on the face's location in the robot's visual field. The tracking algorithm tends to minimize the distance between the position of the face in the robot's visual field and a relevant point in the image which in the case shown is the center of the visual field ($X_t = 0, Y_t = 0$).

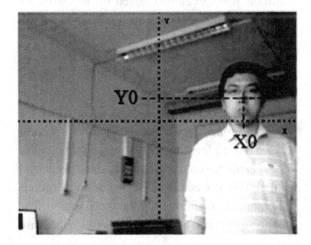 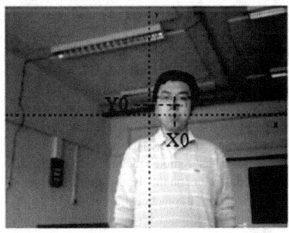

Figure 8. Face detection and image processing stages. Head Pose estimation requires face detection and image preprocessing, in particular edge detection, to the data acquired from the robot's camera.

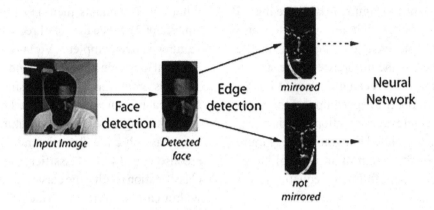

reported in Figure 7. We generate control commands with a simple proportional action as described in the following equation:

$$C_{yaw} = k_{yaw}(X_0 - X_t)$$
$$C_{pitch} = k_{pitch}(Y_0 - Y_t), \tag{14}$$

where k_{yaw} and k_{pitch} represent the strength of the proportional control action relative to the yaw and pitch angles of the robot's head.

Head Pose Estimation

Currently, several commercial methods are available for estimating a person's head pose (e.g. the face API, face.com). These methods are usually optimized for situations where a person is sitting behind his desktop computer: high-resolution images of nearby faces. In robotics the image quality is typically limited: (1) Small humanoid robots like Aldebaran's Nao typically have limited processing capacity available, because of the weight limitations. As a result the cameras do not have optics to improve image quality. (2) It is possible to process images on a fast remote machine, but this poses strong constraints on the bandwidth of the wireless connection. In practice, only low-resolution images are transmitted with sufficient refresh rate.

(3) The robot, if autonomous, operates in hugely varying lighting conditions. Close inspection of Figure 7, for example, reveals large color differences although the scenes are very similar to the human eye. (4) In addition, a robot walking on the floor is never very close to the user. Thus, a person's face only covers a small part of the image (see Figure 7). The first two constraints are of a technical nature and can be remedied with more expensive equipment. The third and fourth constraints, however, are due to the different role a humanoid robot has when interacting with a person. Thus, an improved method is required to estimate head pose from a limited amount of visual information. Because of these reasons a neural network-oriented head pose estimation solution was developed based on the work by Voit et al. (2007), and further developed by van der Pol et al. (2011).

Image Processing

The image patch of the face as detected by the Viola and Jones' face detection method is preprocessed before feeding it into the neural network. The images are converted to black and white and scaled and cropped to 30 pixels wide and 90 pixels tall images containing only the facial region. The image aspect ratio is rather tall than wide, because this way the image contains only parts of the face even when it is rotated over large angles. After adjusting the image to the appropriate size, the image is filtered with a Laplacian edge detection filter. This filter is a translation- and rotation-invariant detector for contrast. The Laplacian works like an edge detector and behaves similarly to a combination of Sobel filters or a Canny edge detector. To be able to average over different data gathered from the neural network, the mirrored image is also passed through to the next level, as well as a cutout one pixel bigger in all directions and its mirrored image.

Neural Network

Neural networks are computationally expensive to train, but they are efficient after training. Other solutions often detect certain features within the face to calculate the rotation by using a three-dimensional model of the head (Murphy-Chutorian & Trivedi, 2009). This requires a high-resolution image for robust detection of the fine textures that define these features. Therefore, it also becomes very dependent on the lighting condition. Our neural network-based approach does not use these features, nor does it rely on a three-dimensional model.

The network is trained using a training set of 5756 images some of them coming from the database created by Gourier, Hall and Crowley (2004). Multiple networks are trained for both yaw (horizontal orientation) and pitch (vertical orientation) angles (see Figure 9). The Levenberg-Marquardt training method is used to train the two-layer, feed-forward neural network. Note that the roll angle cannot be estimated by the neural network, because the face detection algorithm is only able to detect faces that are upright. While training neural networks, especially complex ones like this, chances are high to end up at a local minimum. Therefore, the output of different networks is likely to differ. This method uses the best ten networks to increase performance. The output of the networks is multiple estimates of pitch and yaw angles. After averaging over all of these values the estimate of the head pose is retrieved (see Figure 10).

Caveats

Currently, the utilized face detection cascade for Viola and Jones' object (2002) detection is limited to detect a face when most of the face is visible excluding side views. Consequently, our method currently works for faces rotated less than 90 degrees from looking straight at the camera, in any direction. Head pose is only an estimator of gaze direction as human gaze is also determined by the orientation of the eyes. Nonetheless, head pose is a good estimator for gaze, because when people attend to something for some time they always turn their heads. The performance of our method in real life heavily depends on the training set that was used because the variation of, say, lighting condition in the training set affects the robustness against varying lighting conditions during implementation. We find an average error for the yaw estimation of about 7 degrees. For the pitch angle the average error is approximately 11

Figure 9. Pitch, yaw, and roll angles

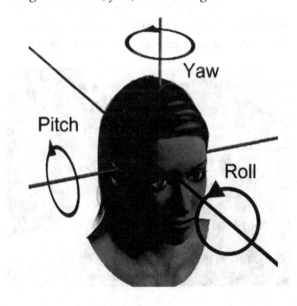

Figure 10. Two-layer feed forward neural network for head pose estimation. Image shows individual networks that combined give an estimate of the yaw angle of rotation.

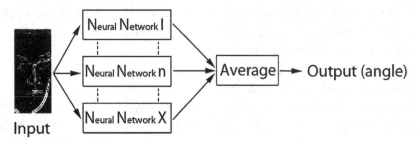

degrees. These results are sufficiently precise for social interaction and they show that our method works from a low vantage point with varying lighting conditions and low resolution images.

CASE STUDY

Effective assistance refers to both, the ability of the robot to interact with the environment and the ability of the robot to interact with the user as those two aspects are tightly coupled. Think about a robot that is able to move safely in a cluttered environment but that is not able to know where its human companion is or where (s)he is looking at. Then the inclusion of such a robot in ambient

homes would not provide any benefit because it would not be able to communicate with the person and remind the user to check some health parameters. On the other hand, in case of a static robotic assistant, the user would be obliged to go to the robot. But in this situation how can a robot function as a reminder assistant if the person herself needs to be active and remember to approach the robot?

These two examples motivate that navigation capabilities of a mobile robot and its ability to acquire information about the user are coupled and they are essential for designing robotics applications for ambient homes. Effectiveness of interaction with the environment and the user can only be achieved by the acquisition of real time

Figure 11. The case study represents tests of the KSERA system done in the Schwechat (Vienna) Nursing Home. For privacy reason we have modified the faces to be unrecognizable.

Figure 12. The actions of various systems components are coordinated with finite element state machines. The figure represents the action sequence that the robot needs to accomplish for going to a person and interacting with her.

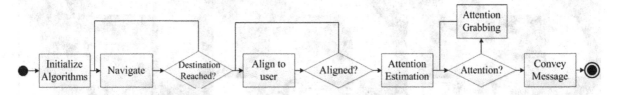

feedback on the environment's and the user's conditions. This feedback can be obtained by the joint application of the computer vision methods presented in the previous sections. Therefore this section presents a case study related to the introduction of a humanoid robot in an AAL application and provides a concrete example of how the computer vision techniques introduced in the previous sections can be integrated for employing a robot in an assisting environment (Figure 11). The robot should navigate autonomously towards a person, establish eye contact, check whether the user is focused on the robot and then ask the user to perform the measurement of his oxygen level. The test case refers to currently ongoing testing of the KSERA system in Schwechat Vienna.

In our case study, we run experiments with different users and one of our test cases is illustrated in Figure 12. This test case consists of two parts: (1) navigate safely towards a person and (2) human robot interaction using nonverbal communication. The event flow for achieving points (1) and (2) is represented as a state machine diagram. At first all the components are initialized, and then the robot navigates towards the person until it reaches its target point. At that moment it tries to establish eye contact with the user modifying its body pose. It then checks the focus of attention of the user and if the user is paying attention it asks the user to measure his oxygen level. We start by letting the robot approach a localized person. Two groups of particles with different setups for person and robot localization are initialized at

Figure 13. Simultaneous person and robot tracking. Left: initial position, Right: robot's movement towards a person.

Figure 14. Robot's movements towards a person. After moving for a short distance the robot stops in a proper location for approaching the user as identified by (Torta et al., 2011).

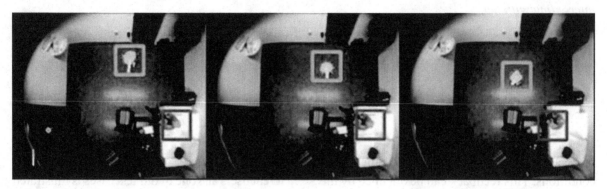

random position in the image space at first (see Figure 13 left).

The person is first localized using motion information. The weights of particles close by the user increase and the particle cloud converges to a single position (Figure 13 right). Meanwhile, the shape feature as well as the color histogram adapts to store features of the localized user, which enables the system to localize the person even when motion is missing. The person localization has been tested and evaluated according to the CLEAR MOT Metrics (Keni & Rainer, 2008). Since only a single person is tracked in the system, based on our goal design, the frame number of misses m and of false positives f_p has been

counted and the multiple object tracking accuracy (MOTA) has been calculated. The test results are shown in Table 1 and for details please see (Yan, Weber, & Wermter, 2011).

For the robot's particles a precise feed-forward motion model has been built according to the robot's behavior. When the robot moves, the particles move also with different orientation. The particles with wrong orientation will fly away from robot's position and only the particles with correct angles can survive which helps the system to estimate robot's orientation. The robot moves from its initial position to the final position in front of the user based on the navigation Equation (8) and Equation (9). Readers interested

Figure 15. Face of the person in the robot's visual field before and after the robot's alignment. Direct eye contact can be built after the alignment.

Table 1. Experimental results (Yan, Weber, & Wermter, 2011)

Name	Total Frame	m	fp	MOTA (%)
Person moving scenario 1	2012	19	22	97.96
Person moving scenario 2	2258	169	12	91.98
Person moving and sitting scenario 1	1190	78	21	91.68
Person moving and sitting scenario 2	980	22	130	84.18
Change environment scenario 1	1151	89	30	89.66
Change environment scenario 2	1564	157	141	80.94
Change light condition in night scenario	160	17	59	52.5
Change light condition in day scenario	540	0	3	99.45
Distracter person scenario 1	1014	48	35	91.81
Distracter person scenario 2	700	57	26	88.14
Distracter person scenario CLEAR 07	2122	188	52	88.68
Total	13691	844	531	89.96

in more complex navigation trajectories can refer to videos reported on the KSERA website http://www.ksera-project.eu. Personal space models have been defined with HRI psychophysical tests and are reported in Torta, Cuijpers, Juola and van der Pol (2011).

Once the robot reaches its target point, its final orientation might be inappropriate for initiating the interaction, as can be seen in Figure 14, because the robot is not able to establish eye contact with the user. In this case the robot can look for the user's face using the algorithm reported in section "Human Robot Interaction" with face detection and head pose estimation and adjusts its body's orientation and head angle to make eye contact with the user. As can be seen in Figure 15, at the beginning the face of the user is not centered in the robot's visual field, but after applying the face tracking algorithm the robot centers the user's face thus aligning to him. The person's attention can be monitored by applying the head pose estimation method as described in section "Head Pose Estimation". If the person is not paying attention to the robot, the robot will generate actions to grab user's attention until the user focuses on the robot. Then the robot conveys a message and the test case ends.

CONCLUSION AND FUTURE RESEARCH DIRECTIONS

The chapter gives an overview of vision algorithms used in a typical scenario of ambient assisted living. We have focused our attention on robot's navigation towards a simultaneously localized person and on human-robot interaction. We discussed challenges for robots in ambient homes as well as the benefits of computer vision for these applications compared to systems with different sensors. A hybrid probabilistic algorithm is described for localizing the person based on different visual cues. The model is to some extent indicative of a human's ability of recognizing objects based on different features. When some of the features are strongly disturbed, detection recovers by the integration of other features. The particle filter parallels an active attention selection mechanism, which allocates most processing resources to positions of interest. It has a high performance of detecting complex objects that move relatively slowly in real time.

We described a sound method for face detection, the Viola and Jones method. We used the coordinates of the user's face in the visual field for correcting the robot's orientation for facing the user. Moreover, we illustrated a head pose estimation method based on the use of multiple neural networks. Once trained, the head pose estimation is computationally inexpensive, requires only low quality images and is robust to non-optimal lighting conditions. This makes our approach, compared to other methods for head pose estimation, especially useful in robotics applications for ambient assisted living.

A case study has been included that provides a concrete example of how the computer vision techniques can be integrated for employing a robot in an assisting environment. The experiments have been conducted and the evaluation shows that intelligent computer vision algorithms using a distributed sensory network (camera and robot) can be merged for achieving more robust and effective robot behavior and improve human-robot interaction in an AAL environment significantly. Our research focus is currently on the localization of a single person and robot navigation based on computer vision technology. However, in a real home situation, multiple persons may appear in a room at the same time and the robot should be able to distinguish them. Hence, in future research we will attempt to extend our model for localizing multiple persons at the same time using the ceiling mounted camera. A sophisticated person recognition model would also be employed in this case to distinguish the target person from the rest based on visual cues obtained from the robot's camera. Intention recognition would be another interesting direction for improving human-robot interaction and for defining the robot's proactive behavior. Fundamental information for intention estimation can come from the ceiling camera and the person's localization method. Another approach is to add a neural-based model for facial emotion recognition so as to understand whether the user is happy or sad and then adapt the robot's interactive behavior. In addition, the elaboration of a novel robot navigation method without camera calibration would be a useful improvement of robotic applications in domestic environments. Camera calibration is essential to eliminate the distortion effect of the camera lens and to ensure the quality of coordinate transformation from the camera view to the real world, but makes the system hard to install by persons without professional knowledge. Therefore, we are considering a model based on neural planning that can learn room mapping from the person's spatial knowledge and plan the robot's movement based on the learned map.

In general, our results show that - although many solutions exist to particular detailed problems like face recognition, navigation and localization - robotic applications in domestic environments like AAL require a level of integration that currently does not exist. This research is only the first step in addressing this issue.

ACKNOWLEDGMENT

The research leading to these results is part of the KSERA project (http://www.ksera-project.eu) funded by the European Commission under the 7th Framework Programme (FP7) for Research and Technological Development under grant agreement n°2010-248085.

REFERENCES

Althaus, P., Ishiguro, H., Kanda, T., Miyashita, T., & Christensen, H. (2004). Navigation for human-robot interaction tasks. *2004 IEEE International Conference on Robotics and Automation, ICRA'04*, Vol. 2, (pp. 1894-1900).

Arkin, R. C. (1998). *Behavior-based robotics*. Cambridge, MA: MIT press.

Ba, S., & Odobez, J. (2004). A probabilistic framework for joint head tracking and pose estimation. *17th International Conference on Pattern Recognition, ICPR 2004*, Vol. 4, (pp. 264-267).

Bahadori, S., Iocchi, L., Leone, G., Nardi, D., & Scozzafava, L. (2007). Real-time people localization and tracking through fixed stereo vision. *Applied Intelligence*, *26*, 83–97. doi:10.1007/s10489-006-0013-3

Bay, H., Tuytelaars, T., & Van Gool, L. (2006). Surf: Speeded up robust features. In Leonardis, A., Bischof, H., & Pinz, A. (Eds.), *Computer Vision - ECCV 2006* (*Vol. 3951*, pp. 404–417). Lecture Notes in Computer Science. doi:10.1007/11744023_32

Belhumeur, P., Hespanha, J., & Kriegman, D. (1997). Eigenfaces vs. fisherfaces: recognition using class specific linear projection. *IEEE Transactions on Pattern Analysis and Machine Intelligence*, *19*(7), 711–720. doi:10.1109/34.598228

Bernardin, K., Gehrig, T., & Stiefelhagen, R. (2008). Multi-level particle filter fusion of features and cues for audio-visual person tracking. In Stiefelhagen, R., Bowers, R., & Fiscus, J. (Eds.), *Multimodal Technologies for Perception of Humans* (*Vol. 4625*, pp. 70–81). Lecture Notes in Computer Science. doi:10.1007/978-3-540-68585-2_5

Bicho, E. (1999). *Dynamic approach to behavior-based robotics*. PhD thesis, University of Minho.

Brooks, A. G., & Arkin, R. C. (2006). Behavioral overlays for non-verbal communication expression on a humanoid robot. *Autonomous Robots*, *22*, 55–74. doi:10.1007/s10514-006-9005-8

Comaniciu, D., Ramesh, V., & Meer, P. (2000). Real-time tracking of non-rigid objects using mean shift. *IEEE Computer Society Conference on Computer Vision and Pattern Recognition*, Vol. 2, (pp. 2142-2150).

Dautenhahn, K. W., Koay, K., Nehaniv, C., Sisbot, A., Alami, R., & Siméon, T. (2006). How may I serve you? A robot companion approaching a seated person in a helping context. *1st ACM SIGCHI/SIGART Conference on Human-Robot Interaction*, (pp. 172-179).

Freund, Y., & Schapire, R. (1995). A desicion-theoretic generalization of on-line learning and an application to boosting. In Vitányi, P. (Ed.), *Computational Learning Theory* (*Vol. 904*, pp. 23–37). Lecture Notes in Computer Science. doi:10.1007/3-540-59119-2_166

Frintrop, S., Königs, A., Hoeller, F., & Schulz, D. (2010). A component-based approach to visual person tracking from a mobile platform. *International Journal of Social Robotics*, *2*, 53–62. doi:10.1007/s12369-009-0035-1

Gourier, N., Hall, D., & Crowley, J. (2004). Facial features detection robust to pose, illumination and identity. *2004 IEEE International Conference on Systems, Man and Cybernetics*, Vol. 1, (pp. 617-622).

Hall, E. (1966). *The hidden dimension*. New York, NY: Doubleday.

Harris, C., & Stephens, M. (1988). A combined corner and edge detector. *Alvey Vision Conference*, Vol. 15, (p. 50). Manchester, UK.

Hecht, F., Azad, P., & Dillmann, R. (2009). Markerless human motion tracking with a flexible model and appearance learning. *IEEE International Conference on Robotics and Automation, ICRA '09*, (pp. 3173-3179).

Hootman, J., & Helmick, C. (2006). Projections of US prevalence of arthritis and associated activity limitations. *Arthritis and Rheumatism*, *54*(1), 226–229. doi:10.1002/art.21562

Hyvärinen, A., & Oja, E. (2000). Independent component analysis: Algorithms and applications. *Neural Networks*, *13*(4-5), 411–430. doi:10.1016/S0893-6080(00)00026-5

Jolliffe, I. (2005). Principal component analysis. In Everitt, B., & Howell, D. (Eds.), *Encyclopedia of statistics in behavioral science*. New York, NY: John Wiley & Sons, Ltd. doi:10.1002/0470013192.bsa501

Kaelbling, L., Littman, M., & Cassandra, A. (1998). Planning and acting in partially observable stochastic domains. *Artificial Intelligence, 101*(1-2), 99–134. doi:10.1016/S0004-3702(98)00023-X

Kahn, J. P., Kanda, T., Ishiguro, H., Gill, B. T., Ruckert, J. H., Shen, S., et al. (2012). Do people hold a humanoid robot morally accountable for the harm it causes? *The Seventh Annual ACM/IEEE International Conference on Human-Robot Interaction* (pp. 33-40). New York, NY: ACM.

Kalal, Z., Matas, J., & Mikolajczyk, K. (2010). P-N learning: Bootstrapping binary classifiers by structural constraints. *IEEE Computer Society Conference on Computer Vision and Pattern Recognition*, (pp. 49-56).

Kaplan, F., & Hafner, V. V. (2006). The challenges of joint attention. *Interaction Studies: Social Behaviour and Communication in Biological and Artificial Systems, 7*, 135–169. doi:10.1075/is.7.2.04kap

Kemmotsu, K., Koketsua, Y., & Iehara, M. (2008). Human behavior recognition using unconscious cameras and a visible robot in a network robot system. *Robotics and Autonomous Systems, 56*(10), 857–864. doi:10.1016/j.robot.2008.06.004

Kendon, A. (1967). Some functions of gaze-direction in social interaction. *Acta Psychologica, 26*, 22–63. doi:10.1016/0001-6918(67)90005-4

Keni, B., & Rainer, S. (2008). Evaluating multiple object tracking performance: The CLEAR MOT metrics. *EURASIP Journal on Image and Video Processing, 2008*, 10.

Kirby, R. (2010). *Social robot navigation*. PhD Thesis, Robotics Institute, Carnegie Mellon University, Pittsburgh, PA.

Kleinke, C. (1986). Gaze and eye contact: A research review. *Psychological Bulletin, 100*(1), 78. doi:10.1037/0033-2909.100.1.78

Lambert, D. (2004). *Body language*. London, UK: Harper Collins.

Lanz, O., & Brunelli, R. (2008). An appearance-based particle filter for visual tracking in smart rooms. In Stiefelhagen, R., Bowers, R., & Fiscus, J. E. (Eds.), *Multimodal Technologies for Perception of Humans* (*Vol. 4625*, pp. 57–69). Lecture Notes in Computer Science. doi:10.1007/978-3-540-68585-2_4

Lawrence, S., Giles, C., Tsoi, A., & Back, A. (1997). Face recognition: A convolutional neural-network approach. *IEEE Transactions on Neural Networks, 8*(1), 98–113. doi:10.1109/72.554195

Lienhart, R., & Maydt, J. (2002). An extended set of Haar-like features for rapid object detection. *2002 International Conference on Image Processing*, Vol. 1, (pp. 900-903).

Louloudi, A., Mosallam, A., Marturi, N., Janse, P., & Hernandez, V. (2010). *Integration of the humanoid robot nao inside a smart home: A case study*. The Swedish AI Society Workshop.

Lowe, D. (2004). Distinctive image features from scale-invariant keypoints. *International Journal of Computer Vision, 60*, 91–110. doi:10.1023/B:VISI.0000029664.99615.94

Mita, T., Kaneko, T., & Hori, O. (2005). Joint haar-like features for face detection. *10th IEEE International Conference on Computer Vision*, Vol. 2, (pp. 1619-1626).

Monteiro, S., & Bicho, E. (2010). Attractor dynamics approach to formation control: Theory and application. *Autonomous Robots, 29*, 331–355. doi:10.1007/s10514-010-9198-8

Muñoz-Salinas, R., Aguirre, E., & Garcá-Silvente, M. (2007). People detection and tracking using stereo vision and color. *Image and Vision Computing, 25*(6), 995–1007. doi:10.1016/j.imavis.2006.07.012

Murphy-Chutorian, E., & Trivedi, M. (2009). Head pose estimation in computer vision: A survey. *IEEE Transactions on Pattern Analysis and Machine Intelligence, 31*(4), 607–626. doi:10.1109/TPAMI.2008.106

Mutlu, B., & Forlizzi, J. (2008). Robots in organizations: The role of workflow, social, and environmental factors in human-robot interaction. *The 3rd ACM/IEEE International Conference on Human Robot Interaction* (pp. 287-294). New York, NY: ACM.

Mutlu, B., Forlizzi, J., & Hodgins, J. (2006). A storytelling robot: Modeling and evaluation of human-like gaze behavior. *6th IEEE-RAS International Conference on Humanoid Robots*, (pp. 518 -523).

Nakauchi, Y., & Simmons, R. (2002). A social robot that stands in line. *Autonomous Robots, 12,* 313–324. doi:10.1023/A:1015273816637

O'Grady, M., Muldoon, C., Dragone, M., Tynan, R., & O'Hare, G. (2010). Towards evolutionary ambient assisted living systems. *Journal of Ambient Intelligence and Humanized Computing, 1,* 15–29. doi:10.1007/s12652-009-0003-5

OECD. (2007). *OECD demographic and labour force database.* Organisation for Economic Co-operation and Development.

Oskoei, A., Walters, M., & Dautenhahn, K. (2010). *An autonomous proxemic system for a mobile companion robot.* AISB.

Osuna, E., Freund, R., & Girosi, F. (1997). Training support vector machines: An application to face detection. *IEEE Computer Society Conference on Computer Vision and Pattern Recognition,* (p. 130).

Pacchierotti, E., Christensen, H., & Jensfelt, P. (2007). Evaluation of passing distance for social robots. *The 15th IEEE International Symposium on Robot and Human Interactive Communication* (pp. 315-320). IEEE.

Piccardi, M. (2004). Background subtraction techniques: A review. *2004 IEEE International Conference on Systems, Man and Cybernetics,* Vol. 4, (pp. 3099-3104).

Ramanan, D., Forsyth, D., & Zisserman, A. (2007). Tracking people by learning their appearance. *IEEE Transactions on Pattern Analysis and Machine Intelligence, 29,* 65–81. doi:10.1109/TPAMI.2007.250600

Salah, A., Morros, R., Luque, J., Segura, C., Hernando, J., & Ambekar, O. (2008). Multimodal identification and localization of users in a smart environment. *Journal on Multimodal User Interfaces, 2,* 75–91. doi:10.1007/s12193-008-0008-y

Schöner, G., Dose, M., & Engels, C. (1995). Dynamics of behavior: Theory and applications for autonomous robot architectures. *Robotics and Autonomous Systems, 16*(2-4), 213–245. doi:10.1016/0921-8890(95)00049-6

Siino, R. M., & Hinds, P. (2004). *Making sense of new technology as a lead-in to structuring: The case of an autonomous mobile robot* (pp. E1–E6). Academy of Management Proceedings.

Sisbot, E. A., Marin-Urias, L. F., Broquère, X., Sidobre, D., & Alami, R. (2010). Synthesizing robot motions adapted to human presence. *International Journal of Social Robotics, 2,* 329–343. doi:10.1007/s12369-010-0059-6

Steg, H., Strese, H., Loroff, C., Hull, J., & Schmidt, S. (2006). *Europe is facing a demographic challenge Ambient Assisted Living offers solutions.*

Syrdal, D. S., Lee Koay, K., & Walters, M. L. (2007). A personalized robot companion? - The role of individual differences on spatial preferences in HRI scenarios. *RO-MAN 2007 - The 16th IEEE International Symposium on Robot and Human Interactive Communication* (pp. 1143-1148). IEEE.

Torta, E., Cuijpers, R., Juola, J., & van der Pol, D. (2011). Design of robust robotic proxemic behaviour. In Mutlu, B. A., Ham, J., Evers, V., & Kanda, T. (Eds.), *Social Robotics* (*Vol. 7072*, pp. 21–30). Lecture Notes in Computer Science Berlin, Germany: Springer. doi:10.1007/978-3-642-25504-5_3

Triesch, J., & Malsburg, C. (2001). Democratic integration: Self-organized integration of adaptive cues. *Neural Computation*, *13*(9), 2049–2074. doi:10.1162/089976601750399308

van der Pol, D., Cuijpers, R., & Juola, J. (2011). Head pose estimation for a domestic robot. *The 6th International Conference on Human-Robot Interaction, HRI '11*, (pp. 277-278).

Viola, P., & Jones, M. (2002). Robust real-time face detection. *International Journal of Computer Vision*, *57*(2), 137–154. doi:10.1023/B:VISI.0000013087.49260.fb

Voit, M., Nickel, K., & Stiefelhagen, R. (2007). Neural network-based head pose estimation and multi-view fusion. In Stiefelhagen, R., & Garofolo, J. (Eds.), *Multimodal Technologies for Perception of Humans* (*Vol. 4122*, pp. 291–298). Lecture Notes in Computer Science. doi:10.1007/978-3-540-69568-4_26

Voit, M., Nickel, K., & Stiefelhagen, R. (2008). Head pose estimation in single- and multi-view environments - results on the clear'07 benchmarks. In Stiefelhagen, R., Bowers, R., & Fiscus, J. (Eds.), *Multimodal Technologies for Perception of Humans* (*Vol. 4625*, pp. 307–316). Lecture Notes in Computer Science. doi:10.1007/978-3-540-68585-2_29

Walters, M., Dautenhahn, K., Boekhorst, R., Koay, K., Syrdal, D., & Nehaniv, C. (2009). An empirical framework for human-robot proxemics. In Dautenhahn, K. (Ed.), *New frontiers in human-robot interaction* (pp. 144–149).

Weber, C., & Wermter, S. (2007). A self-organizing map of sigma-pi units. *Neurocomputing*, *70*(13-15), 2552–2560. doi:10.1016/j.neucom.2006.05.014

West, G., Newman, C., & Greenhill, S. (2005). Using a camera to implement virtual sensors in a smart house. *From Smart Homes to Smart Care: International Conference on Smart Homes and Health Telematics,* Vol. 15, (pp. 83-90).

Yamaoka, F., Kanda, T., Ishiguro, H., & Hagita, N. (2010). A model of proximity control for information-presenting robots. *IEEE Transactions on Robotics*, *26*(1), 187–195. doi:10.1109/TRO.2009.2035747

Yan, W., Weber, C., & Wermter, S. (2011). A hybrid probabilistic neural model for person tracking based on a ceiling-mounted camera. *Journal of Ambient Intelligence and Smart Environments*, *3*(3), 237–252.

Yücel, Z., Salah, A., Merigli, C., & Mericli, T. (2009). Joint visual attention modeling for naturally interacting robotic agents. *24th International Symposium on Computer and Information Sciences*, (pp. 242-247).

Zivkovic, Z., & Krose, B. (2004). An EM-like algorithm for color-histogram-based object tracking. *IEEE Computer Society Conference on Computer Vision and Pattern Recognition*, Vol. 1, (pp. 798-803).

Chapter 16
An Integrated Framework for Robust Human–Robot Interaction

Mohan Sridharan
Texas Tech University, USA

ABSTRACT

Developments in sensor technology and sensory input processing algorithms have enabled the use of mobile robots in real-world domains. As they are increasingly deployed to interact with humans in our homes and offices, robots need the ability to operate autonomously based on sensory cues and high-level feedback from non-expert human participants. Towards this objective, this chapter describes an integrated framework that jointly addresses the learning, adaptation, and interaction challenges associated with robust human-robot interaction in real-world application domains. The novel probabilistic framework consists of: (a) a bootstrap learning algorithm that enables a robot to learn layered graphical models of environmental objects and adapt to unforeseen dynamic changes; (b) a hierarchical planning algorithm based on partially observable Markov decision processes (POMDPs) that enables the robot to reliably and efficiently tailor learning, sensing, and processing to the task at hand; and (c) an augmented reinforcement learning algorithm that enables the robot to acquire limited high-level feedback from non-expert human participants, and merge human feedback with the information extracted from sensory cues. Instances of these algorithms are implemented and fully evaluated on mobile robots and in simulated domains using vision as the primary source of information in conjunction with range data and simplistic verbal inputs. Furthermore, a strategy is outlined to integrate these components to achieve robust human-robot interaction in real-world application domains.

INTRODUCTION

Mobile robots are increasingly being used in real-world application domains such as surveillance, navigation and healthcare due to the availability of high-fidelity sensors and the development of state of the art algorithms to process sensory inputs. As we move towards deploying robots in our homes and offices, i.e., domains with a significant amount of uncertainty, there is a need for enabling robots to learn from sensory cues and limited feedback from non-expert human partici-

DOI: 10.4018/978-1-4666-2672-0.ch016

pants. Human-robot interaction (HRI) poses many challenges such as autonomous operation, safety, engagement, robot design and interaction protocol design (Tapus, Mataric, & Scassellati, 2007). The focus of this chapter is on robust autonomy using sensory cues and high-level feedback from non-expert human participants. Many algorithms have been developed for autonomous operation based on sensory inputs, and for learning from manual training and domain knowledge. Real-world domains characterized by partial observability, non-deterministic action outcomes and unforeseen dynamic changes make it difficult for a robot to operate without any human feedback. At the same time, human participants may not have the expertise and time to provide elaborate and accurate feedback in complex domains (Fong, Nourbakhsh, & Dautenhahn, 2003; Thrun, 2004). Recent research has hence focused on enabling a robot to acquire human feedback when needed and merge human inputs with the information extracted from sensory cues. However, these algorithms require elaborate domain knowledge or fail to model the unreliability of human inputs, limiting their use to simplistic simulated domains or specific real-world applications (Knox & Stone, 2010; Rosenthal, Veloso, & Dey, 2011).

As an illustrative example, consider the robots in Figure 1(left) deployed to interact with humans in offices and homes. Such real-world domains are characterized by unforeseen dynamic changes, e.g., existing objects move, novel objects are introduced and the environmental factors change unpredictably. Assume that sensory cues consist primarily of vision (monocular and stereo) in conjunction with range data and simplistic verbal inputs. Also assume that the robots do not manipulate domain objects and do not have physical contact with humans. Each robot is equipped with core algorithms to process sensory cues with varying levels of reliability and computational complexity. Non-expert human participants provide limited high-level feedback in the form of simplistic verbal inputs that reinforce the robot's actions or resolve ambiguities identified by the robot. Although it is not feasible to process all inputs or model the entire domain and still respond to dynamic changes, each robot has to exploit relevant sensory cues to operate reliably. Given such a scenario, this chapter focuses on the following key questions:

- How to best enable a robot to adapt learning, sensing and processing to different scenarios and participants?

Figure 1. (Left) examples of robot platforms relevant to the research described in this chapter; (right) integrated framework that uses the dependencies between learning, adaptation and interaction to achieve synergetic autonomy in real-world HRI

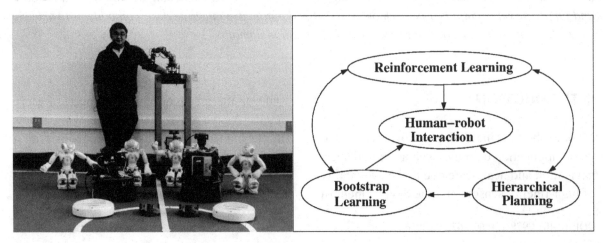

- How to best enable a robot to seek limited high-level feedback from non-expert human participants, and robustly merge human inputs with the information extracted from sensory inputs?

While sophisticated algorithms have been developed for the learning, adaptation and interaction challenges in isolation, the integration of these subfields to enable robust HRI remains an open problem, even as it presents new opportunities to address the existing challenges in the subfields (AAAI Symposium, 2012). This chapter describes a novel probabilistic framework that seeks to answer the questions listed above by jointly addressing the associated learning, adaptation and interaction challenges. The framework is composed of the following components:

- **Bootstrap Learning:** Robots use sensory cues to autonomously and incrementally learn probabilistic graphical models of environmental objects. These learned models enable robots to detect and adapt to unforeseen changes.
- **Hierarchical Planning:** A novel hierarchical decomposition of partially observable Markov decision processes enables robots to automatically adapt learning, sensing and information processing to each of a wide range of tasks.
- **Reinforcement Learning:** Robots acquire high-level feedback from non-expert humans (based on need and availability) and merge the information extracted from human feedback with the information extracted from sensory inputs.

As shown in Figure 1 (right), these components inform and guide each other, e.g., planning and limited human feedback can constrain learning to objects and events relevant to the task at hand, while learning can help automate planning. The remainder of the chapter is organized as follows. The next section motivates the integrated framework for HRI by discussing related work in learning, planning and interaction. Instances of the individual components are then described in the context of visual inputs in the section and subsections following. These instantiations are accompanied by experimental results of evaluating the corresponding algorithms in simulated domains and on wheeled and humanoid robots deployed in indoor domains. Furthermore, this chapter outlines a strategy to integrate the individual components to achieve the desired target of robust human-robot interaction.

RELATED WORK

The proposed framework uses vision as a major source of information. This section motivates the integrated framework for robust human-robot interaction by discussing the limitations of existing algorithms for vision-based learning, planning and interaction.

Vision-Based Learning and Planning

A robot vision system typically includes segmentation, recognition and scene understanding. Computer vision research has produced many algorithms for segmentation (Caselles, Kimmel, & Shapiro, 1997; Comaniciu & Meer, 2002; Felzenswalb & Huttenlocher, 2004), and many robot domains use labeled data to map pixels to color labels. Algorithms have also been developed to characterize (and hence recognize) objects using local image gradients (Lowe, 2004; Mikolajczyk & Schmid, 2004); appearance models (Arashloo & Kittler, 2011; Fergus, Perona, & Zisserman, 2003); hierarchical decomposition of parts (Fidler, Boben, & Leonardis, 2008); or visual cortical mechanisms (Serre, Wolf, Bileschi, Riesenhuber, & Poggio, 2007). Recent research in computer vision has also provided algorithms that use contextual cues learned from images for object recognition (Li,

Parikh, & Chen, 2011). Many robot applications use these algorithms in conjunction with temporal cues and 3D range input (e.g., Kinect, RGB-D cameras, Lidar) for object recognition and scene understanding (Lai, Bo, Ren, & Fox, 2011). However, many of these algorithms are computationally expensive, sensitive to environmental changes, or require extensive domain-specific information.

Sensitivity to environmental factors is a major challenge to the use of visual features, e.g., object recognition algorithms based on image gradients or color distributions are sensitive to illumination, and stereo maps are sensitive to texture-less surfaces (Hartley & Zisserman, 2004). Algorithms developed to provide robustness to changes in environmental factors such as illumination typically require prior knowledge of illuminations and object properties and are computationally expensive (Finlayson, Hordley, & Hubel, 2001; Lammens, 1994). Since a mobile robot has to deal with unexpected changes, robot vision algorithms tend to track changes in visual feature distributions and update parameters of learned models (e.g., mixture of Gaussians) over time (Sridharan & Stone, 2007; Thrun, 2006). However, adaptation to unforeseen changes continues to be a major challenge to the use of robots in the real-world.

In parallel to the research on vision-based learning, many algorithms have been developed for automatic speech recognition and understanding using grammars and probabilistic sequential reasoning methods (Brick & Scheutz, 2007; Guedon, 2005; Rabiner, 1989), resulting in many HRI applications. However, these algorithms require significant prior knowledge, cannot adapt to dynamic changes or do not build strong associations between language and other modalities for human-robot interaction.

A mobile robot in real-world application domains such as offices and homes cannot observe the entire domain or process all sensory inputs. Planning algorithms have hence been developed to sequence sensing and information processing operators based on high-level goals. Modern AI planning algorithms that relax the limiting constraints of classical algorithms (Ghallab, Nau, & Traverso, 2004) have been used in many applications (Brenner & Nebel, 2009; Petrick & Bacchus, 2004; Talamadupula, Benton, Kambhampati, Schermerhorn & Scheutz, 2010). Probabilistic planning algorithms have also been designed for tasks such as visual gesture and object recognition (Li, Bulitko, Greiner, & Levner, 2003). In parallel, active vision algorithms have been developed for sensor placement and multi-target tracking (Kreucher, Kastella, & Hero, 2005), submodular functions have been used for sensor placement (Krause, Singh, & Guestrin, 2008) and visual target recognition has been posed as an information maximization task (Butko & Movellan, 2008). However, many of these methods require manual supervision and many visual planning tasks are not submodular. In recent years, partially observable Markov decision processes (POMDPs) have been used to plan sensory processing for behavior control, navigation and grasp planning on robots (Brook, Ciocarlie, & Hsiao, 2011; Hoey et al., 2010). Although good performance has been achieved using a hierarchy in POMDPs and other planning formulations (Marthi, Russell, & Wolfe, 2009; Pineau, Montemerlo, Pollack, Roy, & Thrun, 2003), a large portion of the data for hierarchy and model creation has to be manually encoded. To enable planning in complex domains, recent work has focused on integrating knowledge representation and logical reasoning (Chen et al., 2010; Galindo, Fernandez- Madrigal, Gonzalez, & Saffioti, 2008), and on switching between classical and probabilistic planning for robot applications (Gobelbecker, Gretton, & Dearden, 2011; Kaelbling & Lozano-Perez, 2011). Researchers have also explored tractable representations for hierarchical POMDPs (e.g., dynamic Bayes nets and factored MDPs) (Theocharous, Murphy, & Kaelbling, 2004; Toussaint, Charlin, & Poupart, 2008), but these algorithms are computationally expensive for complex application domains.

Human-Robot Interaction

Developments in sensory input processing algorithms and cognitive architectures (Anderson et al., 2004; Scheutz, Schermerhorn, Kramer, & Anderson, 2007) have aided the use of robots and software agents in a wide range of applications such as human-computer interaction, elderly care and interaction with autistic children (Canemero, 2010; Robins et al., 2004). Research consortia are focusing on cognitive human-robot interaction (CogX Project, 2011), where information from different cues (e.g., vision and speech) are bound based on predetermined rules. Researchers are also integrating computational cognitive models, multiple spatial representations and sensory cues to enable human-robot collaboration, e.g., in a reconnaissance task (Kennedy et al., 2007). However, adaptive visual processing, speech understanding, knowledge representation and optimal use of human inputs are still open challenges to natural HRI (Cantrell, Scheutz, Schermerhorn, & Wu, 2010).

Two broad design approaches typically characterize HRI efforts: biologically inspired design mimics social behavior and uses theories in life sciences and social sciences, while functional design builds computational models to match the domain's social interaction needs. The limited applicability of existing HRI design guidelines causes designers to develop context-specific guidelines and evaluation methods for each domain (Thrun, 2004). An appealing approach is to analyze domain needs and use social exchange concepts to guide HRI design choices (Lawler, 2001; Wagner & Arkin, 2008).

HRI researchers have developed sophisticated algorithms for enabling a robot to operate autonomously based on sensory cues. Some algorithms use computational models of social interactions between humans, modeling the perceived outcomes of the association and the evolving short-term and long-term constraints (Kleinberg & Tardos, 2008). Research shows that robots learn better when they consider social and environmental cues in addition to mimicking the actions of a partner (Cakmak, DePalma, Arriaga, & Thomaz, 2010). Similarly, research on interactions between robots and toddlers shows that the credibility of interactions is a major contributor to the social significance assigned to a robot (Meltzoff, Brooks, Shon, & Rao, 2010). Recent research also indicates that a socially assistive robot can use verbal and visual feedback to positively impact intrinsic motivation of elderly to perform physical or cognitive tasks (Fasola & Mataric, 2010). There has also been considerable work on using embodied relational (virtual) agents in health care (Bickmore, Schulman, & Yiu, 2010; Rizzo, Parsons, Buckwalter, & Kenny, 2010). However, a key limitation of these (existing) algorithms is that they predominantly use manually-encoded domain knowledge in specific applications, and the use of robots in complex real-world domains continues to be an open challenge.

In parallel to the work on autonomous learning from sensory cues, significant research has been performed to enable a robot to learn from human demonstrations (Argall, Chernova, Veloso, & Browning, 2009; Grollman, 2010; Zang, Irani, Zhou, Isbell, & Thomaz, 2010). These approaches build mathematical models based on research in related fields such as control theory, biology and psychology, and theories of human learning and social interactions among humans. However, extensive manual training requires participants with substantial knowledge of the domain and the robot's capabilities. Although humans can provide useful information about tasks and domain, it is typically difficult for human participants to possess the expertise and time to provide elaborate and accurate feedback in complex domains.

Widespread use of robots in the real-world requires the ability to interact with non-experts (Clarkson & Arkin, 2006; Yanco, Drury, & Scholtz, 2004). In recent times, there has been some work in agent domains and on robots to use limited high-level human feedback when it is available or necessary, e.g., the CoBot that seeks human help to navigate to desired locations (Rosenthal et al., 2011), or the reinforcement learning-based TAMER framework that combines human and environmental feedbacks in simulated game domains (Knox & Stone, 2010). However, existing methods require elaborate prior knowledge of the specific task and domain or do not model the unreliability of human inputs.

Summary: Existing learning and planning algorithms have enabled the use of robots in specific applications, but they make strong assumptions regarding the task and domain, require extensive manual feedback and are computationally expensive. Existing methods for HRI have predominantly focused on teaching the robot to perform specific tasks or on enabling the robot to learn from sensory cues. Although it is intractable for the robot to learn complex models of all domain objects and events in all scenarios, it has to use the relevant information and respond in real-time to dynamic changes. Our framework addresses these challenges by exploiting the dependencies between learning, adaptation and interaction. As a result, mobile robots are able to incrementally learn object models, adapt sensing and information processing to the task at hand, and acquire and use high-level inputs from non-expert human participants based on need and availability.

INTEGRATED FRAMEWORK FOR HRI

The framework described in this chapter seeks to achieve robust HRI by enabling robots to operate autonomously when possible, acquiring and utilizing feedback from non-expert human participants based on need and availability. Consider the illustrative example in the introduction, where a mobile robot equipped with sensors and information processing algorithms is deployed in an office. High-level human feedback is in the form of simplistic verbal inputs that provide positive or negative reinforcement of the robot's actions, or make a choice from multiple options posed by

the robot. The integrated framework consists of three components. The next subsection describes a bootstrap learning algorithm that enables a mobile robot to learn layered graphical models of objects and adapt to changes. The following subsection describes a hierarchical planning algorithm that uses partially observable Markov decision processes to automatically adapt sensing, learning and processing to the task at hand. Finally, an augmented reinforcement learning algorithm that enables a mobile robot to acquire and robustly merge high-level human feedback with the information extracted from sensory cues is described. As stated earlier, each section also describes how the individual algorithms in the integrated framework inform and guide each other. Furthermore, the last subsection illustrates the software architecture for the integrated framework in the context of the learning and planning algorithms described in previous subsections.

Bootstrap Learning

Figure 2 (left) shows an instance of bootstrap learning for visual inputs, where the mobile robot autonomously and incrementally: (a) learns the domain map and layered graphical object models; (b) uses the map and object models to learn visual feature models; and (c) uses the visual feature models to detect and adapt to unforeseen dynamic changes.

Existing simultaneous localization and mapping (SLAM) algorithms are used by the robot to learn and revise the domain map (Davison, Reid, Morton, & Stasse, 2007; Grisetti, Stachniss, & Burgard, 2006). Human input can be used to provide semantic labels to locations in the map. Learning of object models is then based on the observation that many real-world objects tend to possess a unique characteristics (e.g., colors and parts) and trace well-defined motion patterns, although these characteristics and patterns are not known in advance. In addition, given a learned map of the domain, the interesting objects are typically those that can move. Candidate objects in the images are hence identified using motion cues, i.e., by tracking local image gradient features (used in visual SLAM) and clustering the features based on relative velocity. Next, discriminative and descriptive local, global and temporal visual cues with complementary properties are extracted from these candidate image regions to populate the object models. For instance, in Figure 2 (right), gradient features, connection potentials between gradient features, graph-based image segments and color distributions are the features under consideration. The second layer of the object model represents a higher level of abstraction for robustness, e.g., relative spatial arrangement of local gradients, neighborhood relationships of connection potentials between gradient features (using Markov random fields), part-based models of image segments and second-order image statistics of color distributions. These learned models are revised incrementally over subsequent images and used to recognize stationary or moving objects.

An instance of this bootstrap learning approach was used to learn models for objects in different categories, e.g., box, book, airplane, robot, car and human, with about 5-6 different models learned for subcategories within each category. These experiments were conducted over a set of approximately 1000 images, which included images captured by the wheeled robots in Figure 1 (left) and images from computer vision benchmark datasets (e.g., Pascal VOC 2006). The robot autonomously learned models for moving objects and used the models to recognize stationary and moving objects in subsequent images. Experimental results indicate a high classification accuracy of 90% averaged over all categories (and subcategories within a category). Classification errors correspond to images where a sufficient number of unique features were not detected or matched with the learned object models due to motion blur or the fact that some images provide long-shots of the objects (Li & Sridharan, 2012; Li, Sridharan & Zhang, 2011). Figure 3 shows an example of using the learned object models to recognize a target object against a complex background (blue box on book shelf). Merging probabilistic evidence provided by individual components of the learned object models regarding occurrence of the corresponding objects in the image regions enables the robot to exploit the complementary properties of different visual cues. As a result, the robot robustly recognizes the target object (blue box in this example) in the appropriate image region.

The learned map and object models are used to maintain revised models of the underlying visual features. For instance, distributions of im-

Figure 2. (Left) visual feature models, environmental map and object models bootstrap off of each other to incrementally refine the individual models; (right) layered graphical models with belief propagation are used to represent domain objects

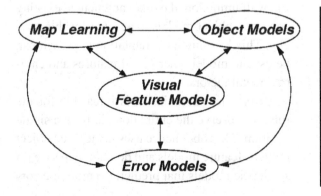
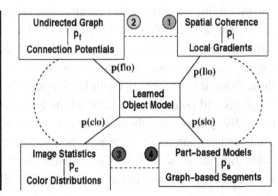

Figure 3. (Left) test image of a box in a complex background; (center) individual match probabilities—the best subcategories within each category are shown along the x-axis; and (right) net match probabilities across different object categories—merging evidence from different components of the learned object models results in robust object recognition

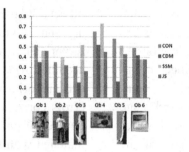
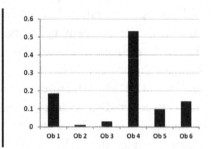

age gradients are organized as histograms to characterize objects, while color distributions are modeled as Gaussian mixture models. Statistical bootstrap tests are used to automatically determine suitable models for different visual feature distributions. The feature models are then revised incrementally using pixels from images of objects recognized using the learned object models. When changes in object configuration or environmental factors (e.g., illumination) cause unforeseen changes in feature distributions, the learned feature models and object models are used to correlate sensor values to these factors based on the hypothesis that images from an environmental state (e.g., specific illumination) have measurably similar distributions in relevant feature spaces. A representation is learned automatically for environmental factors using visual features. The learned feature models and representations are then used to detect and adapt to changes in the corresponding environmental factors. Instances of this learning and adaptation approach has been implemented and evaluated on legged and wheeled robots that autonomously learned color distributions from domain objects (with known/learned positions and color labels) and used the learned color distribution models to detect and adapt to illumination changes (Sridharan & Stone, 2009; Sridharan & Stone, 2007). Color distributions were modeled as Gaussian mixture models and

histograms, and each illumination was represented by: a mapping from pixel values (of images in that illumination) to object color labels, probability density functions (pdfs) in color space and a distribution of distances between these pdfs. When minor illumination changes caused a drift in color distributions and a slow decay of capabilities such as segmentation, the robot automatically extracted image pixels corresponding to known objects to revise the color distribution models. When sudden (or large) illumination changes caused large shifts in color distributions, the average distance between color distributions from the new illumination and the learned color distributions were well outside the range of the expected distribution of distances. If the change was to an illumination that had been modeled, the robot smoothly transitioned to using the corresponding color models for subsequent operations. On the other hand, if the change corresponded to a new illumination, the robot augmented existing color and object models to account for this novelty. This learning and adaptation approach can be used to model other visual features and environmental factors.

In practical domains, it is infeasible for the robot to observe the entire domain from a single position. The robot hence uses the learned object models, feature models and domain map to learn stochastic models that predict: (a) motion errors for different motion patterns; and (b) the likelihood of learning feature models in different locations. Motion planning then simultaneously maximizes the probability of learning desired feature (and object) models and minimizes localization errors (Ghallab, Nau & Traverso, 2004). For instance, the color distribution learning and illumination adaptation approach described above used such a motion planning algorithm to enable the robot to plan motion sequences that placed the robot in the vicinity of objects suitable for learning the color distributions (Sridharan & Stone, 2007). Furthermore, the bootstrap learning algorithm has been used to fuse stereo (visual) cues with range information on wheeled robots operating in indoor and outdoor environments (Murarka, Sridharan, & Kuipers, 2008; Sridharan & Li, 2009).

Real-world application domains are likely to contain a large number of objects that can be represented using different features, making it a challenge to learn appropriate object models (Hoiem, Efros, & Hebert, 2007). The integrated framework will address this challenge using the relationships between components of the framework. Bootstrap learning will thus be made feasible using: (a) planning to identify relevant features for characterizing domain objects relevant to the task at hand (hierarchical planning); and (b) human feedback for reinforcement and disambiguation (augmented reinforcement learning). Conditional probability distributions can also be learned to model relationships between and within the layers of the object models, building richer object descriptions. Furthermore, the learning and adaptation algorithms can be revised to work with other visual (or non-visual) sensory cues.

Hierarchical Planning

In large, complex domains, a robot cannot process all sensory inputs or learn models for all domain objects. At the same time, robust operation in such domains requires that the robot make best use of all the relevant information. Based on

evidence in animals and robots (Horswill, 1993; Land & Hayhoe, 2001), an appealing approach is to retain capabilities for many tasks, direct sensing to relevant locations, and consider the reliability and complexity of available algorithms to automatically determine the sequence of algorithms appropriate for any given task. This objective can be posed as a planning task and as an instance of probabilistic sequential decision-making. More specifically, this section describes the use of partially observable Markov decision processes (POMDPs) to tailor sensing and information processing to the task at hand. POMDPs elegantly model the partial observability and non-determinism of robot application domains. However, POMDP formulations of large real-world domains soon become intractable due to the exponential state explosion of such domains and the high computational complexity of even approximate POMDP solvers (Ong, Png, Hsu & Lee, 2010). A novel hierarchical decomposition is hence incorporated—Figure 4 (left) shows an instance for visual sensing and information processing. For a specific task, the high-level (HL) POMDP computes the sequence of 3D scenes to be analyzed. For a chosen scene, the intermediate-level (IL) POMDP analyzes snapshots (e.g., images) of the scene by choosing a sequence of salient regions of interest (ROIs) to be examined. Each ROI is modeled as a lower-level (LL) POMDP that computes the best sequence of algorithms to be applied on the ROI. Belief propagation between levels of the hierarchy and generation of suitable POMDP models in all levels of the hierarchy occurs autonomously at run-time. Furthermore, the hierarchy is augmented with a communication layer (CL) that enables each robot to merge the information extracted from sensory cues with the information communicated by one or more teammates (Zhang & Sridharan, 2012).

Consider the task of visually locating a human or an object in an office with multiple rooms. The HL-POMDP represents the 3D area as a 2D occupancy grid, which forms the state space. Since

Figure 4. (Left) layered POMDP hierarchy for visual sensing and information processing on a team of robots; (right) visual search based on constrained convolutional policies is more efficient than ad-hoc heuristic search strategies

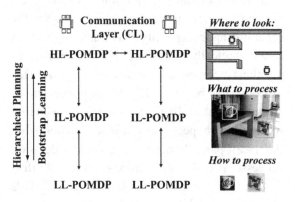

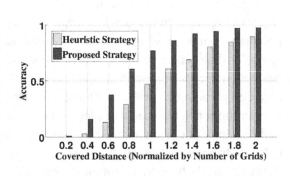

the true underlying state cannot be observed with certainty, a probability distribution over the grid represents the current belief and any prior knowledge about the resident's location (i.e., the belief state). The HL-POMDP's actions cause a robot to move to specific grids and analyze 3D scenes. Planning involves finding the best policy that maps belief states to stochastic action choices. The challenge is that application domains can result in large state spaces and these state spaces can change in response to domain changes. For efficient operation over large areas, shift and rotation symmetries of visual search are exploited to learn a convolutional policy kernel from the policy for a grid map of a small region. The policies for grid maps of large areas are then generated automatically (at run-time) by performing an inexpensive convolution operation with the learned policy kernel. Action utilities in the HL-POMDP are modeled as the expected information gain, i.e., the reduction in entropy of the corresponding belief vectors. In addition, the observation functions of the HL-POMDP are computed automatically based on the learned observation functions of lower levels of the hierarchy. During plan execution, the computed policy is used to repeatedly choose an action and update beliefs based on the observed outcome. A key benefit of this approach is that domain map

changes (e.g., objects are moved or doors are closed) are addressed automatically by suitably re-weighting the computed policy. In addition, the cost of robot motion is modeled by re-weighting the learned policy to trade-off distance of travel against the likelihood of finding the desired targets. Figure 4 (right) shows results where a robot located targets in a 15×15 simulated grid— each point in the figure is the average over 1000 trials, with the convolutional policy computed from a 5×5 policy kernel. As seen in Figure 4 (right), for any desired accuracy (along the y-axis), convolutional policies locate target objects much faster than heuristic (i.e., greedy) search policies.

Once the robot moves to a chosen grid-cell, it analyzes snapshots (e.g., images) of the scene. To locate a target, ROIs in a snapshot, shown enveloped in green rectangles in Figure 4 (left), can be processed using a wide range of visual operators based on bootstrap-learned (object) models, e.g., object recognition operators that use gradient features or parts. However, the POMDP in the joint space of image ROIs soon becomes intractable, e.g., there are approximately 50000 states for just three ROIs and two actions with six outcomes (each). The POMDP hierarchy partially ameliorates this state explosion challenge by modeling each ROI with an LL-POMDP, and using an

IL-POMDP to select the ROI to be analyzed further using the corresponding LL policies. The IL-POMDP hence controls the application of relevant processing algorithms to examine all the ROIs in an image of the chosen scene. The IL-POMDP model parameters (e.g., reward specification and observation functions) are generated automatically at run-time based on the corresponding LL policies and propagated belief. Similarly, relevant LL-POMDP models are learned automatically for any image ROI using bootstrap learning and minimal human supervision. Furthermore, each robot probabilistically merges current beliefs with the beliefs communicated by teammates, enabling the team of robots to collaborate robustly (despite unreliable communication) to achieve a shared objective, e.g., find one or more target objects in the domain.

Instances of the IL and LL of the hierarchy have enabled robots to collaborate with humans to jointly manipulate and converse about table-top objects (Sridharan, Wyatt & Dearden, 2010; Sridharan, Wyatt & Dearden, 2008). Instances of the entire hierarchy have enabled mobile robots to locate target objects in dynamic indoor domains such as offices (Zhang & Sridharan, 2012; Zhang & Sridharan, 2011). These experiments indicate that the hierarchical planning algorithm significantly reduces planning time in comparison to the POMDP in the joint space of all ROIs, as shown in Figure 5 (left). The hierarchical planning approach is also as efficient as state of the art contingency planners while providing substantially higher reliability (Sridharan, Wyatt & Dearden, 2010). Furthermore, robots are able to share information to collaborate robustly with teammates despite unreliable sensing and communication. As shown in Figure 5 (right), belief sharing in conjunction with the POMDP hierarchy enables the team to identify targets with high accuracy in a much smaller number of action steps in comparison to an ad-hoc collaboration strategy.

In real-world application domains with large state spaces and dynamic changes, automated planning and decision-making is a challenge. This challenge will be addressed using the hierarchical planning algorithm in conjunction with other components of the integrated framework. Mobile robots will then be able to adapt learning and planning to the domain and the corresponding tasks by: (a) representing and revising domain knowledge, performing logical reasoning and acquiring information from other robots or humans (e.g., feedback in the form of reinforcement and disambiguation) (Zhang, Bao & Sridharan, 2012); and (b) identifying relevant objects that need to be learned and features that will capture the most information about these objects.

Augmented Reinforcement Learning

For widespread deployment of robots to interact with humans in real-world domains, robots equipped with the learning and planning algorithms described above still need a strategy to acquire and use limited feedback from non-expert human participants. This objective poses two questions: (Q1) how best to robustly merge high-level human input with the information extracted from sensory cues? and (Q2) when and how should human feedback be acquired?

Tasks that require an agent or a robot to learn from repeated interactions with the environment can be posed as a Reinforcement learning (RL) problem. RL is a well-established computational approach, where the desired task is modeled as a Markov decision process (MDP) and an agent repeatedly performs actions to receive a state estimate and a reward signal (Sutton & Barto, 1998). The RL framework has been used in many application domains to enable agents and robots to learn suitable action policies (i.e., mapping from states to actions). As stated earlier, we consider high-level feedback from non-expert human participants—for ease of explanation, this section only considers positive or negative reinforcement of actions, e.g., yes/no feedback. Including human feedback H in the RL framework is a challenge

Figure 5. (Left) hierarchical POMDP significantly reduces planning time in comparison to the POMDP over the joint state space of all image ROIs; (right) merging beliefs obtained by processing sensory cues with the communicated beliefs enables a team of robots to localize targets accurately while traveling a much smaller distance than with an ad-hoc probabilistic collaboration strategy

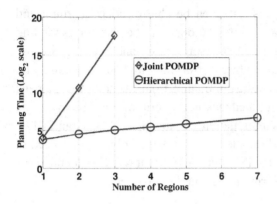 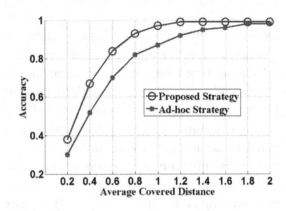

because H may not fit in the same range as environmental feedback R obtained from sensory inputs. In addition, H may be in response to a set of past (or even future) states and actions. Figure 6 (left) shows the augmented reinforcement learning (ARL) approach that is used to answer Q1, i.e., to robustly merge R and H. In the absence of human feedback, the robot uses the standard RL formulation, i.e., a baseline RL algorithm is used to learn an action policy by observing the effects of actions performed in various states. When human feedback is available, the robot uses automatically-computed performance measures (e.g., time for task completion) to bootstrap off of the two feedback signals and incrementally revise their relative contributions to the action choice policy. Specifically, the robot estimates parameters of a function that merges R and H such that the actions chosen by the resultant policy maximize the performance measure(s): $\text{argmax}_{a \in A} f(R,H)$. For ease of explanation, consider the weighted linear combination function: $\text{argmax}_{a \in A} \{w_r R + w_h H\}$ in the fully observable simulated game domains of Tetris and Keepaway soccer—see Figure 6 (center) and Figure 6 (right). The objective in Tetris is to maximize

episode length by dropping blocks such that they complete and hence clear lines. In multiagent Keepaway, the objective is to maximize episode length by enabling keepers to retain ball possession from the takers. In the absence of human feedback, the agent(s) in these domains learn a policy from R (using the baseline RL algorithm) and invoke the top N action policies proportional to their relative ability to maximize episode length. When human feedback is available, the weights corresponding to feedback signals (w_h and w_r) are continuously and incrementally revised based on the degree of match between H and R, and their relative ability to maximize episode length. The individual feedback signals are thus merged to provide the overall action choice policy. Figure 7 shows results of experiments in the Tetris and Keepaway domains, using high-level feedback from four human participants 2-5 times per episode (the yes/no feedback signals are mapped to real-valued rewards). Figure 7 (left) shows the result of using a weighted linear combination function in the Tetris domain, using policy gradient (Sutton & Barto, 1998) as the baseline RL algorithm. The ARL approach to merge R and H significantly increases the episode length in comparison to using R or H (not shown in Figure) individually (Sridharan, 2011).

Next, Figure 7 (right) shows experimental results in the 3 vs. 2 Keepaway domain, using the SMDP version of *Sarsa(λ)* (Stone, Sutton & Kuhlmann, 2005) as the baseline algorithm. This domain changes too quickly for instantaneous human feedback of the agents' action choices. A gamma distribution is hence learned experimentally to model typical human response times. This distribution is used for credit assignment over past states and actions. As seen in Figure 7 (right), using the ARL approach significantly increases episode duration in comparison to the individual feedback signals. In addition, using the learned gamma function for credit assignment further improves the episode duration. Furthermore, when different humans participating in the experimental trials provide intentionally incorrect feedback, the agents are able to recover by revising the weights of feedback signals (Sridharan, 2011).

Since the ARL approach uses belief distributions computed in hierarchical planning to estimate state, Q2 is answered using information-theoretic measures and bootstrap learning algorithms. The state with the maximum belief is considered to be the true state and the entropy of belief distribu-

tions is used as a measure of uncertainty. Asking for human input is modeled as a sensing action that is sequenced to maximize information gain (Zhang, Bao & Sridharan, 2012), similar to the topmost level of the POMDP hierarchy described in the hierarchical planning section.

The ARL approach has been described (above) in the context of simulated agent domains. RL typically requires knowledge of state and an estimate of the transition and reward functions—these are not readily available in robot application domains. The integrated framework will address this challenge by defining rewards based on information gain and global performance measures, states based on the belief states used in hierarchical planning, and transition functions based on bootstrap learning and limited domain knowledge. Bootstrap learning will also provide the models necessary to estimate the likelihood of obtaining relevant information through different query types. Furthermore, the ARL algorithm will be used in conjunction with an algorithm that learns associations between visual and verbal object descriptions, enabling simplistic natural language interactions between humans and agents (Swaminathan & Sridharan, 2011). The robot will thus be able to initiate and sustain interactions with appropriate human participants.

Figure 6. (Left) augmented reinforcement learning enables the robot to bootstrap off of high-level human feedback and environmental feedback in the form of sensory inputs; (center) single-agent Tetris domain; and (right) multiagent 3vs.2 Keepaway domain

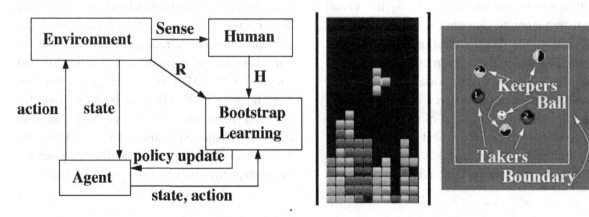

Figure 7. Results of proof of concept experiments in simulated domains: (left) single-agent Tetris domain; and (right) multiagent Keepaway domain. Merging human and environmental feedbacks significantly increases the episode length in comparison to individual feedback mechanisms. Probabilistic credit assignment over past states and actions further improves performance.

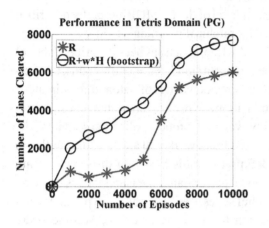 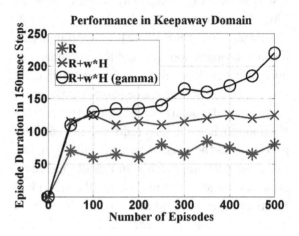

Integration Overview

Finally, consider the architecture that integrates the bootstrap learning, hierarchical planning and augmented reinforcement learning algorithms described above. To enable modular software development, all algorithms were implemented using the popular Robot Operating System (ROS) (Quigley et al., 2009). Figure 8 presents a subset of the architecture that focuses on visual bootstrap learning and hierarchical (probabilistic) planning—the corresponding graph was generated by the ROS command-line option: <rxgraph>. The individual nodes are described below.

The hierarchical planning algorithm is placed within the vs_planner node, while the visual bootstrap learning algorithm is placed within the vs_vision node. Communication between nodes is achieved by publishing topics, i.e., by passing messages. The vs_vision node repeatedly processes input images to learn relevant visual object models. The learned object models are used to recognize objects in subsequent images, populating the <v_pack> package that is sent to the vs_planner node. This package contains the ID of

each detected object, in addition to the distance and bearing of the object (relative to the robot) and a (probability) measure of the uncertainty associated with the observation of the object. These observations are used to perform belief updates within the planning module, as described in the hierarchical planning section. Belief updates occur: (1) when the robot arrives at a desired grid cell and processes one or more images of the scene—belief updates consider presence and absence of the target object; or (2) when the robot detects the target by processing images while moving to a desired grid cell. After the belief update, the planner node sends the coordinates of any desired grid cell to the movement control node move_base (in the goal message) and waits for a response. The move_base node receives the current domain map from the map_server (which can also perform simultaneous localization and mapping—SLAM) and laser range information from hokuyo_node, which contains the driver for the laser range finder. The move_base node also receives navigation goals (if any) from humans through navigation_goals, in addition to pose and odometry information from amcl and erratic_ base_driver respectively. The erratic_base_driver provides the robot-specific coordinate frames (in tf) and the driver for the specific robot platform used in these experiments (e.g., the erratic wheeled robot in Figure 1). The amcl node performs localization using particle filters to provide the pose estimate. The move_base node uses A* search to find a path to the desired grid cell and provides linear and angular velocity commands to the robot's driver (in cmd_vel). These commands result in the robot's motion and one of three responses: arrived, canceled or not-arrived. The arrived response is received when the robot reaches the desired location, while the canceled response represents unexpected cancellation of the motion command. The not-arrived response is usually the result of a dynamic change in the environment, e.g., closing a door makes an office unavailable to the robot. Additional nodes are used (in a similar manner) to create instances of other algorithms, e.g., for augmented reinforcement learning using human feedback.

This architecture has been successfully implemented and used on wheeled robots deployed in indoor office domains (Zhang & Sridharan, 2012). The modular architecture makes it easy to revise specific algorithms (e.g., for autonomous learning or planning) and to use the architecture on other robot platforms in different application domains. Furthermore, other nodes can be added (as and when required) to create instances of algorithms that augment the existing components in the integrated framework. Results of recent experimental trials, including some images and video demos can be viewed online: http://www.cs.ttu.edu/~smohan/RobotAssist.html

CONCLUSION AND FUTURE RESEARCH DIRECTIONS

This chapter described a novel integrated (probabilistic) framework that jointly addressed the learning, adaptation and interaction challenges associated with robust human-robot interaction in real-world application domains. This framework consists of three components: (1) a bootstrap learning algorithm that enables mobile robots to autonomously learn layered graphical models of environmental objects, and to detect and adapt to unforeseen domain changes; (2) a hierarchical planning algorithm based on partially observable Markov decision processes that enables a team of robots to collaborate robustly by sharing beliefs and automatically adapting sensing and information processing to the task at hand; and (3) an augmented reinforcement learning algorithm that enables robots to acquire limited high-level

Figure 8. ROS-based framework for integrating different components. Interaction between hierarchical planning, visual bootstrap learning and other control modules is illustrated.

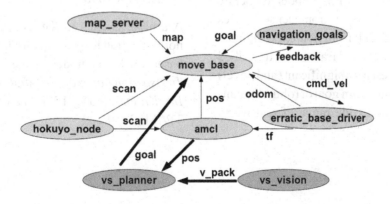

feedback from non-expert human participants, and to robustly merge human feedback with the information extracted from sensory cues. Instances of these algorithms have been implemented in a modular software architecture and evaluated on mobile robots and simulated agents interacting with non-expert human participants in indoor office domains and multiagent game domains.

As stated in the introduction (and illustrated in the previous section), the integrated framework enables the individual algorithms to inform and guide each other, posing novel challenges and providing new opportunities to address the tough challenges in the individual fields. Future work will investigate the full integration of the bootstrap learning, hierarchical planning and augmented reinforcement learning algorithms. For instance, planning will be used to choose the objects and events relevant to the tasks that need to be performed, and to identify features suitable for modeling these objects, e.g., to select visual features that provide the most information about the objects of interest. Similarly, bootstrap learning will be used to autonomously learn the model parameters required to automate hierarchical planning in complex domains. Future work will also focus on building richer object descriptions by integrating visual and verbal cues, resulting in natural language interactions between robots and humans.

As mobile robots are increasingly deployed to interact with humans in real-world application domains such as homes and offices, there is a pressing need for enabling robots to operate autonomously by learning from sensory cues and high-level feedback from non-expert human participants. The integrated framework described in this chapter represents a significant (and novel) step towards this long-term goal of robust human-robot interaction in a wide range of real-world application domains.

ACKNOWLEDGMENT

The author thanks Shiqi Zhang, Xiang Li, Batbold Myagmarjav and Mamatha Aerolla for their help in implementing the algorithms, performing the experimental trials and gathering the results reported in this chapter. The author is also grateful to the following colleagues for many discussions that have guided the research reported in this chapter: Peter Stone, Ian Fasel, Benjamin Kuipers, Jeremy Wyatt, Richard Dearden, Nick Hawes and Aaron Sloman. The research reported in this chapter was supported in part by the ONR Science of Autonomy award N00014-09-1-0658.

REFERENCES

AAAI. (2012). *AAAI Spring Symposium on Designing Intelligent Robots: Reintegrating AI* (SS02). Retrieved October 2011 from http://www.aaai.org/Symposia/Spring/sss12symposia.php#ss02

Aboutalib, S., & Veloso, M. (2010, October). Multiple-cue object recognition in outside datasets. *IEEE International Conference on Intelligent Robots and Systems (IROS)* (pp. 4554-4559). Taipei.

Anderson, J. R., Bothell, D., Byrne, M. D., Douglass, S., Lebiere, C., & Qin, Y. (2004). An integrated theory of the mind. *Psychological Review, 111*(4), 1036–1060. doi:10.1037/0033-295X.111.4.1036

Arashloo, S. R., & Kittler, J. (2011). Energy normalization for pose-invariant face recognition based on MRF model image matching. *IEEE Transactions on Pattern Analysis and Machine Intelligence, 33*, 1274–1280. doi:10.1109/TPAMI.2010.209

Argall, B., Chernova, S., Veloso, M., & Browning, B. (2009). A survey of robot learning from demonstration. *Robotics and Autonomous Systems, 57*(5), 469–483. doi:10.1016/j.robot.2008.10.024

Bickmore, T., Schulman, D., & Yiu, L. (2010). Maintaining engagement in long-term interventions with relational agents. *Journal of Applied Artificial Intelligence; special issue on Intelligent Virtual Agents, 24*(6), 648-666.

Brenner, M., & Nebel, B. (2009). Continual planning and acting in dynamic multiagent environments. *Journal of Autonomous Agents and Multiagent Systems, 19*(3), 297–331. doi:10.1007/s10458-009-9081-1

Brick, T., & Scheutz, M. (2007, March). Incremental natural language processing for HRI. In *ACM/IEEE International Conference on Human-Robot Interaction (HRI)* (p. 263-270). Washington DC, USA.

Brook, P., Ciocarlie, M., & Hsiao, K. (2011). Collaborative grasp planning with multiple object representations. In *International Conference on Robotics and Automation (ICRA)*.

Butko, N. J., & Movellan, J. R. (2008). *I-POMDP: An infomax model of eye movement*. In International Conference on Development and Learning (ICDL).

Cakmak, M., DePalma, N., Arriaga, R., & Thomaz, A. (2010). Exploiting social partners in robot learning. *Autonomous Robots, 29*, 309–329. doi:10.1007/s10514-010-9197-9

Canemero, L. (2010). *The HUMAINE Project*. Retrieved October 2010 from http://emotion-research.net/

Cantrell, R., Scheutz, M., Schermerhorn, P., & Wu, X. (2010). *Robust spoken instruction understanding for HRI*. In ACM/IEEE International Conference on Human-Robot Interaction (HRI).

Caselles, V., Kimmel, R., & Shapiro, G. (1997). Geodesic active contours. *International Journal of Computer Vision, 22*(1), 61–79. doi:10.1023/A:1007979827043

Chen, X., Ji, J., Jiang, J., Jin, G., Wang, F., & Xie, J. (2010, May 10-14). *Developing high-level cognitive functions for service robots*. In International Conference on Autonomous Agents and Multiagent Systems (AAMAS). Toronto, Canada.

Clarkson, E., & Arkin, R. C. (2006). *Applying heuristic evaluation to human-robot interaction systems* (Tech. Rep.). Georgia Institute of Technology GIT-GVU- 06-08.

Cog, X. (2011). *Cognitive systems that self-understand and self-extend*. Retrieved October 2011 from http://cogx.eu/

Comaniciu, D., & Meer, P. (2002). Mean shift: A robust approach towards feature space analysis. *IEEE Transactions on Pattern Analysis and Machine Intelligence, 24*(5), 603–619. doi:10.1109/34.1000236

Davison, A. J., Reid, I. D., Morton, N. D., & Stasse, O. (2007). MonoSLAM: Real-time single camera SLAM. *IEEE Transactions on Pattern Analysis and Machine Intelligence, 29*(6), 1052–1067. doi:10.1109/TPAMI.2007.1049

Fasola, J., & Mataric, M. (2010, August). *Robot motivator: Increasing user enjoyment and performance on a physical/cognitive task*. In International Conference on Development and Learning (ICDL). Ann Arbor, USA.

Felzenswalb, P. F., & Huttenlocher, D. P. (2004). Efficient graph-based image segmentation. *International Journal of Computer Vision, 59*(2), 167–181. doi:10.1023/B:VISI.0000022288.19776.77

Fergus, R., Perona, P., & Zisserman, A. (2003). *Object class recognition by unsupervised scale-invariant learning*. In International Conference on Computer Vision and Pattern Recognition (CVPR).

Fidler, S., Boben, M., & Leonardis, A. (2008). *Similarity-based cross-layered hierarchical representation for object categorization.* In International Conference on Computer Vision and Pattern Recognition (CVPR).

Finlayson, G., Hordley, S., & Hubel, P. (2001, November). Color by correlation: A simple, unifying framework for color constancy. *IEEE Transactions on Pattern Analysis and Machine Intelligence, 23*(11), 1209–1221. doi:10.1109/34.969113

Fong, T., Nourbakhsh, I., & Dautenhahn, K. (2003). A survey of socially interactive robots. *Robotics and Autonomous Systems, 42*(3-4), 143–166. doi:10.1016/S0921-8890(02)00372-X

Galindo, C., Fernandez-Madrigal, J.-A., Gonzalez, J., & Saffioti, A. (2008). Robot task planning using semantic maps. *Robotics and Autonomous Systems, 56*(11), 955–966. doi:10.1016/j.robot.2008.08.007

Ghallab, M., Nau, D., & Traverso, P. (2004). *Automated planning: Theory and practice.* San Francisco, CA: Morgan Kaufmann.

Gobelbecker, M., Gretton, C., & Dearden, R. (2011). *A switching planner for combined task and observation planning.* In National Conference on Artificial Intelligence (AAAI). San Francisco, USA.

Grisetti, G., Stachniss, C., & Burgard, W. (2006). Improved techniques for grid mapping with Rao-Blackwellized particle filters. *IEEE Transactions on Robotics, 23*(1), 34–46. doi:10.1109/TRO.2006.889486

Grollman, D. (2010). *Teaching old dogs new tricks: Incremental multimap regression for interactive robot learning from demonstration.* Unpublished doctoral dissertation, Department of Computer Science, Brown University.

Guedon, Y. (2005). Hybrid Markov/semi-Markov chains. *Computational Statistics & Data Analysis, 49,* 663–688. doi:10.1016/j.csda.2004.05.033

Hartley, R., & Zisserman, A. (2004). *Multiple view geometry in computer vision* (2nd ed.). Cambridge University Press. doi:10.1017/CBO9780511811685

Hoey, J., Poupart, P., Bertoldi, A., Craig, T., Boutilier, C., & Mihailidis, A. (2010). Automated handwashing assistance for persons with dementia using video and a partially observable Markov decision process. *Computer Vision and Image Understanding, 114*(5), 503–519. doi:10.1016/j.cviu.2009.06.008

Hoiem, D., Efros, A., & Hebert, M. (2007). Recovering surface layout from an image. *International Journal of Computer Vision, 75*(1), 151–172. doi:10.1007/s11263-006-0031-y

Horswill, I. (1993). Polly: A vision-based artificial agent. In *National Conference on Artificial Intelligence (AAAI)* (pp. 824-829).

Kaelbling, L., & Lozano-Perez, T. (2011). *Domain and plan representation for task and motion planning in uncertain domains.* In IROS 2011 Workshop on Knowledge Representation for Autonomous Robots.

Kennedy, W., Bugajska, M., Marge, M., Adams, W., Fransen, B., Perzanowski, D., et al. (2007). Spatial representation and reasoning for human-robot interaction. In *Twenty-Second Conference on Artificial Intelligence (AAAI)* (pp. 1554-1559). Toronto, Canada.

Kleinberg, J., & Tardos, E. (2008, May 17-20). *Balanced outcomes in social exchange networks.* In ACM Symposium on Theory of Computing.

Knox, W., & Stone, P. (2010, May). *Combining manual feedback with subsequent MDP reward signals for reinforcement learning*. In International Conference on Autonomous Agents and Multiagent Systems (AAMAS).

Krause, A., Singh, A., & Guestrin, C. (2008). Near-optimal sensor placements in Gaussian processes: Theory, efficient algorithms and empirical studies. *Journal of Machine Learning Research, 9*, 235–284.

Kreucher, C., Kastella, K., & Hero, A. (2005). Sensor management using an active sensing approach. *IEEE Transactions on Signal Processing, 85*(3), 607–624.

Lai, K., Bo, L., Ren, X., & Fox, D. (2011, May 9-13). *Sparse distance learning for object recognition combining RGB and depth information*. In International Conference on Robotics and Automation (ICRA). Shanghai, China.

Lammens, J. M. G. (1994). *A computational model of color perception and color naming*. Doctoral dissertation, Computer Science Department, State University of New York at Buffalo, NY.

Land, M. F., & Hayhoe, M. (2001). In what ways do eye movements contribute to everyday activities? *Vision Research, 41*, 3559–3565. doi:10.1016/S0042-6989(01)00102-X

Lawler, E. J. (2001). An affect theory of social exchange. *American Journal of Sociology, 107*(2), 321–352. doi:10.1086/324071

Li, C., Parikh, D., & Chen, T. (2011, November 6-13). *Extracting adaptive contextual cues from unlabeled regions*. In International Conference on Computer Vision (ICCV). Barcelona, Spain.

Li, L., Bulitko, V., Greiner, R., & Levner, I. (2003). *Improving an adaptive image interpretation system by leveraging*. In Australian and New Zealand Conference on Intelligent Information Systems.

Li, X., & Sridharan, M. (2012, June 5). *Vision-based autonomous learning of object models on a mobile robots*. In Autonomous Robots and Multirobot Systems (ARMS) Workshop at the International Conference on Autonomous Agents and Multiagent Systems (AAMAS), Valencia, Spain.

Li, X., Sridharan, M., & Zhang, S. (2011, May 9-13). *Autonomous learning of vision-based layered object models on mobile robots*. In International Conference on Robotics and Automation (ICRA). Shanghai, China.

Lowe, D. (2004). Distinctive image features from scale-invariant keypoints. *International Journal of Computer Vision, 60*(2), 91–110. doi:10.1023/B:VISI.0000029664.99615.94

Marthi, B., Russell, S., & Wolfe, J. (2009, August). *Angelic hierarchical planning: Optimal and online algorithms*. Technical Report UCB/EECE-2009-122, EECS Department, University of California Berkeley.

Meltzoff, A., Brooks, R., Shon, A., & Rao, R. (2010, October-November). Social robots are psychological agents for infants: A test of gaze following. *Neural Networks, 23*(8-9), 966–972. doi:10.1016/j.neunet.2010.09.005

Mikolajczyk, K., & Schmid, C. (2004). Scale and affine invariant interest point detectors. *International Journal of Computer Vision, 60*(1), 63–86. doi:10.1023/B:VISI.0000027790.02288.f2

Murarka, A., Sridharan, M., & Kuipers, B. (2008). *Detecting obstacles and drop-offs using stereo and motion cues for safe local motion*. In International Conference on Intelligent Robots and Systems (IROS), Nice, France.

Ong, S. C., Png, S. W., Hsu, D., & Lee, W. S. (2010, July). Planning under uncertainty for robotic tasks with mixed observability. *The International Journal of Robotics Research, 29*(8), 1053–1068. doi:10.1177/0278364910369861

Petrick, R., & Bacchus, F. (2004). *Extending the knowledge-based approach to planning with incomplete information and sensing.* In International Conference on Automated Planning and Scheduling (ICAPS).

Pineau, J., Montemerlo, M., Pollack, M., Roy, N., & Thrun, S. (2003). Towards robotic assistants in nursing homes: Challenges and results. *Robotics and Autonomous Systems. Special Issue on Socially Interactive Robots, 42*(3-4), 271–281.

Quigley, M., Conley, K., Gerkey, B., Faust, J., Foote, T., Leibs, J., et al. (2009). *ROS: An open-source robot operating system.* In ICRA Workshop on Open Source Software.

Rabiner, L. R. (1989). A tutorial on hidden Markov models and selected applications in speech recognition. *Proceedings of the IEEE, 77*(2), 257–286. doi:10.1109/5.18626

Rizzo, A., Parsons, T., Buckwalter, G., & Kenny, P. (2010, March 21). *A new generation of intelligent virtual patients for clinical training.* In IEEE Virtual Reality Conference. Waltham, USA.

Robins, B., Dautenhahn, K., Boekhorst, R., Billard, A., Keates, S., & Clarkson, J. (2004). *Effects of repeated exposure of a humanoid robot on children with autism.* Springer-Verlag. doi:10.1007/978-0-85729-372-5_23

Rosenthal, S., Veloso, M., & Dey, A. (2011, August). *Learning accuracy and availability of humans who help mobile robots.* Twenty-Fifth Conference on Artificial Intelligence (AAAI), San Francisco, USA.

Scheutz, M., Schermerhorn, P., Kramer, J., & Anderson, D. (2007). First steps towards natural human-like HRI. *Autonomous Robots, 22*(4), 411–423. doi:10.1007/s10514-006-9018-3

Serre, T., Wolf, L., Bileschi, S., Riesenhuber, M., & Poggio, T. (2007, March). Robust object recognition with cortex-like mechanisms. *IEEE Transactions on Pattern Analysis and Machine Intelligence, 29*(3). doi:10.1109/TPAMI.2007.56

Sridharan, M. (2011, December 18-21). *Augmented reinforcement learning for interaction with non-expert humans in agent domains.* In International Conference on Machine Learning Applications (ICMLA), Honolulu, Hawaii.

Sridharan, M., & Li, X. (2009). *Learning sensor models for autonomous information fusion on a humanoid robot.* In IEEE-RAS International Conference on Humanoid Robots. Paris, France.

Sridharan, M., & Stone, P. (2007). *Global action selection for illumination invariant color modeling.* In International Conference on Intelligent Robots and Systems (IROS). San Diego, USA.

Sridharan, M., & Stone, P. (2009). Color learning and illumination invariance on mobile robots: A survey. *Robotics and Autonomous Systems, 57*(6-7), 629–644. doi:10.1016/j.robot.2009.01.004

Sridharan, M., Wyatt, J., & Dearden, R. (2008, September). *HiPPo: Hierarchical POMDPs for planning information processing and sensing actions on a robot.* In International Conference on Automated Planning and Scheduling (ICAPS). Sydney, Australia.

Sridharan, M., Wyatt, J., & Dearden, R. (2010). Planning to see: A hierarchical approach to planning visual actions on a robot using POMDPs. *Artificial Intelligence, 174*, 704–725. doi:10.1016/j.artint.2010.04.022

Stone, P., Sutton, R., & Kuhlmann, G. (2005). Reinforcement learning for RoboCup soccer keepaway. *Adaptive Behavior, 13*, 165–188. doi:10.1177/105971230501300301

Sutton, R. L., & Barto, A. G. (1998). *Reinforcement learning: An introduction*. Cambridge, MA: MIT Press.

Swaminathan, R., & Sridharan, M. (2011, December). *Towards Natural human-robot interaction using multimodal cues*. Technical Report, Department of Computer Science, Texas Tech University.

Talamadupula, K., Benton, J., Kambhampati, S., Schermerhorn, P., & Scheutz, M. (2010). Planning for human-robot teaming in open worlds. *ACM Transactions on Intelligent Systems and Technology, 1*(2), 14:1-14:24.

Tapus, A., Mataric, M., & Scassellati, B. (2007, March). The grand challenges in socially assistive robotics. *Robotics and Automation Magazine. Special Issue on Grand Challenges in Robotics, 14*(1), 35–42.

Theocharous, G., Murphy, K., & Kaelbling, L. P. (2004). Representing hierarchical POMDPs as DBNs for multi-scale robot localization. In *International Conference on Robotics and Automation (ICRA)*.

Thrun, S. (2004). Toward a framework for human-robot interaction. *Human-Computer Interaction, 19*(1), 9–24. doi:10.1207/s15327051hci1901&2_2

Thrun, S. (2006). Stanley: The robot that won the DARPA grand challenge. *Field Robotics, 23*(9), 661–692. doi:10.1002/rob.20147

Toussaint, M., Charlin, L., & Poupart, P. (2008). Hierarchical POMDP controller optimization by likelihood maximization. In *Proceedings of the 24th Annual Conference on Uncertainty in AI (UAI)*.

Wagner, A., & Arkin, R. (2008). Analyzing social situations for human-robot interaction. *Interaction Studies: Social Behaviour and Communication in Biological and Artificial Systems, 9*(2), 277–300. doi:10.1075/is.9.2.07wag

Yanco, H. A., Drury, J. L., & Scholtz, J. (2004). Beyond usability evaluation: Analysis of human-robot interaction at a major robotics competition. *Human-Computer Interaction, 19*(1), 117–149. doi:10.1207/s15327051hci1901&2_6

Zang, P., Irani, A., Zhou, P., Isbell, C., & Thomaz, A. (2010). Using training regimens to teach expanding function approximators. In *International Joint Conference on Autonomous Agents and Multiagent Systems (AAMAS)* (pp. 341-348).

Zhang, S., Bao, F. S., & Sridharan, M. (2012, June 5). *ASP-POMDP: Integrating non-monotonic logical reasoning and probabilistic planning on mobile robots*. In Autonomous Robots and Multirobot Systems (ARMS) Workshop at the International Conference on Autonomous Agents and Multiagent Systems (AAMAS), Valencia, Spain.

Zhang, S., & Sridharan, M. (2011, August 7-8). *Visual search and multirobot collaboration on mobile robots*. In International Workshop on Automated Action Planning for Autonomous Mobile Robots. San Francisco, USA.

Zhang, S., & Sridharan, M. (2012, June 4-8). *Active visual search and collaboration on mobile robots*. In International Conference on Autonomous Agents and Multiagent Systems (AAMAS). Valencia, Spain.

Section 5
Vision Control

Chapter 17
Collaborative Exploration Based on Simultaneous Localization and Mapping

Domenec Puig
Rovira i Virgili University, Spain

ABSTRACT

This chapter focuses on the study of SLAM taking into account different strategies for modeling unknown environments, with the goal of comparing several methodologies and test them in real robots even if they are heterogeneous. The purpose is to combine them in order to reduce the exploration time. Indubitably, it is not an easy work because it is important to take into account the problem of integrating the information related with the changes into the map. In this way, it is necessary to obtain a representation of the surrounding in an efficiently way. Furthermore, the author is interested in the collaboration between robots, because it is well-known that a team of robots is capable of completing a given task faster than a single robot. This assumption will be checked by using both simulations and real robots in different experiments. In addition, the author combines the benefits of both vision-based and laser-based systems in the integration of the algorithms.

INTRODUCTION

During the last decades the robotics field has provided solutions to problems that were difficult to solve some years ago or simply performs the tasks in faster and simplest way that facilitates the life for human beings. Thus, when a problem has to be solved implies that challenges have to work out. In this sense, one of the most well-known challenges in the robotics field is related to Simultaneous Localization and Mapping problem (SLAM).

Many applications depend on a team of robots being able to navigate in real world environments without human intervention. In order to navigate or path plan, robots often need to consult the representation of their surroundings and sometimes it is the only way to achieve their goal.

Therefore, when a robot is located in an unknown environment, the SLAM solutions conduct to build a map while at the same time the robot is able to determine its location using the map. Thus, to give a solution to the SLAM problem

DOI: 10.4018/978-1-4666-2672-0.ch017

we need to perform the task of integrating the information related with the changes into the map and it becomes more difficult in the design part, as we also have to choose appropriately the tools that were used in the real world.

In this sense, the SLAM problem has become central for the researchers during the last years. For this reason, different proposals have been presented in order to solve it, such as: Extended Kalman Filter, Rao-Blackwellized, FastSLAM, etc. In which the probability theory plays an important role in the whole research related to SLAM.

Until now, we can think that SLAM is only about theoretical part. However, it can be applied in different domains such as indoors, outdoors, underwater or even in the air. However, it leads us to exploration tasks conducting to discover unknown spaces during the time. In this sense, one point that we want to cover in this chapter is related to compare the advantages of both single and multi-robot exploration, but it is well-known that increasing the number of robots also increases the complexity for managing those tasks in an optimal way. It is interesting for several reasons, for instance, tasks may be inherently too complex for a single robot to accomplish, or performance benefits can be gained from using a team of robots, or several robots give us more flexibility and more fault-tolerance than having a single robot for each separated task.

In addition, we want to show that is more advantageous a coordinated exploration in contrast to an uncoordinated exploration. Different benefits can be found: overlapping information which can help to compensate for sensors uncertainty, robots can localize themselves more efficiently, increased robustness or higher quality solutions. However, a set of aspects to consider during the tasks that must be taken into account are: risk of possible interferences between robots, limited communication range, limited energy, computation and mobility or dynamic events.

Furthermore, during an exploration task we expect that robots acquire as much information as possible, this meaning to get both idiothetic information (that corresponds to odometry) and allothetic information (provided by laser range finders, sonars or vision) and, then, combine them for a good representation of the surrounding. Consequently, in order to build reliable maps and to navigate for long periods of time the allothetic information must compensate for idiothetic information drift, while idiothetic information must allow perceptually aliased allothetic information to be disambiguated. Of course, for archive a mathematical model of the sensors, we have to take into account that we use both camera and laser based systems and, then, processing the information for representing the results in an understandable way.

On the other hand, it is necessary to bear in mind the two fundamental paradigms for modeling robot environments: the grid-based (metric) paradigm and the topological paradigm. We want to compare the advantages and disadvantages of both. In addition, solve the assignment problem and compare well-known algorithms, as for instance the frontier cells, based on structure environment using the segmentation of the space or using clustering, etc.

In this chapter, we want to give to the reader a framework related to Simultaneous Localization and Mapping including the dependent tasks: motion control, exploration and improvement of the pose estimated along the time for obtaining a consistent representation of the environment. Moreover, we will discuss the factors that have to bear in mind in the real world, and finally offer a set of results in both simulation and real world for making a comparison between the most well-known approaches. In this sense, we will give a global idea about the existing work and the future of this important research line.

The remainder of this chapter is organized as follows. In the next section, the framework of the Simultaneous Localization and Mapping and artificial vision will be presented, including the mathematical part of two well-known approaches, the former Extended Kalman Filter and the latter

Rao-Blackwellized. In addition, the approaches for extraction of vision based features of the environment are presented, and also an experimental part for camera calibration. In the following section, the different paradigms for the representation of the environment can be seen, too, how to work with the available information given by the sensors. This chapter then presents a study about the multi-robot exploration, including the levels of coordination and its advantages and drawbacks is shown. Moreover, two main tools are explained: the first one related to the Voronoi diagram and the second one to the Hungarian method for the assignment problem. Later, the experimental part is presented, including both simulation and real robots. Finally, in the last section a discussion about the results obtained in the previous sections will be done, including the future research directions.

BACKGROUND

It is well-known that scientists have proposed solutions for the SLAM problem (Chli, 2009). In this sense, SLAM asks if is possible for a mobile robot to be placed in an unknown location, within an unknown environment, and for the robot to incrementally build a consistent map of this environment while simultaneously determining its location within this map. Therefore, it becomes in a difficult challenge. For this reason, SLAM has been one of the notable successes of the robotics community over the last decade. It has been formulated and solved as a theoretical problem in a number of different forms. In addition, SLAM has also been implemented in different domains from indoor robots, to outdoor, underwater and airborne systems.

Furthermore, the great majority of work has focused on improving computational efficiency while ensuring consistent and accurate estimates for the map and vehicle pose (e.g., [Bailey et al., 2006], [Estrada & Tardos, 2005]). However, there has also been much research on issues such

as non-linearity, data association and landmark characterization, all of which are vital in achieving a practical and robust SLAM implementation. Nowadays, improvements and new proposals about SLAM with different points of view have been emerging.

General SLAM Problem

In this section, we are focused in the general SLAM problem, representing in this way the mathematical formulation of it. It is useful for a good understanding of the algorithms that we will describe in the next sections.

Figure 1 shows that, given an environment including a set of landmarks which location is assumed to be known, it is possible to consider a mobile robot moving through the environment taking relative observations of the landmarks using sensors located on the robot in order to determine its position and orientation in space by processing the information obtained by means of those sensors.

At a time instant t, the following quantities are defined:

x_t: The state vector describing the location and orientation of the vehicle.

u_t: The control vector, applied at time $t - 1$ to drive the vehicle to a state x_t at time t.

m_i: A vector describing the location of the ith landmark whose true location is assumed time invariant.

z_{it}: An observation taken from the vehicle of the location of the ith landmark at time t. When there are multiple landmark observations at any one time or when the specific landmark is not relevant to the discussion, the observation will be written simply as z_t.

In addition, the following sets are also defined:

$X_{0:t} = x_0, x_1, ..., x_t = X_{0:t-1}, x_t$: The history of vehicle locations.

Figure 1. The essential SLAM problem. Observations are made between the true robot and landmark locations (taken from Durrant-Whyte & Bailey, 2006).

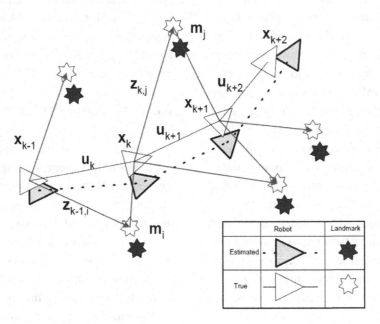

$U_{0:t} = u_1, u_2, \ldots, u_t = U_{0:t-1}, u_t$: The history of control inputs.

$m = m_1, m_2, \ldots, m_n$: The set of all landmarks.

$Z_{0:t} = z_1, z_2, \ldots, z_t = Z_{0:t-1}, z_t$: The set of all landmark observations.

Probabilistic SLAM

In probabilistic form, the SLAM problem requires that the probability distribution (Durrant-Whyte & Bailey, 2006), $P(x_{t:1}, m | Z0: t-1, U0: t-1)$, be computed for all times t. This probability distribution describes the joint posterior density of the landmark locations and vehicle state (at time t) given the recorded observations and control inputs up to and including time t together with the initial state of the vehicle. In general, a recursive solution to the SLAM problem is desirable. Starting with an estimate for the distribution $P(x_{t:1}, m | Z0: t-1, U0: t-1)$ at time $t-1$, the joint posterior, following a control u_t and observation z_t, is computed using the Bayes Theorem. This computation requires that a

state transition model and an observation model are defined describing the effect of the control input and observation respectively.

Online and Full SLAM Problem

From a probabilistic perspective, there are two main strategies for the SLAM problem, which are both of equal practical importance. The first is known as the online SLAM problem that involves estimating the posterior over the momentary pose along with the map. Many algorithms for online SLAM are incremental: they discard past measurements and controls once they have been processed (Thrun et al., 2006).

The second strategy is called the full SLAM problem, which seek to calculate a posterior over the entire path along with the map, instead of just the current pose. This subtle difference between online and full SLAM has ramification in the type of algorithm that can be brought to bear. In particular, the online SLAM problem is the result of the integrating out past poses from the full SLAM problem.

In online SLAM, these integrations are typically performed one-at-a-time. They cause interesting changes of the dependency structures in SLAM. In addition, an example of online SLAM is the well-known Extended Kalman Filter. In contrast, an example of full SLAM is FastSLAM technique.

SLAM Solutions

SLAM is considered a complex problem because to localize itself a robot needs a consistent map and for acquiring the map the robot requires a good estimate of its location. This mutual dependency among the pose and the map estimates makes the SLAM problem hard and requires searching for a solution in a high-dimensional space. For these reasons, a lot of methods have been established for providing a solution to this problem. However, in this section we are focused in two main approaches: Rao-Blackwellized Filter and the Kalman Filter.

Rao-Blackwellized

Murphy, Doucet and colleagues (Doucet et al., 2000; Murphy, 1999) introduced Rao-Blackwellized particle filter as an effective means to solve the SLAM problem. The key idea of this approach is to first use a particle filter to estimate the trajectory of the robot. Afterwards, this estimated trajectory is used to compute a posterior about the map of the environment. The main problem of Rao-Blackwellized approaches is their complexity, measured in terms of the number of particles required to build an accurate map (Stachniss, 2009). Reducing this quantity is one of the major challenges for this family of algorithms. Additionally, the re-sampling step is problematic since it can eliminate good state hypotheses. This effect is also known as the particle depletion problem (Doucet, 1998).

The key idea of Rao-Blackwellized mapping is to separate the estimation of the trajectory from the map estimation process. Moreover, Rao-Black-

wellized mapping uses a particle filter similar to Monte Carlo localization (MCL) (Dellaert et al., 1998). In contrast to MCL, the Rao-Blackwellized particle filter for mapping maintains an individual map for each sample. Each map is built given the observations z1:t and the trajectory x1:t represented by the corresponding particle.

Extended Kalman Filter

The Kalman Filter (KF) was defined by Rudolph E. Kalman, who in 1960 published his famous paper describing a recursive solution to the discrete-data linear filtering problem (Kalman, 1960). The KF is essentially a set of mathematical equations that implement a predictor-corrector type estimator that is optimal in the sense that it minimizes the estimated error covariance (when some presumed conditions are met). Since the time of its introduction, the KF has been the subject of extensive research and application, particularly in the area of autonomous or assisted navigation. This is likely due in large part to advances in digital computing that made the use of the filter practical, but also to the relative simplicity and robust nature of the filter itself. Rarely do the conditions necessary for optimality actually exist, and yet the filter apparently works well for many applications in spite of this situation.

However, KF is highly efficient in the computational sense. It is optimal for linear Gaussian systems, but most robotics systems are non-linear. For this reason, the Extended Kalman Filter (EKF) has been the de-facto approach to the SLAM problem for nearly fifteen years, due to it is an extension of KF but it works with non-linear systems.

The EKF-SLAM solution is very well-known and inherits many of the same benefits and problems as the standard EKF solutions to navigation or tracking problems. Four of the key issues are the following:

1. **Convergence:** In the EKF-SLAM problem, convergence of the map is manifest in the

monotonic convergence of the determinant of the map covariance matrix, and all landmark pair sub-matrices, toward zero. The individual landmark variances converge toward a lower bound determined by initial uncertainties in the robot position and the observations.

2. **Computational Effort:** The observation update step requires that all landmarks and the joint covariance matrix be updated every time an observation is made. Naively, this means computation grows quadratically with the number of landmarks. There has been a great deal of work undertaken in developing efficient variants of the EKF-SLAM solution and real-time implementations with many thousands of landmarks have been demonstrated (Guivant & Nebot, 2001; Kim & Sukkarieh, 2007).

3. **Data Association:** The standard formulation of the EKF-SLAM solution is especially fragile to incorrect association of observations to landmarks (Neira & Tardos, 2001). The loop-closure problem, when a robot returns to re-observe landmarks after a large traverse, is especially difficult. The association problem is compounded in environments where landmarks are not simple points and indeed look different from different view-points.

4. **Non-linearity:** EKF-SLAM employs linear models of non-linear motion and observation models and so inherits many caveats. Non-linearity can be a significant problem in EKF-SLAM and leads to inevitable, and sometimes dramatic, inconsistency in solutions (Julier & Uhlmann, 2001). Convergence and consistency can only be guaranteed in the linear case.

Vision-Based SLAM

Vision is an extremely important sense for both humans and robots, providing detailed information about the environment. A robust vision system should be able to detect relevant objects and provide an accurate representation of the world to higher level processes. The vision system must also be highly efficient, allowing a resource limited agent to respond quickly to a changing environment.

Notwithstanding, the sensors used in most of the SLAM algorithms are traditionally sonars and laser range scanners because they directly retrieve depth information. Therefore the inclusion of the new landmark in its estimated 3D position on the map is straightforward. In recent years, several methods capable of performing SLAM with purely visual information have started to appear. However, in images where the landmark depth information is not directly accessible, the landmarks map location must be obtained from the reconstruction of two or more views (Hartley & Zisserman, 2003).

For obtain a representation of the environment it is necessary to use features of the robot surroundings, but it is well-known that using pure range data for object recognition are attractive. But, 2D laser rangers (LIDAR systems) are definitely not sufficient to identify objects, due they are restricted to a single plane. In addition, its price and working condition make them still difficult to use in mobile robots.

Recently, researchers have found that cameras are a convenient way to acquire a 3D image in different scenarios. We can use a simple webcam or a stereo camera with the goal to build the environment and give a solution to the SLAM problem. However, in order to do that, we need to extract the key points that contain useful information for instance recognition.

Furthermore, time-of-flight cameras allow the RGB video and range imaging to be integrated together (which is referred to as "RGBD," in which "D" refers to a "depth" or "distance" channel) (Iddan & Yahav, 2001). Thus, a time-of-flight camera (ToF camera) is a range imaging camera system that resolves distance based on the known speed of light, measuring the time-of-flight of a light signal between the camera and the subject for each point of the image. The time-of-flight camera is a class of scannerless LIDAR, in which the entire scene is captured with each laser or light pulse, as opposed to point-by-point with a laser beam such as in scanning LIDAR systems.

Features in Computer Vision

The increasing availability of standard cameras in the last years has opened a large amount of alternatives in SLAM research. Cameras offer the advantages of full 3D data, low cost, and compactness as well as the intuitive "human-like" appeal of purely visual sensing. Perhaps, more importantly, they provide seemingly limitless potential for the extraction of detail, both geometric and photometric, and for semantic and other higher level scene interpretation. Furthermore, it is just the large amount of information provided by cameras that put in place new challenging opportunities for solving SLAM problems.

Also, it is well-known that systems using single cameras, stereo cameras, or omnidirectional cameras, often in combination with odometry or inertial sensing, have demonstrated reliable and accurate localization and mapping and shown that cameras are highly viable sensors for SLAM.

In order to carry out SLAM using vision, features from the environment have to be extracted that can be used as landmarks for SLAM. These features have to be stable and observable from different viewpoints and angles. Many types of environment features have been used for vision-based SLAM. However, for achieve a feature-based SLAM approach encompasses the following four basic functionalities:

1. **Environment feature selection:** It consists in detecting in the perceived data, features of the environment that are salient, easily observable and whose relative position to the robot can be estimated. This process depends on the kind of environment and on the sensors the robot is equipped with:

it is a perception process that represents the features with a specific data structure.

2. **Relative measures estimation:** Two processes are involved here: (observation) estimation of the feature location relatively to the robot pose from which it is observed; (prediction) estimation of the robot motion between two feature observations. This estimate can be provided by sensors, by a dynamic model of robot evolution fed with the motion control inputs, or thanks to simple assumptions, such as a constant velocity model.

3. **Data association:** The observations of landmarks are useful to compute robot position estimates only if they are perceived from different positions: they must imperatively be properly associated (or matched), otherwise the robot position can become totally inconsistent.

4. **Estimation:** This is the core of the solution to SLAM: it consists in integrating the various relative measurements to estimate the robot and landmarks positions in a common global reference frame. The stochastic approaches incrementally estimate a posterior probability distribution over the robot and landmarks positions, with all the available relative estimates up to the current time.

Stereo vs. Monocular Vision

SLAM must put into play robust and scalable real-time algorithms that fall into three main categories: estimation (in the prediction-correction sense), perception (with its signal processing), and decision (strategy). The problem is now well understood: the whole decade of the nineties was devoted to solve it in 2D with range and bearing sensors, and big progress was achieved in the estimation side, to the point that researchers claim that the subject is approaching saturation (which is some kind of optimality obtained by evolutionary mechanisms) and recent research has focused on the perception side.

The best example of this trend is vision-based SLAM, and specially monovision SLAM, where bearings-only measurements reduce observability, thus delaying good 3D estimates. Davison (2003) showed real-time feasibility of monovision SLAM with affordable hardware, using the original Extended Kalman Filter (EKF) SLAM algorithm. Also, in the work of (Sola et al., 2005) can be seen that undelayed landmark initialization (i.e. mapping the landmarks from their first, partial observation) was needed when considering singular trajectories of the camera.

In this direction, an interesting work (Eade & Drummond, 2006) has recently appeared that uses the constant time FastSLAM2.0 algorithm (Montemerlo et al., 2003). Stereo-vision SLAM has also received considerable attention. The ability to directly obtain 3D measurements allows us to use the best available SLAM algorithms and obtain very good results with little effort in the conceptual parts.

Good works on stereo SLAM usually put the accent on advanced image processing, which may require highly specialized programming (for instance, the real time construction and querying of big data bases and the hardware implementation of robust feature trackers). The drawback of stereo-based systems is that they strongly depend on precise calibrations to be able to extend their naturally limited range of 3D observability.

Although, researches proposed a lot of solutions that show the benefits of putting together both mono and stereo vision SLAM:

1. Important objects for reactive navigation, which are close to the robot, are rapidly mapped with stereo.
2. Good orientation and localization is achieved with bearings-only measurements of remote landmarks.
3. In addition, updates can be performed on any landmark if it is only visible from one camera.
4. Also, precise previous calibration of the stereo rig extrinsic parameters of the cameras is no longer necessary, due to the fact that dynamic self-calibration of these stereo rig extrinsic parameters can be incorporated, thus making such an intrinsically delicate sensor more robust and accurate.

BUILDING THE ENVIRONMENT

The robots can move in different environments that depends on the situation that they can be found, in this sense, the robots must be equipped with appropriate sensors in order to act in the best way in different situations. The simple case is that a robot can move in a static environment but in a real situation the environment is usually dynamic.

Furthermore, one of processes required for SLAM is map learning, which consists of memorizing the data acquired by the robot during exploration in a suitable representation. In addition, they have to be able response in outdoor and indoor places.

Localization and map-learning are interdependent processes - as using a map to localize a robot requires that the map exists, while building a map requires the position to be estimated relative to the partial map learned so far. On the contrary, path-planning is a rather independent process that takes place once the map has been built.

Available Information

Perception in robots is based on multiple sensors that provide information on their environment. Thus, there are two types of sources related with the information: idiothetic and allothetic. The former is related with the information that corresponds to odometry, and the latter corresponds to the information that is provided by laser range finders, sonars or vision. There are two main uses of these information sources:

1. They may be used to directly recognize a place or a situation. In this case, any cue, such as sonar time-of-flight or color may be used.

2. They may be converted to information expressed in the 2D space related to the idiothetic data thanks to a metric model of the corresponding sensors. With such a metric model, it is possible to infer the allothetic cues that would be sensed in unvisited places, or to infer the relative positions of two places in which allothetic information has been gathered (Filliata & Meyer, 2003).

Based on the combination of both, they are endowed with the capacity of constructing a topological or metric map, localizing themselves with this map and planning paths to reach goals. One consequence of integrating these two kinds of information is that they can compensate for each other to some extent. Allothetic cues may compensate the cumulative errors resulting from the use of idiothetic information. On the contrary, idiothetic cues may serve to disambiguate distinct locations that appear the same to the robot's sensors. Such mechanisms however, do not necessarily reflect an actual adaptive response to environmental changes but rather apply to planned situations occurring in a stable environment. Adaptation to unexpected changes requires more sophisticated interactions.

The drawbacks and advantages of these two sources of information are complementary. Indeed, the main issue raised by idiothetic information is that, because it involves an integration process, it is subject to cumulative error. This leads to a continuous decrease in quality, and therefore such information cannot be trusted over long periods of time. On the contrary, the quality of allothetic information is stationary over time, but it suffers from the perceptual aliasing problem, i.e., the fact that, for a given sensory system, two distinct places in the environment may appear the same.

Consequently, in order to build reliable maps and to navigate for long periods of time the allothetic information must compensate for idiothetic information drift, while idiothetic information must allow perceptually aliased allothetic information to be disambiguated.

Representation of the Environment

The problem of acquiring models is difficult and far from being solved. The following factors impose practical limitations on a robot's ability to learn and use accurate models:

1. **Sensors:** They often are not capable of directly measuring the quantity of interest. For example, cameras measure color, brightness and saturation of light, whereas for navigation one might be interested in assertions such as "there is a door in front of the robot."

2. **Perceptual limitations:** The perceptual range of most sensors, such as ultrasonic transducers, or cameras, is limited to a small range around the robot. To acquire global information, the robot has to actively explore its environment.

3. **Sensor noise:** Sensor measurements are typically corrupted by noise. Often, the distribution of this noise is not known.

4. **Drift/slippage:** Robot motion is inaccurate. Unfortunately, odometric errors accumulate over time. For example, even the smallest rotational errors can have huge effects on subsequent translational errors when estimating the robot's position.

5. **Complexity and dynamics:** Robot environments are complex and dynamic, making it principally impossible to maintain exact models and to predict accurately.

6. **Real-time requirements:** Time requirements often demand that internal models must be simple and easily accessible.

Recent research has produced two fundamental paradigms for modeling robot environments: the grid based (metric) paradigm and the topological paradigm (Thrun & Bucken, 1996).

Grid Based Paradigm

The simplest method for building a grid based map calls upon an incremental scheme that consists of estimating the robot's position using the current map and in subsequently locally updating the map around this position whenever new information is acquired by the robot. The term incremental stems from the fact that the way new information is added to the map cannot be reconsidered afterwards, even if this information turns out to be incorrect. For example, if the robot's position is wrongly estimated, any update of the map will be incorrect. If the robot later recovers its true position, it cannot take this information into account to cancel the effects of wrong past map updates. This incremental scheme may be problematic, for example, when large cycles must be covered in the environment. Upon closing such cycle, a lot of information is gathered about the robot's previous positions but cannot be used, and the corresponding map-learning strategies often fail in such circumstances.

Occupancy Probability

Grid maps discretize the environment into equally sized cells. Each cell represents the area of the environment it covers. It is assumed that each cell is either free or occupied by an obstacle. Occupancy grids store for each cell c a probability $p(c)$ of being occupied by an obstacle. Taking as a reference the algorithm presented by Moravec and Elfes which computes the occupancy probability $p(m)$ for the grid map m. It takes into account a sequence of sensor observations $z_{1:t}$ obtained by the robot at the positions $x_{1:t}$ and seeks to maximize the occupancy probability for the grid map. One assumption in the algorithm of Moravec and Elfes is that the different cells are independent. Therefore, the probability of a map m is given by the product over the probabilities of the individual cells:

$$p(m) = \prod_{c \in m} p(c) \qquad (1)$$

Topological Paradigm

Topological maps are built on top of the grid-based maps. The key idea is simple but effective: Grid-based maps are decomposed into a small set of regions separated by narrow passages such as doorways. These narrow passages, which are called critical lines, are found by analyzing a skeleton of the environment. The partitioned map is mapped into an isomorphic graph, where nodes correspond to regions and arcs connect neighboring regions. This graph is the topological map. The precise algorithm works as follows:

1. **Thresholding:** Initially, each occupancy value in the occupancy grid is thresholded. Cells whose occupancy value is below the threshold are considered free-space (denoted by F). All other points are considered occupied (denoted by C).
2. **Voronoi diagram:** For each point in free-space $(x, y) \in$ F, there is one or more nearest point(s) in the occupied space C. We will call these points the basis points of (x, y), and the distance between (x, y) and its basis points the clearance of (x, y). The Voronoi diagram is the set of points in free-space that have at least two different (equidistant) basis-points (Choset, 1996; Garrido, Moreno, & Blanco, 2009).
3. **Critical points:** The key idea for partitioning the free space is to find critical points." Critical points (x, y) are points on the Voronoi diagram that minimize clearance locally. In other words, each critical point (x, y) has the following two properties: (a) it is part of the Voronoi diagram, and (b) the clearance of

all points in a neighborhood of (x, y) is not smaller.

4. **Critical lines:** Critical lines are obtained by connecting each critical point with its basis points. Critical points have exactly two basis points (otherwise they would not be local minima of the clearance function). Critical lines partition the free-space into disjoint regions.

5. **Topological graph:** The partitioning is mapped into an isomorphic graph.

Based on the previous theory, Figure 2 shows a graphical representation of each part of the algorithm.

The importance of integrating metric and topological maps for scaling up mobile robot operation has long been recognized. Among the first to propose this idea were (Elfes, 1987) and (Chatila & Laumond, 1985). Elfes devised algorithms for detecting and labeling occupied regions in occupancy maps, using techniques from computer vision. He also proposed building large-scale

topological maps, but he did not devise an algorithm for doing so.

In brief, both paradigms are used with the same importance, depending on the needs that want to cover. In this sense, as a comparison can be seen a table that shows the advantages and disadvantages of both.

On the other hand, as stated in (Van Zwynsvoorde, Simeon, & Alami, 2001), pure geometric representations may not be well suited for robotic navigation in large scale environments. New models combine topological and metrical information to give compact and efficient representations in large scale environments. Also, when dealing with large environments, the acquisition of huge amount of range data is a time consuming task. Then, it is necessary to devote efforts to register different types of sensor data, which have different projections, resolutions and scaling properties.

Furthermore, although perception plays a key role in object search within robotics, it is also very important to use semantic knowledge to narrow

Figure 2. Extracting topological maps: (a) metric map, (b) Voronoi diagram, (c) critical points, (d) critical lines, (e) topological regions, and (f) the topological graph (Thrun, 1997)

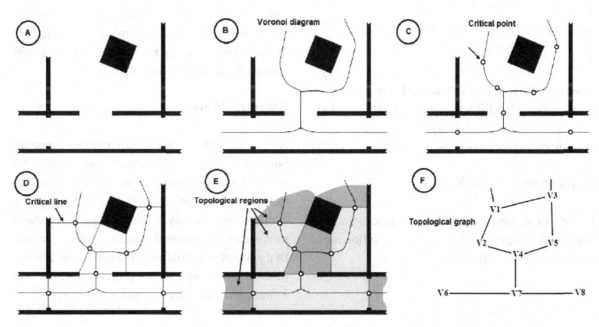

down the search space in large-scale environments. Pruning the search space is mainly important because of two reasons, first, robot perception is computational expensive, and second, if robots have no clue about potential object locations, objects will only be found by employing exhaustive search methods or by chance (Saito, et al., 2011).

Sensor Model for a Laser Range Finder

During the experimentation part, we used different kinds of sensors. However, laser range finder is a crucial tool for testing the algorithms because itself includes the model. In the case of grid-based maps, each cell c that is covered by the nth beam $z_{t,n}$ of the observation z_t and whose distance to the sensor is shorter than the measured one, is supposed to be unoccupied. The cell in which the beam ends is supposed to be occupied. The function $dist(xt, c)$ refers to the distance between the sensor and the center of the cell c. This can be formulated as:

$$p(c \mid Z_{t,n}, x_t) =$$
$$\begin{cases} p_{prior}, z_{t,n} \ is \ a \ \max imum \ range \ reading \\ p_{prior}, \ c \ is \ not \ \cov ered \ by \ z_{t,n} \\ p_{occ}, \ \mid z_{t,n} - dist(x_t,c) \mid < r \\ p_{free}, z_{t,n} \geq dist(x_t,c), \end{cases}$$

$$(2)$$

where r is the resolution of the grid map. Furthermore, it must hold $0 \leq p_{free} < p_{prior} \leq p_{occ} \leq 1$. In this case, we used the Hokuyo Range Finder for both modeling and experimental part.

Sensor Model for Sonar

In order to obtain reliable range data from a Sonar Sensor to remove chronic errors, the following steps are performed:

1. **Thresholding:** Range readings above a certain maximum R_u, are discarded. We observe that sonar readings caused by specular reflections are often near the maximum range of the device (R_{max}). With R_u slightly below R_{max}, many of these readings are discarded. The system becomes slightly myopic, but the overall quality of the map improves. Very large open spaces are detected by analyzing the set of distance values obtained from the sonar.

2. **Averaging:** Several independent readings from the same sensor at the same position are averaged. The Sonar readings are subject to error not only from reflections but also from other causes such as fluctuations in the effective sensitivity of the transducer. As a result, readings show certain dispersion. Averaging narrows the spread.

3. **Clustering:** A set of readings from one sensor at a given position sometimes shows a clustering of the data around two different mean values. This happens when different readings are being originated by objects at staggered distances.

A range reading is interpreted as providing information about space volumes that are probably empty and somewhere occupied. The functions are based on the range reading and on the spatial sensitivity pattern of the sonar.

Camera Model

As it is well-known, a camera performs a mapping between the 3D world (object space) and a 2D image. Thus, all cameras modeling central projection are specializations of the general projective camera. The anatomy of this most general camera model can be examined using the tools of projective geometry. In this way, the geometric entities of the camera, such as the projection centre and image plane, can be computed quite simply from

Table 1. Advantages ($\triangle$) and disadvantages (∇) of grid-based and topological approaches to map building

Grid based (metric) approach	Topological approach
$\triangle$ Easy to build, represent, and maintain	$\triangle$ Permits efficient planning, low space complexity (resolution depends on the complexity of the environment)
$\triangle$ Recognition of places (based on geometry) is non-ambiguous and view point-independent	$\triangle$ Does not require accurate determination of the robot's position
$\triangle$ Facilitates computation of shortest paths	$\triangle$ Convenient representation for symbolic planner/problem solver, natural language
∇ Planning inefficient, space consuming (resolution does not depend on the complexity of the environment)	∇ Difficult to construct and maintain in large-scale environments if sensor information is ambiguous
∇ Requires accurate determination of the robot's position	∇ Recognition of places often difficult, sensitive to the point of view
∇ Poor interface for most symbolic problem solvers	∇ May yield suboptimal paths

its matrix representation. Specializations of the general projective camera inherit its properties, for example their geometry is computed using the same algebraic expressions.

The specialized models fall into two major classes: Finite Cameras, Cameras at Infinity.

Finite Cameras

This kind of cameras is the most specialized and simplest camera model, which is the basic pinhole camera by considering the central projection of points in space onto a plane (Hartley & Zisserman, 2003).

Under the pinhole camera model, a point in space with coordinates $\chi = (X, Y, Z)_T$ is mapped to the point on the image plane where a line joining the point χ to the centre of projection meets the image plane.

Cameras at Infinity

This category corresponds to cameras with centre lying on the plane at infinity. We can find two types of cameras within this category: the affine,

and the non-affine cameras (we are focused in the first one, which is the most common in this classification).

In this sense, the affine camera is called in this way because points at infinity are mapped to points at infinity. The optical centre, since it lies on the principal plane, must also lie on the plane at infinity. From this we have:

1. Conversely, any projective camera matrix for which the principal plane is the plane at infinity is an affine camera matrix.
2. Parallel world lines are projected to parallel image lines. This follows because parallel world lines intersect at the plane at infinity, and this intersection point is mapped to a point at infinity in the image. Hence the image lines are parallel.

Camera vs. Laser

In comparison with a laser, the range image generation process resides on the use of the colinearity equations to project the points of the cloud over the image plane. Furthermore, the information obtained by both sensors has to be treated in different ways. Of course, a camera has been seen as a tool more powerful than a laser due to the benefits that the former can offer. However, if we combine the benefit of both sensors, the final representation of the environment that can be obtained would be potentially better than if only one sensor is used. Next table shows the advantages and disadvantages of both sensors.

MULTI-ROBOT EXPLORATION: ASSIGNMENT OF ROBOTS TO REGIONS

During this chapter, we are focused in two main topics: the former related with the advantage to use a multi-robot system in comparison with a single robot, the latter related with the assignment problem. In this sense, we will prove that a collaborative work is able to complete a set of goals in a fast and effective way.

On the other hand, since collaborative work involves multiple robots, it is interesting for several reasons. For instance, a task may be inherently too complex for a single robot to accomplish. Therefore, performance benefits can be gained from using a team of robots. Also, using several robots provides more flexibility and increases the fault-tolerance with respect to the case when a single robot is responsible for each separated task.

Multi-Robot Exploration

The problem of exploring an unknown environment and constructing a map is central to mobile robotics. The ability to build an internal representation of the environment is also critical to most intelligent agents. Explicit definition of exploration, include: investigation of unknown regions or travel undertaken to discover what a place is like or where it is. In this way, we want to build an accurate representation of the environment based on the navigation/exploration of a team of robots, of

course as we have mentioned previously, it is not an easy task and it becomes more complex when we involved multi-robot exploration. However, an exploration action can involve a sequence of control actions. In brief, the goal of exploration process is to cover the whole environment in a minimum amount of time.

On the other hand, the action by a team of robots implies a cooperative behavior. According to (Noreils, 1993), robot cooperation is a situation in which several robots operate together to perform some global task that either cannot be achieved by a single robot, or whose execution can be improved by using more than one robot, thus obtaining higher performances.

Furthermore, when dealing with heterogeneous robot teams with different sensing capabilities, multiple choices must be considered in designing a solution to a given exploration problem, based upon cost, robot availability, flexibility in robot use, etc. Designing an optimal robot team for a given application prior to deployment requires a significant amount of analysis and consideration of the tradeoffs in alternative strategies. The idea of the optimal team design is to engineer the best robots for a particular application in advance, and then apply those robots to the application with a certain solution strategy in mind.

Thus, it is necessary to develop general techniques that enable any collection of heterogeneous robots to reorganize into subteams as needed depending upon the requirements of the application tasks and on the available robots and their resources. An additional assumption is that with a large team of heterogeneous robots, different combinations of robots will be able to solve certain tasks in different ways. The desired objective is to achieve a high degree of flexibility and fault tolerance in the team solution through the ability of the robots to autonomously reconfigure into subteams (Parker, 2003).

Of course, we are focused in exploration based on a team of robots for all the advantages that we want obtain:

Table 2. Advantages (↑) and disadvantages (↓) between laser and camera

SCANNER LASER	CAMERA
↓Not accurate extraction of lines	↑High accuracy in the extraction of lines
↓Not visible junctions	↑Visible junctions
↓Color information available on low resolution	↑Color information on high resolution
↑Straightforward access to metric information	↓Awkward and slow access to metric information
↑High capacity and automatization in data capture	↓Less capacity and automatization in data capture
↓Data capture not immediate. Delays between scanning stations. Difficulties to move the equipment	↑Flexibility and swiftness while handling the equipment
↑Ability to render complex and irregular surfaces	↓Limitations in the renderization of complex and irregular surfaces
↓High cost	↑Low cost
↑Not dependent on lighting conditions	↓Lighting conditions are demanding
↓3D model is a cloud without structure and topology	↑The 3D model is accessed as a structured entity, including topology if desired

1. Accomplish a single task faster than a single robot.
2. Overlapping information, which can help to compensate for sensors uncertainty.
3. Localize themselves more efficiently.
4. Increased robustness.
5. Higher quality solutions.
6. Completion of tasks impossible for single robots.

However, we can find many obstacles to effective coordination, such as:

1. Risk of possible interferences between them.
2. Limited communication range.
3. They use longer detours that may be necessary in order to avoid collisions and covered large space.
4. Dynamic events.
5. Limited energy, computation and mobility.

In addition, when robots work together as a team, the members that perform each task should be the ones that promise to use the least resources to do the job. The performance of multi-robot in redundancy and co-operation contributes to task solutions with a more reliable, faster, or cheaper way. Many practical and potential applications, such as unmanned aerial vehicles (UAVs), spacecraft, autonomous underwater vehicles (AUVs), ground mobile robots, and other robot-based applications in hazardous and/or unknown environments can benefit from the use of it. In this sense, robots have to learn from, and adapt to their operating environment and their counterparts.

Most approaches to multi-robot exploration proceed in the following way. First, a set of potential target locations or target areas is determined. Secondly, target locations are assigned to the individual members of the team. The robots then approach those target locations and include their observations obtained along the paths into a map. This process is repeated, until the environment has completely been explored. A stopping criterion can be given by a threshold (Stachniss, 2009).

In this way, recent strategies, e.g. (Puig, Garcia, & Wu, 2011), for coordinated multi-robot exploration have been proposed. Thus, the aforementioned work applies a global optimization strategy based on clustering to guarantee a balanced and sustained exploration of big workspaces. The algorithm optimizes the on-line assignment of robots to targets, keeps the robots working in separate areas and efficiently reduces the variance of average waiting time on those areas. The latter ensures that the different areas of the workspace are explored at a similar speed, thus avoiding that some areas are explored much later than others, something desirable for many exploration applications, such as search and rescue.

Coordinated Exploration

In this section, we are focused in the levels of coordination, the advantages and drawbacks of each one. However, if we think in the simplest way that a team of robots can explore an environment, then we address to an uncoordinated way, because each robot is simply focused in exploring without taking into account important factors such as redundancy, limited communication, optimization, etc. Of course, a good coordination has to bear in mind these factors in order to obtain the best results.

In addition, with an uncoordinated way we can obtain undesirable situations yielded by the lack of coordination. For instance, a robot may choose the same exploration view point selected by other robots or, at least, the map's region that a robot can sense may overlap the sensed regions by other robots (see Figure 3-A). As well as, a robot may select an exploration viewpoint for which its chosen trajectory is blocked by other robot (see Figure 3-B), or a robot's sensor may be occluded by other robots located within its sensory field of view (see Figure 3-C).

In contrast to this, in a real world we expected to avoid these situations in order to obtain an acceptable result, based on the previous, a coordination exploration is the best option. In this sense, next we are focused in the framework related to coordinated exploration.

As a general way, the framework in coordinated multi-exploration, so as to improve collective performance can be presented as:

1. **Sharing information:** Most of the work related with multi-robot exploration (MRE) has been devoted to the definition of different distributed architectures that perform the interaction between the behaviors of individual robots. Furthermore, most of the work about communication has addressed the communication structure, neglecting another important dimension: the communication content.

2. **Robotic Mapping:** Robotic mapping addresses the problem of acquiring spatial models of physical environments with mobile robots equipped with range sensors. In this sense, sensors that have always a limited range, are subject to occlusions and yield noisy data. Therefore, robots have to navigate through the environment and build the map iteratively. It worth to mention that one of the challenges include the sensor modeling, the representation and the exploration.

3. **Exploration and active/passive sensing:** When a robot or a team of robots explore an unknown environment to build a map, the main goal is to acquire as much new information as possible with every sensing cycle, so as to minimize the mission time.

Levels of Coordination

Following, we can find three main levels related with the coordination, each one depends of the kind of situation that we want to focus:

- No exchange of information.
- **Implicit coordination:** Sharing a joint map.
  - Communication of the individual maps and poses.
  - Central mapping system.

- **Explicit coordination:** Determine better target locations to distribute the robots.
  - ○ Central planner for target point assignment.

As it can be seen, the first level is the simplest one, the robots do not worried about be agree with the other members, they are only worried about to explore using the sensors that they have (for instance, ultrasonic sensors). Of course, as we mentioned previously it has many drawbacks: explore the same place, no optimization etc. For this reason, this kind of coordination is not usually used.

On the other hand, as projects and teams grow in size and complexity, tasks and member dependencies become more numerous, diverse and complex, thus increasing the need for team coordination. Coordinated teams manage these dependencies effectively using a number of explicit and implicit mechanisms and processes. Teams coordinate explicitly using task programming mechanisms (e.g., schedules, plans, procedures, etc.) or by communicating. We call these mechanisms "explicit" because team members use them purposely to coordinate.

However, teams can also coordinate implicitly (i.e., without consciously trying to coordinate) through team cognition, which is based on shared knowledge that the team members have about the

task and about each other. This shared knowledge helps the team members to understand what is going on with the task, and also helps them to anticipate what is going to happen next, and which actions team members are likely to take, thus helping them become more coordinated. In particular, this is the anticipation of the actions and needs of team members and task demands, and dynamic adjusting of their own behavior accordingly, without prior plan of activity or communicating with each other [Wittenbauma et al., 1996].

Of course, this kind of coordination offers us the optimal solution when we want to explore with a team of robots, because the task can be distributed for avoiding redundancy and, in consequence, we obtain a better solution for the task of building the map of the space. In addition, having communication between robots adds the advantage of planning the best robot distribution. Thus, for instance, if a robot breaks, then another robot may continue with the same task, and complete it without repeating information.

As an example, in Figure 4 can be seen in a simple way the advantages that will be obtained with an explicit robot coordination. In it can be seen the exploration of two robots, with each color representing the route that was taken by each one. In the first image can be seen redundancy in the exploration in comparison with the second one,

Figure 3. Undesirable situation in uncoordinated exploration

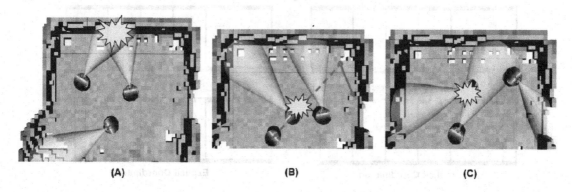

(A)   (B)   (C)

which shows a better distribution of the work. However, the best kind of coordination finally depends of the application and the final goal.

Another way of explicit coordination comes from the multi-robot coalition formation. Solutions to the coalition formation problem have many potential applications, especially in situations where tasks are located at considerable distances from one another and teams of robots need to be dispatched to different locations to autonomously complete their designated tasks (Vig & Adams, 2005).

A formation in multi-robot systems involves the coordination of a group of robots to get into and maintain a formation with a certain shape (e.g., wedge, chain, etc.), in order to solve problems in a variety of application fields, such as: rescue operations, landmine removal, remote terrain and space exploration, control of arrays of satellites and unmanned aerial vehicles.

Furthermore, during the last recent years, swarm robotics has evolved as a new approach to the coordination of multi-robot systems, which consist of large numbers of mostly simple physical robots. It is supposed that a desired collective behavior emerges from the interactions between the robots and interactions of robots with the environment. Two key characteristics in swarm robotics are miniaturization and low-cost. These are the constraints in building large groups of

robotics. Thus, the simplicity of the individual team member should be emphasized.

In contrast, for strongly centralized systems, the global information including the environment as well as the locations of all the robots is shared. It is typical for small number of robots in well structured environments and it is not robust to dynamic environments or failures in communications and other uncertainties.

In summary, the main characteristics of explicit coordination are: has proved efficiency in work coordination; conducts to the optimal solution; requires a good knowledge of the work in progress and an effective process tracking; allows interoperation by interconnecting their processes; is complicated in the designing part. However, the implicit coordination: does not allow an important analysis in modeling; it is not currently efficient to support interpretability processes; it is dynamic and flexible because the process is not really modeled, and in consequence, it can be changed easily.

The Assignment Problem

In this section, we present two of the main methodologies to solve the assignment problem that we will test in our experiments. The former is the Voronoi diagrams, which is one of the most fundamental methods related to computational

Figure 4. Comparison between implicit and explicit coordination in the same environment

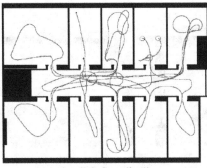

Implicit Coordination

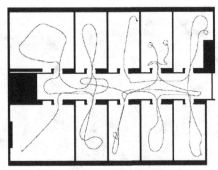

Explicit Coordination

geometry for the decomposition of a given space. The latter is the Hungarian method, which consists of a combinatorial optimization algorithm that solves the assignment problem in polynomial time.

Voronoi Diagram

This method decomposes a set of objects in a spatial space to a set of polygonal partitions. Formally, for any set of objects in a 2D or 3D space, a polygonal shape surrounds the object such that approximately any point in the polygon is closer to its generated object than any other generated object.

Two points are considered to be neighboring if the segment joining them does not intersect any other segment smaller than itself. This definition of neighborhood ensures that the subdivision generated is planar. Such a planar subdivision is called a Voronoi Diagram (also Dirichlet diagram or Thiessen diagram). The division segments are called the Voronoi edges, and the points of intersection of these edges, the Voronoi centers. Each planar face of the subdivision is called a Voronoi region and each Voronoi region can be associated uniquely with a point of the input set.

Among the different algorithms for the construction the Voronoi diagram, the divide-and-conquer approach is one of the most recognized, and it is detailed next.

Divide-and-Conquer Approach

A widely used method to design fast algorithms is divide and conquer, which allows to construct a Voronoi diagram for a set, S, of n points in time $O(n\log n)$.

It is based on the computation of the convex hull of a set of points (i.e., is the smallest convex set that includes all the points) in two or more dimensions. The process in 2D consists of dividing the set S of points into two equal sized sets L and R, such that all points of L are to the left of the most leftmost points in R. Recursively find

the convex hull of L and R. To merge the left hull and the right hull it is necessary to find the two upper and lower common tangents. The upper common tangent can be found in linear time by scanning around the left hull in a clockwise direction and around the right hull in an anti-clockwise direction. The two tangents divide each hull into two pieces. Finally, the edges belonging to one of these pieces must be deleted. As mentioned above, this procedure can be generalized to any number of dimensions.

The recurrence relation established by a divide-and-conquer approach is in the form,

$$T(n) = 2 \times T(\frac{n}{2}) + f(n) \qquad (3)$$

where, $f(n)$ is the time taken to merge the two half solutions at each step. Thus, to achieve $O(n\log n)$ time complexity, the merge operation must be performed in $O(n)$ time. Thus, a divide-and-conquer algorithm essentially deals with an efficient method of merging, while the divide step can be done easily in $O(n)$ time by dividing the points in to two halves, L and R, about the median x coordinate.

Thus, the set S has been divided in to two equal halves, L and R, with L containing all points to the left of the median x coordinate and R containing all points to the right. Further, assuming that, by recursion, the Voronoi diagrams of L and R, $V(L)$ and $V(R)$ have been constructed. For the base cases, where S contains 2 or 3 points, the Voronoi diagram can be made by perpendicular bisector construction. By visual inspection, notice that we can create a chain about the joint of L and R, which is the set of points equidistant from a point in L and a point in R. Importantly, the line segments in this chain are the new edges of the Voronoi diagram $V(S)$, i.e. $V(L \cup R)$. And, the edges of $V(L)$ and $V(R)$ remain in $V(L \cup R)$ except for those which are intersected by this chain (for more details see [Aurenhammer,1991]).

The Hungarian Method

The Hungarian method is a combinatorial optimization algorithm which solves the assignment problem in polynomial time and that anticipated later primal-dual methods.

In this algorithm the input is a cost table established according to the cost needed for completing different exploration tasks, and the output is an equivalent cost table in which the array of zero needed for a complete assignment constitutes an optimal assignment. The main idea of the algorithm is to modify the cost table until there is at least one zero in every column or row of the table, so as to find a complete assignment scheme according to the zeros. This scheme is an optimal assignment when it is applied to the cost matrix for the total cost in this scheme is the least, and the algorithm can be always converging on an optimal solution in finite steps. The basic theory of this algorithm is that when a constant is added or subtracted to any row (column), the optimal assignment won't change.

Given an $n \times n$ matrix $R = (R_{ij})$ of positive integers, find the permutation $j_1, ..., j_n$ of the integers $1, ..., n$ that maximizes the sum $r_{ij} + ... + r_{rnjn}$. Find non-negative integers $u_1, ..., u_n$ and $v_1, ..., v_n$ subject to

$$u_i + v_j \geq r_{ij} (i, j = 1, ..., n) \qquad (4)$$

that minimize the sum $u_1, ..., u_n + v_1, ..., v_n$. A set of non-negative integers satisfying (4) will be called a cover (or an adequate budget) and the positions (i,j) in the matrix for which equality holds are said to be marked (or qualified in the associated Simple Assignment Problem); otherwise (i,j) is said to be blank. A set of marks is called independent if no two marks from the set lie in the same line (row or column). If the largest number of independent marks that can be chosen is m, then m lines can be chosen that contain all of the marked positions.

If a cover for R is given, a largest set of independent marks is found; if this set contains n marks then the marked (i, j) constitute the desired assignment. If the set contains less than n marks then a set of less than n lines containing all of the marked (i, j) is used to improve the cover.

Figure 5. (a) Shows the division of S into L and R about median x coordinates, while (b) shows the individual Voronoi diagrams V(L) and V(R) of L and R overlaid together.

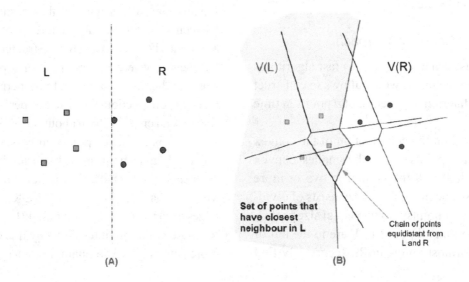

The following algorithm applies the above theorem to a given $n \times n$ cost matrix to find an optimal assignment (Junger et al., 2010).

- **Step 1:** Subtract the smallest entry in each row from all the entries of its row.
- **Step 2:** Subtract the smallest entry in each column from all the entries of its column.
- **Step 3:** Draw lines through appropriate rows and columns so that all the zero entries of the cost matrix are covered and the minimum number of such lines is used.
- **Step 4:** Test for Optimality: (i) If the minimum number of covering lines is n, an optimal assignment of zeros is possible and we are finished. (ii) If the minimum number of covering lines is less than n, an optimal assignment of zeros is not yet possible. In that case, proceed to Step 5.
- **Step 5:** Determine the smallest entry not covered by any line. Subtract this entry from each uncovered row, and then add it to each covered column. Return to Step 3.

EXPERIMENTS

Different sets of experiments have been carried out in order to illustrate the application of the theoretical basis presented in the previous sections. This section is organized as follows. First at all, a comparison including the advantages and drawbacks between a single and a team of robots will be presented in two domains: uncoordinated and coordinated exploration. After that, the calibration of the odometry will be performed in order to avoid the error in the final result. Also, the extraction of some features related to the data set acquired by the stereo camera will be shown. Following, a consistent map created in both, indoors and outdoors domains will be presented. Finally, a graphical representation about the advantages and disadvantages related with each of the SLAM approaches will be done.

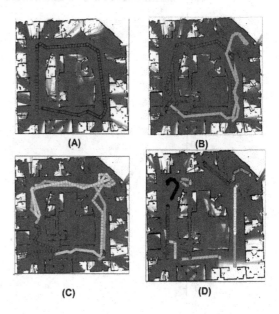

Figure 6. Uncoordinated exploration with an incremental team of robots

Based in the previous, we simulated the exploration in an unknown environment with a single robot and with an incremental team of robots (2; 4; 6; 8; 10; 12; 14). The objective was to observe the decrease in the exploration time in both coordinated and uncoordinated exploration. Also, we presented two techniques for the segmentation of the environment. First, a partition of the environment is done by the Voronoi graph with the goal of generating targets, then we determine the set of frontiers and compute the cost for reaching the subsegment with each robot: the matrix that is generated is solved based on the Hungarian Method, and that process is repeated until all the segments are covered. The results will be compared with other approaches.

The main algorithm that we propose for the segmentation of the environment is shown below:

- Determine segmentation $S = s_1, ..., s_n$ of the map.
- Determine the set of frontier targets for each segment.

Figure 7. Comparison between uncoordinated and coordinated exploration

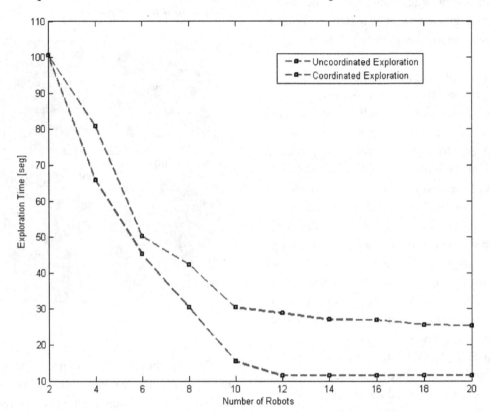

- Compute for each robot i the C_s^i for reaching each map segment $s \in S$.
- Discount cost C_s^i if robot i is already in segment s.
- Assign robots to segments using the Hungarian Method.
- For all segments s do.
- Assign robots to frontier targets in s w.r.t path costs using the Hungarian method.
- End for.

Figure 6 shows an uncoordinated exploration with 2, 4 and 10 robots. Thus, notice that undesirable situations are produced as, for example, when robots choose the same exploration direction selected by other robots, or if the trajectory is blocked by another robot and the robot's sensor is occluded by the other robots located within its sensory field of view. In addition, redundant work is done and, in consequence, the exploration time is increased. Of course, it is not an optimal solution (i.e. it is the worst situation related with a team of robots).

In Figure 7 we present the results obtained in simulation, based on a map of 308×309 cells. In this sense, we tested the algorithms taking the same reference for all of them (using Player-Stage libraries, and simulating Pioneer robot AT3, Hokuyo laser URG and Bumblebee stereo camera). Thus, in the case of uncoordinated exploration (Figure 7 - red line) we can notice a decrease in exploration time, but as we mentioned previously a redundant work can be done. It means that the solution for this task was inefficient (of course not optimal). In the real world it is desirable to avoid this kind of situation.

In contrast, Figure 7 - (blue line) shows the results based on the simplest coordination algorithm (frontier cells) that in brief: it determines the frontiers cells; then compute for each robot the cost for reaching each frontier cell; after that, choose the robot with the optimal overall evaluation and assign the corresponding target point to it; as next step, a reduction of the utility of the frontier cell visible is done. As a result of it, we can see that the exploration time was reduced, and redundant work was avoided.

From the results depicted above, it is clear that the coordinated exploration is more advantageous than the uncoordinated.

Without a doubt, if we move to real world, we should face to different problems that we must solve in order to obtain a good result, as a clear example, the odometry, which is sensitive to errors due to the integration of the velocity measurements over time to give position estimates. Fast and

accurate data collection, equipment calibration, and processing are required in most cases with the goal of use it in an effectively way.

Moreover, an experiment related to the correction of the odometry can be seen in Figure 8: the ground truth given by the black sequence, but the result obtained by the real odometry of the robot can be seen in

the trajectory drawn by the blue sequence, after applying the correction we obtained a great approximation to the real one given by the red sequence. Indubitably, we can notice that it is necessary this correction in order to obtain an acceptable result. On the contrary, if we put together both idiothetic and allothetic information the final result will be not the expected.

In this sense, Figure 9 illustrates a real sequence using the Pioneer AT3 robot. Thus, we

Figure 8. Correction of the odometry

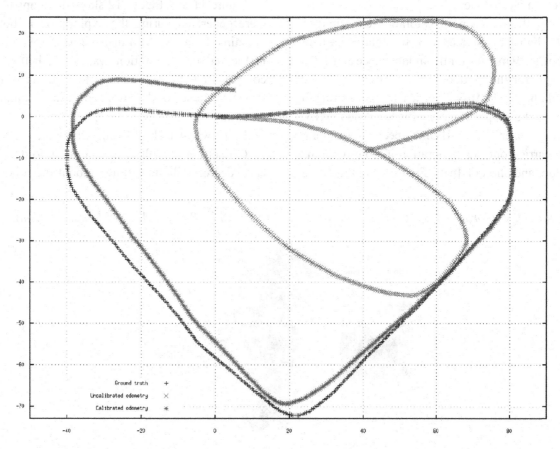

Figure 9. Odometry correction during a real sequence

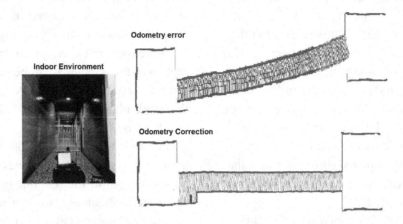

present an image with the odometry error and the correction in an indoor space. Before to process the information we verify the odometry error and we can notice that a correction was necessary. For the sequence, the Pioneer used the Hokuyo laser and a stereo camera, and some sequences with a webcam were done.

For the real-world exploration experiments with the robot Pioneer AT3, we obtained a data set of 1539 images for the indoor space and 1154 images for the outdoor space. Furthermore, we used the CARMEN libraries for generating a consistent map using the Pioneer robot AT3, Hokuyo laser URG and Bumblebee stereo camera. We carried out the exploration in both domains indoor and outdoor. In the Figure 9 and Figure 10

can be seen the readable logs generated by the information acquired during the exploration. We can see the representation of the information related to odometry, detection of the laser, camera, etc. After that, the correction for the odometry error was done.

Figure 11 and Figure 12 show the maps that were generated during the exploration in both domains. It can be seen that the result obtained in the indoor exploration was much better in comparison to the outdoor, due to the decrease of the performance of the laser sensor in the output domains.

Using Rao-Blackwellized particle filter based on (Grisetti et al., 2007) it is possible to drastically decrease the uncertainty about the robot's

Figure 10. Exploration task in the outdoor domain carried out by Pioneer AT3, without processing

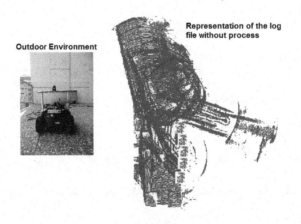

Figure 11. Consistent map generated by the exploration of Pioneer 3AT in an indoor domain

pose in the prediction step of the filter (see Figure 13). For that reason, we take both raw laser range data and odometry given by the Pioneer robot. In addition, it seriously reduces the problem of particle depletion.

As we mentioned previously, the coordinated exploration has advantages in comparison with uncoordinated exploration. Of course, there are many approaches for multi-robot exploration, and some of them are based on the segmentation of the environment (e.g., [Solanas & Garcia, 2004; Wurm et al., 2006] proposed different ways for coordination). The first one is related with the *K*-means method, while in contrast, the second is based on both *K*-means and Voronoi diagrams, and the third approach is using Voronoi diagrams but with the advantage of solving the assignment problem with the Hungarian method (we are focused in it).

Based on the segmentation of the environment, we can find some advantages and disadvantages: using *K*-means adds many conditions that in real world are not realistic, as for instance, the size and shape of the map is predetermined and based on indoor applications. Also, the proposition related to combine both *K*-means and Voronoi diagrams is not a realistic application, due to they are not based in the uncertainty model.

In comparison, solving the segmentation of the environment based on Voronoi diagrams, and solving the assignment problem with the Hungarian Method, it is a realistic solution and the computational cost is decreased. In addition, an optimal solution can be found. We are focused in it, and we show next the advantages of this approach. For the experiments, we used our own

data set and two additional data sets (the first related to Intel lab provided by Dieter Fox and the second related to the Newell-Simon Hall in Carnegie Mellon provided by Nick Roy).

Figure 13 and Figure 14 show the results we obtained by comparing the previous approaches. In particular, in Figure 13 can be seen different steps of the exploration of an indoor scenario performed by our proposed process for multi-robot coordinated exploration, based on the segmentation of the environment and assigning each segment to a robot according to the Hungarian Method. Moreover, Figure 14 shows the comparison between the three approaches tested in this work, by comparing the decrease of the exploration time when the number of robots increases. As it can be seen, the proposed coordination strategy based on segmentation was the best in comparison with the rest, due to the solution of the assignment targets is solved in an optimal way. Of course, the frontier based coordination is a good approach, but it does not consider the structure of the environment, and thus, it can assign one region to more than one robot.

CONCLUSION AND FUTURE RESEARCH

In this chapter, we presented a framework for simultaneous localization and mapping based on an active coordination of a team of robots. We

Figure 12. Final map generated by the exploration task in an outdoor domain

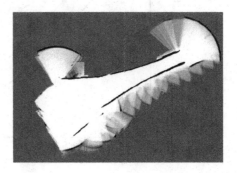

Figure 13. Multi-robot coordination based on the proposed segmentation of the environment

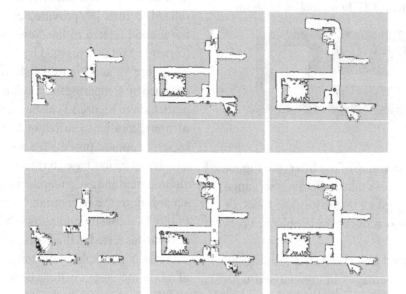

Figure 14. Comparison (in simulation time) between uncoordinated exploration, simple coordination exploration and coordination based on segmentation exploration

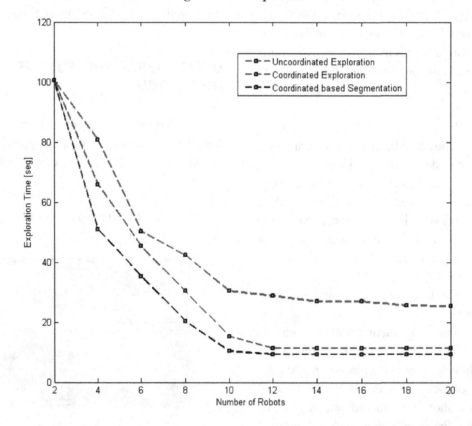

worked in both simulation and real-world with the goal to make a consistent representation of the environment. In this way, we used different algorithms for improving the distribution of the tasks, achieving the reduction in the exploration time.

Real experiments were performed by using the Pioneer Robot AT3, which was equipped with different sensors (stereo camera, webcam, and laser). Moreover, we used the CARMEN libraries for acquiring both idiothetic and allothetic information (and, thus, create the data set useful for the experiments). In addition, while we worked in real-work environments, we had to bear in mind different factors such as: odometry calibration, camera calibration, adjust the laser information and, then, solve the problem related with the integration of the information after each exploration step. Also, we took into account the fact that, for compensating the odometry error, we used the information provided by the laser and we reduced it with the least squares estimation. After that, we were able to create a consistent representation of the environment.

We compared the multi-robot exploration based in frontiers with the uncoordinated exploration. Without doubt, the best results were obtained by the first one, because the second produced undesirable situations yielded by the non-coordination, such as: a robot may choose the same exploration view point selected by other robots, a robot trajectory may be blocked by other members, etc. In comparison, frontier cells gave a better distribution of the work avoiding redundant work.

Furthermore, we proved that multi-robot exploration by means of a coordination based on the segmentation of the environment is better than the two previously mentioned algorithms. The main reason is due to the fact that, since we solve the assignment problem by using the Hungarian Method, we assure that the result will be the optimal. Also, the assignation of each segment only will be done to one robot.

As a future work, we want to address some problems that have not been tackled in this work. In particular, we will work in distinguishing flexible objects that can be found in the way (for instance, if the robot find a curtain it can be able to access in this space and do not take it as a solid object). In addition, we expect to integrate the visual-odometry correction instead of the laser-odometry correction. On the other hand, we will advance towards the use of the laser to create a consistent 3D representation of the environment, instead of the current 2D representation, and compare it with the 3D representation based on the features extracted from the environment, which were given by a stereo camera, in order to finally study the possible integration of both sensors for improving the 3D reconstruction of the environment.

REFERENCES

Aurenhammer, F. (1991). *Voronoi diagrams: A survey of a fundamental geometric data structure*. ACM. doi:10.1145/116873.116880

Ayache, N., & Faugeras, O. (1988). Building, registrating and fusing noisy visual maps. *The International Journal of Robotics Research*, 7(6), 45–65. doi:10.1177/027836498800700605

Bailey, T., Nieto, J., Guivant, J., Stevents, M., & Nebot, E. (2006). Consistency of the ekf-slam algorithm. *Proceedings of the IEEE/RSJ Conference on Intelligent Robots and Systems (IROS)*.

Chatila, R., & Laumond, J. (1985). Position referencing and consistent world modeling for mobile robots. In *Proceedings of the IEEE International Conference on Robotics and Automation*, (pp. 138-143).

Chli, M. (2009). *Applying information theory to efficient SLAM*. Ph.D. thesis, Imperial College London.

Choset, H. (1996). *Sensor based motion planning: The hierarchical generalized Voronoi graph*. PhD thesis, California Institute of Technology.

Davison, A. (2003). Real-time simultaneous localisation and mapping with a single camera. In *Proceedings of the International Conference on Computer Vision*, Nice.

Dellaert, F., Fox, D., Burgard, W., & Thrun, S. (1998). *Monte Carlo localization for mobile robots*. In IROS.

Doucet, A. (1998). *Technical report, Signal Processing Group*. Dept. of Engineering, University of Cambridge.

Doucet, A., de Freitas, J., Murphy, K., & Russel, S. (2000). Raoblackwellized particle filtering for dynamic Bayesian networks. *Proceedings of the Conference on Neural Information Processing Systems (NIPS)*, (pp. 176-183).

Durrant-Whyte, H. (1987). Uncertain geometry in robotics. *IEEE Transactions on Robotics and Automation, 4*(1), 23–31.

Durrant-Whyte, H., & Bailey, T. (2006). Simultaneous localization and mapping (slam): Part i. *Robotics and Automation Magazine, IEEE, 13*(2), 99–110. doi:10.1109/MRA.2006.1638022

Durrant-Whyte, H., Rye, D., & Nebot, E. (1996). *Localization of automatic guided vehicles*. The 7th International Symposium (ISRR).

Eade, E., & Drummond, T. (2006). Scalable monocular slam. *Computer Vision and Pattern Recognition, 1*, 469–476.

Elfes, A. (1987). Sonar based real world mapping and navigation. *IEEE Journal on Robotics and Automation, 3*(3), 249–265. doi:10.1109/JRA.1987.1087096

Estrada, C., & Tardós, J. D. (2005). Hierarchical slam: real-time accurate mapping of large environments. *IEEE Transactions on Robotics, 21*(4), 588–596. doi:10.1109/TRO.2005.844673

Filliata, D., & Meyer, J. (2003). Map-based navigation in mobile robots: I. A review of localization strategies. *Cognitive Systems Research, 4*(4), 243–282. doi:10.1016/S1389-0417(03)00008-1

Garrido, S., Moreno, L., & Blanco, D. (2009). Exploration of 2D and 3D environments using Voronoi transform and fast marching method. *Journal of Intelligent & Robotic Systems, 55*(1), 55–80. doi:10.1007/s10846-008-9293-7

Grisetti, G., Stachniss, C., & Burgard, W. (2007). Improved techniques for grid mapping with Raoblackwellized particle filters. *IEEE Transactions on Robotics, 23*(1), 34–46. doi:10.1109/TRO.2006.889486

Guivant, J., & Nebot, E. (2001). Optimization of the simultaneous localization and map-building algorithm for real-time implementation. *IEEE Transactions on Robotics and Automation, 17*(3), 242–257. doi:10.1109/70.938382

Hartley, R., & Zisserman, A. (2003). *A multiple view geometry in computer vision*. Cambridge University Press. doi:10.1017/CBO9780511811685

Iddan, G. J., & Yahav, G. (2001). 3D imaging in the studio (and elsewhere…). *Proceedings of the Society for Photo-Instrumentation Engineers, 4298*, 48–55. doi:10.1117/12.424913

Julier, S., & Uhlmann, J. (2001). A counter example to the theory of simultaneous localization and map building. In *Proceedings of the IEEE International Conference on Robotics and Automation*, (pp. 4238-4243).

Junger, M., Liebling, T., Naddef, D., Newhauser, G., & Wolsey, L. (2010). *50 years of integer programming, 1958-2008*. Berlin, Germany: Springer-Verlag. doi:10.1007/978-3-540-68279-0

Kalman, R. E. (1960). A new approach to linear filtering and prediction problems. *Journal of Basic Engineering, 82*, 35–45. doi:10.1115/1.3662552

Kim, J., & Sukkarieh, S. (2007). Real-time implementation of airborne inertial-SLAM. *Robotics and Autonomous Systems, 55*, 62–71. doi:10.1016/j.robot.2006.06.006

Montemerlo, M., Thrun, S., Koller, D., & Wegbreit, B. (2003). Fastslam 2.0: An improved particle filtering algorithm for simultaneous localization and mapping that provably converges. *In Proceedings of International Conference on Artificial Intelligence (IJCAI)*.

Moravec, H. (1980). *Obstacle avoidance and navigation in the real world by a seeing robot rover*. Technical Report CMU-RI-TR-3, Carnegie-Mellon University.

Murphy, K. (1999). Bayesian map learning in dynamic environments. *Proceedings of the Conference on Neural Information Processing Systems (NIPS)*, (pp. 1015-1021).

Neira, J., & Tardos, J. (2001). Data association in stochastic mapping using the joint compatibility test. *IEEE Transactions on Robotics and Automation, 17*(6), 890–897. doi:10.1109/70.976019

Noreils, F. R. (1993). Toward a robot architecture integrating cooperation between mobile robots: Application to indoor environment. *The International Journal of Robotics Research, 12*(1), 79–98. doi:10.1177/027836499301200106

Parker, L. E. (2003). The effect of heterogeneity in teams of 100+ mobile robots. In Schultz, A., Parker, L. E., & Schneider, F. (Eds.), *Multi-robot systems: From swarms to intelligent automata* (*Vol. 2*, pp. 205–215).

Puig, D., Garcia, M. A., & Wu, L. (2011). A new global optimization strategy for coordinated multi-robot exploration: Development and comparative evaluation. *Robotics and Autonomous Systems, 59*(9), 635–653. doi:10.1016/j.robot.2011.05.004

Saito, M., et al. (2011). Semantic object search in large-scale indoor environments. *Proceedings of IROS*, 2011 San Francisco, CA, USA.

Smith, R., & Cheesman, P. (1987). On the representation and estimation of spatial uncertainty. *The International Journal of Robotics Research, 5*(4), 56–68. doi:10.1177/027836498600500404

Sola, J., Monin, A., Devy, M., & Lemaire, T. (2005). *Undelayed initialization in bearing only slam*. IEEE International Conference on Intelligent Robots and Systems.

Solanas, A., & Garcia, M. A. (2004). Coordinated multi-robot exploration through unsupervised clustering of unknown space. In *Proceedings of IEEE/RSJ International Conference on Intelligent Robots and Systems*, (pp. 717-721).

Stachniss, C. (2009). *Robotic mapping and exploration*, Vol. 55. STAR Springer tracts in advanced robotics. Springer.

Thrun, S., & Bucken, A. (1996). Integrating grid-based and topological maps for mobile robot navigation. *Proceedings of the Thirteenth National Conference on Artificial Intelligence*, Portland, Oregon.

Thrun, S., & Bucken, A. (1997). Learning maps for indoor mobile robot navigation. *Technical report*, 1-60.

Thrun, S., Burgard, W., & Fox, D. (2006). *Probabilistic robotics*. Cambridge, MA: The MIT Press.

Van Zwynsvoorde, D., Simeon, T., & Alami, R. (2001). Building topological models for navigation in large scale environments. *Proceedings of ICRA, 4*, 4256–4261.

Vig, L., & Adams, J. A. (2005). A framework for multi-robot coalition formation. *Proceedings of the 2nd Indian International Conference on Artificial Intelligence*.

Wittenbauma, G., Stasser, G., & Merry, C. (1996). Tacit coordination in anticipation of small group task completion. *Journal of Experimental Social Psychology, 32*(2), 129–152. doi:10.1006/jesp.1996.0006

Wurm, K. M., Stachniss, C., & Burgard, W. (2006). Coordinated multi-robot exploration using a segmentation of the environment. *IEEE/RSJ International Conference on Intelligent Robots and Systems* (pp. 1160-1165).

ENDNOTE

[1] This work was partly supported by the Spanish Government through project Avanza Competitividad I+D+I, TSI-020100-2010-970.

Chapter 18
An Embedded Vision System for RoboCup

Pedro Cavestany-Olivares
University of Murcia, Spain

Juan José Alcaraz-Jiménez
University of Murcia, Spain

David Herrero-Pérez
University of Murcia, Spain

Humberto Martínez Barberá
University of Murcia, Spain

ABSTRACT

In this chapter, the authors describe their vision system used in the Standard Platform League (SPL), one of the official leagues in RoboCup competition. The characteristics of SPL are very demanding, as all the processing must be done on board, and the changeable environment requires powerful methods for extracting information and robust filters. The purpose is to show a vision system that meets these goals. The chapter describes the architecture of the authors' system as well as the flowchart of the image process, which is designed in such a manner that allows a rapid and reliable calibration. The authors deal with field features detection by finding intersections between field lines at frame rate, using a fuzzy-Markov localisation technique. Also, the methods implemented to recognise the ball and goals are explained.

INTRODUCTION

In the past few years we have witnessed a significant development within the field of robotics. In particular, there has been a great research on mobile platforms. This research has had to deal with many challenges, as computing resources on board usually are limited and most algorithms require outputs to be in real time. In addition to this, the main exteroceptive sensor of average mobile robots is a low-quality camera, which has

led researchers to seek robust and simple filters so that noise is ruled out and imprecisions dimmed. Our work aims at contributing on this research.

In this chapter we describe our vision system used in the Standard Platform League (SPL) which is one of the official leagues in RoboCup competition. All the code, examples and implementations that we present in this work have been developed for *Los Hidalgos* team, which has participated in the 2010 and 2011 editions of the RoboCup in association with L3M team, and in several international competitions.

DOI: 10.4018/978-1-4666-2672-0.ch018

In the SPL, platform and environment are standard. The standard platform is the Nao robot, a humanoid robot provided by Aldebaran Robotics. Nao is a light, compact, humanoid robot, fully programmable and easy to operate. The Nao version in which our software has been implemented has a biped configuration based in 21 degrees of freedom, which enables it to develop a great mobility. It is 57 cm tall and weighs 4.5 Kg. Concerning the computational architecture, the robot is equipped with an x86 AMID Geode 500MHz processor, 256 MB of SD-RAM and 2GB of flash memory. There are two communication interfaces in the robot: Ethernet and Wi-Fi 802.11g. The operative system is Open, an open Linux based distribution. The main exteroceptive sensors of the Nao are two non-stereo cameras, placed on the forehead and on the mouth of the robot, provided with VGA resolution. The first point to undertake is whether it is worthwhile using both cameras of the Nao, alternatively, or only one, and which one. Switching the camera takes at least 2 sec, and in every switch it is necessary to recalibrate the parameters of each one. We decided to use only one camera. The mouth camera is able to see both the feet of the robot and far away objects, thanks to the head joints. Since the robot has to see the ball when is near, so that it may get ready to kick the ball, and the goals, that usually are relatively far, this is the camera that we work with.

The standard environment is given by the soccer field, and the colour of the involved objects. Thus, player uniforms and goals are coloured: the blue goal belongs to the blue team, and the yellow goal to the red team. The ball is orange. Therefore, colour segmentation and its recognition is critical. For example, due to lighting conditions, orange could easily be mistaken for yellow. Our segmentation approach has been developed to avoid these confusions and to permit a quick and good calibration. Our recognition algorithms are designed to face as many situations of sighting the different objects as possible, especially in the case of goals. However, goals are not frequently perceived in game conditions, because robots are constantly looking at the floor when seeking the ball. Moreover, natural landmarks over the field are constantly perceived, i.e. field lines, which can be used to update robot localisation more frequently. For these field lines to be successfully used there are two issues that must be addressed: a robust field features detection in real time, and a robust localisation able to manage such information. We are using a fuzzy-Markov self-localisation technique, in which the robot location is modelled as a belief distribution on a 2½ D possibility grid (Buschka et al., 2000). This formalism allows us to represent and track multiple possible positions where the robot might be. Furthermore, it only requires an approximate model of the sensor system and a qualitative estimate of the robot displacement.

According to the league rules, all the processing must be done on board. For practical reasons, it has to be performed in real time, which prevents us from using time consuming algorithms. Nao camera provides an image every 33 ms, and our algorithms should be able to process every image within this time. As SPL allows any change in the software scope, we will try to avoid the middleware provided by Aldebaran, which slows down the image process.

We follow the ThinkingCap architecture (Martínez-Barberá & Saffiotti, 2000), a two-layer architecture which clearly reflects a cognitive separation of modules. From the conceptual point of view, modules are arranged by the nature of their processing tasks. From a software point of view, interfaces are clear and well defined, so that replacing or improving modules is not a difficult task.

Specifically, the objectives of this chapter are:

- To describe a fully functional vision system integrated in ThinkingCap architecture on a limited mobile robotic platform.

- To explain in detail our segmentation method, which significantly eases the calibration task.
- To present a fuzzy-Markov self-localisation technique, based on field features recognition algorithm.
- To describe robust object recognition methods and their filters implemented in a simple fashion so that the frame rate time premise is achieved.

This chapter is organised as follows. The next section describes the vision system architecture and flowchart. The section following devoted to explaining how world objects are detected by the vision module. In the first part of this section the process of colour coded objects perception is described. In the second part the algorithm for detecting field lines is explained. Once the information has been extracted from the processed images it is possible to treat these data in order to get knowledge; the chapter tackles this stage of our system in the Knowledge Extraction section. The end of the chapter exposes experimental results carried out along our research. In the last section, we evaluate our system and propose future work ahead.

ARCHITECTURE AND VISION MODULE

This section is devoted to the architecture implemented on the Naos and more specifically, the vision system architecture and the communication with the hardware device.

ThinkingCap Architecture

Many approaches have been adopted among Robocup teams for the software architecture. rUNSWift has implemented a task-hierarchy for multi-agent teams of 4 Naos, giving as a result a fault-tolerant network-centric architecture. The

software architecture of CMurfs, instead, is centred on a shared memory system, with different modules that communicate one another. Another paradigm is used by Upennalizers. This team has an architecture based on Lua language, with compiled *c* files that call the lower layers of the robot.

The Nao uses the layered architecture shown in Figure 1. This is a variant of the ThinkingCap architecture, which is a framework for building autonomous robots jointly developed by Örebro University and the University of Murcia (Martínez-Barberá & Saffiotti, 2000). We outline below the main elements of this architecture:

- The lower layer (commander module or CMD) provides an abstract interface to the sensori-motor functionalities of the robot. CMD accepts abstract commands from the upper layer, and implements them in terms of actual motion of the robot effectors. In particular, CMD receives set-points for the desired displacement velocity $<v_x, v_y, v_\theta>$, where v_x, v_y are the forward and lateral velocities and v_θ is the angular velocity, and translates them to an appropriate walking style by controlling the individual leg joints.
- The middle layer maintains a consistent representation of the space around the robot (Perceptual Anchoring Module, or PAM), and implements a set of robust tactical behaviours (Hierarchical Behaviour Module, or HBM). PAM acts as a short-term memory of the location of the objects around the robot: at every moment, the PAM contains an estimate of the position of these objects based on a combination of current and past observations with self-motion information. For reference, objects are named Ball, Net1 (own goal) and Net2 (opponent goal). PAM is also in charge of camera control, by selecting the fixation point according to the current perceptual needs (Saffiotti & LeBlanc, 2000). HBM

Figure 1. The ThinkingCap architecture

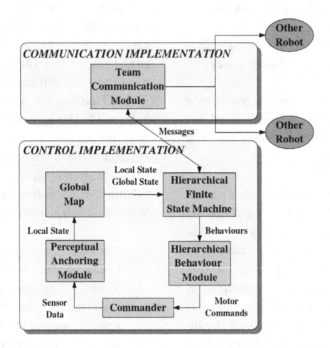

performs a set of navigation and ball control behaviours.

- The higher layer maintains a global map of the field (GM) and makes real-time strategic decisions based on the current situation (situation assessment and role selection is performed in the HFSM, or Hierarchical Finite State Machine). Self-localization in the GM is based on fuzzy logic (Buschka, Saffiotti & Wasik, 2000; Herrero, Martínez-Barberá & Saffiotti, 2004), as it will be explained later on. The HFSM implements a behaviour selection scheme based on finite state machines.
- Radio communication is used to exchange position and coordination information with other robots (via TCM, or Team Communication Module).

This architecture provides effective modularisation as well as clean interfaces, making it easy to develop independently different parts. Furthermore, its distributed implementation allows the execution of each module in a computer or robot indifferently. For instance, the low level modules can be executed on-board a robot and the high level modules can be executed off-board, where some debugging tools are available. However, a distributed implementation generates serious synchronisation problems. This causes delays in decisions and robots cannot react fast enough to dynamic changes in the environment. For this reason we have favoured the implementation of mixed mode architecture: at compilation time it is decided whether it will be a distributed version or a monolithic one (where the whole architecture is a thread and the communication module is another, see Figure 1).

Vision System Architecture

This section is devoted to the communication with the hardware device and to the vision system architecture, which is fitted within PAM. Firstly the access to the Nao camera is explained. Secondly, the vision system architecture is expounded.

Image Capture

The Nao camera provides images through a middleware developed by Aldebaran, called *Naoqi*. *Naoqi* consists of a main broker (an executable that listens to commands on an IP address and port) and some modules linked to it containing a set of robotic functions. The module provider of the camera images is called VIM (Video Input Module). In order to obtain an image, the processing image modules must subscribe to VIM. Due to the flexibility required in *Naoqi* (to offer the same accessing interface from different applications) many of these operations are not optimised and do not provide enough throughput for competition requirements. To improve this drawback, our image acquisition system has been developed to access directly to the camera images through the V4L2 (Video for Linux 2) driver, avoiding the use of VIM. The V4L2 code has been taken from the code released by German team B-Human. V4L2 is configured in streaming mode: the images are stored at camera frame rate in a number of predefined buffers in the kernel space. When an image is requested to be processed, V4L2 returns a pointer to the last filled buffer and the image is copied into the application user space. The streaming mode allows the continuous image acquisition whenever free buffers are available and returns always the most recent image captured. The results of this improvement are shown in the experimental results section.

Vision Flowchart

As Figure 2 shows, the vision flowchart has two branches: the coloured objects recognition path (above) and the field features detection path (bottom). Here both paths are described in detail. The starting point of the coloured objects recognition path is (Wasik & Saffiotti, 2002), where it is expected that the system receives the image already segmented by hardware. As our system receives raw images, the segmentation becomes another stage of the process.

The colour based vision system works as follows. When an image in YUV 422 colour space is received, it is created a segmented image through a look-up table (LUT) of 7 bits. This LUT is automatically generated by means of a YUV-HSV space colour conversion, in the following manner: the position of every input in the LUT corresponds to the coordinates of a colour in YUV colour space, whose values are converted to HSV. After that, the HSV value obtained is segmented according to segmentation thresholds. The result of this segmentation is a label which fills the corresponding input of the LUT. Therefore, by accessing to the LUT position indicated by the YUV coordinates, the segmented colour for these coordinates is returned as a result.

In the following stage, the colour labelled image is used to generate blobs. A blob is a set of pixels of the same colour grouped together because of their proximity. In the recognition phase, the blobs that do not satisfy a set of constraints (such as the constraint given by the horizon view) are ruled out by several filters. Finally, blobs that correspond to real objects are processed to estimate the distance from the local coordinate reference system. The resulting output of this tube is the set of currently sensed features of the environment, called Local Perceptual State or LPS, which are used by the control module of the robot, HBM.

In the field features detection path, the first step after taking a camera image is computing the horizon line, based on the pose of the robot and its head. The idea is to limit the search for field lines to the horizontal plane, thus saving computing time and avoiding false positives. Once this is done a Sobel filter to extract borders is applied to the brightness channel of the YUV image. By applying super-sampling to the border-filtered image from the bottom up to the horizon line, transitions of the form non-border (black) → border (white) → non-border (black) are stored and considered for further processing.

The following stage is shared with the coloured objects recognition path, where the YUV image is converted to the HSV colour space, as we have previously explained. Next, non-field line transitions are filtered out from the candidate list. These candidate transitions are checked with the HSV colour segmented image to detect which ones have been produced by the field lines. As a result we obtain segments that only correspond to field lines. The fifth step of this path is finding intersections between the labelled segments to produce field lines corners. The following stage consists of grouping these corners into labelled field features, according to the type of intersection found. In the corners detection section, these steps are further elaborated.

In the last step the distance and orientation from the robot to each detected field feature are computed. The pose of the camera is computed by using the joint angles of the legs and the head. The pixel from the image that represents a field feature is projected onto the field, subject to the constraint that the feature is on the horizontal ground plane. Then the projection point is used to compute the corresponding distance and orientation.

WORLD OBJECTS RECOGNITION AND FEATURES DETECTION

Here the algorithms developed to extract characteristics out of segmented images are described. Both the procedure followed in order to identify colour coded objects and the technique used for detecting field features are fully explained. Each subsection is devoted to the corresponding process.

Colour Based Vision

In order to detect the coloured objects of the standard environment of SPL, the images have to be colour segmented. This segmentation has to be robust, which means that it has to give good results under normal conditions and acceptable results under changing conditions, such as blurred images or changes in lighting conditions.

In the common literature colour segmentation methods are classified into Thresholding, Edge-based and Region-based methods, sometimes used in combination (Sonka, Hlavak & Boyle, 1996). Thresholding techniques assume that pixels whose colour value is within a certain range belong to the same object (Sahoo, Soltani, Wong & Chen, 1988). Edge-based methods rely on the assumption that pixel values change suddenly at the edge between two regions (Palmer, Dabis & Kittler, 1996). Finally, region-based methods assume that adjacent pixels in the same region have similar colour value. This similarity depends on the selected homogeneity criterion, usually based on some threshold value.

Colour Thresholding

A key point in the colour thresholding is the correct choice of the colour space. Colours have to be calibrated in this abstract mathematical model. Wasik & Saffiotti (2002) define the bounds of the colours by using basic shapes in each slice of the division of the Y channel, in the YCbCr colour space. In other words, the calibration consists of specifying the bounds of the colours for each slice of the Y channel. These bounds are defined using one prism (Cb-Cr) for each colour in every Y slice. This is a hard and slow task because sometimes it is not possible calibrate properly some colours, when Y division is not fine enough. In (Bruce, Balch & Veloso, 2000), the YCbCr colour space is fully tessellated by software into a 3D grid. The calibration process consists of defining a grid map for each colour. This method is also hard since the colour regions are tuned by selecting each cell of the grid map for each colour. In both cases, the process to calibrate the bounds of the colours in this space (YCbCr) is critical and laborious. It leads to difficult choices between accepting many false positives and discarding many true negatives.

Regardless of all this disadvantages, thresholding in YCbCr is quite used in RoboCup domain

Figure 2. The vision system flowchart

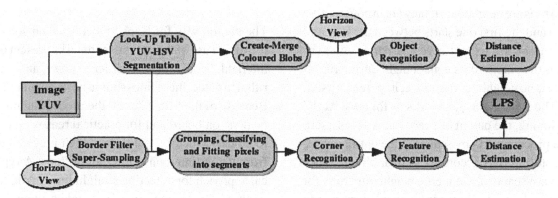

for performance reasons, as Nao robots provide images in this colour space. Some works (Palma-Amestoy, Guerrero, Vallejos & Ruiz del Solar, 2006) aim to solve the problem of wrong colour definitions under lighting changes and very close colours in this space by using ambiguous colours (or soft colours).

In this work, colours are defined in the HSV colour space which contains colour information as human visual sensation. This shows quite convenient to calibrate the colours because they are clearly separated in this colour space, since Hue channel represents the perception of the nuance, i.e. the colour. Thus, it is possible to fit these regions by using basic shapes, like prisms. In the RoboCup scenario, from a practical point of view, this implies that HSV calibration is one order faster than YUV calibration. This allows us calibrating the robots 10 minutes prior to matches. Something not many teams can do.

Therefore, in this colour space, colour regions can be bounded together only using prisms. Given that there is no problem of intersections among these prisms, calibration process is drastically simplified. We only have to define few pixels of each colour. These pixels will be used to define the prism whose internal region is the colour definition. Furthermore, the method gains robustness: we can be less restrictive on calibrating the

S and V channels, since the colour information is represented in the Hue channel.

The complexity of the YCbCr-HSV conversion is solved by using the already mentioned look-up table (LUT) of variable resolution. In our application, we use a resolution of 7 bits as trade-off between memory used and saved computational cost. This LUT is automatically filled in every calibration. As we have explained in the Vision Flowchart section, for every input of the LUT an YCbCr-HSV conversion is carried out. Then, if the HSV coordinates fall within a defined prism, the content of the corresponding input is filled with the segmented colour associated to that prism.

Blob Creation

The blob creation is undertaken by using the Run-length encoding (RLE) method. RLE is a compression technique used in images where objects are captured in flat colours, and where there are few colours at stake. (Bruce et al., 2000) uses this technique for labelling images, and our algorithm is based on it.

The RLE method treats the image as a set of lines. At every line runs of consecutive pixels in the same colour are detected, and stored by means of the first pixel and the number of pixels that make the run up. In this work the segmented image is sorted out into blobs using this procedure, but adding the concepts of neighbour runs and parent. Two runs are neighbours if they are in consecutive lines and the first one starts between the start and the end of the second one. The parent of every run is the parent of its upper neighbouring run. If there is no neighbour run above, its parent is itself.

The RLE algorithm works as follows. At the beginning, the parent of every run is set as itself. The process starts by scanning the entire image, and storing runs according to their colour. Once runs are scanned, we merge neighbours runs for every colour. This union is obtained by assigning the parent of the upper one to the parent of the bottom one. A filter is applied in order to get only one parent per set of runs. This procedure allows us to define a set of runs through its parent.

At this point we are able to create blobs. A blob is created for every run that the parent of which is itself. The rest of runs are added to the blob of its parent. If two blobs of the same colour are nearer than a gap value, they are merged. The gap is fixed for each colour, and its value is configurable. Figure 3 shows an example where we can observe the colour segmented image (a) and blobs using RLE (b). The blobs are depicted using one rectangle with a point, which represents the blob's centre that is calculated as the centre of gravity of the labelled pixels.

Corners Detection

The natural way for Naos to localise themselves is to make use of the natural landmarks present on the field, i.e. the lines. In order to do so, an algorithm to detect these lines should be implemented. Because of the LPS rules all the processing must be done on board and for practical reasons it has to be performed in real time, which prevents us from using time consuming algorithms. A typical approach for detecting straight lines in digital images is the Hough Transform and its numerous variants. These variants have been developed to try to overcome the major drawbacks of the standard method, namely, its high time complexity and large memory requirements. rUNSWift is currently relying their recognition system on edge-features, by developing a sensor model that will provide multiple hypotheses for the new localisation module. Their research is focused on foveated vision and virtual saccades to maximise the scare computational resources available on the Nao. Cmurf, in turn, uses scanline-based object detection for all objects. Nao Devil team adresses line detection by applying clustering heuristic based on positions and relative distance to generate line fragments. B-Human also scans lines in order to build regions and cluster them into features.

Figure 3. Coloured objects detection: (a) colour segmented image, (b) blobs using RLE, and (c) objects recognised after filters

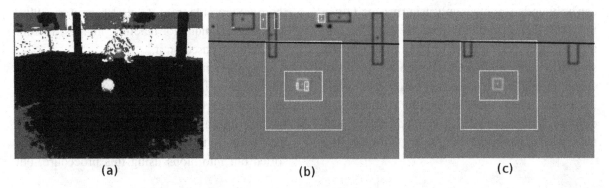

(a)   (b)   (c)

Instead of using the field lines as references for self-localization, we decided to use the corners produced by the intersection of the field lines. The main reasons for using corners are that they can be labelled depending on the type of intersection and they can be tracked more appropriately. In addition, detecting corners can be more computationally efficient than detecting lines. Also, this technique provides a robust approach in order to avoid false positives, like other robots, (which are white). As it has been outlined in the vision flowchart section, first a Sobel filter (Figure 4(b)) on the horizon plane is applied in order to extract borders. Next, the non-field transitions are ruled out by using the HSV colour space segmentation (Figure 4(c)). In particular we consider the following transition types:

- **Carpet-to-line:** Transition from carpet (green) pixels to line (white) pixels.
- **Line-to-carpet:** Transition from line (white) pixels to carpet (green) pixels.
- **Carpet-to-goal:** Transition from carpet (green) pixels to goal pixels (blue or yellow).

These labelled transitions are grouped together into sets of transitions that belong to the same straight segment. These segments are obtained using the Recursive Iterative End Point Fit Algorithm (RIEPFA) (Duda & Hart, 1973). RIEPFA groups a set of points into segments by evaluating the distance of the points to candidate end segment points, which are recursively divided into smaller segments until the fit criteria is hold (point to segment distance). Segments are further evaluated to meet a minimum number of points and maximum point-to-point distance criteria. The segments that are not rejected by the previous criteria are then labelled according to the transition pixels that originated the segment.

Now it is possible to find intersections among these segments in order to find field line corners, as Figure 4 (a) shows. The intersections are evaluated according to the segments labelling,

segments orientation, and end points proximity. Valid intersections are then labelled as following:

- **Closed-intersection:** Intersection between carpet-to-line or line-to-carpet segments, with an angle less than 180 degrees.
- **Open-intersection:** Intersection between carpet-to-line or line-to-carpet segments, with an angle greater than 180 degrees.
- **Goal-intersection:** Intersection between a carpet-to-line or line-to-carpet segment and a carpet-to-goal segment. The angle is not taken into account.

Once the intersections have been found, we group them into labelled field features, rejecting those that do not fall into any of the following categories:

- **Type C:** An open-intersection nearby of a closed-intersection. This feature can be found in the corners of the goal keeper area.
- **Type T-field:** Two closed-intersections. This feature can be found in the intersection of the field lines perimeter.
- **Type T-goal:** A closed-intersection nearby of a goal-intersection. This feature can be found in the intersection between the goal lines and the corresponding goal.

The resulting corner-features, together with the goals, are used for localizing a robot in the field. In the Fuzzy Self-Localisation section, we show how to represent the uncertainty associated to these features, and how to use these features in our fuzzy localization technique in order to obtain an estimate of the robot position.

KNOWLEDGE EXTRACTION

This section is devoted to obtain information out of the features detected in the previous section. First, the methods used to identify the ball and

Figure 4. Image from goalkeeper position: (a) sub-sampling and intersections, (b) Sobel filter of the YUV brightness channel, and (c) colour segmentation of the HSV image

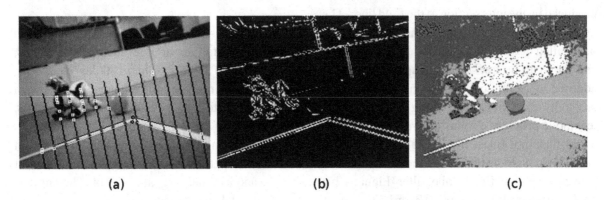

(a)   (b)   (c)

goals and to estimate their location with respect to the robot are described. Next, the fuzzy self-localisation technique developed is presented.

Coloured Objects Recognition

Here we present the methods that we apply to recognize the ball and goals. We have previously found blobs with the same colour as the objects that we are looking for. The first step is to filter out objects which are not of our interest. A horizon filter is used for the whole image and specific feature based filters for each object. Finally, we explain how distance estimations are performed.

The approach commonly adopted in the SPL at this regard is to use scan line-based object detectors, as CMurfs and NaoDevils do. B-Human is a representative example. In goal recognition, this team scans along the horizon and when a blue or yellow segment is found, a vertical scanning is performed and then a perpendicular scan along the vertical just created vector. Austin Villa applies a recursive method that merges predicted measurements with current observations. Some teams have novel approaches: HTWK tries to avoid the calibration problem, and recognises objects by applying in real time the knowledge of the shape of the objects. Upennalizers, instead, recognises objects from statistics, such as the bounding box,

the centroid location and the chord lengths in the region. Our system is designed in order to make the most of the structures created in the previous steps, like blobs and the green background.

Horizon Filter

The main filter that we set is the horizon line. This line is created by using the pixels segmented in green. In a column of the image, the green pixels are searched from top to bottom. We store the height where the first green pixel is found. This scanning is run by doing a subsampling of 5 columns step. Then, a linear regression of least squares to the stored points is performed.

In order to improve the robustness of this filter, we apply a hysteresis to the horizon line. The hysteresis raises the horizon a fixed number of pixels, in order to be conservative in the rejection of blobs. Due to the differences between ball recognition and goal recognition, different hysteresis is applied for each object. In Figure 3 the black line that crosses the picture (b) from one side to the other is the horizon line of (a) plus the hysteresis used in the goal recognition method.

As the linear regression just focuses on the top pixels of the field, it is not necessary a very accurate calibration for green colour. In addition to this, the linear fitting assures a good tolerance to noise. This filter defines the area where we should search for objects. All objects above the horizon are rejected.

Ball Recognition

The ball recognition tries to estimate the location of the ball with respect to the robot, i.e. the distance away and the horizontal angle from the local coordinate reference system.

In addition to the horizon line, two more filters are applied for the recognition of the ball to guarantee a faithful estimation of the centre of the ball. First, we check if the size of the blob is big enough. This filter allows us not to mistake noise for the ball. Secondly, a density test of orange pixels within the blob is carried out. If the blob candidate passes the test, we will be able to locate properly the centre of the ball. Figure 3 shows the recognition process for the ball: (a) depicts how the ball is seen in the segmented image, in (b) all the blobs created appear, and (c) shows the recognized blob for the ball.

Goal Recognition

The purpose of goal recognition is to estimate the distance, angle and orientation of the goal centre in relation to the robot, as it is shown in Figure 5 (a).

Goal recognition is a task more complex than ball recognition. First of all, we must decide which blobs belong to each goal. After that, it is necessary to be able to take information from the selected blobs. We could get the distance, angle and orientation if we just knew the location of the base of both goal posts. Therefore, special attention must be given to the recognition of the goal posts.

The core of this method lies in considering that each goal post is an object, and that the goal is made up of two posts. If the robot is looking for a goal, and it sees a post, it should still search for the other one. So we do not use the upper part of

posts. The cut-off point is defined by the horizon line, with the hysteresis. Thanks to this hysteresis the beginning of posts are covered, even though when the goal is located at a high distance. Our method will only work with this part of the posts. Thus, we filter out all the false positives which are in the background and simplify the recognition algorithm.

In order to make the recognition accurate this algorithm uses the previous location of the goal, along with the odometry of the robot. It is also true that we miss the chance to get information about the top part of the goal, but with this method we avoid to deal with difficult problems of filtering, such as ruling out objects in blue/yellow which are near the goal or overlapping it. Furthermore, this lack of information is not very relevant, due to the high frequency of the image process.

Goalposts Recognition algorithm works as follows: Once the bottom part of the posts has been spotted, the previous location of each goalpost is updated, taking into account the odometry. In the case of a yellow blob, a further filter is carried out: we check if that yellow blob is part of the ball, since yellow and orange are very similar. Now, all the blobs which were goal candidates have been reduced up to a maximum of two (the two goalposts) and they are merged into one blob. Then, we focus on a square of 3 pixels wide located on the centre of the resulting blob. If we find no pixel belonging to the blob inside this square, we are seeing two posts, since it is not possible that the width of a post reaches the centre of the blob, as the blob is made up of two posts. Otherwise, we are only seeing one post. If there are two, each side of the blob is scanned, and for each side, the bottom pixel must be searched. In the case of only one, we have to decide what post is, by comparing the distance between this post and the position of each goalpost calculated at the beginning. We assign the just seen post to the closer one.

Another benefit that we obtain from this algorithm is the treatment of occlusions. Since the method is able to recognize a post being alone, and the upper part of the goal is ruled out from the recognition task, we do not need the goal to be centred in the image, and avoid dealing with most of typical occlusions, such as seeing a post and half crossbar.

Figure 3 shows how this method works. From the segmented image (a), all the blobs of interest are created (b). Then, the yellow blobs next to the ball and the objects above the horizon are deleted (c). Now, only blobs of the goal and the ball remain in the picture. Thus, we reject false positives and get more accurate estimations. The robot may not know what the side of each goal post is, but this is irrelevant because the consequences of this possible error might only mean that the robot locates itself outside the field, and we already know that is not true.

Distance Estimations

Once we have located the centre of the ball or the base of each goal post, it is possible to calculate the distance from the corresponding object to the robot.

First, we study the simplest situation, like in Figure 5 (b), where legs are straight and torso is vertical. By considering the vertical and horizontal view angle and the position of the camera over the floor, the distance and the angle to the object can be estimated through simple trigonometric operations. However, the robot is usually walking, with the torso leaning, legs bent, in many possible poses. Therefore, images may come from around the axis, and camera location changes continuously.

The position of the camera is worked out by computing the kinematic transformations through all the joints involved. The image turn is solved by turn transformations of the image. The problem will be completed by taking into account the height over the floor of each object (the radius for the ball, zero for the base of the goal posts). Finally,

locating the centre of the goal and its orientation will result from the measurement of distance to each post.

Fuzzy Self-Localisation

In this section the method implemented to provide our robots with a self-localisation is exposed. First, a fuzzy notation for the location of the observed corners and the estimated pose of the robot is described. By means of this notation we develop our self-localisation system, aiming to be robust against imprecision, unreliability and ambiguity.

Fuzzy Locations

Location information may be affected by different types of uncertainty, including vagueness, imprecision, ambiguity, unreliability, and random noise. An uncertainty representation formalism to represent locational information should be able to represent all of these types of uncertainty and to account for the differences between them. Fuzzy logic techniques are attractive in this respect (Saffiotti, 1997). It can be argued that Monte Carlo approach is the most common in the Robocup. However, we find several reasons why Fuzzy systems are more convenient here: (i) only an approximate sensor model is required, (ii) several facets of location uncertainty can be represented, and (iii) ambiguities in the sensor information are directly represented, thus avoiding having to solve the data association problem separately. In addition, it is proven that Fuzzy systems deal with outliers better than stochastic approaches. Herrero et al. (2010, 2011) develop further this point.

We can represent information about the location of an object by a fuzzy subset μ of the set $\mathbf{X}$ of all possible positions (Zadeh, 1978). For instance, $\mathbf{X}$ can be a 6-D space encoding the (x, y, z) position coordinates of an object and its (θ, φ, η) orientation angles. For any $\mathbf{x} \in \mathbf{X}$, we read the value of $\mu(\mathbf{x})$ as the degree of possibility that the object is located at $\mathbf{x}$ given the available information.

Figure 5 (c) shows an example of a fuzzy location, taken in one dimension for graphical clarity. This can be read as "the object is believed to be approximately at θ, but this belief might be wrong". Note that the unreliability in belief is represented by a uniform bias b in the distribution, indicating that the object might be located at any other location. Total ignorance in particular can be represented by the fuzzy location $\mu(\mathbf{x}) = 1$ for all $\mathbf{x}\ \mathbf{X}$.

Representing the Robot Pose from Observations

Following (Buschka et al., 2000), we represent fuzzy locations in a discrete format in a position grid: a tessellation of the space in which each cell is associated with a number in [0, 1] representing the degree of possibility that the object is in that cell. In our case, we use a 3D grid to represent the robot's belief about its own pose, that is, its (x, y) position plus its orientation θ.

This 3D representation has the problem of having a high computation complexity, both in time and space. To reduce complexity, we adopt the approach proposed by Buschka et al. (2000). Instead of representing all possible orientations in the grid, we use a 2D grid to represent the (x, y) position, and associate each cell with a trapezoidal fuzzy set $\mu_{x,y} = (\theta, \Delta, \alpha, h, b)$ that represents the uncertainty in the robot's orientation. Figure 5 (c) shows this fuzzy set. The θ parameter is the centre, Δ is the width of the core, α is the slope, h is the height and b is the bias. The latter parameter is used to encode the unreliability of our belief as mentioned before.

For any given cell (x, y), $\mu_{x,y}$ can be seen as a compact representation of a possibility distribution over the cells $\{(x, y, \theta) \mid \theta\ [-\pi, \pi]\}$ of a full 3D grid. The reduction in complexity is about two orders of magnitude with respect to a full 3D representation (assuming an angular resolution of one degree). The price to pay is the inability to handle multiple orientation hypotheses on the same (x, y) position - but we can still represent multiple hypotheses about different positions. In our domain, this restriction is acceptable.

An important aspect of our approach is the way to represent the uncertainty of observations. Suppose that the robot observes a given feature at time t. The observed range and bearing to the feature is represented by a vector. Knowing the position of the feature in the map, this observation induces in the robot a belief about its own

Figure 5. Notation used in measurements: (a) target values for the goal distance estimation, (b) optimal situation when estimating distance to the goal, and (c) fuzzy set representation of an angle measurement θ

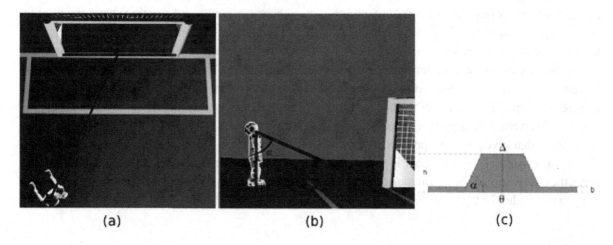

(a)   (b)   (c)

Figure 6. Distance measurements: (a) error measurements in ball detection, and (b) error measurements in goal detection

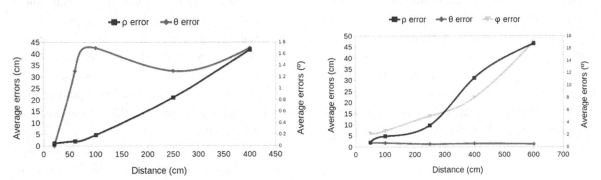

position in the environment. This belief will be affected by uncertainty, since there is uncertainty in the observation.

In our domain, we consider three main facets of uncertainty. First, imprecision in the measurement, i.e. the dispersion of the estimated values inside an interval that includes the true value. Imprecision cannot be avoided since we start from discrete data (the camera image) with limited resolution. Second, unreliability, i.e. the possibility of outliers. False measurements can rise from a false identification of the feature, or from a mislabelling. Third, ambiguity, i.e. the inability to assign a unique identity to the observed feature as features (e.g., corners) are not unique. Ambiguity in observation leads to a multi-modal distribution for the robot's position.

All these facets of uncertainty can be represented using fuzzy locations. For every type of feature, we represent the belief induced a time t by an observation by a possibility distribution $S_t(x,y,\theta \mid)$ that gives, for any pose (x, y, θ), the degree of possibility that the robot is at (x, y, θ) given the observation. This distribution constitutes our sensor model for that specific feature.

The shape of the distribution $S_t(x, y, \theta \mid)$ depends on the type of feature. In the case of goal observations, this distribution is a circle of radius in the (x, y) plane, blurred according to the amount of uncertainty in the range estimate. An example of this belief distribution is depicted in Figure 7 (a). We only show the (x, y) projection of the possibility distributions for graphical clarity.

Note that each circle has a roughly trapezoidal section. The top of the trapezoid (core) identifies those values which are fully possible. Any one of these values could equally be the real one given the inherent imprecision of the observation. The base of the trapezoid (support) identifies the area where we could still possibly have meaningful values, i.e. values outside this area are impossible given the observation. In order to account for unreliability, then, we include a small uniform bias, representing the degree of possibility that the robot is "somewhere else" with respect to the measurement.

As Figure 7 (a) shows, the $S_t(x, y, \theta \mid)$ distribution induced by a corner-feature observation is the union of several circles, each centred on a feature in the map, since our simple feature detector does not give us a unique ID for corners. It should be noted that the ability to handle ambiguity in a simple way is a distinct advantage of our representation. This means that we do not need to deal separately with the data association problem, but this is automatically incorporated in the fusion process. Data association is one of the unsolved problems in most current self-localization techniques, and one of the most current reasons for failures.

Fuzzy Self-Localisation

As it is said, in order to model the world Monte Carlo particle filter variants are the most popular amongst Robocup teams. However, a few of them are carrying out new approaches with good results. rUNSWift team is experimenting with multi-robot-multi-modal Kalman filtering. Upennalizers team addresses this issue with a pose estimation algorithm that incorporates a hybrid Rao-Blackwellized respresentation that reduces computational time, while still providing a high level of accuracy. Nao Devils Team, instead, make use of Bayesian filters in combination with concepts based on multi robot SLAM.

Our approach to feature-based self-localization extends the one proposed by Buschka et al. (2000), who relied on unique artificial landmarks. Buschka's approach combines ideas from the Mar-

kov localization approach proposed by (Burgard, Fox, Hening & Schmidt, 1996) with ideas from the fuzzy landmark-based approach technique proposed by (Saffiotti & Wesley, 1996).

The robot's belief about its own pose is represented by a distribution G_t on a 2½ D possibility grid as described in the previous section. This representation allows us to represent, and track, multiple possible positions where the robot might be. When the robot is first placed on the field, G_0 is set to 1 everywhere to represent total ignorance. This belief is then updated according to the typical predict-observe-update cycle of recursive state estimators as follows.

- **Predict:** When the robot moves, the belief state G_{t-1} is updated to G_t using a model of the robot's motion. This model performs a translation and rotation of the G_{t-1} distribution according to the amount of motion, followed by a uniform blurring to account for uncertainty in the estimate of the actual motion.

- **Observe:** The observation of a feature at time t is converted to a possibility distribution S_t on the 2½ grid using the sensor model discussed above. For each pose (x, y, θ), this distribution measures the possi-

Figure 7. Self localisation results: (a) belief induced by the observation of a type T-field feature, (b) belief on the goal location of the goalkeeper, and (c) position error of this belief

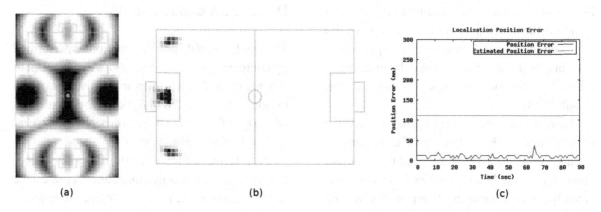

(a)   (b)   (c)

bility of the robot being at that pose given the observation.

- **Update:** The possibility distribution S_t generated by each observation at time t is used to update the belief state G_t by performing a fuzzy intersection with the current distribution in the grid at time t. The resulting distribution is then normalized.

If the robot needs to know the most likely position estimate at time t, it does so by computing the centre of gravity of the distribution G_t. A reliability value for this estimate is also computed, based on the area of the region of G_t with highest possibility and on the minimum bias in the grid cells. This reliability value is used, for instance, to decide to engage in active re-localization behaviour. In practice, the predict phase is performed using tools from fuzzy image processing, like fuzzy mathematical morphology, to translate, rotate and blur the possibility distribution in the grid (Burgard et al., 1996). The intuition behind this is to see the fuzzy position grid as a grey-scale image.

For the update phase, we update the position grid by performing point-wise intersection of the current state G_t with the observation possibility distribution $S_t(\cdot|r)$ at each cell (x, y) of the position grid. For each cell, this intersection is performed by intersecting the trapezoid in that cell with the corresponding trapezoid generated for that cell by the observation. This process is repeated for all available observations. Intersection between trapezoids, however, is not necessarily a trapezoid. For this reason, in our implementation we actually compute the outer trapezoidal envelope of the intersection. This is a conservative approximation, in that it may over-estimate the uncertainty but it does not incur the risk of ruling out true possibilities.

There are several choices for the intersection operator used in the update phase, depending on the independence assumptions that we can make about the items being combined. In our case, since the observations are independent, we use

the product operator which reinforces the effect of consonant observations.

EXPERIMENTAL RESULTS

In order to validate the efficiency and robustness of our vision system, three types of experiments have been conducted: the time processing, the quality of the measurements of distance and our self-localisation method.

Time Processing

A critical factor is the capability to process an image as fast as possible. Our challenge is to process images in less than the period of execution of PAM (40 ms). Time is considered in the three possible resolutions given by the Nao camera: VGA (640 * 480), QVGA (320 *240) and QQVGA (160 * 120). Table 1 shows the resulting processing times using VIM and our image acquisition system in these resolutions.

Capture's time in VIM is so long due to the conversion performed within this stage. As table 1 shows, the benefits of using our image acquisition system are considerable and increase with the resolution, apart from the independence from *Naoqi* obtained. As the process using V4L2 in QVGA resolution takes less than the period of the thread of Vision, this is the resolution that we have worked with.

Distance Measurements

The validation of distance estimations is performed by comparing the provided values by the robot to the true ones. These values are ρ (distance) and θ (angle) for the ball. For the goal, φ (orientation) is added (see Figure 5 (a)). The respective measurements are depicted in Figure 6.

According to Figure 6 (a), in short distances the distance measurements are quite accurate. Due to limitations as the non-linearity of the trigono-

Table 1. Comparison of processing times using VIM and V4L2-based acquisition system

Processing Times (ms)	VGA		QVGA		QQVGA	
	VIM	V4L2	VIM	V4L2	VIM	V4L2
Image capture	46	0.04	11	0.04	9	0.04
Segmentation	79	71.5	18.1	18.1	4.3	8.7
Blob forming	9.1	9.4	2.4	2.6	0.7	0.7
Recognition	0.3	0.3	0.2	0.2	0.1	0.1
TOTAL	134	81	32	21	14	10

metric operations, the rough resolution and the gives of joints, the error increases with the distance. However, the error in angle measurements remains small among all the distances, thanks to the less dependence on hardware for this value. The calibration in HSV colour space guarantees the ball recognition even when it is far away, in distances such as 400 cm.

Figure 6 (b) shows similar results to Figure 6 (a). Since two locations have to be estimated (the base of both goalpost), errors in ρ are accumulated, and grow faster with the distance to the goal. As the case of the ball, errors in θ remain small with the distance. Errors in φ estimation are given, mainly, by the influence of the rough resolution, which distorts the measurements in high distances.

It is remarkable the good results obtained in short distances (50 cm, 100 cm), where the location of the goal as a whole is determined as a combination of two estimations, one for each post. The measurements have been taken through estimations with and without occlusions, in order to simulate real situations.

Self-Localisation

In order to validate the field feature detection process described in this chapter, a localisation experiment has been performed. For convenience, the platform used for this experiment is the robot AIBO, the former standard platform in Robocup. In this experiment the robot has been placed in the goal area, facing more or less to the opposite

goal. This is the typical goalkeeper position. For a goalkeeper localisation is critical because many behaviours depend on the absolute position. If localisation fails, the robot would start wandering around, possibly leaving the goal clear to the other team. The localisation process is initialised with a belief distributed along the whole field, i.e. it does not know its own location. Then the robot starts scanning its surroundings by moving its head from left to right. As soon as a feature is perceived, it is incorporated into the localisation process. The experiment corresponds to a RoboCup scenario, in which the goals and the field lines are the only perceptual sources.

In the experiments we compare the estimated position with the real position. In order to measure the real robot position we use an external vision system. This is composed of an overhead camera with a wide angle lens mounted on an aluminium structure at 2.5 metres over the floor. The robot wears a coloured mark which allows computing both the position and orientation of the robot. Because of the high distortion of the lens, the accuracy of the position is of ±2 centimetres and ±5 degrees.

The experiment starts with the belief distributed along the whole field and the robots are left for some seconds scanning the surroundings. The resulting belief is shown in Figure 7 (b). The highlighted box placed in front of the goal corresponds to the estimate of the uncertainty of the robot's position (obtained by finding the bounding box of the highest possibility area) and the arrow corresponds to the estimate of the robot's position (obtained by defuzzification of the fuzzy belief with the centre of gravity).

The reduction of the uncertainty conditions very much the accuracy of the localisation process, that is, less uncertainty usually leads to better accuracy. Figure 7 (c) shows the absolute position error over time. The absolute error is computed taking the difference between the estimated position (on-board) and the measured position (overhead camera). The estimated absolute error is computed from the uncertainty of the belief distribution and the estimated position (on-board). Because the belief distribution is a grid, the minimum estimated absolute error corresponds to the tessellation size, which corresponds to 100 mm in these experiments.

CONCLUSION

In this chapter we have described the vision system used by *Los Hidalgos* team for RoboCup, in particular the official field of the Standard Platform League (SPL), and how it fits within the whole system developed on our Naos, which happens to be an instance of the ThinkingCap architecture.

The image processing is executed every 40 ms in QVGA resolution, due to hard real time constraints. In order to meet this deadline, a V4L2 driver is used. Segmentation and blob-forming methods are approached in robustness and efficiency terms. The more suitable algorithms among those presents in the state of art have been chosen for designing the full vision system.

A key point of the presented approach is the colour segmentation step in HSV, because it simplifies drastically the calibration process and provides more robustness under slight changes in lighting conditions. This is an important point in robot competitions and demonstrations setups. For example, in RomeCup 2010 our team was able to complete a calibration in 10 minutes on average, while other teams spent much more time.

An object recognition method has been presented. This method has been designed so that two purposes are fulfilled: to determine where the ball and goals are, regardless of the location of the robot on the field, and to filter out possible false positives in the background.

The typical approach for self-localisation in SPL is the recognition of field lines. However, in this domain motion estimates are highly unreliable, observations are uncertain, accurate sensor models are not available, and real time operation is of essence. The described technique in this chapter is based on the recognition of field line intersections, keeping into main the constraints of the SPL: on-board processing and the resolution of the camera. The computational burden of the process is low. In addition, most of the efforts of this technique are in the direction of avoiding false positives, which can lead to wrong localisation.

In order to validate the usability of the proposed vision process, the detected features are incorporated in a localisation filter. Without loss of generality, we use a fuzzy-Markov grid in which we have modelled the perception and its associated uncertainty of different types of field line intersections. This localisation filter provides an effective solution to the problem of localisation. We have presented an experiment to show the performance of the localisation process.

Our method needs unique goals to cope with the natural symmetry of the field. The standard field might move from unique goals to equally coloured-goals. Once the use of field lines and unique goals localisation is common in the leagues, there will be necessary any other means to break with the ambiguity, which are left as future work.

REFERENCES

Bruce, J., Balch, T., & Veloso, M. (2000). Fast and inexpensive color image segmentation for interactive robots. *Proceedings of the IEEE International Conference on Intelligent Robots and Systems* (pp. 2061-2066). Takamatsu, Japan.

Burgard, W., Fox, D., Hennig, D., & Schmidt, T. (1996). Estimating the absolute position of a mobile robot using position probability grids. In *Proceedings of the National Conference on Artificial Intelligence*.

Buschka, P., Saffiotti, A., & Wasik, Z. (2000). Fuzzy landmark-based localization for a legged robot. In *Intelligent Robots and Systems* (pp. 1205-1210). Takamatsu, Japan.

Duda, R., & Hart, P. (1973). *Classification and scene analysis*. John Wiley and Sons.

Herrero-Pérez, D., & Martínez-Barberá, H. (2011). Fuzzy mobile-robot positioning in intelligent spaces using wireless sensor networks. *Sensors (Basel, Switzerland)*, *11*, 10820–10839. doi:10.3390/s111110820

Herrero-Pérez, D., Martínez-Barberá, H., LeBlanc, K., & Saffiotti, A. (2010). Fuzzy uncertainty modeling for grid based localization of mobile robots. *International Journal of Approximate Reasoning*, *51*, 912–932. doi:10.1016/j.ijar.2010.06.001

Herrero-Pérez, D., Martínez-Barberá, H., & Saffiotti, A. (2004). Fuzzy self-localization using natural features in the four-legged league. In *Proceedings of the International RoboCup Symposium*, Lisbon, Portugal.

Martínez-Barberá, H., & Saffiotti, A. (2000). *ThinkingCap-II architecture*. Retrieved November 20, 2011, from http://robolab.dif.um.es/tc2/

Palma-Amestoy, R., Guerrero, P., Vallejos, P., & Ruiz-del-Solar, J. (2006). *Context-dependent color segmentation for Aibo robots*. In 3rd IEEE Latin American Robotics Symposium, Santiago de Chile, Chile.

Palmer, P. L., Dabis, H., & Kittler, J. (1996). Performance measure for boundary detection algorithms. *Computer Vision and Image Understanding*, *63*(3), 476–494. doi:10.1006/cviu.1996.0036

Saffiotti, A. (1997). The uses of fuzzy logic in autonomous robot navigation. *Soft Computing*, *1*(4), 180–197. doi:10.1007/s005000050020

Saffiotti, A., & LeBlanc, K. (2000). Active perceptual anchoring of robot behavior in a dynamic environment. *In Proceedings of the IEEE International Conference on Robotics and Automation* (pp. 3796-3802). San Francisco, USA.

Saffiotti, A., & Wesley, L. P. (1996). Perception-based self-localization using fuzzy locations. In *Proceedings of the 1st Workshop on Reasoning with Uncertainty in Robotics* (pp. 368-385). Amsterdam, NL.

Sahoo, P. K., Soltani, S., Wong, A. K. C., & Chen, Y. C. (1988). A survey of thresholding techniques. *Computer Vision Graphics and Image Processing*, *41*(2), 233–260. doi:10.1016/0734-189X(88)90022-9

Sonka, M., Hlavac, V., & Boyle, R. (1996). *Image processing analysis and machine vision*. International Thomson Computer Press.

Wasik, Z., & Safiotti, A. (2002). Robust color segmentation for the RoboCup domain. *Proceedings of the International Conference on Pattern Recognition*, Quebec City, Quebec, CA.

Zadeh, L. A. (1978). Fuzzy sets as a basis for a theory of possibility. *Fuzzy Sets and Systems*, *1*, 3–28. doi:10.1016/0165-0114(78)90029-5

Chapter 19
Visual Control of an Autonomous Indoor Robotic Blimp

L. M. Alkurdi
University of Edinburgh, UK

R. B. Fisher
University of Edinburgh, UK

ABSTRACT

The problem of visual control of an autonomous indoor blimp is investigated in this chapter. Autonomous aerial vehicles have been an attractive platform for a wide range of applications, especially since they don't have the terrain limitations the autonomous ground vehicles face. They have been used for advertisements, terrain mapping, surveillance, and environmental research. Blimps are a special kind of autonomous aerial vehicles; they are wingless and have the ability to hover. This makes them overcome the maneuverability constraints winged aerial vehicles and helicopters suffer from. The authors' blimp platform also provides an exciting platform for the application and testing of control algorithms. This is because blimps are notorious for the uncertainties within their mathematical model and their susceptibility for environmental disturbances such as wind gusts. The authors have successfully applied visual control by using a fuzzy logic controller on the robotic blimp to achieve autonomous waypoint tracking.

INTRODUCTION

A blimp is a special kind of lighter-than-air air-ship; it does not have a rigid skeleton supporting its balloon. Blimp and airship automation has recently emerged as an attractive field of research due to their properties.

Unmanned aerial vehicles in general have advantages over unmanned ground vehicles. They are able to reach locations where it is hard for ground vehicles to reach due to hazards or terrain limitations. They also have the advantage of a larger field of view making them able to survey and collect data of a larger area of terrain at a given instance. Unmanned aerial vehicles are also faster and have better maneuverability.

Blimps also have advantages over winged unmanned aerial vehicles and helicopters. Blimps

DOI: 10.4018/978-1-4666-2672-0.ch019

have much safer failure degradation. They are able to hover over one area for a long time, achieve low altitude flights and do not suffer from maneuverability constraints. They also have minimal vibration and do not influence the environment they are in. The properties previously mentioned make them ideal for data collection, exploration, monitoring and research applications. They take off and land vertically. This means that they can be easily deployed with no need for a runway, which makes them attractive as platforms for rescue operations or as communication beacons when communication is cut-off from a certain area. Other attractive properties include long flight durations and low energy consumption as they depend on buoyancy to achieve vertical position. The blimp's relatively slow speed makes it also an attractive platform for computationally expensive algorithms that need many state updates such as simultaneous localization and mapping (SLAM).

Blimps have been studied as a viable platform for rapidly deployable communication beacons (Flahpour et al., 2009), advertisements and atmospheric data collection and analysis. They are also attractive for military operations such as surveillance and rapid equipment deployment. Blimps serve as an option for providing images and information about regions which have suffered natural catastrophes. Map building and localization of targets have also been studied through the work of LAAS/CNRS (Hygounenc, Soueres, & Lacroix, 2004). Astro-explorations are also an application studied by the Jet Propulsion laboratory at NASA (Kampke & Elfes, 2003).

Blimp Used

The Surveyor blimp "YARB" (Yet Another Robotic Blimp), which is a 66" helium blimp, was employed in this project. This robotic blimp is driven by three motors, two propellers and a third vectoring motor. The onboard electronics include a Blackfin processor, color camera and a Matchport

wireless LAN interface. A network camera fixed to the ceiling provides the images for the image processing algorithms. The processing is done on a laptop, and the motor commands are sent to the blimp via wireless LAN interface. Testing was performed on a Toshiba satellite pro laptop with i5-520 CPU with 4GB DDR3 SDRAM.

The blimp is 1.68m long and has a maximum diameter of 0.76m giving it a fineness ratio (length/diameter) of 2.2. It has a volume of 0.26m^3 and a total lift capacity of 0.3kg given that the lighter than air gas used is helium. While hydrogen is a cheaper alternative that provides more lift capacity for the same volume, helium remains the safer choice.

The blimp platform under study has a few drawbacks making its control rather challenging. The most challenging aspects of the control problem are modeling the dynamics of the blimp and accounting for uncertainties. Examples of uncertainties include disturbances in the form of temperature and pressure variations that could vary the size of the blimp's envelope and vary the buoyancy, or disturbances such as wind gusts. Another problem faced in this project is that the blimp's envelope leaks helium varying its buoyancy from one test run to the other. Airship dynamics are also notoriously hard to control due to large moment of inertia (Khoury & Gillett, 2002). Furthermore, the blimp's lack of an internal rigid frame structure makes its envelope susceptible to expansion and contraction due to acceleration, pressure and temperature variations, adding uncertainty to the blimp's dynamic model. Signal latency has also been observed in our platform as well as delay in control signals.

Therefore, the blimp is indeed a hard platform to control; and just like any controller, for successful operation, an input of positional state is essential. The states we are interested in regarding blimp control are: position in three dimensional space, vertical and horizontal velocities, angular position and finally angular velocity. The space the blimp flies in is shown in Figure 1. In practice this represents the space inside the Informatics Forum at the Edinburgh University.

Ideally, a global positioning system (GPS) would be the most suitable choice for the inputs. However; because the blimp operates indoors, the GPS signal might not be accurate enough or not available. Sensors in the form of accelerometers and inertial navigation systems are also not available to this platform as blimps have the disadvantage of limited payload. Payload, when speaking about blimps, is a function of envelope size, and with small blimps limited payload means limited amount of sensors the blimp can be equipped with. This means that the variety of information that the blimp's controller can be fed is limited.

As such, we employ a ceiling camera looking down on the blimp's operation area to guide the blimp and provide positional state. Figure 2.a shows the image as seen from the ceiling camera and the virtual waypoints and the line that blimp should ideally follow. Figure 2.a also shows sources of wind gust that influence the behavior of the blimp. They are labeled as doors, stairs and elevators. The setup of the camera system is shown in Figure 2.b.

Following the analysis of each image frame captured from the ceiling camera, the absolute parameters of position in 3D space, linear velocity, orientation and angular velocity are computed. They are transformed into relative parameters by calculating the deviation from preset values. The relative positions represent error signals and can be then fed into the fuzzy control algorithm to obtain motor output.

In this chapter we discuss how we have successfully applied a vision system to extract and locate the robotic blimp. Visual servoing was applied to the blimp by way of a fuzzy logic controller to achieve indoor autonomous waypoint tracking. The vision system as well as the fuzzy logic controller were successful at achieving the objective of indoor waypoint controller and proved to be robust against environmental disturbances. It is worth noting that given the described setup, this system is intended for indoor applications; especially that the blimp's size cannot handle outdoor wind as well as the need of a ceiling camera for the localization and control of the blimp in this current stage. The blimp has an onboard camera that future stages of research will employ for localization and control.

BACKGROUND

This section introduces previous research done on airships and blimps. The first subsection mentions the major airship platforms and the control

Figure 1. The space blimp flies in

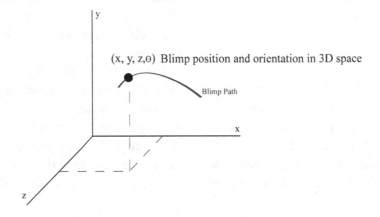

Figure 2. a) An image as seen from the ceiling camera and the virtual waypoints as well as the sources of error; b) the setup of the camera system

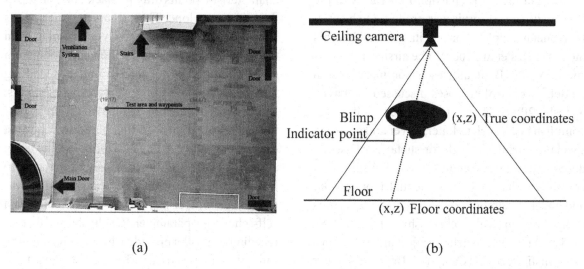

(a)   (b)

algorithms used. The second subsection introduces vision algorithms applied for blimp and airship control. The third subsection focuses on fuzzy logic control schemes applied on airships and blimps.

Airships in the Literature

This subsection aims to provide a summary of major airship platforms and discusses the control aspects used in each of these projects. Reviews on airship platforms as well as other UAV can be found in (Avenant, 2010), (Liu, Pan, Stirling, & Naghdy, 2009) and (Ollero & Merino, 2004).

The University of Stuttgart's project "Lotte" is a 15m airship, with a volume of 107m³ and has a maximum payload of 12kgs. It has been a platform for many research projects such as aerodynamic research (Lutz, Funk, Jakobi, & Wagner, 2002). The dynamical characteristics of the ship have been modeled using system identification techniques (Kornienko, 2006). The control relies on a number of sensor inputs such as GPS (global positioning system) information for position tracking as well as electronic compasses. The accelerations are calculated using inertial measuring units and the helium temperature and pressure is calculated

and compensated against. A full description of the sensors used in this project can be found in (Kungl, Schlenker, Wimmer, & Kröplin, 2004) and a discussion on the controllers used is given in (Wimmer et al., 2002).

The LAAS-CNRS airship "Karma" is 8m, has a volume of 15m³ and a maximum payload of 3.5kg (Hygounenc et al., 2004). This platform had been developed for high resolution terrain mapping and the controllers are built to execute planned trajectories by using the blimp sensor input and detecting special ground elements (Lacroix, Soueres, Hygounenc, & Berry, 2003). Positioning is done through vision, where two successive frames are analyzed to get a position update. The controller assumes decoupling between longitudinal and lateral planes. Once the airship achieves the desired longitudinal position the lateral controller starts to achieve path following. The airship's control algorithm involves geometric and dynamic models whose constraints are taken into account by employing backstepping techniques (Hygounenc & Soueres, 2003). This airship platform has been used to apply SLAM (simultaneous localization and mapping) techniques successfully as discussed in (Hygounenc et al., 2004).

The Titan Aerobot project developed at the NASA and the Jet Propulsion Laboratory at the University of California was proposed to be used for planetary exploration on Titan, one of Saturn's moons (Elfes et al., 2005). The airship is an Airspeed AS-800B, it utilizes a nonlinear airship model for control purposes discussed in (Payne & Joshi, 2004). The controllers were built to accomplish tasks such as loiter, hover and cruise. A special controller is built for subtasks of ascent, descent, turning and altitude control. A full list of controllers that include sequential-loop-closure and linear-quadratic-regulator control algorithms is discussed in (Kulczycki, Joshi, & Hess, 2006).

The AURORA Airship project at the Autonomous Institute of CTI Campinas, Brazil, is another important airship platform that has been used for environmental monitoring missions, investigations of airship dynamic models and visual servoed guidance (Azinheira et al., 2002). The control of this airship is discussed in (Ramos et al., 2001) where the airship makes use of a proportional integral (PI) controller and a proportional derivative (PD) controller to follow a path trajectory by outputting a heading angle. The problem of hover control has been also been investigated using this platform and is discussed in (Azinheira, Ramosb, & Buenob, 2000) where image processing is used to provide an offset from the desired position which then is input into the controller to account for the positional deviation.

Blimp Vision Systems

This subsection introduces work done on vision systems used for blimp feedback and control. The vision systems introduced cover both ceiling cameras as well as on board cameras.

Visual feedback control using receding horizon control, also known also model predictive control, applied to unmanned planar blimp system is presented in (Kawai, Hirano, Azuma, & Fujita, 2004). The camera is fixed on a ceiling looking down on

the test area and blimp. The camera identifies two characteristic points drawn (black dots) on top of the blimp envelope. These characteristics points have different sizes and can always be seen from the ceiling camera. The position and orientation of the blimp can be deduced by image processing techniques applied to the characteristics points. Visual feedback control using two different controllers were applied on this platform. The first controller was a receding horizon controller using a control Lyapunov function as terminal cost to stabilize the system. The second controller was a linear parameter using carrying system with a self-scheduling parameter. Results showed that the receding controller provided the better results. The same platform was used in (Kawai, Kitagawa, Izoe, & Fujita, 2003) for PD (proportional derivative) control. A dynamical model is derived. Feedback linearization techniques are applied to achieve a visual linearizing feedback PD controller.

In (van der Zwaan, Bernardino, & Santos-Victor, 2000) an onboard vision system is applied on a small-sized, indoor blimp. Visual control is applied to this platform to achieve station keeping and docking; the main objective here is to maintain the blimp at a certain 3D location and orientation in reference to a specific landmark. Information about the blimp's pose and location is extracted through the vision system; this information is then used to attain visual servoing. The visual algorithm applied makes use of an initial image patch set by the user. Subsequent blimp movements would then distort this reference image. Image registration is then applied through minimization of the error function (sum-of-squared-differences) to calculate optical flow. The measurements obtained by the image processing are then passed to PD controllers so as to achieve the objective of station keeping and docking. The PD parameters were set experimentally. The height of the blimp was maintained by comparing the initial's patch area and the current tracked patch area. Results showed success in tracking the object and achieving station keeping and docking. The system had difficulties with lateral movements of the blimp due to air currents, as this blimp does not have lateral degrees of freedom to control. This required using rotation of the blimp which introduced oscillatory behavior because the rotation of the blimp is not performed around the camera optical axis.

Visual navigation of robotic airships has been investigated in (Xie, Luo, Rao, & Gong, 2007). An onboard camera extracts and tracks natural landmarks (buildings in a city) and uses them as visual beacons for localization and control. The system makes use of geographical information systems (GIS) to extract the geometric information of these extracted visual beacons. An algorithm is then applied to obtain orientation and position of the airship by comparing the given geometric data of the beacons (by the GIS) with those extracted by the vision system. Visual feedback is then passed to an optimal fuzzy flight control system to keep the airship on a predefined track. Genetic algorithms were applied for the optimization of the controller.

In (Azinheira et al., 2002) visual servo control of a hovering outdoor robotic airship is discussed. An onboard camera is used to identify a circle on the ground and a ball floating above it. This represents the hovering location. The circle-ball configuration was used as it has an interesting property of decoupling longitudinal and lateral dynamics which simplifies the design of the controller. Once these landmarks are extracted, their properties are transformed into visual signals. An image Jacobian is built using these visual signals and used for visual servoing. The algorithm includes airship dynamics and uses optimal control design to obtain motor command signals. Experiments have been set up to test environments of no wind disturbances, slight wind gusts and finally strong winds and gusts. The results show that stabilization of the airship was maintained by the optimal controller even with the existence of harsh wind conditions.

Blimp Control

This subsection introduces major work done on proportional integral derivative (PID) and fuzzy controllers applied to blimps. While the main focus will be on projects relating to fuzzy controllers, we will mention ongoing work in the form of intelligent controllers such as model predictive controllers and reinforcement learning controllers that have shown very good results when applied to blimps.

Acquiring the blimp dynamical model is a first step of studying controller design. A general dynamical model for blimps is presented in (Gomes & Ramos, 2008). In this work, a platform for controller design and simulation research is presented through a complete physical and dynamical model of the blimp.

Classical control methods in the form of PID control have the advantage of simple implementation and reliability, however it can be computationally expensive to model the system and tune its parameters. Work on a PD controller can be seen in (Azinheira et al., 2002). This project employed a dynamical model controlled by a PD error controller that gets feedback from an onboard camera that sends feedback signals to the controller. PID control has been also been applied to landing of a blimp by Toshihiko Takaya in (Tayaka, Minagawa, Yamamoto, & Ohuchi, 2006), using orbital control. Another platform is presented in (Hygounenc et al., 2003) where a PID controller is used for altitude and horizontal positions of the blimp.

In the work of Falahpour et al. in (Falahpour et al., 2009), a fuzzy logic controller was compared with a PID controller. System model and dynamics were derived in order to apply PID control. The model accounted for air friction and random wind gusts. The PID controller was able to achieve the desired position and could cope with gusts of wind of varying direction and force. The fuzzy controller used three membership functions for the inputs and five membership functions for the outputs. There are four error inputs (plane position, orientation and angular speed) giving eighty one control rules. The defuzzification method used was the weighted average method. Results from the comparison showed better performance with the fuzzy controller in terms of less oscillation and faster convergence speed. This result was obtained under MATLAB simulation of the second order balloon dynamic model.

Gonzalez et al. in (Gonzalez, Burgard, Sanz, & Fernandez, 2009) applied a fuzzy altitude controller on a low-cost autonomous indoor blimp and compared it to PID control. Vertical and horizontal controllers were decoupled much like our system. The system also employs a fuzzy collision avoidance controller where a PID collision avoidance controller failed to provide satisfactory results. PID altitude control parameters were experimentally calculated using the Zieger-Nichols method and showed good results in stable, undisturbed environments but showed large oscillations in environments with disturbances. The fuzzy logic controller has two inputs of velocity and vertical position error and employed five membership functions for each input, actuation output had nine membership functions. The Fuzzy controller out-performed the PID controller in practical tests especially in environments with wind disturbances. The same platform had a fuzzy collision controller that uses five membership functions for velocity and three membership functions for positional error. The fuzzy controller showed the desired behavior while the PID had oscillatory behavior and was judged to be inadequate.

In (Jian-guo & Jun, 2008) altitude control of an autonomous airship is investigated with the use of fuzzy logic. Seven triangular membership functions were used for the positional error as well as for the speed of the blimp. The controller has two fuzzy logic subsystems, one that calculates the current error and the other calculates the predicted error. Each one is activated depending on the blimp's altitude. This compound controller was designed in this way to be robust to disturbances.

The results showed that the blimp was able to achieve and maintain the required altitude as well as being robust to parametric perturbations and disturbances.

Backstepping control, model-predictive control and reinforcement learning control of autonomous blimp navigation are prominent control methodologies currently receiving much attention. Important reviews and introductory material to the field for control of autonomous airships is given in (Liu et al., 2009) and (Ollero & Merino, 2004).

BLIMP EXTRACTION

In this section we discuss the methods investigated to detect the pixels that belong to the blimp object. These pixels are an important region of interest (ROI) for the extraction of positional states. Our setup has the advantage of a fixed camera with a fixed viewpoint. Ideally, we would like to subtract the current frame from the fixed viewpoint to detect changes in the blimp's position. However; several problems have to be dealt with. These problems come in the form of dynamically changing ambient light, varying sun positions, varying sunlight intensity, the cloudy nature of the city of Edinburgh, shadows and finally dynamically changing reflections and glares that are produced on the background. Other problems the detector has to deal with are external objects being introduced in to the background such as people, furniture and stationary objects. More serious problems come in the form of white tables that can be easily be mistaken for the blimp.

Two detection methods will be discussed in this section. First we will investigate detection with background subtraction using principle component analysis (PCA) techniques as discussed in (Majecka, 2009). The second method involves using colored thresholding as discussed in (Ntelidakis, 2010).

Blimp Detection: Background Modeling Using PCA

In (Majecka, 2009), different methods were used to model the background image seen from the ceiling camera; these methods are discussed in this subsection.

The first method of modeling the background image was achieved by calculating the mean of fifty images that were obtained in different lighting conditions.

The second method applied the idea of constantly updating the mean background model with the most recently captured image. A weighted sum is used to combine these two images. This idea proved to be robust to changes in overall lighting conditions. However; as this method relies on weights, it can be difficult to optimally fine tune, and achieve the correct results.

The third technique used chromaticity coordinates to represent the mean image. This image would then be subtracted from a current image also represented by its chromaticity coordinates. A dynamically chosen threshold was then applied to detect the blimp. This technique proved robust against reflections, however; it failed in the case of shadows.

Finally background modeling using PCA techniques was applied. This method has been most successful in extracting objects from the background. It is discussed with more detail in the following paragraphs.

PCA was used to model the background image as seen from the ceiling camera in the Informatics Forum. Fifty images were obtained in different lighting conditions. This image-set contained images with glares, light reflections, shadows and all major disturbances that can occur in the background image, examples are shown in Figure 3.a and 3.b. A background model was produced using PCA from these fifty images. The background model is composed of the mean image as well as the eigenvectors that represent the significant principal components (directions of the most

variation in the image), these were retrieved by way of eigendecomposition. To extract newly introduced objects to the background in any new image, this new image is projected onto the space characterized by the most significant eigenvectors and then projected into the original image space. This projected image is then compared to the original image. Foreground objects resulting from the previous operation would represent objects alien to the background. They can be extracted using binary thresholding.

To extract the blimp, we only used the first eigenimage F of the Principle Component Model and the mean image seen by the ceiling camera M. The projection of the current image C onto the first eignimage would then be $P = (C - M) * F$. This projection is then used to estimate the background by the following equation $B = P*F + M$. Detection can then be done by applying $O = abs$ $(C - B)$. Finally, a binary threshold is applied on O to extract the blimp blob.

The blob (blimp) is next used as a mask on the current frame to obtain a ROI image of the blimp itself. Next, a new histogram is built of this ROI to find a threshold to extract the indicator point (black marker) on the white blimp. That marker is then used to obtain the orientation of the blimp by applying the *arctan* function to the center of gravity of the blimp and the indicator point's center of gravity.

The steps of the algorithm are presented next:

- Obtain grayscale image of the current frame.
- Apply the PCA method on the current to obtain the difference image (DI).
- Build the DI's histogram.
- Apply smoothing to the histogram and find the optimum threshold value (valley between two peaks).
- Threshold the DI image to obtain a binary image.
- Apply morphological open operation to remove noise and small artifacts.

Figure 3. a) Reflections on the background image; b) shadows and lighting variations on the background image; c) mean background image; d) first principal component (figure copied from (Majecka, 2009) by permission of author)

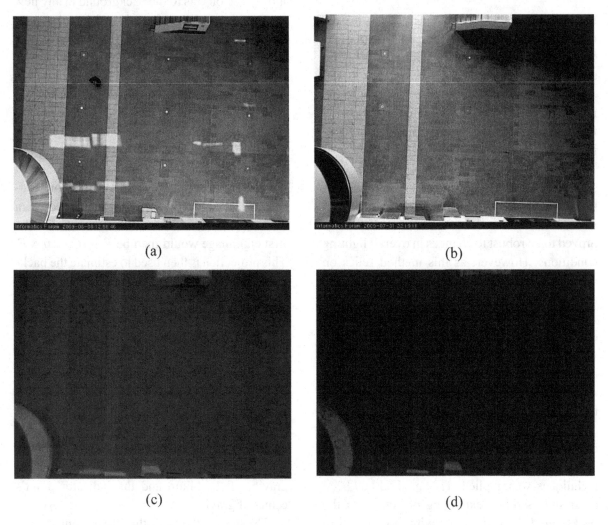

(a)　　　　　　　　　　　　　　　(b)

(c)　　　　　　　　　　　　　　　(d)

- Label remaining blobs.
- Calculate the area of the blobs and assume the largest to be the blimp.
- Create a mask from the identified blimp blob.
- Fill all gaps in the mask.
- Create an inverted image of the original current frame, and apply the mask to it to obtain an ROI consisting of the blimp.
- Build the ROI's histogram.
- Apply smoothing to the histogram and find the optimum threshold (valley between

two peaks) value for the ROI. This value will separate the blimp from its marker point.
- Apply thresholding to obtain binary image, foreground will represent the marker.
- Apply morphological open operation to re-move noise.
- Label remaining blobs.
- Calculate the area of the blobs and assume the largest to be the blimp's marker.

This algorithm proved very fast, however; it still suffered from failures under certain lighting conditions, and whenever the blimp flew over white patches of the floor. Thus, there was a need to devise a different algorithm for the detection of the blimp.

Blimp Detection: Using Color Threshold

This algorithm that was adapted from (Ntelidakis, 2010) makes use of the fact that the blimp is the whitest object in the image frame. As such, this approach becomes robust to reflections from the sun or sudden changes in ambient light brightness. This is because these changes are never bright enough or intense enough to become as intense as the white blimp. However; other white objects in the background are extracted, therefore a mask is applied to mask out the stairs and the visible desk area.

The steps of this algorithm are presented in the following points:

*   Obtain image of the current frame. This is shown in Figure 4.a.
*   Blur the current frame to remove false positives.
*   Convert the image to grayscale. This is shown in Figure 4.b.
*   Use a high threshold (200) to transform the grayscale image into a binary image. This makes sure that only the whitest objects are seen as white pixels while others become black pixels in the binary image. This step is shown in Figure 4.c.
*   Apply morphological operation of erosion to remove salt noise.
*   Apply morphological operation of dilation; this is to repair any damage done by the erosion in the previous step. This step is important because we want to maintain the pixels that represent the blimp marker, as its detection is important to calculate

the orientation of the blimp. The result is shown in Figure 4.d.

*   Apply the mask to remove the stairs area in the background as it is too white, as shown in Figure 4.e.
*   Label the blobs and calculate their sizes.
*   Pick the largest blob and assume it is the blimp.
*   Give elliptical properties to the blimp and find its center of mass, minor and major axes and size.
*   Use the blimp as an ROI, as shown in Figure 4.f.
*   Find the biggest blobs inside that ROI.
*   Pick the blob with the smallest compactness.
*   Find its center of mass.
*   Estimate the blimp orientation based on this center of mass, and the center of mass of the blimp.

This method then returns the important values of blimp's center of mass coordinates, the blimp's orientation, major and minor axes of the elliptical shape fitted to the blimp and finally the area of the blimp as seen from the ceiling camera.

Now that we have this information, we can calculate the blimp's position in 3D space. The next section discusses the methods used to calculate these parameters.

Results

The results discussed in this section are obtained after an analysis of 30 recorded flights. The background subtraction using PCA method was not reliable as mentioned earlier, failing 40% of the time. The results were worse when the blimp went across white objects in the background, the algorithm would then fail 100% of the time. The color threshold method however, was successful 100% of the time during the hours of the morning till the afternoon. It performs poorly in poor lighting conditions and requires adjustment of the threshold parameters to get correct results. A future

Figure 4. a) Current image; b) image blurred and converted to greyscale; c) binary image (threshold =200); d) binary image after erosion and dilation. e) masked image; f) ROI for black marker identification (figure adapted from Ntelidakis, 2010)

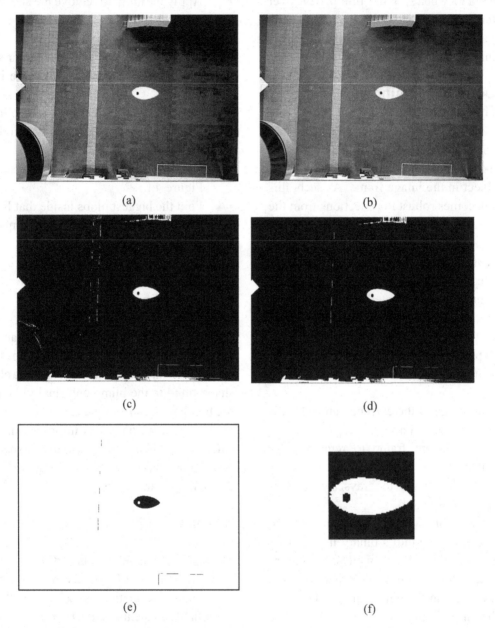

modification could be allowing these parameters to vary dynamically such as to account for the changing lighting conditions. The color threshold method also proved to be robust to shadows and the sun's reflection and other failures experienced in the first algorithm. This method also proved

faster; taking only 0.3 seconds compared to 0.5 seconds obtained by the background subtraction using PCA method.

For validation, the physical start and end points were known and the software was tested against them. The software was always successful in returning the correct coordinates in feet. For the height validation, the length of the rope given to the blimp was also a physically measurable metric, and it was always around 15 feet; the software was successful in returning that figure as well.

BLIMP'S VISUAL ODOMETRY

In this section we discuss the calculations involved in transforming the pixel information obtained from the image frame obtained to real word information; vital to the controller performance (Ntelidakis, 2010).

In the last section we have obtained:

x_t: Planar position (x_{im}, z_{im}) in the current image frame t in pixels.

theta: Angular orientation of the blimp (yaw) in radians.

R_{im}: The length of the major axis of the blimp (from the fitted ellipsoid).

A: Area of the blimp in pixels.

The information we aim to obtain is the blimp's position in three dimensional space, vertical and horizontal velocities, angular orientation (yaw) and finally angular velocity. We are only interested in the 4 degrees of freedom (3D position and angular orientation (yaw)). We assume that the roll and pitch parameters of the blimp are always maintained due to the blimp's nature. In all the calculations discussed in this section the distortion of the ceiling camera is neglected.

The following subsections discuss the calculation of altitude, planar position, planar velocity, and vertical velocity.

Calculating Altitude, Position, and Velocities

To calculate the location of the blimp, we make use of consecutive frames of the extracted blimp as seen in Figure 5.a. We also make use of the fact that the blimp is flying inside an irregular pyramid

with a height of 77.58 feet; this is the camera's height from the floor. The irregular pyramid's base is a rectangle 52x39 feet-squared. This is shown in Figure 5.b. Estimating the blimp's planar position and height is very important in this case; this is because the blimp changes planes inside this pyramid as it flies inside it. Each of these planes has different widths and lengths. After we calculate the blimp's position inside the irregular pyramid, we apply calculations to project the blimp's position onto the ground coordinates. These positions are used to calculate planar and vertical velocities.

VISUAL SERVOING CONTROLLER

Now that we have the inputs required for the controller function we can input them into a controller to issue the appropriate motor commands for waypoint tracking. As discussed earlier, the blimp is a challenging platform to control. Moreover; it is being constantly attacked by several environmental disturbances. In this section we discuss the controllers used in our project and introduce the methodology used to compare between them (Alkurdi, 2011). First we discuss the experimental setup used in our project.

Experimental Setup

We used waypoint tracking as an experiment for measuring the performance of the controllers applied to the blimp. The setup can be seen in Figure 2.a. The blimp is expected to travel between the two red dots in a straight line. A way to measure the actual performance is shown in Figure 6 below. As the blimp flies in 3D space, we calculate the (x,y,z) coordinates through the vision system and store them at each time step. We calculate the perpendicular distance of the each data point from the line it should be following. This is shown in Figure 6.a. Over all the time steps we would obtain a surface in 3D space. This surface is shown in Figure 6.b. Ideally, for perfect control, we would like to minimize this surface area to zero. However; this is not possible because of a set of inaccuracies within the blimp model and external disturbances. The following subsection discusses these errors.

Sources of Error

The blimp does not have a pressure sensor to calculate the contained helium pressure, so each day of testing the blimp will contain a different amount of helium. The gondola's position also differs at the beginning of each testing day. This has the effect of changing the center of mass for the blimp and could alter the performance slightly.

During the test run, factors such as people passing, elevators running, and doors opening in the Informatics Forum ground level will induce gusts that drive the blimp in a certain direction, pushing it away off its course.

The blimp has a lot of inertia when moving in a certain direction, thus changing the direction requires some time. It then becomes important that no series of faulty motor commands are sent to the blimp, as this will cause the blimp to stray

from its path and it requires more time to set it back onto its course.

The software for the blimp calculates the area of the blimp as seen from a network camera fixed at the ceiling of the Informatics Forum looking down on the testing area. The area is calculated to get a height estimation, and since this is done every frame, a lot of noise is introduced into the recorded height of the blimp, as lighting conditions might change from one captured frame to another.

The blimp is held by two ropes at each end of its envelope. The amount of rope length given to the blimp affects the weight the blimp is carrying and thus would affect its final height. The amount of rope length can vary throughout the test run, and can affect the results.

Other issues that affect the blimp during its run is change of temperature in the testing area. This will lead to a change in pressure and change of the height of the blimp.

These errors are expected to be dealt with via the controller function such that a correct motor command is sent to the blimp to overcome these disturbances and set the blimp back to its correct path. The controllers applied to our robotic blimp are discussed in the next subsection.

Figure 5. a) Binary image of the extracted blimp b) the irregular pyramid that the blimp is flying inside

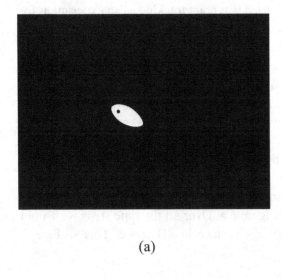

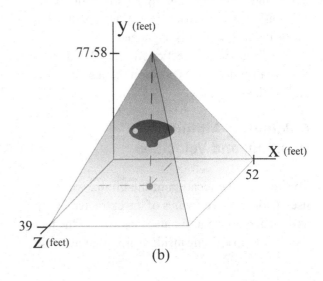

(a)   (b)

Blimp Control

Three different linguistic controllers were applied to our blimp platform. The aim was to avoid computationally expensive mathematical algorithms for the realization of real time control. The first of these controllers is a 3-parameter state based controller. The second controller studied was a 4-parameter state based controller. Finally a fuzzy logic controller was studied.

The 3-parameter state based controller determines the current, predicted and the rate of change of the blimp's position for position control and it also calculates the current, predicted and the rate of change of the blimp's orientation for heading control. If the three parameters of say, position, are in a certain state, then a certain hardcoded motor impulse command is applied in the current time step. For more information please refer to (Alkurdi, 2011).

The 4-parameter state based controller is basically the same. However; an additional parameter of history is added. The idea was to study the effect of past state to further enhance the blimp's understanding of its position.

The fuzzy logic controller was introduced to study the effect of having a continuous range of motor commands issued from the controller; rather than having hardcoded impulse commands as in the previous two controllers. The fuzzy logic control system employed two sub controllers, one for heading and one for position. The heading controller has an input parameter of angle that uses 7 triangular membership functions, and a parameter of angular velocity that uses 3 triangular membership functions. The output used 5 triangular membership functions. The position controller has an input parameter of distance from goal that uses 5 triangular membership functions, and an input parameter of planar velocity that uses 3 triangular membership functions. The output used 3 triangular membership functions. The rule base is determined through our experience of the blimp's performance. Full design parameters are discussed in (Alkurdi, 2011).

Results

The fuzzy logic controller outperformed the other two controllers. Many test runs were conducted

Figure 6. a) The perpendicular distance of the each data point from the line it should be following; b) the surface obtained in 3D space

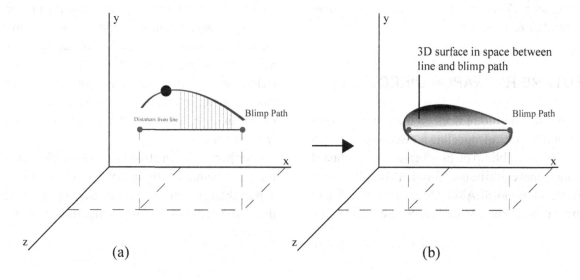

Figure 7. a) A typical recorded path in 3D; b) average path of the blimp's path under fuzzy control

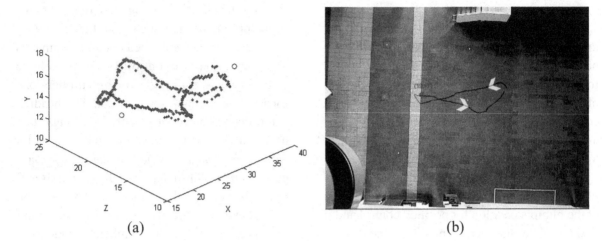

(a)   (b)

to test the performance of each controller and on average the flying area in 3D was smaller for fuzzy logic (area was 1923 feet²) than the other two controllers (the area was 2714 feet² and 2300 feet² for the 3 and 4 parameter controller respectively). The fuzzy logic controller improved the performance of the blimp by 29% compared to the 3-parameter state based controller, and by 16% compared to the 4-parameter state based controller. The fuzzy logic controller was most successful at keeping the blimp on the required path compared to the other two controllers.

In terms of time, the fuzzy logic controller (3.35 minutes) performed its task the fastest (*versus* 3.55 and 4.27 seconds for the 3 and 4 parameter controller respectively).

FUTURE RESEARCH DIRECTIONS

In this section future work on the vision, control and mathematical modeling is discussed. In terms of vision, the blimp platform carries an on-board camera that could be used for SLAM applications. Work on mono-SLAM techniques can be found in (Davison, Reid, Molton, & Stasse, 2007). The application of such algorithms can complement or replace the current localization algorithms.

As for the control algorithms, further tuning to the fuzzy controller would enhance the performance of the robotic blimp. This could be done manually by changing the number and/or shape of the membership functions. The ranges the membership functions cover can also be manipulated to tune the fuzzy logic controller. The rule base can always be changed to alter the behavior of the controller and is an important factor in tuning the controller. Finally, the defuzzification method can be changed, where the center of area method that is used throughout this project provides a weighted average output to the motors, a more/less aggressive method could be used. Some other defuzzification methods that can be tested are: Mean of maxima, leftmost maximum and rightmost maximum. Automatic tuning; however, can be done in the form of ANFIS (adaptive neural fuzzy inference system) (Kitt, Chekima, & Dhargam, 2011).

Further work can also be done through obtaining a mathematical model of the robotic blimp and applying computer simulations to understand the best way to apply control algorithms on this platform.

CONCLUSION

The research showed that the visual feedback was effective for blimp control even under unpredictable indoor conditions that included varying lighting conditions and gusts arising from door openings, people walking and elevators. Additional research showed that the fuzzy logic controller was more effective than a 3-parameter state table approach or a 4-parameter state table approach.

The continuous range of outputs the fuzzy logic controller provides to the motors, as well as the ability to fuzzify the inputs into more than one membership function have made a difference in the performance of the blimp. These factors have rendered the fuzzy logic controller to be the better performing controller. The 3-parameter state based controller and the 4-parameter state based controller both had fixed outputs to depending on what location the blimp was located in the map; theses outputs amounted to 5 possibilities. As stable as that is, it would reduce the ability of the blimp motor actions and make the motion less smooth than desirable. The 4-parameter state based controller performed better than the 3-parameter state based controller because of its ability to exclude faulty assumptions about the blimps position by predicting the blimps future position. This would remove many ambiguities and help the blimp avoid issuing any faulty commands that could take the blimp in an undesired position; a situation that needs more time to recover from.

A typical path by the blimp is a run between the two waypoints; they are shown as circles in Figure 7.a. A typical recorded path in 3D space is shown in Figure 7.a, the red dots represent the outbound path, the blue dots represent the inbound path. Figure 7.b shows the average path of 20 runs of the fuzzy logic controlled blimp. These figures show the successes of the image processing and visual control techniques and the range of their coverage.

REFERENCES

Alkurdi, L. (2011). *Fuzzy logic control of a robotic blimp*. Master's thesis, University of Edinburgh, U.K.

Avenant, G. C. (2010). *Autonomous flight control system for an airship*. Master's thesis, Stellenbosch University, South Africa.

Azinheira, J. R., de Paivab, E. C., Ramosb, J. J. G., & Buenob, S. S. (2000). Hovering control of an autonomous unmanned airship. In *Proceedings of the 4th Portuguese Conference on Automatic Control*, (pp. 430–435).

Azinheira, J. R., Rives, P., Carvalho, J. R. H., Silveira, G. F., de Paiva, E. C., & Bueno, S. S. (2002).Visual servo control for the hovering of an outdoor robotic airship. In *Proceedings of the IEEE International Conference on Robotics and Automation ICRA '02*, Vol. 3, (pp. 2787–2792).

Davison, A. J., Reid, I. D., Molton, N. D., & Stasse, O. (2007). MonoSLAM: Real-time single camera SLAM. *Pattern Analysis and Machine Intelligence*, *29*(6), 1052–1067. doi:10.1109/TPAMI.2007.1049

Elfes, A., Montgomery, J. F., Hall, J. L., Joshi, S. S., Payne, J., & Bergh, C. F. (2005, September). *Autonomous flight control for a titan exploration aerobot*. Paper presented at the 8th International Symposium on Artificial Intelligence, Robotics and Automation in Space. Munich, Germany.

Falahpour, M., Moradi, H., Refai, H., Atiquzzaman, M., Rubin, R., & Lo Presti, P. G. (2009, October). *Performance comparison of classic and fuzzy logic controllers for communication airships*. Paper presented at the IEEE/AIAA 28th Digital Avionics Systems Conference DASC '09, Orlando, FL, U.S.A.

Gomes, S. B. V., & Ramos, J. G., Jr. (1998). Airship dynamic modeling for autonomous operation. In *Proceedings of the IEEE International Robotics and Automation Conference,* Vol. 4, (pp. 3462–3467). Leuven, Belgium.

Gonzalez, P., Burgard, W., Sanz, R., & Fernandez, J. L. (2009). Developing a low-cost autonomous indoor blimp. *Journal of Physical Agents, 3*(1), 43–52.

Hygounenc, E., Jung, I., Soueres, P., & Lacroix, S. (2004). The autonomous blimp project of LAAS-CNRS: Achievements in flight control and terrain mapping. *The International Journal of Robotics Research, 23,* 473–511. doi:10.1177/0278364904042200

Hygounenc, E., & Soueres, P. (2003). Lateral path following GPS-based control of a small-size unmanned blimp. In *Proceedings of the IEEE International Conference of Robotics and Automation ICRA '03,* Vol. 1, (pp. 540–545).

Jian-guo, G., & Jun, Z. (2008, December). *Altitude control system of autonomous airship based on fuzzy logic.* Paper presented at the 2nd International Symposium on Systems and Control in Aerospace and Astronautics, Shenzhen, China.

Kampke, T., & Elfes, A. (2003). Optimal aerobot trajectory planning for wind-based opportunistic flight control. *IEEE/RSJ International Conference on Intelligent Robots and Systems (IROS 2003),* Vol. 1, (pp. 67–74).

Kawai, Y., Hirano, H., Azuma, T., & Fujita, M. (2004). Visual feedback control of an unmanned planar blimp system with self-scheduling parameter via receding horizon approach. In *Proceedings of the IEEE International Conference on Control Applications, Vol. 2,* (pp. 1063-1069).

Kawai, Y., Kitagawa, S., Izoe, S., & Fujita, M. (2003, August). *An unmanned planar blimp on visual feedback control: Experimental results.* Paper presented at the SICE 2003 Annual Conference, Fukui, Japan.

Khoury, G. A., & Gillett, J. D. (2002). *Airship technology.* Cambridge, UK: Cambridge University.

Kitt, W. W., Chekima, A., & Dhargam, J. A. (2011). Design and verification of an ANFIS based control system for an autonomous airship on a simulation environment. *Global Journal of Computer Application and Technology, 1,* 119–128.

Kornienko, A. (2006). *System identification approach for determining flight dynamical characteristics of an airship from flight data.* PhD thesis, University of Stuttgart, Germany.

Kulczycki, E. A., Joshi, S. S., & Hess, R. A. (2006). *Towards controller design for autonomous airships using SLC and LQR methods.* In AIAA Guidance Navigation and Control Conference and Exhibit.

Kungl, P., Schlenker, M., Wimmer, D. A., & Kröplin, B. H. (2004). Instrumentation of remote controlled airship "LOTTE" for in-flight measurements. *Aerospace Science and Technology, 8*(7), 599–610. doi:10.1016/j.ast.2004.06.004

Lacroix, S., Jung, I.-K., Soueres, P., Hygounenc, E., & Berry, J.-P. (2003). The autonomous blimp project of LASS/CNRS: Current status. *Experimental Robotics VIII, 5,* 487–496. doi:10.1007/3-540-36268-1_44

Liu, Y., Pan, Z., Stirling, D., & Naghdy, F. (2009). Control of autonomous airship. In *Proceedings of the IEEE International Conference on Robotics and Biomimetics* (pp. 2457–2462).

Lutz, T., Funk, P., Jakobi, A., & Wagner, S. (2002, July).*Summary of Aerodynamic Studies on the LOTTE Airship*. Paper presented at the 4th International Airship Convention and Exhibition. Cambridge, UK.

Majecka, B. (2009). *Statistical models of pedestrian behavior in the Forum*. Master's thesis, University of Edinburgh, U.K.

Ntelidakis, A. (2010). *Using a blimp to build an interior model of the forum*. Master's thesis, University of Edinburgh, U.K.

Ollero, A., & Merino, L. (2004). Control and perception techniques for aerial robotics. *Annual Reviews in Control, 28*(2), 167–178. doi:10.1016/j.arcontrol.2004.05.003

Payne, J., & Joshi, S. S. (2004). *6 degree-of-freedom non-linear robotic airship model for autonomous control*. Unpublished technical report, University of California. Oakland, California, U.S.

Ramos, J. J. G., de Paiva, E. C., Azinheira, J. R., Bueno, S. S., Bergermana, M., Ferreira, P. A. V., & Carvalho, J. R. H. (2001). Lateral/directional control for an autonomous unmanned airship. *Aircraft Engineering and Aerospace Technology, 73*(5), 453–458. doi:10.1108/EUM0000000005880

Takaya, T., Kawamura, H., Minagawa, Y., Yamamoto, M., & Ohuchi, A. (2006). PID landing orbit motion controller for an indoor blimp robot. *Artificial Life and Robotics, 10*(2), 177–184. doi:10.1007/s10015-006-0385-9

Van der Zwaan, S., Bernardino, A., & Santos-Victor, J. (2000). Vision based station keeping and docking for an aerial blimp. In *Proceedings of the International Conference on Intelligent Robots and Systems,* Vol. 1, (pp. 614-619). Takamatsu, Japan.

Wimmer, D., Bildstein, M., Well, K. H., Schlenker, M., Kungl, P., & Kröplin, B.-H. (2002, October) *Research airship "lotte" development and operation of controllers for autonomous flight phases*. Paper presented at the International Conference on Intelligent Robots and Systems. Lausanne, Switzerland.

Xie, S.-R., Luo, J., Rao, J.-J., & Gong, Z.-B. (2007). Computer vision-based navigation and predefined track following control of a small robotic airship. *Acta Automatica Sinica, 33*(3), 286–291. doi:10.1360/aas-007-0286

ADDITIONAL READING

Alkurdi, L. (2011). *Fuzzy logic control of a robotic blimp*. Master's thesis, University of Edinburgh, U.K.

Avenant, G. C. (2010). *Autonomous flight control system for an airship*. Master's thesis, Stellenbosch University, South Africa.

Bekiroglu, K. (2010). *Vision based control of an autonomous blimp with actuator saturation using pulse width modulation*. Master's thesis, Northeastern University, U.S.A.

Fukao, T., Fujitani, K., & Kanade, T. (2003). Image-based Tracking Control of a Blimp. In *Proceedings of the 42nd IEEE Conference on Decision and Control,* Vol. 5, (pp. 5414-4519). Maui, Hawaii, U.S.A.

Liu, Y., Pan, Z., Stirling, D., & Naghdy, F. (2009). Control of autonomous airship. In *Proceedings of the IEEE International Conference on Robotics and Biomimetics* (pp. 2457–2462).

Metelo, F. M., & Campos, L. R. G. (2003). *Vision based control of an Autonomous Blimp (VIDEOBLIMP). Undergraduate final year project.* Portugal: Universidade Técnica de Lisboa.

Moutinho, A. (2007). *Modeling and nonlinear control for airship autonomous flight.* PhD thesis, Universidade Técnica de Lisboa, Portugal.

Ollero, A., & Merino, L. (2004). Control and perception techniques for aerial robotics. *Annual Reviews in Control, 28*(2), 167–178. doi:10.1016/j.arcontrol.2004.05.003

KEY TERMS AND DEFINITIONS

Blimp: A lighter-than-air airship that does not have an exterior frame to hold its envelope.

Fuzzy Logic Control: Applying fuzzy logic mathematics to control a given system. This ultimately leads to a control system that uses a linguistic framework that an experienced user can understand, rather than a mathematical framework of transfer functions.

Object Detection: In image processing, object detection is the operation of deciding whether a desired object is in a given image (and often also includes location of the object or extracting the pixels that belong to the desired object).

Thresholding: In image processing, thresholding is applying an pixel intensity value(s) to discriminate foreground objects from background objects (e.g. pixels brighter than the threshold belong to the foreground object).

Visual Odometry: Extracting the robot's pose and location by the use of vision sensors.

Visual Servoing: Applying vision sensors and algorithms to extract feedback parameters necessary for robotic motion control.

Waypoint Tracking: A robot is said to have an objective of waypoint tracking if its goal is to achieve sequential tracking of specific points in 3D space.

Section 6
Visual Attention

Chapter 20
Selective Review of Visual Attention Models

Juan F. García
Universidad de León, Spain

Francisco J. Rodríguez
Universidad de León, Spain

Vicente Matellán
Universidad de León, Spain

ABSTRACT

The purpose of this chapter is both to review some of the most representative visual attention models, both theoretical and practical, that have been proposed to date, and to introduce the authors' attention model, which has been successfully used as part of the control system of a robotic platform. The chapter has three sections: in the first section, an introduction to visual attention is given. In the second section, relevant state of art in visual attention is reviewed. This review is organised in three areas: psychological based models, connectionist models, and features-based models. In the last section, the authors' attention model is presented.

VISUAL ATTENTION

Attention can be defined as the cognitive process by which human beings focus on a certain aspect of the environment while ignoring all others, or as the management of the resources that allow for information processing (Anderson, 2004). Talking to someone while ignoring other conversations around or driving properly while at the same time ignoring distracting elements adjacent to the road are examples of attention. Attention is one of the most studied topics in psychology and neuroscience.

Attention may be influenced by many channels of information (like animals obtain information from their different senses), referring to the case when attention works with visual information channels as "visual attention". Since this is the only

DOI: 10.4018/978-1-4666-2672-0.ch020

source of information considered in this chapter, even when we refer to it generically as "attention" it will always be visual in nature.

There are several reasons why attention is useful for any system, whether biological or artificial. The most obvious is to reduce input data thus freeing up resources used, resources which could then be employed in other system tasks. It is also interesting the ability to simplify a complex problem into simpler ones. Additionally, using a mechanism of attention, the influence of information that may cause distractions can be suppressed or at least reduced.

Attention in Robotics

Robotic vision systems are often structured as a system of analysis at one end and a response system in the other. The analysis part generates a model of the environment using the images obtained from the same and the response part uses these models to plan or perform an action. Vision in this case acts as a pre-action stage, calculating all the features that the planning system might need.

Thanks to attention, instead of processing all pixels in each image, with the amount of resources that would imply, only those which presumably contain information relevant to the current task are selected. That is, the system adapts to extract from the images only the information it really needs.

Attention seen as a means to select information for tasks, as if it were a filter, is also very useful in the field of robotics. Autonomous robots should be able to interact with complex environments and scenarios, simultaneously maintaining different goals for each of their possible behaviours, the latter being understood as sets of actions that seek a goal.

Attention in this case can guide robots behaviour, restricting the amount of processed visual information to what is relevant to the tasks at hand and ignoring other elements (distractors). It also provides a coordination mechanism, because

selection of stimuli can serialise the actions of the current behaviour.

The latest trends in robotics to solve problems related to attention go through the application of bio-inspired models (Frintrop *et al.*, 2005a; Itti & Koch, 2000; Navalpakkan & Itti, 2006; Torralba *et al.*, 2006; Tsotsos *et al.*, 1995). It is important to note that although algorithms used are biologically inspired their goal is not to model biological systems.

Active Vision

Active vision is defined as the perceptual system's ability to both select and analyse specific parts of a visual scene and to modify the parameters, both external and internal, of image creation (fixation point, focal length, …) (Bajcsy, 1986; Ballard, 1990; Krotkov, 1987)

It is the task of active vision system to determine the most attractive regions of each scene (attention) and move the point of fixation to specific regions, both within and outside the field of view (gaze control). Active vision is only possible by combination of many components: visual attention, gaze control, selection of data obtained from different sources, depth calculations made from disparity estimation between images, and hand-eye coordination (Backer *et al.*, 2001; Brunnström *et al.*, 1994).

These techniques depend on the task to be performed and require real-time solutions. Since the 80, robotic platforms with mobile cameras have emerged to meet these constraints (Jenning & Murray, 1997; Vieville *et al.*, 1995; Zangenmeister, 1996), specifically used to validate the conceptual models implicit in the system.

Gaze Control

Gaze control is the process of directing the fixation point through a scene to allow for the development of perceptual, cognitive and behavioural activity (Henderson, 2003). Figure 1 shows an example of the variations of the fixation point in an image during the visual search process. Circles represent the points of fixation (the larger the radius, the longer the fixation time) and lines represent saccades.

The properties that determine where to position the fixation point vary depending on which part of perception considered. The task performed is of great importance in the attention case, while gradients in space or time (areas of high contrast or movement) influence preatentive perception. Yarbus (Yarbus, 1969) pioneered the study of movement of the fixation point in terms of the information required for a given task.

There are plenty of studies in the field of gaze control. An example is the use of objects recognition in a given scene as input to validate a gaze control system, as in (Brunnström *et al.*, 1994). In this case, the gaze control system follows the edges and lines in the image. The model starts by checking and classifying a particular corner or junction which is especially remarkable. This point is then connected to other corners or intersections to define more complex shapes.

Another example of gaze control is the work of Maki et al. (Maki *et al.*, 1996), in which gaze control is specifically designed to track moving objects. The most important aspect of it is the time-variant visual attention system it implements. If the system is in " follow mode",the focus is exclusively on the object pursued while a mask is applied to the rest of the scene. while in "search mode", the attention is directed toward a new moving object. Changes in attention between different moving objects are managed through an internal algorithm based on different criteria. This model of gaze control has been an inspiration for later works (Lopez *et al.*, 2006, 2007; Maki *et al.*, 2000) in which various modifications like the inclusion of new inputs such as depth or motion (horizontal flow) are incorporated.

In the field of human-robot interaction active vision system with gaze control and monitoring functions have also been successfully developed. The Milva project (Boehme *et al.*, 1998), a mobile robot reacts to nonverbal forms of interaction such as gestures or mime to meet transportation demands. In this case, the active vision system allows for the detection of hands that are fixated by the gaze control system and then analysed to determine the type of action requested.

In the same research line, Waldherr et al. (Waldherr *et al.*, 2000) developed a service robot equipped with gaze control which could identify and follow a person in an office environment. The operator guides the robot to specific areas that need cleaning and can also instruct it on garbage collection tasks using gestures (or static postures) performed with his arms.

In summary, most gaze control system operate guided by the image and the model. Moreover, in most cases this model is unique and defines a single behaviour (search for a particular object type or property). Also noteworthy is that the most common bottom-up component in these models is based in filtering operations.

Active Vision without Visual Attention

Active vision system can also be developed without the use of visual attention. The alternative to this would be designing an appropriate selection mechanism for each particular problem, modeling the environment and determining the methods needed to identify and select relevant aspects of that model (Rowley *et al.*, 1995; Wren *et al.*, 1997).

The major disadvantage of this option is the reimplementation of the selection mechanism required for each system and for each model of the environment. The effect of each discrepancy between the model and the modeled environment would also need to be individually established.

In contrast, the use of a visual attention system eases the separation between the main parts of the

Figure 1. Fixation point variations through a visual scene

attention control system and the modules that have to be adapted to each specific problem.

Conceptual Foundations of Attention

The study of attention in humans from the point of view of psychology has allowed characterisation of the basic concepts involved in this process. These theoretical concepts are the basis for most theoretical, qualitative and computational attention systems proposed to date.

Focus of Attention

The focus of attention of a system can be defined as the region, object or item in which the system's perceptual resources are concentrated. Humans can change their focus of attention either by moving the fixation point or by focusing on one part of the field of vision. In the first scenario we speak of "overt attention", and in the second one we speak of "covert attention" (Wright & Ward, 2008).

Changes in covert attention are up to four times faster than overt changes. This speed difference can be used to analyse a potential fixation point to conclude whether it is worth or not to move our eyes to that position.

Over the years, different paradigms have been proposed describing how humans focus attention (Milanese, 1990). In the zoom metaphor, cognitive resources can be either distributed to the entire field of view (wide angle), or focused on a portion (zoom lens).

The reflector metaphor (see Figure 2) (Julesz, 1991) assumes that a light moves through the scene, illuminating certain areas for detailed processing while other parts remain obscure and are ignored. The size and resolution of the light source may vary depending on circumstances. Even now is discussed whether this point of light moves continuously or intermittently, jumping from one area to another (LaBerge *et al.*, 1997).

Early and Late Selection

In the human perceptual processes, there is a distinction between early and late selection. Early selection analyses the full field of view, registering environmental features, while late selection integrates features into proto-objects. Both are running continuously, so that the latter uses the results of the former.

The question about in which of these phases attention come into play has led to a longstanding debate. Some authors argue that attention is involved during early selection stage, being its function to reduce the amount of input data (Posner *et al.*, 1980). Proponents of this theory claim that the visual perceptual system has limited capacity and attention acts as a filter for it, selecting only the information that must be processed (Broadbent, 1958; LaBerge, 1995). We speak here of attention for resource management.

Other authors claim that human perceptual capacity is huge and does not require a previous stage of selection of information (Allport, 1987; Neumann *et al.*, 1986). They argue that attention is involved in late selection (which is why they also call it "attentive perception") and is not present during early selection (which they call "preatentive perception"), being its function to remove distracting elements from the scene so that only items of interest influence the actions

and decisions (Pashler & Badgio, 1985). In this case we speak of attention for action.

The two theories can be integrated if information selection performed by attention is considered to take place in different time depending of the task to be fulfilled and the available resources.

The terms preattentive and attentive perception are used nowadays to refer to early and late selection respectively even though it is considered that attention is actually present in both processes.

Unit of Selection

The unit of selection, defined as that which attention can identify (eg, a particular object or just the space it occupies) is one of the key aspects of attention.

In psychology research, there is disagreement on whether to consider the space or the objects as unit of selection (Baylis & Driver, 1993).

Since objects always occupy a space, is difficult to distinguish whether selection affects the object or the space in which it is located. Some authors have found evidence for the existence of pure selection of objects (Blaser *et al.*, 2000; Hübner

Figure 2. Reflector metaphor

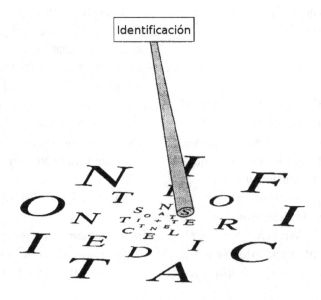

& Backer, 1999), but despite of that space it is still considered the unit of selection.

It is nevertheless established that a particular space or region is selected because the objects or features it contains, so in many cases we are really working with the space occupied by an object.

ATTENTION MODELS

Visual attention is studied by a multitude of disciplines, from medicine to engineering through psychology and neurology. Each of them presents its own conceptual models (abstract representations) of visual attention. This section contains the most relevant proposals in these fields.

Multiscale Pyramids

Some attention models, like Tsotsos's Selective tuning model (Tsotsos *et al.*, 1995) or Itti's Saliency–based visual attention model (Itti *et al.*, 1998a,b; Itti & Koch, 2000, 2001) are based on the use of so-called multi-scale pyramid representation of visual information.

Visual information can be organised in a multiscale pyramid (Anderson & Essen, 1987; Burt, 1988; Tsotsos, 1987; Uhr, 1972), with various levels containing images of different resolution and dimension of the same scene (see Figure 3). The higher up the pyramid, the lower resolution and dimesion images have. Each level of the pyramid is connected to its immediately upper and lower levels.

Psychological and Theoretical Models

Most studies in this field focus on the empirical aspect of attention, providing quality attention control models at best (qualitative models seek to represent concepts, defining and describing their qualities or actions without performing quantifications).

Guided Search

One of the visual attention models most widely applied in psychology nowadays is Wolfe's Guided search, originally published in 1989 (Wolfe *et al.*, 1989) and now in its fourth revision, published in 2007 (Wolfe, 2007). It is also very interesting because of its similarity with some control of attention models for computer vision.

Wolfe's work tries to model the visual search process and it is based on Treisman's feature integration model (Treisman & Gelade, 1980), perhaps the leading model of visual search, which seeks to explain the distinction between serial and parallel searches with a two-stage model. A fairly limited, "preattentive", parallel stage of processing is followed by a more sophisticated, serial stage. Treisman holds that only basic features such as color, size, and orientation can support parallel search, whereas all other stimuli require a serial search.

In Guided search's experimental psychological paradigm, a subject is asked about the presence or absence of any particular object in a scene they are witnessing. The search time (the time it takes the subject to find the object) may depend on the number of distracting elements present, and gives an indication of how fast one can focus attention on a target.

Search tasks can be divided into parallel, with search times independent of the number of distracting elements, and in series, with search times linearly increasing according to the given distractions.

Guided Search starts with the extraction of a number of basic features, including orientation and colour. The value of these features at each point of the observed scene are stored in so-called feature maps. These maps are finally reduced to a single activation map called "saliency map" by weighted sum. This map redirects attention to its "hills" while positions already visited are stored in an "inhibition map". The efficiency of this model is largely given by the relationship between the activation levels of both interesting and distracting elements appearing in the activation map.

This model has similarities with an older one dating from 1985 published by Koch and Ullman (Koch & Ullman, 1985), which also presents a stage for parallel extraction of features and uses a master map of attention to represent saliency. A WTA (winner-takes-all) process is applied on this map and its maximum are then obtained by means of an inhibition map (see Figure 5).

Connectionist Models

Connectionist models are named for its hierarchical structure of interconnected components. Sometimes, these models integrate two different aspects of attention as a different way of seeing the same problem: finding objects of unknown position and identification of an object present in a known position (Deco, 2000; Hamker, 2000).

The main concern of connectionist models is the transformation of the most salient part of the observed scene in a frame of reference (Backer *et al.*, 2001). These transformations are associated with a translation and, in some cases, with the use of a size-independent mappig. For active vision systems this approach may not be of great importance since it is possible to perform a

translation-independent mapping just by fixating the striking object.

The following are some of the most important connectionist models.

Selective Tuning

In 1995, Tsotsos et al. (Tsotsos *et al.*, 1995) published their Selective tuning model (known in some sources as inhibitor beam model) to select those regions with a greater visual information load. This method tries to be biologically feasible (in fact it was initially hypothesised for the vision of primates), but also applies to any computer problem that requires selection of information in images, including those present in robotics. It can also be adapted to tasks that require a specific type of attention.

The model begins with the idea of attention as a means to optimise visual search tasks. It defines a hierarchy of WTA processes, embedded in the visual processing pyramid discussed earlier, which progressively and continuously refines visual information received (see Figure 6).

Using an input pattern, a winner process of the WTA is obtained at the top level (the region of the image at the upper level that is closest to the input pattern is chosen). This process will be the overall winner and will spread through a an inhibitor ray to lower levels of the hierarchy until reaching the lowest one, determining a new local winner for each level. Only those processes connected with the winner process of the immediately upper level are chosen at each level.

At the end of the process, the bottom layer has a set of units that correspond to the region of the input image with more visual information. To change the focus of attention, it is enough to inhibit the previously chosen region and obtain a new representation by the same process.

Later in 2005, Tsotsos et al. apply their model of selective tuning to a real world attention task: the detection of motion in an image sequence (Tsotsos *et al.*, 2005). In this case, the application

Figure 3. Multiscale pyramid

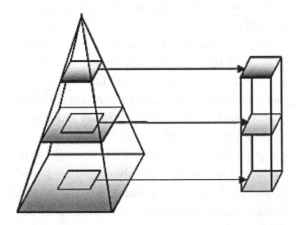

Figure 4. Wolfe's guided search (Wolfe et al., 1989)

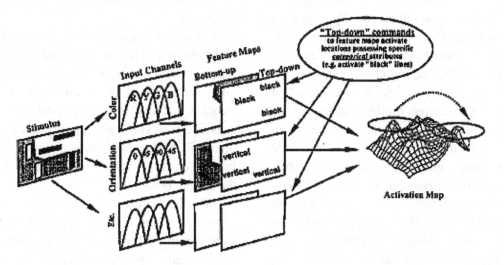

tion of the model allows for the identification of movement patterns in the images, incorporating to such identification the specific attention processes of selective tuning.

In 2009, Tsotsos et al. present an alternative model to the selective tuning as an attention-control method that can detect patterns of movement: the direction maps (Gryn *et al.*, 2009). With this new model, results obtained are applied in the field of surveillance and security. Specifically, the method was tested on videos recorded by seven different traffic or safety cameras.

This technique is based on detecting motion patterns by abstracting from the intensity of the pixels of the image to explicit representations that capture the movement in time and space. This spatio-temporal movement through regions of interest is captured in so-called Direction maps. These maps abstract video information, assigning two-dimensional vectors that represent the local direction of movement to regions quantified in space and time (see Figure 7).

Direction maps are obtained directly from video sequences in real time and are subsequently used to create templates to be used to identify or compare specific movement patterns (called

patterns of interest) in subsequent image sequences.

Figure 8 contains a representation of the technical approach followed by the authors. In this image, previous filtering steps necessary to create space and time quantified regions can be further appreciated.

Routing Circuit

Olshausen et al. in 1995 described a neuronal model for the representation of visual objects regardless of their position (Olshausen *et al.*, 1995). The work is based on a "dynamic routing" circuit that creates a framework for the position of objects from certain parts of the input image. This allows for the spatial relationship between objects (the location of the objects relative to each other) not to be lost.

The variable resolution pyramid structure of the input images allows the model to scale the object representation to the appropriate size for the frame of reference. Once the frame is built, it is used by the object recognition module (based on correspondence with templates) to identify the elements present in the scene. Scaling and information flow, which determine the regions of the image to be considered relevant, are regulated by interconnected and hierarchically organised control neurons. This flow of information varies depending on which neurons are active, hence the name dynamic routing. For changes in attention and gaze is sufficient to inhibit those neurons that contribute to the current routing pattern.

This model has some bearing on recent work in the field of medicine. Specifically, there are learning studies pertaining to the area of neurology, as the work of Buonomano et al. in 2009 on the influence of stimulation intervals on learning (Buonomano *et al.*, 2009), bio-inspired visual sensors networking (Ghosh *et al.*, 2007), and other MIT studies on neural computation (Miao & Rao, 2007; Uchizawa *et al.*, 2006).

In line with connectionist models, SCAN (Signal Channelling Attentional Network) model of Postma et al. (Postma *et al.*, 1997) is introduced in 1997. It is also based on a routing neural structure organized in levels to which a classifier network is added.

This classifier network (somewhat similar to the classification module of Olshausen's routing circuit) generates a pattern (called expectation pattern) for the type of target object. This pattern is then delivered to all levels of the neuronal structure. If any of the WTA processes of these levels finds similarities with the pattern, it generates a control signal that modulates the routing.

Feature–Based Models

Feature–based models main concern is how to find out the regions of interest in a given scene. They are influenced by the biological model of Koch and Ullman (Koch & Ullman, 1985) and the features integration theory of Treisman et al. (Treisman & Gelade, 1980). They are based on the extraction of features of the environment and their representation on maps called "feature maps". These maps are then integrated into a single saliency map which will be used to guide

the attention to the most conspicuous regions of the image.

Milanese (Milanese, 1993; Milanese *et al.*, 1995) developed his model for extracting salient regions based on some of the concepts previously published model by Koch and Ullman in 1985 (Koch & Ullman, 1985). The work of Milanese uses filter banks to obtain feature maps, conspicuous regions maps and a saliency map, using the latter to create the attention map that will govern the system behaviour.

Initially, feature maps for orientation, curvature, edge and colour contrast are generated from the original image. Then, interesting areas are detected on this maps through a new filtering, storing the results in conspicuous regions maps. These maps, in order to obtain regions of interest (ROI, regions of interest) compact and homogeneous, are combined in the so-called saliency map. In the end, the final attention map is generated through a binarisation process.

Figure 5. Koch and Ullman attention model (Koch & Ullman, 1985)

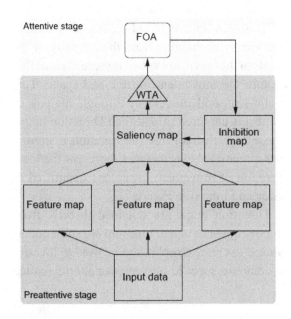

380

Figure 6. Selective tuning model by Tsotsos et al. (1995)

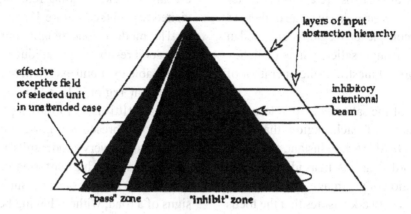

However, this model neither implements gaze changes, nor establishes the relationships that exist with the recognition system.

Work in early 2000 (Itti *et al.*, 1998a; Itti & Koch, 2001; Kadir & Brady, 2001) seeking practical implementation of visual saliency models are inspired by the work of Milanese.

Another example of feature-based model is the model implemented by Giefing et al. in the gaze control module built into their active vision system (Giefing *et al.*, 1992). In this case, feature maps are obtained by correlation of the input image with a certain kind of texture (linear elements, curves, …). These feature maps are combined, after being filtered using DoG filters (Difference of Gaussian, a wavelet mother function of null total sum obtained by subtracting a wide Gaussian from a narrow Gaussian), and their maximum values are entered in a single map of interest (see Figure 9).

According to the similarities between the information previously entered into the system (stored in a semantic network) and the observed features, a hypothesis of possible objects that may be present in the image is generated. Each of these features is considered part of an object (or several objects), and the relative position of each of them is added to the interest maps, while those regions of the image that have already been checked are inhibited in the same map.

Model of Saliency–Based Visual Attention for Rapid Scene Analysis

Inspired by the visual system of primates, Itti et al. develop their model of saliency–based visual attention for rapid scene analysis (Itti *et al.*, 1998a,b; Itti & Koch, 2000, 2001). This model combines in a single a saliency map the features extracted from a series of images of different scales.

The model input are static images defined in RGB colour space, from which a multiscale pyramid with 9 levels is obtained (Greenspan *et al.*, 1994). At each level, the original image resolution is gradually reduced by filtration, achieving scales ranging from 1:1 (level 0, original image) to 1:256 (level 8).

Through a process of differentiation between each scale images, feature maps are extracted for each level of the pyramid (Itti *et al.*, 1998b).

Three types of feature maps are obtained: intensity maps, extracted from an intensity image I calculated as the average of the colour channels R, G and B, colour, obtained from the colour channels, and orientation, obtained from the intensity using Gabor pyramids (Greenspan *et al.*, 1994).

The maps within each group (intensity, colour and orientation) are grouped in so-called conspicuity maps, one for each of the three groups mentioned above, which are denoted by $\bar{I}$, $\bar{C}$, and $\bar{O}$ respectively.

Finally, these three maps are combined, after normalisation to prevent noise in some of them to potentially hide conspicuous regions contained in others, to obtain a single saliency map (saliency map, SM). Figure 10 illustrates the operation of the system.

The purpose of the saliency map is to quantify the "saliency" of each region that falls within the visual field. At each instant, the maximum value recorded in this map identifies the most striking region of the image.

A WTA neural network ensures that the focus of attention is directed to this region of the image while simultaneously an inhibition process (called "inhibition of return", see Figure 10) temporary hides the same region in the saliency map. This allows for dynamic changes in attention, allowing the system to jump to the next region of great-

est salience and avoiding returning to a recently visited region (see Figure 11).

This model is conceptually simple and provides very good results. With a reduced computational load is able to obtain excellent results in identifying salient regions. In addition, in several experiments (Itti *et al.*, 1998b), it proved to be very robust to the presence of noise and, more interestingly, obtained very good results in the recognition of objects with a high information content in an image, such as a car on a mountain road or traffic signs of a road, without having been specifically designed for it.

NAVIS

The NAVIS (Neural Active Vision) visual attention model was developed by B. Mertsching et al.

Figure 7. Movement with orientation in space and time (Gryn et al., 2009)

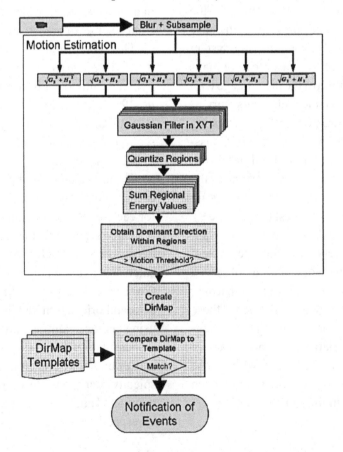

Figure 8. Direction maps implementation (Gryn et al., 2009)

and published in 1999 (Mertsching *et al.*, 1999) and is the basis of several later works (Backer *et al.*, 2001; Backer & Mertsching, 2003; Frintrop *et al.*, 2005b) dating from 2001, 2003 and 2005 respectively.

The operation of the system consists of three distinct phases: first, salient regions of the image are identified and selected computing a series of features. Second, a high-level component performs the control of attention using the available knowledge about the environment and the current state of the system. Finally, behaviour is activated, choosing the most appropriate for the current situation from a predefined set of basic behaviours.

The first stage of the NAVIS system (identification and selection of conspicuous regions) falls into the category of feature–based models. Three feature maps are built: edges orientation, area orientation, and colours, which will allow the system to obtain information on symmetry, eccentricity and colour contrast. The computed features are general-purpose, as was the case with the model proposed by Itti et al. (Itti *et al.*, 1998a,b; Itti & Koch, 2000, 2001), which makes it highly domain-independent, although it is possible to add new specialised features.

Finally, just as happens in other models (Itti & Koch, 2000; Wolfe, 1994), the information converge into a single saliency map that will change the focus of attention. The selection and object tracking is integrated into a single process using a system based on dynamic neural fields (Amari, 1977; Backer *et al.*, 2001).

Model Summary

Table 1 summarises the most relevant attention models introduced in previous sections.

VISUAL ATTENTION MODEL FOR ROBOT CONTROL

The model of attention developed as part of the Nao robot control module fits into feature–based attention models. Specifically, the model presented here extends the practical application that Itti et al. (Itti *et al.*, 1998a,b; Itti & Koch, 2000, 2001) made of the classical model of Koch and Ullman (Koch & Ullman, 1985). Both its reduced computational cost and the use of basic features such as colour, intensity and orientation, often used in the development of tasks in the field of robotics, make this model a suitable candidate to work in this discipline.

The objective of the extended model is improving those areas in which the original had its major drawbacks as well as applying the benefits of attention to robotics field. The use of an attention model allows not only for a reduction of the time required by vision algorithms, but also for an improvement of the system's accuracy, easing decision-making process through the limitation of allowed actions to those which can be applied to the elements contained in the regions selected by the attention model.

Scenario and Main Contributions

Vision and control systems in Robotics are usually implemented in an impulse-analysis-response fashion. Given a visual impulse, The proposed model builds up a so called saliency map using intensity, colour, and orientation contrast information contained in the observed scene. The most relevant regions of the obtained saliency map are then grouped in small clusters called blobs which are sorted according to their attentional relevance.

The most important contributions of the proposed model and the real differences between it and the original are the way colour and orientation information is extracted, as well as the normalisation method employed and the blob–based map scan proposed. Input image size is also reduced by means of a virtual fovea mask, further releas-

ing computational resources while also reducing the influence of distracting peripheral elements.

The model has been validated using the same image benchmarks proposed by Itti (Itti & Koch, 2000; Itti & Koch, 2001; Itti *et al.*, 1998). For better comparison with the original model and to prevent some of its extra features (not present in our model) from slowing it down, an implementation of it using the same programming language and artificial vision libraries employed to build the proposed model has been used instead of the version available at their web http://ilab.usc.edu.

Robotic Platform

Further prove of the model's usefulness for attention–related tasks solving has been obtained by

Figure 9. Active vision system: preatentive module (Giefing et al., 1992)

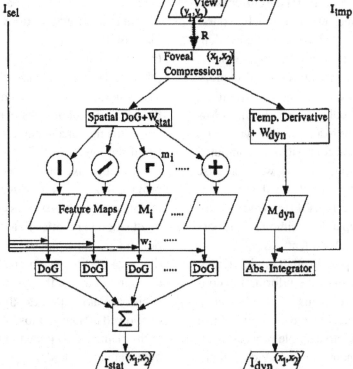

Figure 10. Saliency–based visual attention model (Itti et al., 1998a)

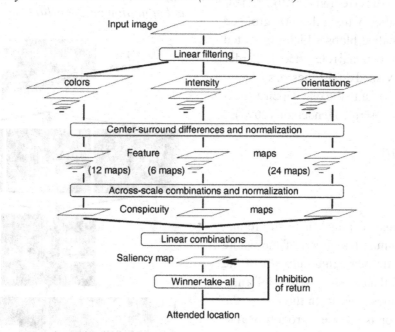

using it for controlling a robotic platform, Nao robot, from Aldebaran Robotics company.

Nao robot (see Figure 12) has two 30 fps video cameras located in the forehead and in the mouth, each one with a maximum resolution of 640x480, but they can neither be used simultaneously nor are capable of stereo vision since their field of view is not overlapped. All control is made on-board using a x86 AMD Geode chip at 500 MHz, 256 MB of SDRAM memory and a standard 1 Gb in flash memory that can be upgraded. Given that all teams use this same platform in the RoboCup SPL competition, software optimisation becomes critical. More than 75% of processor time is used for visual related tasks, which means every little improvement in this area will greatly benefit the whole system.

The model has allowed for successfully completion of the any ball challenge, proposed at RoboCup SPL (Standard Platform League) 2009 (see Experiments section below), a world-wide competition in which every participant competes using this robot, hence its name.

Model

At any given time, the maximum registered in the saliency defines the most important region from an attentive point of view.

To build up the saliency map, three maps are used: an intensity map, a colour map and an orientation map.

The maps assign high values to those areas which stand out in the magnitude they meassure, for instance, the intensity map will assign high values to those areas the intensity (light) of which changes a lot in relation to their surround.

The maps are obtained using the original camera image. To reduce the amount of pixels to be computed, and thus improving performance, a virtual fovea mask which simulates the human eye progressive resolution decrement is used. The further a region is from the centre of the image, the greater the amount of masked pixels will be (masked pixels will not be analysed). The Fovea section shows a more detailed description of this fovea mask.

Once the most relevant regions have been identified within the saliency map, they are grouped in small clusters called blobs which are sorted according to their attentional relevance. The order in which blobs are visited establishes an scanning path of the scene similar to the one which a spectator's eyes would describe Henderson (2003).

Implementation

Fovea

The fovea is a small depression in the retina. It grants the maximum resolution of the whole field of view and despite being only 1% of the retinal area, more than 50% of visual information processed comes from it. In the rest of the retina, the resolution is inversely proportional to the distance to the fovea. It allows sharp vision and tasks dependant on it such as lecture or driving. The term fovea is often used in Robotics to specify that the centre of the image is not treated the same way as the periphery.

Multi resolution sensors (as the fovea, be it real or virtual) are not necessary for the attention system to work. However, there are a couple of reasons (biological and computational) which make interesting to use them:

- The amount of sensor information is reduced in comparison to using the whole field of view at maximun resolution.
- High resolution and wide field of view can be combined as a consequence of the latter.
- Peripheral vision gives only contextual information, allowing for a not so exhaustive process in these areas.

All these advantages can be applied to a robotic vision system, from a hardware or software approach (Tistarelli & Sandini, 1991; van der Spiegel *et al.*, 1989). Hardware solutions are not suitable for us due to our robot being standard, so an algorithmic solution has been chosen.

Figure 11. Saliency maps, focus of attention change and inhibition process (Itti et al., 1998a)

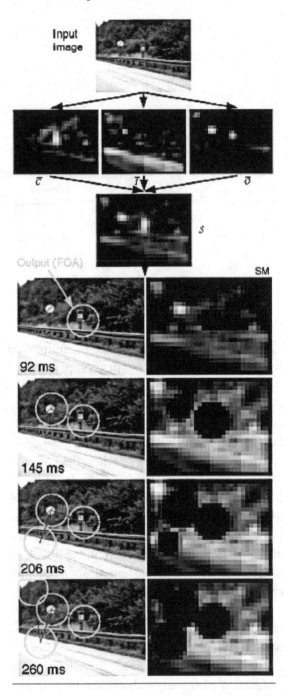

Our solution is based on C. J. Westelius (1995) work, but instead of modifying the original image to create the foveal effect, we use a mask to

Table 1. Attention models

Model	Objective	Main features	Attention control
Classic	Conspicuous regions	Feature maps	Saliency map
Guided Search	Conspicuous regions	Top-down control	Activation map
		Color and orientation	
Selective tuning	Object search	Patterns and WTA	Frame of reference
		Movement patterns	
Routing circuit	Objects tracking	Input patterns	Routing circuit
SCAN	Objects search	Neural structure	Expectation pattern
		Classifier network	
Salient regions extraction	Conspicuous regions	Orientation	Attention map
		Edges	
		Curvature	
Saliency–based attetion	Conspicuous regions	Intensity	Saliency map
		Colour	
		Orientation	
NAVIS	Conspicuous regions	Attractivity maps	Saliency map
	Attention and behaviour	Contrast	
		Eccentricity	
		Simmetry	

grant access only to certain of its pixels. This mask allows for simulating the lower resolution of peripheral areas without slowing down the system with unnecessary filter and subsampling operations.

It is possible to configure the amount, size and resolution reduction ratio of these areas. By default, three areas are used: fovea (full resolution, 40% of the image), parafovea (1:2 reduction, 40% of the image) and perifovea (1:4 reduction, 20% of the image). In case attentional relevance of peripheral areas needs to be further reduced, parafovea and perifovea precision can be halved, using a 1:1:4:16 (original:fovea:parafovea:perifo vea) mask. Figure 13 shows the result (right) of applying a 1:1:2:4 mask (centre) to the original image (left).

Maps

The input of the model are 640x480 pixels static RGB colour images (Figure 14 shows the input image which will be used as an example during the explanation of the map construction process). After the previously explained fovea mask has been applied, these images are used to build multiscale pyramids (Anderson & Essen, 1987) for every map used in the model. Each pyramid has 9 levels and a resolution reduction factor of $1 : 2^n$ for each of them. Level 0 means then no reduction (1:1, original image), while maximum reduction happens at level 8 (1:256).

Intensity Maps

The first step of the model consists of creating a nine level intensity pyramid which represents the "intensity" (luminosity) of each image pixel. Using the original image, a intensity matrix M_I is obtained by combination of the R, G and B channels value:

$$m_I(i,j) = \frac{m_R(i,j) + m_G(i,j) + m_B(i,j)}{3}$$

387

The intensity pyramid is then created using M_I, with $M_I(n)$ being the intensity matrix corresponding to the *nth* level of the pyramid. Using the pyramid, six intensity maps $M_{I(m,n)}$ are obtained by across-scale difference, $\ominus$, which is obtained by interpolation of the maps to the finer scale and point-by-point subtraction.

The amount of maps, the way scales are paired, and the reasons behind this process are the same already explained in Itti *et al.* (1998a)

In a more optimised version of the model, the finnest scale is $n=2$ and not $n=0$ to reduce noise, excessive detail, and the amount of pixels to be computed (160x120 at scale 2 instead of 640x480 at scale 0), improving both performance and robustness.

Finally, an intensity conspicuity map $\bar{I}$, representing those conspicuous locations from an intensity point of view, is generated combining all the previous maps through across-scale addition, $\oplus$, which consists of reduction of each map to scale $n=4$ (40x30 resolution) and point-by-point addition:

$$\bar{I} = \oplus M_{I(m,n)}$$

Figure 15 shows the intensity conspicuity map $\bar{I}$ for the image at Figure 14.

Colour Maps

Four pyramids representing "colour" of each image pixel are created using the normalised R, G and B channels and a yellow channel Y (obtained using the three previous ones).

Two consecutive normalisations are employed: the first normalisation is a necessity of using RGB colour space, in which channels include intensity information, thus, in order to make the result independent to environmental light, they have to be normalised by intensity. To do so, we applied the same formulae used in Walther (2006), which improves the quality of the obtained maps by properly obtaining yellow colour information.

Second normalisation is not related with the colour space and is not present in the original model. It allows for promotion of those chan-

Figure 12. Nao robot (figure copyrighted by Aldebaran Robotics)

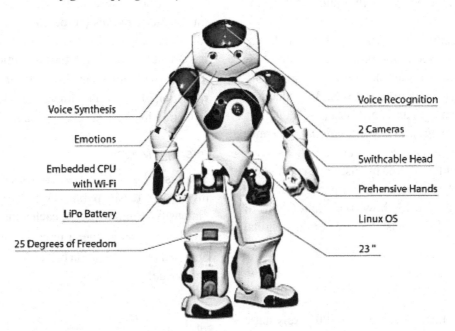

Figure 13. Fovea mask

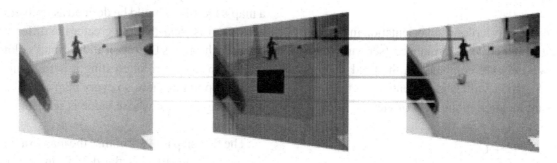

nels whose information is more salient from the very first steps of the attentional algorithm. The method used for channel normalisation is the same used for map normalisation, explained in the Normalisation section. Reasons and advantages behind channel normalisation are reviewed in the Experiments section.

The four colour pyramids are then used to generate a set of 12 colour maps, six for difference between red and green components, $M_{RG(m,n)}$, and six for blue and yellow difference, $M_{BY(m,n)}$, in a similar fashion to the intensity maps.

$M_{BY(m,n)}$ are obtained in a similar way to $M_{RG(m,n)}$ but using the Blue and Yellow components instead.

Finally, a colour conspicuity map $\overline{C}$, representing those conspicuous locations from a colour contrast point of view, is generated combining all the previous maps:

$$\overline{C} = \oplus\left[M_{RG(m,n)} + M_{BY(m,n)}\right]$$

Figure 16 shows the colour conspicuity map $\overline{C}$ for the image at Figure 14.

Orientation Maps

The model builds up a set of orientation maps which are merged in a final orientation map O which represents the location of those elements which stand out from an orientation point of view in comparison to the rest of the objects present in the image. As is the case with colour and intensity, orientation maps are obtained using pyramids.

Four different pyramids are used, one for each of the four orientation θ considered $\left(0°, 45°, 90°, \text{ and } 135°\right)$ by applying a set of Gabor filters directly to the Intensity pyramid levels.

This approach differs from the original (Greenspan *et al.*, 1994; Itti & Koch, 2000) since it doesn't apply Gabor filters to an edge pyramid (which is obtained by either subtracting consecutive levels of the intensity pyramid or by applying an edge extractor kernel to each of its levels). Results obtained this way not only gather orientation information correctly, but are also more noise tolerant and more efficient. The Experiments section further explains this point.

These pyramids are then used to generate a set of 24 orientation maps $M_{O(m,n,\theta)}$, six for each orientation.

Finally, an orientation conspicuity map $\overline{O}$, representing those conspicuous locations from an orientation contrast point of view, is generated combining all the previous maps:

$$\overline{O} = \oplus M_{O(m,n,\theta)}$$

Figure 17 shows the orientation conspicuity map $\overline{O}$ for the image at Figure 14.

Saliency map

Once intensity, colour and orientation maps have being obtained and normalised (see subsection below), they are combined in the final saliency map S which will guide attention to the most relevant location in the field of view:

$$S = \frac{\overline{I} + \overline{C} + \overline{O}}{3}$$

Figure 18 shows the 3D (left) and 2D (right) saliency map S for the image at Figure 14.

The saliency map S is then applied to the original image as obtained by the robot camera, promoting the most relevant locations and hiding the rest. In Figure 19 this process is illustrated: left image is the original coloured image. Central image shows the results of applying the saliency map in Figure 18 to the original image (the darker the area, the less salient it is). Right image shows the regions with higher saliency across the whole map (green rectangles). Please note that the system proposed only tell us "where" to look at (area) and not "what" (object) to look for; the fact that the ball and the goal post are in those areas is a consequence of being the most notorious regions of the image from a colour and intensity point of view.

Normalisation

Before obtaining each conspicuity map $(\overline{I}, \overline{C}, \ and \ \overline{O})$ and the final saliency map S, each individual map has to be normalised. Normalisation is also applied to each individual colour channel before building up the colour pyramids.

The normalisation method used has been called direct current offset–based normalisation. It is an order of magnitude faster than Itti's local iterations method while obtaining results of comparable quality.

The main idea behind this method is the concept of a signal's direct current offset, defined as

its average amplitude. Following this definition, a map's DC offset could be defined as its average activity. It is clear that the greater the difference between the map's DC offset and the map's global maximum, the more interesting it is from an attentional point of view, so every map could have a wighting factor p whose value is set using this difference.

The first step is to normalise the map to a fixed dynamic range ([0,1] is the default interval) so that modality dependant differences are removed (see Itti *et al.* (1998a). After that, it is necessary to obtain the map's maximum M and its DC offset DC_O, defining then the wighting factor $p' = (M - DC_O)^2$ for the map.

Since spatial distribution of a map's activity is also very important to properly promote a map (see Experiments section), four different weighting factors p_i are used instead of one. These fac-

Figure 14. Original input image

tors are calculated by scanning the map using four different scanning routines R starting from each of its four corners: from left to right and from top to bottom (i=1), from right to left and from top to bottom (i=2), from left to right and from bottom to top (i=3) and from right to left and from bottom to top (i=4).

This way, the spatial distribution of map's activity can be determined at the same time it is scanned.

An "activity counter" a_c is used to store the value of saliently interesting pixels. A pixel value is considered interesting (and then added to the counter a_c) if its value falls between $1.5DC_O$ and M.

Since M and DC_O are dynamically updated while scanning the map for a_c computation, the more high activity values are found, the less values are added to a_c. This way, maps in which it takes more time to find the global maximum (which means maximum activity regions are concentrated in smaller areas and thus are more salient) get a higher weighting factor.

This process is followed to obtain each factor p_i, resetting M, DC_O and a_c before starting computation of each of them. The final map's weighting factor p is the average of the four p_i values:

$$p = \frac{p_1 + p_2 + p_3 + p_4}{4}.$$

To summarise the whole process, the normalisation is performed in four consecutive steps:

1. The map is normalised to a fixed dynamic range ([0,1] by default).
2. The map is globally multiplied by:
$$p = \frac{p_1 + p_2 + p_3 + p_4}{4}$$
where
$$p_i = \left(M - \overline{a_c} \right)^2, \forall i \in R$$

Blob–Based Map Scan

The region of the saliency map with maximum activity identifies the most conspicuous region of the visual scene and is chosen as focus of attention of the system. In absence of any other control mechanism, the system will always select the same area over and over again (in case of static scenes). In some cases, this may be an undesired behaviour since it prevents other regions but the most relevant to be explored.

Other authors use an inhibition of return mechanism to solve this problem (Itti *et al.*, 1998a;

Figure 15. Intensity map

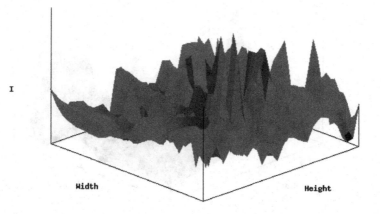

Koch & Ullman, 1985). This technique basically consists of applying an inhibition region of the same size of the focus of attention in all regions already visited, temporary making them non–salient and allowing the next most conspicuous region to be analysed.

An alternative approach based on blobs can improve model efficiency. Blobs are clusters of neighbour pixels which share some kind of property, in this case having a saliency value close enough: those pixels which are spatially next to each other and have an activity within a given range with respect to the maximum of the map are grouped in the same blob.

Several intervals can be used to generate the blobs. For instance, once the saliency map is obtained, blobs can be obtained for three activity interval: [90,100], [75,90), and [60,75) (with the limits of the intervals representing the specified percentage of the map's maximum). Every interval can have none to several blobs.

Each blob has a saliency value, computed as the average saliency of the pixels it contains. This allows all blobs to be sorted according to their saliency values so that regions enclosed by them can subsequently be visited in that order.

Figure 20 shows an example of visual scene scan using blobs and the differences between using one and three intervals to generate them.

Thanks to using this method to scan the saliency map, image processing procedures can be locally and easily applied to regions enclosed in a specific blob without having to deal with the whole image. This can be useful, for instance, if some kind of object not specially salient needs to be found in the current visual scene: almost certainly, one of the saliency blobs generated will contain such an object, so that finding it is just a matter of processing one after another until the desired object is located.

Experiments

The "any ball" RoboCup challenge can been chosen as an example to test the effectiveness of the attention control model in robotics environments.

For this challenge, the robot is placed in the game field along with several random coloured and multi-sized balls. The robot has then a couple of minutes to score the biggest amount of goals possible. Classic colour filter algorithms used for image segmentation are not useful in this scenario, since not only ball colour is unknown, but they can also have the same colour as the ground (green).

In order to solve this challenge both a simulation and a real scenario have been used.

Simulation is so called because only the attention model is implemented, supposing that the robot would be able to kick the ball after knowing

Figure 16. Colour map

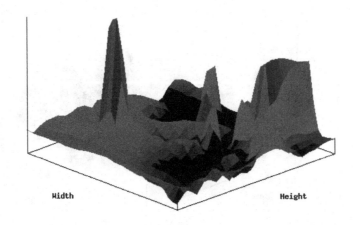

its location. In this case, just finding the location of any ball is enough to consider the test a success.

The term "real scenario" means that ball approaching and kicking are also implemented.

Simulation

To simulate this scenario, the robot is given some pictures of the game field containing a random number of different colour balls (see Figure 21).

The attention model always finds the most salient region in the image, and as long as that region is not dealt with (or inhibited), it will not find any other region. This means that regions chosen as most salient which do not contain any ball must be masked (inhibited), so that others containing a ball can be chosen as focus.

Once a region containing a ball is chosen (see Figure 22), robot should approach to it and try to score a goal by kicking it. This part of the experiment is not implemented during simulation, but it can be assumed that the ball will end up further from the robot than it was when chosen as focus. To simulate that, once a region containing a ball is chosen by the model, it is assumed that the robot could kick it and that specific ball is removed from the next input image for the robot.

With the originally most salient ball no longer present in the field of view (see Figure 23), the saliency map changes and a new most salient region is chosen (see Figure 24). The previously explained process is now repeated: if the new region contains a ball, it is chosen as focus and kicked, otherwise, it is inhibited and the second most salient region is checked, repeating the process until finding a region containing a ball or not finding any at all.

The results obtained are very promising, with a 100% success rate for the images used. Even the regions containing small balls with almost the same colour of the ground are chosen in the last iterations of the algorithm (see Figure 25). It can be easily understood that the colour map (top left image at Figure 26) gives no useful information in this particular case, since the whole field of view is almost of the same colour (except for the white lines of the field). However, the intensity map (top right image at Figure 26) shows strong peaks at those areas containing either shades, which should be minimal except for the one belonging to the ball (due to it being the only object in the field apart from the robot), or different light reflection patterns, as it happens with the region containing the ball since the ball is made of a different material from the ground's. The final

Figure 17. Orientation map

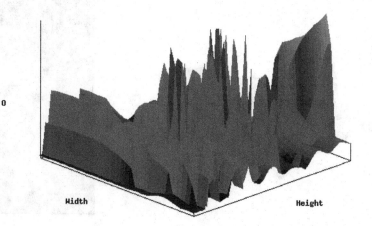

saliency map obtained once again chooses the region containing the ball as the most salient one (see bottom left and bottom right images at Figure 26).

Real Scenario

In this case, the whole process involving scoring a goal is implemented: the attention algorithm is used to locate a target ball to which the robot approaches and kicks in order to score.

Also, with the objective to prove that the model can be applied not only to objects of any colour and texture but also to objects of any shape and size, objects other than coloured balls have been used.

In this case, since objects are not removed from the field, the robot eventually kicks the same object over and over again until it gets away from its field of view. Once this happens, the robots resumes the search for other objects, choosing the most salient one and kicking it, until everything on the field has been cleared.

To put the whole system to the test, three different coloured objects are positioned in the field at random locations. As it may be expected from the results commented in the previous section, the robot successfully chooses every target, one after another, gets close to them and kicks them. A video showing this behaviour along with the attention model input, the real images, and its output, the target locations, can be watched at http://robotica.unileon.es/~jfgars/pubs.

Software Implementation, License, Use, and Distribution

The software implementation of the model presented in this section has been developed in C and C++. It can be downloaded from http://robotica. unileon.es/~jfgars/pubs/atencion.

There are two releases available:

- **Fastpicture:** Stand-alone version of the model.
- **Scaner:** Implementation of the model specific for the Nao robot.

Both the stand-alone and the Nao robot specific code are free software. Redistribute them and/or modify them under the terms of the GNU General Public License, version 2 onwards, as published by the Free Software Foundation.

This code is distributed in the hope that it will be useful, but without any warranty. For

Figure 18. Saliency map

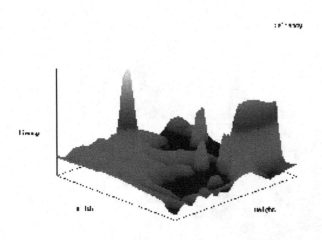

Figure 19. Saliency map applied to the original image

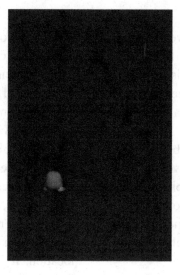

more information, you can download the GNU General Public License from http://www.gnu.org/licenses/gpl.txt.

Conclusions

In this chapter, most relevant visual attention models have been analysed and our own attention model, aimed to work on–board a mobile robot, has been introduced.

The proposed model fits within feature–based models. Specifically, it is based on Itti's visual attention model, chosen as a starting point for our work due to its simplicity and computational efficiency and from which main troubles have been solved.

Its advantages compared to the original model are both qualitative and quantitative, affecting not only the quality of the maps obtained but also the time required to do so, enabling it to operate in a limited hardware environment: the Nao robot, robotic platform whose control has been carried out using this model.

The model has been adapted to make the robot able to identify conspicuous elements located on the RoboCup game field.

The achievement of both tasks, computational optimisation and identification of regions which are useful to the robot, has been simplified thanks to the use of a virtual fovea, which allows for a reduction of the computational needs of the proposed algorithms and eases the precision of the model blurring potentially distracting elements.

Contributions of the present work can be grouped in two main categories: contributions to the theoretical model and contributions to its implementation.

The first ones are modifications of Itti's model theoretical foundations. The second ones are proposals which maintain the ideas of the original model, but differ on a more efficient implementation of them from a computational point of view.

Bellow, contributions in each of these categories are summarised.

Theoretical Model

The main contributions to the theoretical model are usage of a virtual fovea, direct current offset–based normalisation, colour channels normalisation for colour maps computation, and blobs–based scan of the saliency map.

Fovea

The proposed model uses a virtual fovea to reduce the resolution of the scene's peripheral areas. As a result, saliency of the central regions is promoted, while contextual influence of the potentially distracting elements located in peripheral regions of the image is reduced.

This is specially useful for the RoboCup since distracting elements such as the crowd or the robot's feet always appear in the peripheral areas. Since it is virtual, computed by the robot's software which masks the image instead of modifying it, performance loss is almost inappreciable ($\approx$1 ms).

Direct Current Offset–Based Normalisation

A new type of normalisation is used to highlight maps which contain more interesting information from an attentional point of view, direct current offset–based normalisation, which weighs maps depending on the value and the spatial distribution of their activity peaks (maximums).

This involves a qualitative improvement from content–based global non–linear normalisation, which only considers differences between maximum values, which results in erroneous weigh of maps with a reduced amount of maximums close to each other either in absolute value or in spatial location. It is also a quantitative improvement from iterative localised interactions normalisation, which requires a computation time an order of magnitude superior.

Figure 20. From left to right and from top to bottom: original image, saliency map and blob–based map scan. Bottom left image shows the blobs generated considering only the range [90,100], while bottom right image shows the results obtained using three intervals: [90,100], [75,90), and [60,75). The number close to each blob is the order in which it will be visited.

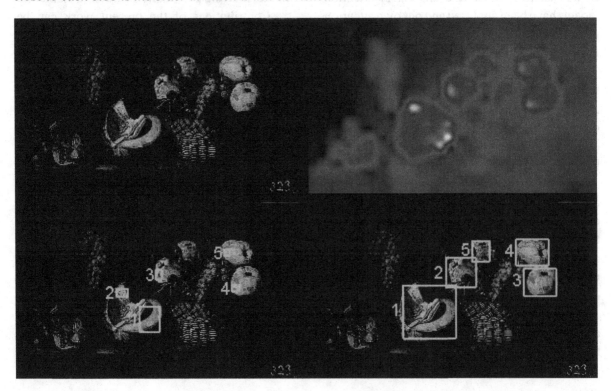

Figure 21. Any ball challenge input image

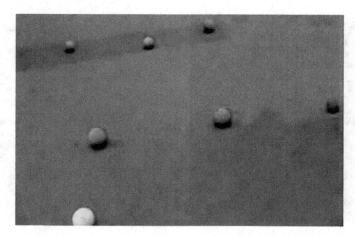

Independent Normalisation of Colour Channels

Each one of the colour channels r, g, b is normalised independently as a previous step to colour maps calculation, so that every channel has a given weight which is proportional to the attentional value of the information it contains.

This way, when computing colour maps, the most conspicuous channel of the pair which makes up the map is properly highlighted, either red or green for RG maps and either blue or yellow for BY maps. As a result, colour double–opponency model is properly implemented.

Using this method, erroneous simplifications in maps creation (like obtaining the opponency using the absolute value of the difference of the considered channels) or employing valid but computationally inefficient approaches (like computing each opponency independently: red–green, green–red, blue–yellow, and yellow–blue) are avoided.

Blobs Based Map Scan

The proposed method consists of creating blobs with the most conspicuous regions of the saliency map, creating then a sorted list with them (ordered from most to least salient) so that most important regions are visited first

This is not an improvement of the original model per se, but a consequence of the model's objectives (working on-board the Nao robot and in the RoboCup environment), since this way each generated blob can be analysed faster, independently and more efficiently in order to look for specific objects like a ball or the keeps.

Implementation

Implementation of the model is a contribution by itself, since the current research is not limited to a theoretical development, but also includes an implementation and empirical validation of the theoretical contributions.

The most relevant implementation improvements are the orientation pyramids computation using Gabor filters directly over the intensity pyramid and half wave rectification applied as a previous step to normalisation.

Orientation Pyramids Computation

In the original model, Itti obtains an edge pyramid (La Place pyramid) using the intensity pyramid by substraction of its consecutive levels. Gabor filters were then applied to this pyramid to build up the four orientation pyramids.

The proposal of the implemented model is leaving out this step, directly applying the filters

Figure 22. Most salient region of the input image

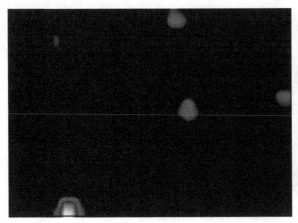 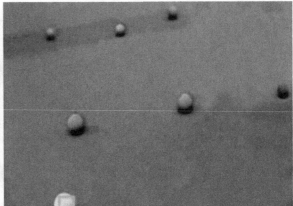

over the intensity pyramid using a kernel which allows for highlighting of thicker oriented elements. Since edge pyramid doesn't have to be computed, performance of the model is improved.

Half Wave Rectification

Half wave rectification as previous step to maps normalisation to is also proposed. This operation is the same than Itti applies to colour channels, discarding those values bellow a given intensity threshold. In the proposed model, rectification is applied to each normalised map, using a certain percentage of the global maximum of the map as threshold.

The most important effect of this rectification is noise reduction and a better smoothing of the obtained maps.

Model Improvements Thanks to Contributions

As it has already been stated in this section, present work contributions result in a both quantitative and qualitative improvement of the original model.

Its application for humanoid robot controlling in the RoboCup environment has also allowed for

Figure 23. Any ball challenge second input image

Figure 24. Most salient regions of the second input image

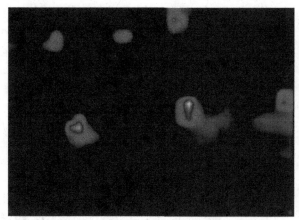 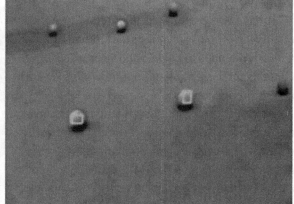

solving certain situations which could barely been handled using other artificial vision techniques more usual in this kind of environment.

Quantitative Improvements

Quantitative improvements result in a faster model: correct colour maps calculation is faster when normalising each channel instead of using separated maps, orientation maps calculation is also faster since edge pyramid is put aside, and, above all, normalisation used is an order of magnitude faster than iterations normalisation used by Itti and a 20% faster than content–based global normalisation, which eases its use in real time.

Ultimately, the model obtained is on average 23% faster than the Itti's version which uses content–based global normalisation, and 75% faster than the version which uses iterations normalisation (Garcia, 2011).

Qualitative Improvements

Qualitative improvements allow the model to generate better quality maps, solving at the same time specific problems of the original model (Garcia, 2011): orientation maps obtained are smoother since Gaussian pyramids (intensity pyramid) are used to generate them and all maps show less noise

Figure 25. Input image containing a ball of the same colour of the ground

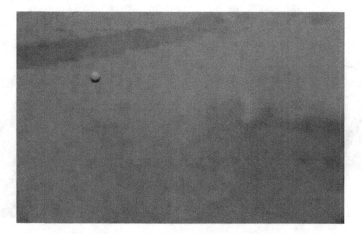

thank to half way rectification as a previous step to normalisation.

It has been established that a map has good quality if it properly identifies the kind of stimulus which is intended to contain and, at the same time, contains the lesser amount possible of noise and non desired elements.

In order to measure this quality in the most objective way, a method to estimate it from the quotient between the map's optimal activity (maximums which the map should contain) and its real activity (maximums which it really contains) has also been proposed.

This way, those maps whose real activity is very close to its optimal activity obtain a higher quality degree (closer to 1).

Having this into account, the quality of normalised maps is comparable to the quality of those obtained by the normalisation methods used by Itti, but with the advantage of not modifying the original maps.

Specific difficulties of content–based global normalisation are also solved.

Application to Robotic Humanoid Control

The application of the model to the RoboCup domain allows for solving certain challenges otherwise complex, specially if only colour filter and segmentation classic techniques are employed.

Attention reduces available actions to those which can be applied to the objects contained within the regions which are selected as salient.

Figure 26. Colour, intensity and saliency map and most salient regions of an input image containing a ball of the same colour of the ground

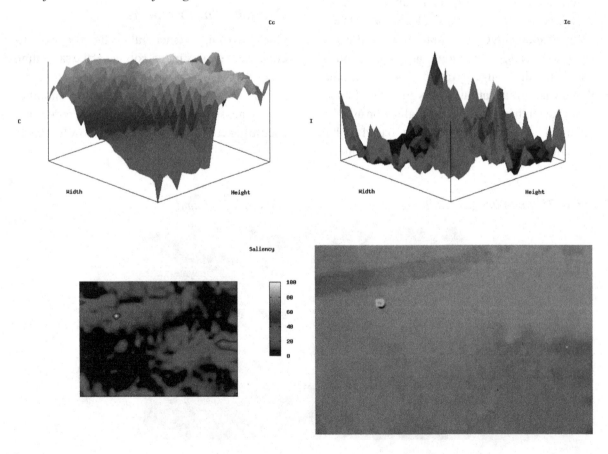

Any ball challenge proposed solution illustrates this. Thanks to the used system it is possible to find in the game field balls of any colour and texture, even those whose colour is very close to the one of the terrain.

Computing visual scene's saliency is less time consuming and has less performance problems (difficulties when dealing with partial occlusions of the objects) than shape recognition techniques (Hough transform).

Limitations

In order for the model to be continuously usable in real time, that is, in order to be able to compute the saliency map corresponding to each frame, it would be necessary to further reduce the computation time required for the algorithm, which is currently around 115 ms for each analysed image using maximum resolution (640×480 pixels) in a XPS M1330 laptop with a Intel Core™ Duo T5850 2.16 GHz processor and 4 GB RAM memory (Garcia, 2011).

Computation time can be reduced down to 42 ms when working with 160×120 images, which is the same as taking level 2 of the pyramid as starting level, and if initial colour and intensity channels and final saliency map are also obtained using the same resolution.

Also, the fact that none of the measured times include the time necessary to obtain the input image for the model has to be taken into account. This may increase computation time from 1.6 ms to 20 ms while working on-board the robot.

For a more detailed analysis of the model and its exprimental results, see (Garcia, 2011).

REFERENCES

Allport, A. (1987). *Selection for action: Some behavioural and neurophysiological considerations of attention and action*. Erlbaum.

Amari, S.-I. (1977). Dynamics of pattern formation in lateral inhibition type neural field. *Biological Cybernetics*, *27*, 77–87. doi:10.1007/BF00337259

Anderson, C., & Van Essen, D. (1987). Shifter circuits: A computational strategy for dynamic aspects of visual processing. In *Proceedings of the National Academy of Sciences USA*.

Anderson, J. R. (2004). *Cognitive psychology and its implications*. Worth Publishers.

Backer, G., & Mertsching, B. (2003). Two selection stages provide efficient object-based attentional control for dynamic vision. In International Workshop on Attention and Performance in Computer Vision.

Backer, G., Mertsching, B., & Bollmann, M. (2001). Data and model driven gaze control for an active vision system. *IEEE Transactions on Pattern Analysis and Machine Intelligence*, *23*, 1415–1429. doi:10.1109/34.977565

Bajcsy, R. (1986). *Passive perception vs. active perception*. In IEEE Workshop on Computer Vision.

Ballard, D. H. (1990). *Animate vision, technical report 329. Technical report*. Computer Science Department, University of Rochester.

Baylis, G., & Driver, J. (1993). Visual attention and objects: Evidence for hierarchical coding of locations. *Journal of Experimental Psychology. Human Perception and Performance*, *19*, 451–470. doi:10.1037/0096-1523.19.3.451

Blaser, E., Pylshyn, Z., & Holcombe, A. (2000). Tracking an object through feature space. *Nature*, *408*(6809), 196–199. doi:10.1038/35041567

Boehme, H.-J., Brakensiek, A., Braumann, U.-D., Krabbes, M., & Gross, H.-M. (1998). Neural networks for gesture-based remote control of a mobile robot. *Proceedings of the IEEE World Congress Computational Intelligence*, Vol. 1, (pp. 372–377).

Broadbent, D. E. (1958). *Perception and communication.* New York, NY: Pergamos Press. doi:10.1037/10037-000

Brunnström, K., Eklundh, J.-O., & Uhlin, T. (1994). Active fixation for scene exploration. *International Journal of Computer Vision, 17,* 137–162. doi:10.1007/BF00058749

Buonomano, D. V., Bramen, J., & Khodadadifar, M. (2009). Influence of the interstimulus interval on temporal processing and learning: Testing the state-dependent network model. *Philosophical Transactions of The Royal Society B. Biological Sciences, 364,* 1865–1873. doi:10.1098/rstb.2009.0019

Burt, P. (1988). Attention mechanisms for vision in a dynamic world. *Proceedings Ninth International Conference on Pattern Recognition,* Beijing, China.

Deco, G. (2000). *A neurodynamical model of visual attention: Feedback enhancement of spatial resolution in a hierarchical system.*

Frintrop, S., Backer, G., & Rome, E. (2005). Goal-directed search with a top-down modulated computational attention system. *Lecture Notes in Computer Science, 3663,* 117–124. doi:10.1007/11550518_15

Garcia, J. F. (2011). *Contributions to visual attention computation and application to the control of a humanoid robot.* PhD thesis, Universidad de Leon.

Ghosh, B. K., Polpitiya, A. D., & Wang, W. (2007). Bio-inspired networks of visual sensors, neurons, and oscillators. *Bio-Inspired Networks of Visual Sensors, Neurons, and Oscillators, 95,* 188–214.

Giefing, G.-J., Janûen, H., & Mallot, H. (1992). *Saccadic object recognition with an active vision system.* In 10th European Conf. Artificial Intelligence.

Greenspan, H., Belongie, S., Goodman, R., Perona, P., Rakshit, S., & Anderson, C. H. (1994). Overcomplete steerable pyramid filters and rotation invariance. *Proceedings of IEEE Computer Vision and Pattern Recognition,* (pp. 222–228).

Gryn, J. M., Wildes, R. P., & Tsotsos, J. K. (2009). Detecting motion patterns via direction maps with application to surveillance. *Computer Vision and Image Understanding, 113,* 291–307. doi:10.1016/j.cviu.2008.10.006

Hamker, F. H. (2000). *Distributed competition in directed attention.*

Henderson, J. M. (2003). Human gaze control during real-world scene perception. *Trends in Cognitive Sciences, 7,* 498–504. doi:10.1016/j.tics.2003.09.006

Hübner, R., & Backer, G. (1999). Perceiving spatially inseparable objects: Evidence for feature-based object selection not mediated by location. *Journal of Experimental Psychology. Human Perception and Performance, 11,* 583–597.

Itti, L., & Koch, C. (2000). A saliency-based research mechanism for overt and covert shifts of visual attention. *Vision Research, 40,* 1489–1506. doi:10.1016/S0042-6989(99)00163-7

Itti, L., & Koch, C. (2001). Computational modelling of visual attention. *Nature Reviews. Neuroscience, 2,* 194–203. doi:10.1038/35058500

Itti, L., Koch, C., & Niebur, E. (1998). Attentive mechanisms for dynamic and static scene analysis. *IEEE Transactions on Pattern Analysis and Machine Intelligence, 20,* 1254–1259. doi:10.1109/34.730558

Itti, L., Koch, C., & Niebur, E. (1998). A model of saliency-based visual attention for rapid scene analysis. *IEEE Transactions on Pattern Analysis and Machine Intelligence, 20,* 1254–1259. doi:10.1109/34.730558

Jenning, C., & Murray, D. (1997). Gesture recognition for robot control. In *Proceedings of the Conference on Computer Vision and Pattern Recognition*.

Julesz, B. (1991). Early vision and focal attention. *Reviews of Modern Physics*, 66(3), 735–772. doi:10.1103/RevModPhys.63.735

Kadir, T., & Brady, M. (2001). Saliency, scale and image description. *International Journal of Computer Vision*, 45, 83–105. doi:10.1023/A:1012460413855

Koch, C., & Ullman, S. (1985). Shifts in selective visual attention: Towards the underlying neural circuitry. *Human Neurobiology*, 4, 219–227.

Krotkov, E. (1987). *Exploratory visual sensing for determining spacial layout with an agile camera system*. PhD thesis, University of Pennsylvenia.

LaBerge, D. (1995). *Attentional processing*. Harvard University Press.

LaBerge, D., Carlson, R. L., Williams, J. K., & Bunney, B. G. (1997). Shifting attention in visual space: Tests of moving-spotlight models versus an activity-distribution model. *Journal of Experimental Psychology. Human Perception and Performance*, 23, 1380–1392. doi:10.1037/0096-1523.23.5.1380

Lopez, M. T., Fernandez, M. A., Fernandez-Caballero, A., Mira, J., & Delgado, A. E. (2007). Dynamic visual attention model in image sequences. *Image and Vision Computing*, 25, 597–613. doi:10.1016/j.imavis.2006.05.004

Lopez, M. T., Fernandez-Caballero, A., Fernandez, M. A., Mira, J., & Delgado, A. E. (2006). Motion features to enhance scene segmentation in active visual attention. *Pattern Recognition Letters*, 27, 469–478. doi:10.1016/j.patrec.2005.09.010

Maki, A., Nordlund, P., & Eklundh, J.-O. (1996). A computational model of depth-based attention. In *Proceedings of the 13th International Conference on Pattern Recognition*.

Maki, A., Nordlund, P., & Eklundh, J.-O. (2000). Attentional scene segmentation: Integrating depth and motion. *Computer Vision and Image Understanding*, 78, 351–373. doi:10.1006/cviu.2000.0840

Mertsching, B., Bollmann, M., Hoischen, R., & Schmalz, S. (1999). The neural active vision system navis. *Handbook of Computer Vision and Applications*, 3, 543–568.

Miao, X., & Rao, R. (2007). Learning the lie groups of visual invariance. *Neural Computation*, 19, 2665–2693. doi:10.1162/neco.2007.19.10.2665

Milanese, R. (1990). *Detection of salient features for focus of attention. Technical report*. Computing Science Center, University of Geneva.

Milanese, R. (1993). *Detecting salient regions in an image: From biological evidence to computer implementation*. PhD thesis, University of Geneva.

Milanese, R., Gil, S., & Pun, T. (1995). Attentive mechanisms for dynamic and static scene analysis. *Optical Engineering (Redondo Beach, Calif.)*, 34, 2428–2434. doi:10.1117/12.205668

Navalpakkan, V., & Itti, L. (2006). An integrated model of top-down and bottom-up attention for optimizing detection speed. In *Proceedings of the 2006 IEEE Computer Society Conference on Computer Vision and Pattern Recognition*.

Neumann, O., van der Hejiden, A. H. C., & Allport, A. (1986). Visual selective attention: Introductory remarks. *Psychological Research*, 48, 185–188. doi:10.1007/BF00309082

Olshausen, B., Anderson, C., & Essen, C. V. (1995). A multiscale dynamic routing circuit for forming size- and position-invariant object representations. *Journal of Computational Neuroscience*, 2, 45–62. doi:10.1007/BF00962707

Pashler, H., & Badgio, P. (1985). Visual attention and stimulus identification. *Journal of Experimental Psychology*, 11, 105–121.

Posner, M., Snyder, C., & Davidson, B. (1980). Attention and the detection of signals. *Journal of Experimental Psychology, 109,* 160–174.

Postma, E. O., van den Herik, H. J., & Hudson, P. T. W. (1997). Scan: A scalable model of attentional selection. *Neural Networks, 10,* 993–1015. doi:10.1016/S0893-6080(97)00034-8

Rowley, H., Baluja, S., & Kanade, T. (1995). *Human face detection in visual scenes. Technical report.* Carnegie Mellon University.

Tistarelli, M., & Sandini, G. (1991). Direct estimation of time-to-impact from optical flow. *IEEE Workshop on Visual Motion,* (pp. 52–60).

Torralba, A., Oliva, A., Castellanos, M. S., & Henderson, J. M. (2006). Contextual guidance of eyes movements and attention in real-world scenes: The role of the global features in object research. *Psychological Review, 113,* 766–786. doi:10.1037/0033-295X.113.4.766

Treisman, A. M., & Gelade, G. (1980). A feature-integration theory of attention. *Cognitive Psychology, 12,* 97–136. doi:10.1016/0010-0285(80)90005-5

Tsotsos, J. (1987). A complexity level analysis of vision. In *Proceedings International Conference on Computer Vision: Human and Machine Vision Workshop,* London, England.

Tsotsos, J., Culhane, S. M., Winky, W., Lay, Y., Davis, N., & Nuflo, F. (1995). Modeling visual attention via selective tuning model. *Artificial Intelligence, 78,* 507–545. doi:10.1016/0004-3702(95)00025-9

Tsotsos, J., Liu, Y., Martinez-Trujillo, J. C., Pomplun, M., Simine, E., & Zhou, K. (2005). Attending to visual motion. *Computer Vision and Image Understanding, 100,* 3–40. doi:10.1016/j.cviu.2004.10.011

Uchizawa, K., Douglas, R., & Maass, W. (2006). On the computational power of threshold circuits with sparse activity. *Neural Computation, 28,* 2994–3008. doi:10.1162/neco.2006.18.12.2994

Uhr, L. (1972). Layered recognition cone networks that preprocess, classify and describe. *IEEE Transactions on Computers, 21,* 758–768. doi:10.1109/T-C.1972.223579

van der Spiegel, J., Kreider, G., Claeys, C., Debusschere, I., Sandini, G., Dario, P., … Soncini, G. (1989). *A foveated retina like sensor using ccd technology.* Kluwe.

Vieville, T., Clergue, E., Enriso, R., & Mathieu, H. (1995). Experimenting with 3-d vision on a robotic head. *Robotics and Autonomous Systems, 14,* 1–27. doi:10.1016/0921-8890(94)00019-X

Waldherr, S., Thrun, S., & Romero, R. (2000). A gesture-based interface for human-robot interaction. *Autonomous Robots, 9,* 151–173. doi:10.1023/A:1008918401478

Walther, D. (2006). *Interactions of visual attention and object recognition: Computational modeling, algorithms, and psychophysics.* PhD thesis, California Institute of Technology, 2006.

Westelius, C. J. (1995). *Focus of attention and gaze control for robot vision.* PhD thesis, Department of Electrical Engineering, Linköping University, Sweden.

Wolfe, J. M. (1994). Guided search 2.0: A revised model of visual search. *Psychonomic Bulletin & Review, 1,* 202–238. doi:10.3758/BF03200774

Wolfe, J. M. (2007). *Guided search 4.0: Current progress with a model of visual search.* Brigham and Womens Hospital and Harvard Medical School.

Wolfe, J. M., Cave, K. R., & Franzel, S. L. (1989). Guided search: An alternative to the feature integration model for visual search. *Journal of Experimental Psychology. Human Perception and Performance*, *15*, 419–433. doi:10.1037/0096-1523.15.3.419

Wren, C., Azarbayejani, A., Darrell, T., & Pentland, A. (1997). Pfinder: Real-time tracking of the human body. *IEEE Transactions on Pattern Analysis and Machine Intelligence*, *19*, 780–785. doi:10.1109/34.598236

Wright, R. D., & Ward, L. M. (2008). *Orienting of attention*. Oxford University Press.

Yarbus, A. L. (1969). *Eye movements and vision*. New York, NY: Plenum.

Zangenmeister, W. (1996). *Designing an anthropomorphic head-eye system*.

KEY TERMS AND DEFINITIONS

Active Vision: Perceptual system's ability to both select and analyse specific parts of a visual scene and to modify the parameters, both external and internal, of image creation.

Attention: Cognitive process by which human beings focus on a certain aspect of the environment while ignoring all others, or as the management of the resources that allow for information processing.

Focus Of Attention: Region, object or item in which the system's perceptual resources are concentrated.

Fovea: Small depression in the retina which grants the maximum resolution of the whole field of view.

Gaze Control: Process of directing the fixation point through a scene to allow for the development of perceptual, cognitive and behavioural activity.

Robocup: International robotics competition founded in 1997 which aims to promote robotics and AI research, by offering a publicly appealing challenge.

Robot: Physical agent capable of properly interacting with its external environment.

Saliency Map: Activation map representing the most conspicuous regions of the environment which is used in certain attention models to establish the focus of attention.

Visual Attention: Attention which works with visual information channels.

Chapter 21
Attentive Visual Memory for Robot Localization

Julio Vega
Rey Juan Carlos University, Spain

Eduardo Perdices
Rey Juan Carlos University, Spain

José María Cañas
Rey Juan Carlos University, Spain

ABSTRACT

Cameras are one of the most relevant sensors in autonomous robots. Two challenges with them are to extract useful information from captured images and to manage the small field of view of regular cameras. This chapter proposes a visual perceptive system for a robot with a mobile camera on board that cope with these two issues. The system is composed of a dynamic visual memory that stores the information gathered from images, an attention system that continuously chooses where to look at, and a visual evolutionary localization algorithm that uses the visual memory as input. The visual memory is a collection of relevant task-oriented objects and 3D segments. Its scope and persistence is wider than the camera field of view and so provides more information about robot surroundings and more robustness to occlusions than current image. The control software takes its contents into account when making behavior or navigation decisions. The attention system considers the need of reobserving objects already stored, of exploring new areas and of testing hypothesis about objects in the robot surroundings. A robust evolutionary localization algorithm has been developed that can use both the current instantaneous images or the visual memory. The system has been programmed and several experiments have been carried out both with simulated and real robots (wheeled Pioneer and Nao humanoid) to validate it.

DOI: 10.4018/978-1-4666-2672-0.ch021

INTRODUCTION

Computer vision research is growing rapidly, both in robotics and in many other applications, from surveillance systems for security to the automatic acquisition of 3D models for Virtual Reality displays. The number of commercial applications is increasing, like traffic monitoring, parking entrance control, augmented reality videogames and face recognition. In addition, computer vision is one of the most successful sensing modalities used in mobile robotics. Cameras have been incorporated in the last years to robots as common sensory equipment. They are very cheap sensors and may provide much information to robots about their environment. However, extracting relevant information from the image flow is not easy. Vision has been used in robotics for navigation, object recognition, 3D mapping, visual attention, robot localization, etc.

Robots usually navigate autonomously in dynamic environments, and so they need to detect and avoid obstacles. There are several sensors which can detect obstacles in robot's path, such as infrared sensors, laser range finders, ultrasound sensors, etc. When using cameras, obstacles can be detected through 3D reconstruction. Recovering 3D-information has been the main focus of the computer vision community for decades. Stereo-vision methods are the classic ones, based on finding pixel correspondences between the two cameras and triangulation, despite they fail with untextured surfaces. Vision depth sensors like Kinect offer now a different technology for visual 3D reconstruction. In addition, structure from motion techniques builds three-dimensional structure of objects by analyzing local motion signals over time, even from only one camera (Richard & Zisserman, 2003).

Moreover, many works have also been presented in vision based navigation and control that generate robot behavior without explicit 3D reconstruction. The temporal occlusions of relevant stimuli inside the images are one hindrance in this approach. The control algorithm should be robust to the lack of time persistence of relevant stimuli in images. This also poses a problem when the objects lie beyond the current field of view of the camera. To solve it, some systems use omnidirectional vision. Others, like humanoids or robots with pantilt units, use mobile regular cameras that can be orientated at will and manage a visual memory of robot surroundings that integrate the information from the images taken from different locations. The visual representation of interesting objects around the robot beyond current field of view may improve the quality of robot's behavior as it handles more information when making decisions. The problem of selecting where-to-look-at at every time, known as gaze control or overt attention (Itti & Koch, 2001; Zaharescu *et al.*, 2005), arises there. Usually the need to quickly explore new areas and the need to reobserve known objects to update their positions, etc. influence that selection. This kind of attention is also present in humans, as we are able to concentrate on particular regions of interest in a scene by movements of the eyes and the head, just by shifting attention to different parts of it. By driving attention specifically to small regions which are important for the task at hand we avoid wasting effort and processing trying fully understand the whole surroundings, and devote as much as possible only to the relevant part.

Another relevant information that can be extracted from images is robot location. Robots need to know their location inside the environment in order to unfold the proper behavior. Using robot sensors (specially vision) and a map, the robot may estimate its own position and orientation inside a known environment. Robot self-localization has proven to be complex, especially in dynamic environments and in those with much symmetry, where sensors values can be similar at different positions.

In this chapter we propose a visual perceptive system for an autonomous robot composed of three modules. First, a short term dynamic visual memory of robot surroundings. It gets images from a mobile camera, extracts edge information and offers a wider field of view and more robustness to occlusions than instantaneous images. This memory stores 3D segments representing the robot surroundings and objects. The memory contents are updated in a continuous coupling with the current image flow. Second, a gaze control algorithm has been developed to select where the camera should look at at every time. It manages the movement of the camera to periodically reobserve objects already stored in the visual memory, explore the scene and test tentative object positions in a time sharing fashion. These two modules working in conjunction build and update an attentive visual memory of the objects around the robot. Third, a visual localization algorithm has been developed that uses current image or the contents of the memory to continuously estimate the robot position. It provides a robust localization estimation and has been specifically designed to bear with symmetries in the environment.

The remainder of this chapter is organized as follows. Second section reviews some of the related works. Third section presents the design of the proposed visual system, its components and connections. The next three sections describe its three building blocks: visual memory component, visual attention algorithm and localization component. The seventh section includes several experiments, both with simulated and real robots, performed to validate our system. Finally some conclusions are summarized.

RELATED WORKS

Many issues have been tackled in the intersection of computer vision and robotic fields: vision-based control or navigation, vision-based map building and 3D representation, vision-based localization, object recognition, attention and gaze control among others. We will review here some examples in the topics most related with the proposed visual perceptive system.

Regarding vision based control and navigation, Remazeilles (Remazeilles *et al.*, 2006) presented

the design of a control law for vision-based robot navigation. The particularity of this control law is that it does not require any reconstruction of the environment, and it does not force the robot to converge towards each intermediary position in its path.

Recently, Srinivasan (Srinivasan *et al.*, 2006) presented a new system to increase accuracy in the optical flow estimation for insect-based flying control systems. A special mirror surface is mounted in front of the camera, which is pointing ahead instead of pointing to the ground. The mirror surface decreases the speed of motion and eliminates the distortion caused by the perspective. In this image objects move slower than in the camera downwards, simplifying the optical flow calculation and increasing its accuracy. Consequently, the system increases the speed range and the number of situations under which the aircraft can fly safely.

Regarding visual map building, representation of the environment and navigation, Badal (Badal *et al.*, 1994) reported a system for extracting range information and performing obstacle detection and avoidance in outdoor environments based on the computation of disparity from the two images of a stereo pair of calibrated cameras. The system assumes that objects protrude high from a flat floor that stands out from the background. Every point above the ground is configured as a potential object and projected onto the ground plane, in a local occupancy grid called Instantaneous Obstacle Map (IOM). The commands to steer the robot are generated according to the position of obstacles in the IOM.

Goldberg (Goldberg *et al.*, 2002) introduced a stereo vision-based navigation algorithm for the rover planetary explorer MER, to explore and map locally hazardous terrains. The algorithm computes epipolar lines between the two stereo frames to check the presence of an object, computes the Laplacian of both images and correlates the filtered images to match pixels from the left image with their corresponding pixels in the right image. The work also includes a description of the navigation module GESTALT, which packages a set of routines able to compute actuation, direction, or steering commands from the sensor information.

Gartshore (Gartshore *et al.*, 2002) developed a map building framework and a feature position detector algorithm that processes images on-line from a single camera. The system does not use any matching. Instead, it computes probabilities of finding objects at every location. The algorithm starts detecting the object boundaries for the current frame using the Harris edge and corner detectors. Detected features are backprojected from the 2D image plane considering all the potential locations at any depth. The positioning module of the system computes the position of the robot using odometry data combined with image feature extraction. Color or gradient from edges and features from past images help to increase the confidence of the object presence in a certain location. Experimental results in indoor environments set the size of the grid cells to 0.25m x 0.25m and the robot moved 0.1m between consecutive images.

As we already mentioned, in autonomous robots it is important to perform a visual attention control. The cameras of the robots provide a large flow of data and the robot needs to select what is interesting and ignore what it is not. There are two approaches to visual attention: covert and overt attention. The first one selects interesting areas inside the image for further processing (Tsotsos *et al.*, 1995; Itti & Koch, 2001), (Marocco & Floreano, 2002). The second one selects interesting areas in the environment surrounding the robot, even beyond the current field of view, and looks at them (Cañas *et al.*, 2008).

One of the concepts widely accepted in this area is the salience map. It is found in (Itti & Koch, 2001), as a covert visual attention mechanism, independent of the particular task to be performed. This bottom-up attention builds in each iteration the conspicuity map for each one of the visual features that attract attention (as color, movement or edge orientations). There are competition dynamics inside each map and they

are merged into a single representative salience map that drives the focus of attention to the area with highest value.

Hulse (Hulse *et al.*, 2009) presented an active robotic vision system based on the biological phenomenon of inhibition of return, used to modulate the action selection process for saccadic camera movements. They argued that visual information has to be subsequently processed by a number of cortical and sub-cortical structures that place it: 1) in context of attentional bias within egocentric salience maps; 2) the aforementioned IOR inputs from other modalities; 3) overriding voluntary saccades and 4) basal ganglia action selection. Thus, biologically there is a highly developed, context specific method for facilitating the most appropriate saccade as a form of attention selection.

The use of a camera in motion to facilitate object recognition was proposed by (Ballard, 1991), and has been used, for example, to distinguish between different forms in the images (Marocco & Floreano, 2002). Arbel and Ferrie presented in (Arbel & Ferrie, 2001) a gaze-planning strategy that moves the camera to another viewpoint around an object in order to recognize it. Recognition itself is based on the optical flow signatures that result from the camera motion. The new measurements, accumulated over time, are used in a one-step-ahead Bayesian approach that resolves the object recognition ambiguity, while it navigates with an entropy map.

Grid based probabilistic localization algorithms were successfully applied with laser or sonar data in small known environments (Burgard & Fox, 1997). They use discretized probability distributions and update them from sensor data and movement orders, accumulating information over time and providing a robust position estimation. Particle filters use sampled probability functions and extend the techniques to larger environments, using them even with visual data as input (Dellaert *et al.*, 1999). At the beginning the maps were provided in advance but in the last years the SLAM techniques tackle localization simultaneously with the map building. There are many particle filter based SLAM techniques. In addition, one of the most successful approaches is Mono-SLAM from Andrew Davison (Newcombe & Davison, 2010; Carrera *et al.*, 2011) based on a fast Extended Kalman Filter for continuous estimation of 3D points and camera position from relevant points in the image.

In (Mariottini & Roumeliotis, 2011) Mariottini and Roumeliotis presented a strategy for active vision-based localization and navigation of a mobile robot with a visual memory where previously visited areas are represented as a large collection of images. It disambiguates the location taking into account the sequence of distinctive images, while concurrently navigating towards the target image.

In (Jensfelt & Kristensen, 2011), Jensfelt and Kristensen presented an active global localization strategy that uses Kalman filtering (KF) to track multiple robot pose hypotheses. Their approach can be used even with incomplete maps and the computational complexity is independent on the size of the environment.

DESIGN

The proposed perceptive system is designed for autonomous robots that use a mobile camera, like that on the head of humanoids or in robots with pan-tilt units. The block diagram of the robot control architecture is showed in Figure 1. The three building blocks (visual memory, gaze control and localization algorithm) have been grouped into two main software components: active_visual_memory and localization. They receive data from robot sensors, like camera and encoders, and extract refined information like description of objects around the robot or robot position. This information is provided to other actuation components like the navigation algorithm or other control units.

First, the active_visual_memory component builds a local visual memory of objects in the robot's surroundings. The memory is built analyzing each camera image looking for relevant objects (like segments, parallelograms, arrows, etc) and updating the object features already stored in the memory like their 3D position. The memory is dynamic and is continuously coupled with camera images. The new frames confirm or correct the object features stored in memory, like their 3D relative position to the robot, length, etc. New objects are introduced in memory when they appear in images and do not match any known object.

This memory has a broader scope than the camera field of view and objects in memory have more persistence than the current image. Regular cameras typically have 60 degrees of scope. This would be good enough for visual control but a broader scope may improve robot responses in tasks like navigation, where the presence of obstacles in the robot's surroundings should be taken into account even if they lie outside the current field of view.

This memory is intended as local and short-term. Relative object positions are estimated using robot's odometry. Being only short term and continuously correcting with new image data there is no much time to accumulate error in the object estimated relative position. Currently the system deals only with objects on the floor plane (ground hypothesis) and uses a single camera. It can be extended to any 3D object position and two cameras.

Second, in order to keep this short term visual memory consistent with reality, the system has mechanisms to properly refresh and update it. The camera is assumed to be mobile, typically mounted over a pan-tilt unit. Its orientation may be controlled and changed during robot behavior at will, and so, the camera may look towards different locations even if the robot remains static. In order to feed the visual memory, an overt attention algorithm has been designed to continuously guide camera movements, choosing where to look at at every time. It has been inserted inside the active_visual_memory component and associates two dynamic values to each object in memory: salience and life (quality). Objects with low life

Figure 1. Block diagram of the proposed visual system

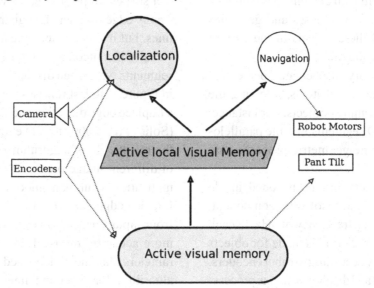

are discarded and objects with high salience are good candidates to look at.

The position of objects already in memory are themselves foci of attention in order to refresh their perceived features. Random locations are also considered to let the robot explore new areas of its surroundings. In addition, new foci of attention may also be introduced to check the presence of some hypothesized objects. For instance, once the robot has seen three vertices of a parallelogram, the position of the fourth one is computed from the visual memory and ordered as a tentative focus of attention for the camera.

Third, a vision based localization algorithm has been developed in the localization component. It uses a population of particles and an evolutionary algorithm to manage them and find the robot position. The health of each particle is computed based on the current image or based on the current contents of the visual memory. The local visual memory provides information about robot's surroundings, typically more than the current instantaneous sensor readings. In this way, the visual memory may be used as a virtual sensor and its information may be used as observations for the localization algorithm. Because of its broader scope it may help to improve localization,

especially in environments with symmetries and places that look like similar according to sensor readings.

These two software components and the three building blocks will be described in detail in the following sections.

LOCAL VISUAL MEMORY

The goal of our visual memory is to do a visual tracking of the various basic objects in the scene surrounding the robot. It must detect new objects, track them, update their relative position to the robot and remove them from the memory once they have disappeared.

Figure 2 shows the main modules of the visual memory building block. The first stage of the visual memory is a 2D analysis, which detects 2D segments present in the current image. These 2D segments are compared with those predicted from the current visual memory 3D contents. The 3D object reconstruction module places relevant 2D segments in 3D space according to the ground-hypothesis (we assume all objects are flat on the floor as simplifying hypothesis). Finally, the 3D memory module stores their position in 3D space, update or merge them with existing 3D segments, calculates perceptual hypotheses and generates new predictions of these objects in the current image perceived by the robot.

The visual memory also creates perceptual hypothesis with the stored items, allowing the system to abstract complex objects. For instance, it groups a set of 3D segments into the parallelogram concept if some geometric properties are hold.

The visual memory has been coded inside the active_visual_memory software component, running iteratively at a frequency of 5Hz. In each iteration a motion prediction is made for objects in the visual memory according to robot encoders, images are acquired and displayed, image processing occurs and the memory contents are updated. To save computing power, robot odometry can be used to trigger the image processing and only analyze frames when the robot has moved away a certain distance or angle from its position when the last image processing was done.

2D Image Processing

The main goal of this module is to extract 2D straight segments as a basic primitive to get object shapes. In prior releases we used the classic Canny edge filter and Hough transform to extract lines, but it was not accurate and robust enough. Usually its outcome was not fully effective, line segments were often disconnected as can be seen in Figure 3. In last releases we replaced this with a Laplace edge detection and the Solis' algorithm (Solis *et al.*, 2009) for 2D extraction, improving the results. Solis' algorithm uses a compilation of different image processing steps such as normalization, Gaussian smooth, thresholding, and Laplace edge detection to extract edge contours from input images. This solution is surprisingly more accurate, robust, faster and with less parameters than the widely used Hough Transform algorithm for detecting lines segments at any orientation and location.

To implement these techniques, we use the OpenCV library. A comparison can be seen in Figure 3 where the Solis algorithm extract many more 2D segments. The detected segments are shown as blue lines. While the Hough approach is able to recognize just a really small set of segments, the Solis one gets most of them. The floor used in this image is a textured surface and so some false positives appear.

Figure 2. Modules of the visual memory

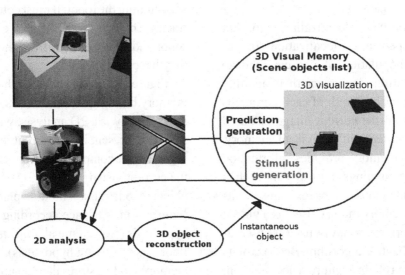

Figure 3. Differences between Canny+Hough (left) and Solis algorithm (right)

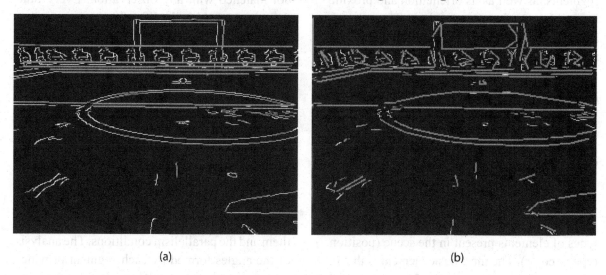

(a)   (b)

Figure 3. Differences between Canny+Hough (left) and Solis algorithm (right)

The 2D analysis system is connected directly to the 3D visual memory contents to alleviate the computational cost of image analysis. Before extracting features of the current image, the system predicts inside the 2D image the appearance of those objects stored in the 3D memory which are visible from the current position. We use our projective geometry library to do this. Each stored 3D visible object is projected on the image plane as shown in Figure 4 (left). The system refutes/ corroborates such predicted segments, comparing them with those coming from the 2D analysis on observed images. This comparison provides three sets of segments, as seen in Figure 4: those that match with observations, those that do not match, and observed segments that are unpredicted, that is, without an homologous in the 3D memory. Matched segments will be used to update the information of their homologous in 3D memory. Unpredicted observed segments will be located in 3D and inserted in the visual memory as new 3D segments.

Reconstruction with 3D Segments

This module is responsible to obtain 3D instantaneous information from 2D segments and objects in the current image. To do this we rely on the idea of ground-hypothesis, assuming all objects are flat on the floor. This simplifying assumption makes more easy to estimate the third dimension from a single camera. This module can be replace with other 3D techniques and full 3D estimation in case of using a stereo pair.

The are four relevant 3D coordinate systems in this approach, as shown in Figure 5. First, the absolute coordinate system whose origin lies somewhere in the world where the robot is moving. Second, the system located at the base of the robot. The robot odometry gives its position and orientation with respect to the absolute system. Third, the system relative to the base of the pan-tilt unit to which the camera is attached to. It has its own encoders for its position inside the robot at any given time, with pan and tilt movements with respect to the base of the robot. And fourth, the camera relative coordinate system, displaced and oriented in a particular mechanical axis from the pan-tilt unit.

Once we have the 3D segments, and before including them on the 3D memory, some postprocessing is needed to avoid duplicates in memory due to noise in the images. This postprocessing compares the relative position between segments, as well as its orientation and proximity, maybe merging some of them. The output is a set of observed 3D segments situated on the robot coordinate system. Figure 6 shows the segments detected in the current image, the segments predicted from the current position, and all of them in the 3D observed scene.

Inserting Segments into the 3D Visual Memory

3D visual memory comprises a dynamic set of lists which stores information about the different types of elements present in the scene (position, type or color). The most basic element is the 3D segment. The visual memory also can establish relationships between them to make up more complex elements such as arrows, parallelograms, triangles, circles or other objects.

As mentioned before, the 2D analysis returns different subsets of segments, as the result of comparison between observed and predicted segments from 3D memory. If a segment is identified in the current image and it does not match the predictions, the system creates a new one in 3D. For matched segments they are located in 3D merged with the 3D segments already stored in the visual memory. They will be nearby segments with similar orientation and so the system combines these segments into a new one taking the longest length of its predecessors, and the orientation of the more recent, as probably it is more consistent with reality (the older ones tend to have more noise due to errors in robot odometry). The 3D segments have an attribute named uncertainty which increases as the time the segment remains in memory and is also taken into account and updated. To make this fusion process computationally lighter, the system has a 3D segment cache of the full 3D segment collection with only the segments close to the robot (in a radius of 4m).

As it will be described later, the 3D segments have an attribute named life which decreases as the time the segment remains in memory and is

not matched with any observation. Every time there is a matching the life of the corresponding 3D segment is increased. If the uncertainty on a 3D segment falls below a given threshold it is deleted from visual memory.

Complex Primitives in Visual Memory

Visual memory manages simple 3D segments and other primitives like parelellograms. The segments and their corresponding vertices are used to detect parallelograms checking the connection between them and the parallelism conditions. The analysis of the angles formed by each segment provides information about how the segments are connected to each other. The visual memory can estimate the position of a possible fourth vertex using the information about edges and the other three vertices. In addition, the parallelogram primitive can be used to merge incomplete or intermittent segments. This capability makes our algorithm robust against occlusions, which occur frequently in the real world.

Figure 7 illustrates an example of occlusion that is satisfactorily solved by our algorithm. The (a) and (b) images show the observation on incomplete parallelograms and include noise. The results of reconstruction of parallelograms can be seen in Figure 7-(a), with the extracted four parallelograms spread on the floor. The robot, after several snapshots with incomplete parallelograms, captures the real ones in 3D avoiding the noise in the observations.

VISUAL ATTENTION

The second building block of the proposed visual perception system is the visual attention. It uses two object attributes: salience and life to decide where to look at at every moment and to forget objects when they disappear from the scene. In addition, this block includes a mechanism to con-

Figure 4. 3D projection on the image plane (left) and matching between predicted and observed segments (right)

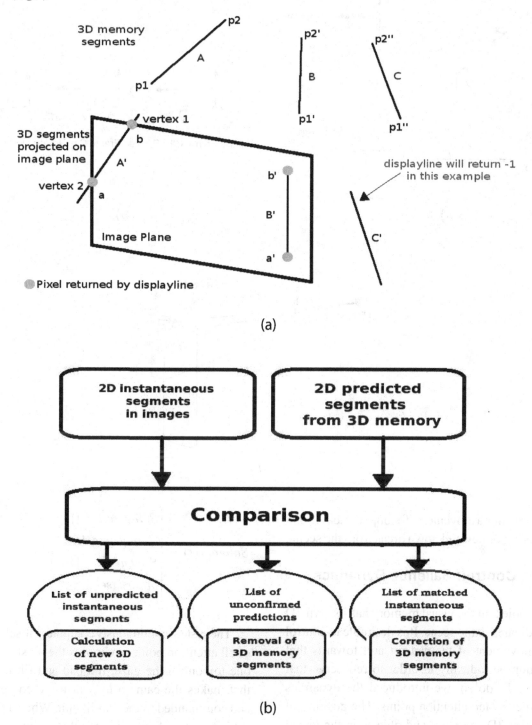

(a)

(b)

Figure 5. 3D coordinate systems used

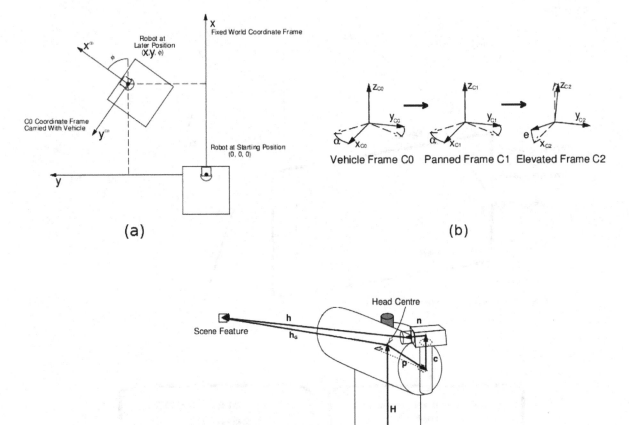

(a)

Vehicle Frame C0   Panned Frame C1   Elevated Frame C2

(b)

(c)

trol the camera movements for object tracking and exploring of new unknown areas from the scene.

Gaze Control: Salience Dynamics

Each object in the visual memory has its own 3D coordinates in the scene. It is desirable to control the movement of the pan-tilt unit towards that position periodically in order to reobserve that object. To do so, we introduced the dynamics of salience and attention points. The position of different 3D segments and objects in the visual memory are attention candidate points. Each one has a salience that grows over time and vanishes every time that element is visited, following the equation:

$$Salience(t) = \begin{cases} Salience(t-1) \\ if\ object\ attended \\ Salience(t-1)+1 \\ otherwise \end{cases}$$

The system continuously computes the salience of all attention points a chooses the most salient one to control the gaze, the pan and tilt orders that makes the camera look at it are computed and commanded to the pantilt unit. When a point is visited, its salience is set to 0. A point that the system has not visited recently calls more attention than one which has just been attended. If the salience is low, it will not be visited now. The

Figure 6. Scene situation, instantaneous image, and 3D scene reconstruction

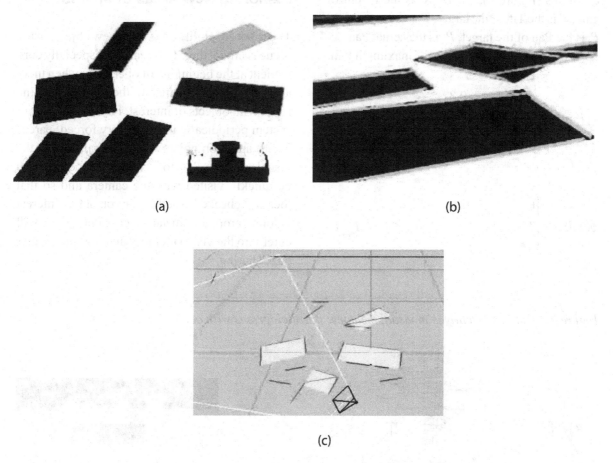

(a)　　　　　　　　　　　　　　　　　(b)

(c)

system is thus similar to the behavior of a human eye, as pointed by biology studies (Itti & Koch, 2005) when the eye responds to a stimulus that appears in a position that has been previously treated, the reaction time is usually higher than when the stimulus appears in a new position.

After a while the most salient point is re-computed again and so the focus of attention is changed, implementing a kind of time sharing gaze control. The designed algorithm allows so to alternate the focus of the camera between the different objects in the scene according to their salience. We assume that an object will be found near the location where it was previously observed the last time. If the object motion were too fast the reobservations would not match with current object location.

In our system all objects have the same slope in the saliency dynamics, the same preference of attention and so all of them are observed with the same frequency. If we assigned different rates of growth of salience we could have different priori-ties for the objects, causing the pan-tilt unit to look more times at objects whose salience grows faster.

Tracking of a Focused Object

When the gaze control chooses the attention point of a given object, the system will look at it for a certain time (3 seconds), tracking it in the case it moves spatially. For this tracking we use two proportional controllers to command the pan and tilt speeds and thus continually keep that object on the image center. The controller follows next equations (Figure 8), where: K_p is the P control gain, T_t is the Tilt of the target, T_a is the actual Tilt, P_t is the Pan of the target, P_a is the actual Pan, M_t is the maximum Tilt and M_p is the maximum Pan.

$$v(Pan) = \begin{cases} 0 & if \quad \in < 0.3 \\ K_p \cdot (P_t - P_a) & if \quad 0.3 \leq \in < M_p \\ M_p & if \quad M_p < \in \end{cases}$$

$$v(Tilt) = \begin{cases} 0 & if \quad \in < 0.1 \\ K_p \cdot (T_t - T_a) & if \quad 0.1 \leq \in < M\_t \\ M_t & if \quad M_t < \in \end{cases}$$

Exploring New Areas of Interest

The robot capability to look for new objects in the scene is interesting. This search is especially convenient at the beginning of operation, when there are many unknowns areas of the scene around the robot with objects of interest. For that search our system periodically inserts (every forcedSearch-Time) attention points with high salience in the visual memory. Due to its high salience they will be quickly visited with the camera and so that location checked whether any object of interest is found around it. In such a case that object will enter into the visual memory and into the regular gaze sharing.

Figure 7. Complex primitives in visual memory: parallelograms with occlusion

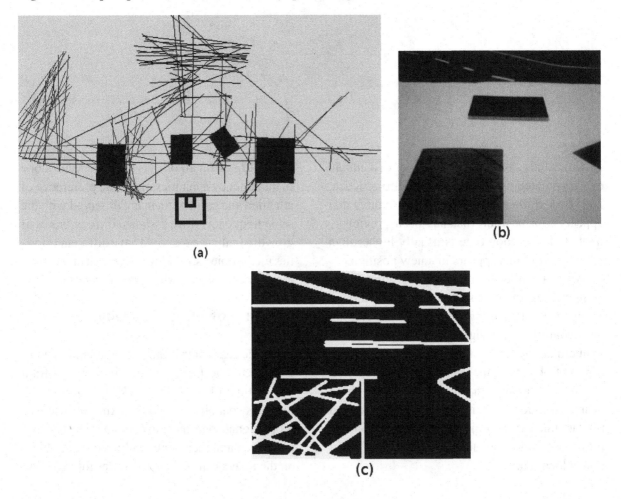

(a)

(b)

(c)

Figure 8. P-controller mechanism

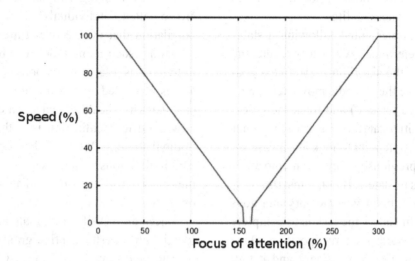

The scanning points can be of two types: random and systematic ones. Random points are distributed uniformly within the pan-tilt range (pan = (-159, + 159), tilt = (-31, +31)). Systematic scanning points follow a regular pattern to finally cover the whole scene around the robot. With them, the system ensures that eventually all areas of the scene will be visited.

There will be a proliferation of points of exploration in the beginning, when there are few objects in memory to reobserve. As the robot discovers objects, the desire to explore new areas will decrease in proportion to the number of already detected objects.

Representation of the Environment: Life Dynamics

As already mentioned in previous sections, the objects may eventually disappear from the scene, and then they should be removed from the memory in order to maintain coherence between the representation of the scene and the reality. To forget such old elements, we have implemented the life dynamics that follows this equation.

$$Life(t) = \begin{cases} \min(MAX_{LIFE}, Life(t-1) + \Delta) \\ if\ object\ observed \\ Life(t-1) - 1 \\ otherwise \end{cases}$$

Life of unobserved 3D segments or objects decrease over time and every time an object or 3D segment and it is observed in the images (just because the gaze control visits it or one near object), its life increases, with a maximum saturation limit. This way when the life of an object exceeds a certain threshold, that means it is still on the scene, whereas when is below it that means it has gone and is deleted from visual memory.

Attention Module Operation

The proposed visual attention module is fully bottom-up. The objects surrounding the robot guide the movements of the camera, just to reobserve them, to track them or to explore the environment looking for them. Periodically the system updates the salience and life attributes of the objects that have already stored in memory following previous equations. It checks whether any of them is already outdated, because its life is below a certain threshold. If not, it increases its salience and reduces its life.

It has been implemented following a state-machine that determines when to execute the different steps of the algorithm: select next goal (state 0), complete the saccadic movement (state 1), analyze image (state 2) and track the object (state 3). In the initial state the system asks whether there is any attention point to look at (in case we have an object previously stored in memory) or not. If so, it goes to state 1. If not, it inserts a new scanning attention point into memory and goes back to state 0. In state 1 the task is to complete the movement towards the target position. Once there, the automata goes to stage 2 and it will analyze whether there are relevant objects in the images or not. After a while it returns to state 0 and starts again.

EVOLUTIONARY VISUAL LOCALIZATION

We have designed a new approach to solve robot self-localization specifically designed to bear symmetries. It is an evolutionary algorithm, a type of meta-heuristic optimization algorithm that is inspired by the biological evolution.

In this kind of algorithms, candidate solutions are so-called "individuals", which belong to a population that evolve over time using genetic operators, such as mutation or crossover. Each individual is a tentative robot position (X, Y, θ) and is evaluated with a quality function which calculates its "health", that is, a measure to know how good its localization is with regard to the optimal solution. We have defined two different health functions, one based on instantaneous measures of robot sensors and another one based on the visual memory contents.

Races are set of individuals around a given location, they perform a fine-grain search around it. The algorithm has N races which compete among each other to be the race containing the best pose estimation. Each race has several associated parameters like the number of iterations without being deleted, the number of iterations containing the best pose estimation, etc.

The main idea of the algorithm consists of keeping several races competing among each other in several likely positions. In case of symmetries from observations, the algorithm will create new races on various positions where the robot might be located. After some iterations, predictably, new observations will provide information to reject most of races and the algorithm will obtain the real

Figure 9. Basic diagram of evolutionary algorithm

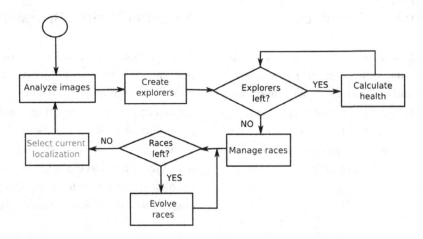

Figure 10. Image analysis before merging (left) and after merging (right)

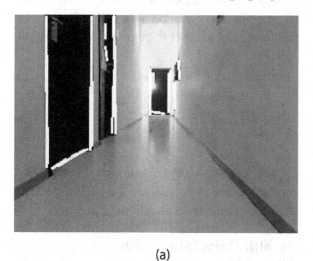

(a)

(b)

robot pose from the best race. On each iteration the algorithm performs several steps:

- Health race calculation using the information obtained after analyzing images.
- **Explorers creation:** We spread randomly new individuals with the aim to find new candidate positions where new races could be created.
- **Races management:** We create, merge or delete races depending on their current state.
- **Races evolution:** We evolve each race by using genetic operators. Besides, if the robot is moving, we update all races taking into account this movement.
- After calculating each race health, we will select one of them to set the current robot pose.

On each iteration of the algorithm there are several steps to be performed to know the current robot pose (Figure 9). They are described in the following sections.

Analyzing Images

When the localization algorithm uses the instantaneous camera images, it first detects lines inside the images following the same technique described in "Local Visual Memory" section. This time the algorithm does not discard segments over the horizon, as it does not require that objects lie on the floor. Some post-processing in the image is performed to clean and refine the segments. The post-processing associates a label to each segment depending on the main colors at both sides of the segment. Segments with unknown type are rejected. After labelling each segment, the algorithm tries to merge segments with the same type if their extremes are very close to each other, joining consecutive segments together (Figure 10). Too small segments are ruled out.

Instead of using these lines directly as input data, the algorithm divides them into sampling points to make the comparison between lines easier, as it will explained in the next section. We created a grid with different cell sizes and we only save a new point where the lines detected intersect with this grid (Figure 11). The size of this grid cells changes because we want to analyze the upper part of the image more deeply than the lower one, since further objects will be at the top of the image and its resolution will be smaller. All these sampling points selected will be the input data to calculate the health of each individual.

Health Calculation from Instantaneous Images

The health of an individual placed at certain location is computed comparing the theoretical set of visible objects and segments from that location (theoretical observation) with objects and segments currently observed (real observation). The more similar the predicted segments and the observed one are, the more likely such location is the correct one.

The theoretical observations are generated ad-hoc for each particular location, projecting lines from the environment map into the camera placed at that location (Figure 12). It contains the lines the robot would see if it were placed at that location (Figure 10 (right)).

For each sampling points in the observed lines the Euclidean distance d_i in pixels to the closest theoretical line with the same label is computed. After calculating d_i for all points, the Individual's

health is computed as the average distance, following the next equation, where N is the number of points and M is set to 50 pixels for a 320x240 image size.

$$H = 1 - \frac{\sum_{i=0}^{N} \frac{d_i}{N}}{M}$$

We will show in "Experiments" section several experiments to analyze the health function behavior in different situations.

Health Calculation with Visual Memory

In case of using the visual memory we don't need to analyze each image, just the current visual memory contents from the active_visual_memory component, that is, the set of 3D segments inside, relative to our robot. So we can't compare lines in image as we did before. Besides, we have to take into account that lines may not be detected completely or they may be divided into several small lines. Thus, to calculate the current health of an individual we cover all lines belonging to the visual memory, for each line we get its extremes

Figure 11. Image grid (left) and points selected (right)

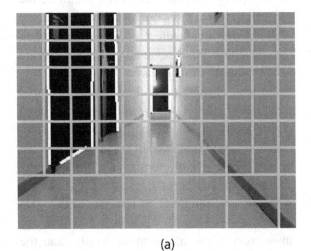

(a)

(b)

Figure 12. Lines detected in current image and theoretical image

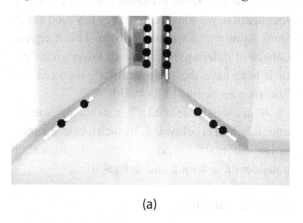

(a)

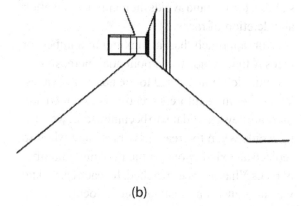

(b)

and calculate the Euclidean distances d_{js} and d_{je} to the closest theoretical line with the same label, that is, similar to health function with instantaneous images but in 3D. After calculating d_{js} and d_{je} for each line, we can calculate the health as follows, where N is the number of lines and M is set to 0.5 meters.

$$H = 1 - \frac{\sum_{i=0}^{N} \left(\dfrac{d_{js} + d_{je}}{2} \right)}{M}$$

Explorer Creation

Explorers are individuals that don't belong to any race and that try to find likely positions with good health. There are two ways to spread explorers around the environment: randomly or following a predesigned pattern of search positions. In order to be truly general and avoid the overfitting of the algorithm we chose the random approach.

When explorers creation is performed, the algorithm creates N explorer inviduals, calculates their health and arrange them according to its it. Then, the best M explorers are promoted to became a candidate to create a new race. The number of explorers created changes dynamically depending on current algorithm status: if the robot is lost,

N is increased, and decreased it if its location is reliable. Besides, since explorers creation its time consuming, when current location is reliable the explorer creations is not executed at each algorithm iteration, but once every certain iterations.

Race Management

In the long term the algorithm manages several races and needs a policy to decide when to create a new race, delete it, or merge two races. The algorithm uses two parameters of each race for that: ''victories'' and ''age''. A race increase its ''victory'' counter in an iteration when its health is higher than that of the rest of races. At the race creation this parameter is set to 0. This number can also decrease if the winner in a given iteration is other race. This parameter is useful to select the race that finally sets the current robot pose estimation in each iteration. The age parameter shows the number of algorithm iterations since its creation. It has been created with two objectives. First, it preserves new races from dying too soon to avoid creating races that will be deleted in the next iteration. When a race is created this race can't be deleted or replaced (although it may be merged) until its age reaches 3. Second, it avoids deleting a race because of wrong sensor information. If a race has had the highest health in an iteration, the algorithm will not delete it at least until 9 iterations after that. It provides some stability to races and avoids the continuous creation and deletion of races.

Our approach has a maximum number or races N that avoids the exponential increasing of computation time related to the number of races. Whenever the explorers creation is executed and there are new candidates to became a race, we have to decide when to create new races and when to replace an existing one. If the maximum number of races N has not been reached, for each candidate we find out if an existing race is located in the same position (X, Y, θ). In such case, we don't try to create a new race, but we assume that the existing race already represents this candidate, and its age is increased. If candidates are innovative enough, we create a new race. If the maximum number of races has been reached, the candidate will replace a victim race when the health of the candidate is greater than that of the victim and the victim's victories parameter falls below 0. If any race can't be replaced, candidates are ignored.

In case two races evolving towards the same localization, we consider that they have led to the same solution, so we merge them. This merging consist of deleting the race with lower victories, or if both have the same victories, we keep the best race according to its health.

A race will be deleted when its victories are 0 and its health is below 0.6. In such a case the race is not in the real robot pose anymore, the algorithm considers it is wrong and deletes it.

Race Evolution

When a race is created from an explorer, all its individuals are created applying a random thermal noise to the explorer who created the race. From then on, in the next iterations, its individuals evolve through three genetic operators: elitism, crossover and mutation. With elitism the algorithm selects the N best individuals of each race, arranges them according to their health. They are saved in the next iteration without any change. With crossover the algorithm randomly selects several

Figure 13. (a) Situation; (b) instantaneous image; (c) short-term memory

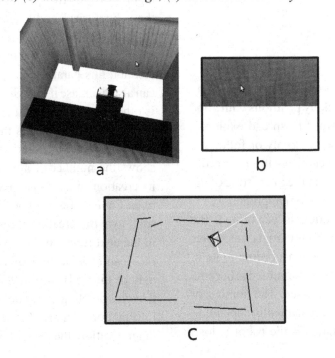

Figure 14. (a) Situation; (b) current onboard image

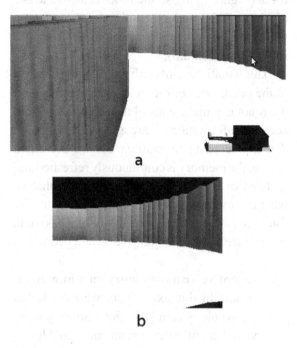

a

b

pairs of individuals calculates their average with their values (*X, Y, θ*) for the next iteration. With mutation the algorithm selects an individual randomly and applies a thermal noise to its position and orientation.

Besides, in case the robot has moved since the last iteration, we apply a motion operator to all races and individuals at the beginning of each iteration using robot odometry. Once all the individuals of each race have evolved, the algorithm calculates the final pose of the race as the average of its elitist, to avoid abrupt changes.

Selecting the Robot Pose Estimation

After evaluating all the existing races, the algorithm chooses one of them in each iteration to be the current pose of the robot. The race selected will be the one with more victories and its pose will determine the calculated robot pose by the algorithm.

The first step is selecting the race with greatest health in the current iteration (R_i). If the race

selected was the same in the previous iteration (R_p), we will increase the victories of R_i and will decrease the victories of the rest. However, if R_i and R_p are different, we only change the races victories if the difference between R_i health and R_p health is greater enough. This distinction is made because we want race changing to be difficult if R_p has been the selected race during a lot of iterations. With this behavior, we try to help the races who have been selected in the previous iterations and we only change the winner race when the health difference is big enough (what would mean that the current localization is wrong).

EXPERIMENTS

To verify our different approaches of visual memory, visual attention and visual localization, we conducted several experiments. Our experimental real platforms were an ActivMedia Pioneer 2DX robot equipped with a Logitech Autofocus camera (2 megapixels) and a Nao Robot from Aldebaran Robotics (v3 model). Besides, we have used Gazebo 0.9 as robot simulator. All our experiments are implemented on C++ with Jderobot robotics software platform, which uses ICE as a middleware.

Attentive Visual Memory Experiments

1. **Robot in the middle of a room:** For the first experiment, the robot is in the middle of a room (see Figure 13-a). Then the robot turns around on itself. Figure 13-b shows a instantaneous view from the room, where robot is able to detect a simple line on the floor. After a few seconds, the robot has turned full circle, having stored all the information about their surrounding environment. Thus, we can see in Figure 13-c, how the short-term memory provides more information than an instantaneous image.

2. **Robot navigating a curve:** This experiment shows how the robot is unable to navigate using only the instantaneous information received from the camera. The situation as shown in Figure 14-a, the robot approaches to a curve area, while navigating through a corridor. If robot used only instantaneous image (Figure 14-b), it would able to see only just some lines in front of itself (Figure 15-a), but with short-term memory it can observe that the path in front of itself is a curve (Figure 15-b). In addition, robot can quickly explore the surrounding environment thanks to the visual attention mechanism, which forces the system to explore unknown areas of the environment.

3. **Robot occlusions:** Here, the situation is presented to solve a temporary occlusion. This happens very often in real environments where there are dynamic objects which can obstruct the robot field of view.

The initial situation is showed in Figure 16-a. After a few seconds, robot has recovered environment information thanks to the short-term memory and the visual attention system (as showed in Figure 16-b).

Then another robot appears, as showed in Figure 17-a, occluding the field of view of our robot (Figure 17-b), so the robot is unable to see anything. This situation continues for some time while the second robot moves away from our robot (Figure 18-a,b).

This situation is solved by our system because of the persistence of the short-term memory, and the robot can make control decisions taking into account information of areas beyond the robot that is occluding its camera. As we presented before, the memory is continuously refreshed and updated over time. If it is inconsistent, that is, what the robot sees does not match with the information stored in memory, the system has some persistence before changing the memory contents.

4. **Attentive visual memory on a humanoid robot:** In this experiment we have tested our whole system including visual memory, visual attention algorithm and visual localization on a real robotics platform such as Nao humanoid robot. The robot is in the middle of our department corridor recovering information about the environment around itself. It is able to move autonomously around the corridor; furthermore, it is moving its neck in order to detect all segments in a few seconds. We can see in Figure 19 some snapshots and the short-term memory built with them.

Figure 15. (a) Information in current field of view; (b) short-term memory

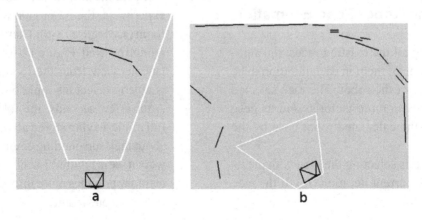

a

b

Figure 16. (a) Situation; (b) short-term memory got after a while

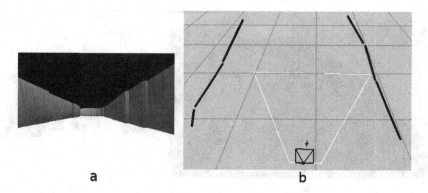

a   b

Visual Localization Experiments

We have performed several experiments to validate the evolutionary algorithm as well, especially with real robots. The algorithm has several parameters to be configured, such as the maximum number of races (10), the number of exploiters (30), and the percentages of the genetic operators previously explained (20% elitism, 20% crossover and 60% mutation). The maximum number of races has been a crucial factor. A high number of races improves the accuracy of the calculated localization, but it also increases the algorithm execution time. We have selected a value, 10, that offers a good balance between efficiency and accuracy.

1. **Typical execution in humanoid robot:** The first experiment has been performed with a real Nao robot travelling through a corridor (Figure 20). It shows how the algorithm is able to follow the real movement of the robot starting on a known position. At first, the robot is located in a known position, afterwards we move the robot around the environment and measure its localization error. The red line in Figure 20 shows the calculated positions, the green line the real robot path, and the brown area is the error measured. The average error has been 11,8 cm and 2,1 grades, the algorithm is able to follow the robot movement even when its instantaneous observations don't provide

Figure 17. (a) Situation; (b) field of view

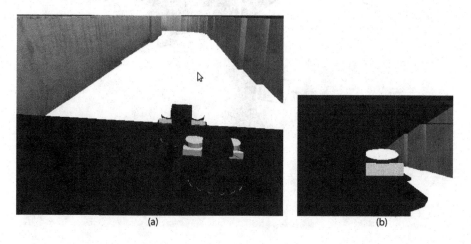

(a)   (b)

Figure 18. (a) Situation; (b) on board current image

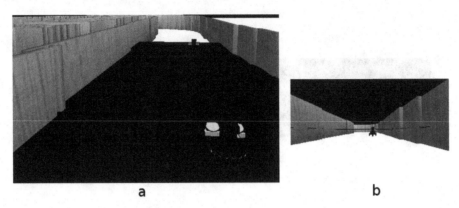

a   b

enough information thanks to robot odometry. Besides, we can emphasize that the trajectory followed by the robot is very stable and is always close to the real location of the robot.

2. **Dealing with symmetries and kidnappings:** The second experiment (Figure 21) shows how the algorithm works with symmetries and kidnappings. At first instant (1.a), we locate the robot in front of a door, so the algorithm creates several races where the robot may be located, the algorithm select one of them (a wrong one) but keeps another one on the right location, then the robot

moves and obtains more information from the world and finally rules the wrong location out and selects the correct one (at 1.b instant). Afterwards, we kidnap the robot to another location (2.a) and it takes a while until the robot changes to the new right location. It happens because the location's reliability changes gradually to avoid changing with false positives, but after a while, it changes to the new position (2.b). A second kidnap is performed (3.a), this case is similar to the first one, at first it selects a wrong localization (very close to the correct one, 3.b), but after some iterations it changes to the correct

Figure 19. Visual memory with 3D segments coming from four images of robot surroundings

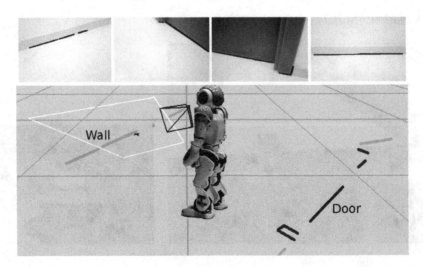

Figure 20. Nao robot travelling a corridor for experiments (left) and estimated localization and position error over time (right)

(a)

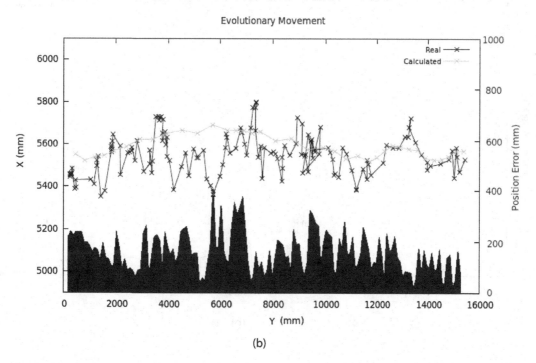

(b)

Figure 21. Position error over time (a) and angle error over time (b)

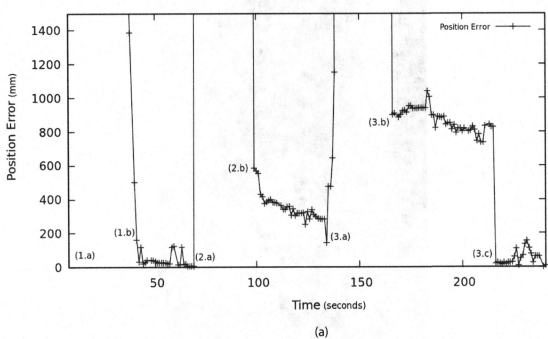

(a)

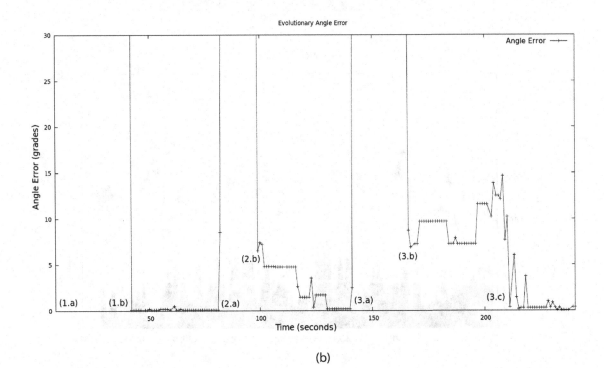

(b)

Figure 22. Image obtained (left) and probabilities calculated with θ equals to 0 radians(right)

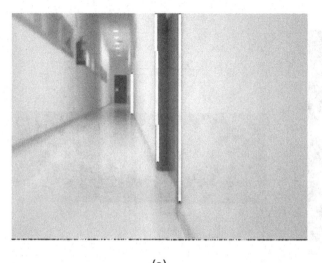

(a)

(b)

one (3.c). The average error after selecting the correct race has been 14,9 cm and 3,2 degrees. We have also measured the time spent until the algorithm calculates a new plausible pose after a kidnapping (recovery time). In this experiment was 29,6 secs.

3. **Health function based on instantaneous images:** To validate the health function, we have implemented a debug mechanism to show graphically the value returned by

the health function in different positions. In Figure 22 we show the value returned in all positions (X, Y), where red areas are the ones with highest probability and white areas with the lowest. As we can see, the position with highest probabilities is plausible with the input image.

In case of symmetries we obtain high probabilities in several positions. In Figure 23 we show

Figure 23. Image obtained in front of a door (left) and probabilities calculated for any θ (right)

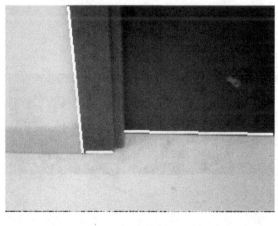

(a)

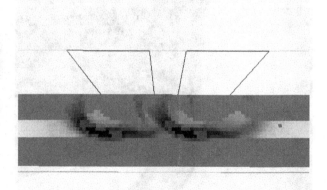

(b)

Figure 24. Image obtained (left) and probabilities calculated without occlusions or false positives for any θ (right)

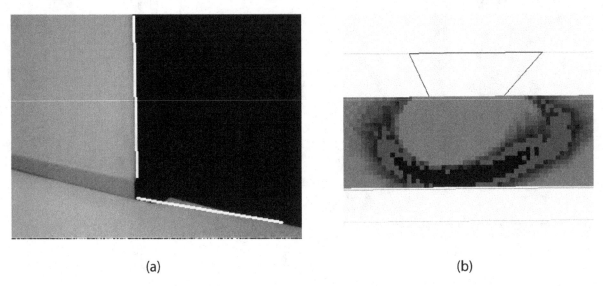

(a)   (b)

what would happen if we obtained an image in front of a door, then we would obtain high probabilities in front of each door of our environment:

The localization algorithm has been tested in more complex scenarios, with occlusions and false positives. In case of occlusions the algorithm keeps a good behavior, since our approach doesn't penalize if an object is not detected in the image

when it should be, and the only objection is a higher probability in some field areas (compare Figure 24 and Figure 25). However, false positives affect negatively to health function and the calculated localization can be totally wrong (compare Figure 24 and Figure 26).

Figure 25. Image obtained (left) and probabilities calculated with occlusions for any θ (right)

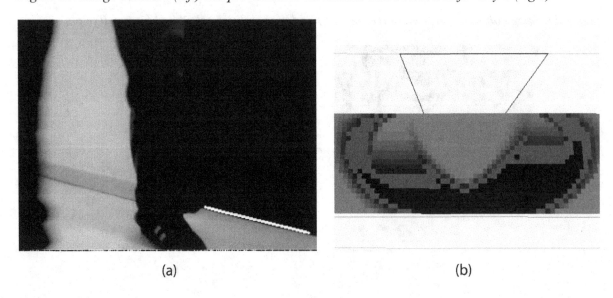

(a)   (b)

Figure 26. Image obtained (left) and probabilities calculated with false positives for any θ (right)

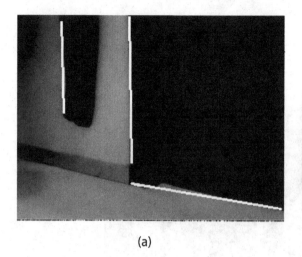

(a)

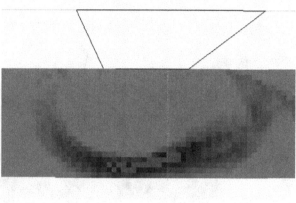

(b)

4. **Health function based on visual memory:** In case of using visual memory instead of instantaneous images for the health function, the calculated values will be similar to previous health function, but we will get two benefits: there will be less symmetries, since we will get more information about the environment, and temporal occlusions won't affect our health function. For instance, Figure 27 shows the health map and position estimation calculated with the visual memory of Figure 19.

CONCLUSION

In this chapter a visual perception system for autonomous robots has been presented. It processes the images from a mobile camera and builds a short term local memory with information about the objects around the robot, even if they lie outside the current field of view of the camera. This visual memory stores 3D segments and simple objects like parallelograms with their associated properties like position, uncertainty (inverse of life), color, etc.. It allows better navigation decisions and even better localization as includes more information than the current image, which can even be temporary occluded.

An overt visual attention mechanism has been created to continuously select where the mobile camera should look at. Using a salience dynamics and choosing the most salient point the system shares the gaze control between the need to reobserve objects on the visual memory and the need explore new areas, providing also Inhibition of Return.

We developed a visual self-localization technique that uses an evolutionary algorithm. It keeps a population of particles to represent tentative robot positions and the particle set evolves as new visual information is gathered or with robot movements. It has been especially designed to bear symmetries, grouping particles into races. There is one race for each likely position and inside it individuals do the fine grain search. It can work both with just the current image or the contents of the visual memory.

This visual perception system has been validated both on real robots and in simulation. The memory nicely represents the robot surroundings using the images from the mobile camera, whose movement is controlled by our attention mechanism. The memory is dynamic but have some persistence to deal with temporary occlusions. The localization works in real time, provides position errors below 15cm and 5 degrees and is robust

Figure 27. Health values for any θ (left) and estimated position (right) calculated with visual memory

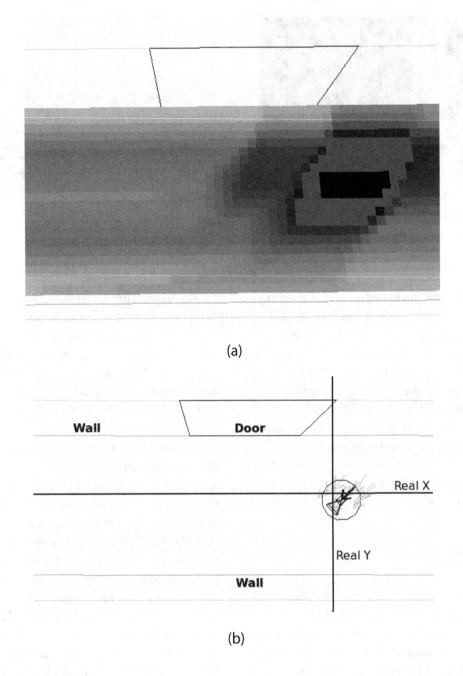

(a)

(b)

enough to recover from kidnappings or estimation errors in symmetric environments.

We are working in extending the system to stereo pairs and dealing with objects at any 3D position, not just the floor. We are also studying how to deal with more abstract objects like tables and chairs into the visual memory. Regarding localization we are working in introducing a monoSLAM EFK for each race of the evolutionary algorithm and improve them to extract localization information from abstract objects, not only 3D points.

REFERENCES

Arbel, T., & Ferrie, F. (2001). Entropy-based gaze planning. *Image and Vision Computing, 19*(11), 779–786. doi:10.1016/S0262-8856(00)00103-7

Badal, S., Ravela, S., Draper, B., & Hanson, A. (1994). A practical obstacle detection and avoidance system. Proceedings of 2nd IEEE Workshop on Applications of Computer Vision, (pp. 97-104).

Ballard, D. H. (1991). Animate vision. *Artificial Intelligence, 48*, 57–86. doi:10.1016/0004-3702(91)90080-4

Burgard, W., & Fox, D. (1997). Active mobile robot localization by entropy minimization. *Proceedings of the Euromicro Workshop on Advanced Mobile Robots*, Los Alamitos, CA, (pp. 155-162).

Cañas, J. M., Martínez de la Casa, M., & González, T. (2008). Overt visual attention inside JDE control architecture. *International Journal of Intelligent Computing in Medical Sciences and Image Procesing, 2*(2), 93-100. ISSN: 1931-308X

Carrera, G., Angeli, A., & Davison, A. J. (2011). Lightweight SLAM and navigation with a multi-camera rig. In *Proceedings of the 5th European Conference on Mobile Robots ECMR 2011,* (pp. 77-82).

Dellaert, F., Burgard, W., Fox, D., & Thrun, S. (1999). Using the CONDENSATION algorithm for robust, vision-based mobile robot localization. In *Proceedings of the Conference on Computer Vision and Pattern Recognition, CVPR-99,* Vol. 2, (pp. 588-594). Fort Collins, CO: IEEE Computer Society.

Gartshore, R., Aguado, A., & Galambos, C. (2002). Incremental map buildig using an occupancy grid for an autonomous monocular robot. *Proceedings of Seventh International Conference on Control, Automation, Robotics and Vision ICARCV,* (pp. 613–618).

Goldberg, S. B., Maimone, M. W., & Matthies, L. (2002). Stereo vision and rover navigation software for planetary exploration. *Proceedings of IEEE Aerospace Conference,* (pp. 2025–2036).

Hulse, M., McBride, S., & Lee, M. (2009). Implementing inhibition of return; embodied visual memory for robotic systems. *Proceedings of the 9th International Conference on Epigenetic Robotics: Modeling Cognitive Development in Robotic Systems,* Lund University Cognitive Studies, Vol. 146, (pp. 189–190).

Itti, L., & Koch, C. (2001). Computational modelling of visual attention. *Nature Reviews. Neuroscience, 2*, 194–203. doi:10.1038/35058500

Itti, L., & Koch, C. (2005). A model of saliency-based visual attention for rapid scene analysis. *IEEE Transactions on Pattern Analysis and Machine Intelligence, 20*(11), 1254–1259. doi:10.1109/34.730558

Jensfelt, P., & Kristensen, S. (2001). Active global localization for a mobile robot using multiple hypothesis tracking. *IEEE Transactions on Robotics and Automation, 17*(5), 748–760. doi:10.1109/70.964673

Mariottini, G. L., & Roumeliotis, S. I. (2011). Active vision-based robot localization and navigation in a visual memory. *Proceedings of International Conference on Robots and Automation.*

Marocco, D., & Floreano, D. (2002). Active vision and feature selection in evolutionary behavioral systems. *Proceedings of International Conference on Simulation of Adaptive Behavior* (SAB-7), (pp. 247-255).

Newcombe, R. A., & Davison, A. J. (2010). Live dense reconstruction with a single moving camera. In *Proceedings of the IEEE Conference on Computer Vision and Pattern Recognition (CVPR)*, 2010, (pp. 1498-1505). San Francisco, California (USA), June 2010.

Remazeilles, A., Chaumette, F., & Gros, P. (2006). 3D navigation based on a visual memory. *Proceedings of Robotics and Automation.*

Richard, H., & Zisserman, A. (2003). *Multiple view geometry in computer vision.* Cambridge University Press.

Solis, A., Nayak, A., Stojmenovic, M., & Zaguia, N. (2009). Robust line extraction based on repeated segment directions on image contours. *Proceedings of the IEEE Symposium on CISDA.*

Srinivasan, V., Thurrowgood, S., & Soccol, D. (2006). An optical system for guidance of terrain following in UAVs. *Proceedings of the IEEE International Conference on Video and Signal Based Surveillance* (AVSS), (pp. 51–56).

Tsotsos, J. K., Culhane, S., Wai, W., Lai, Y., & Davis, N. (1995). Modeling visual attention via selective tuning. *Artificial Intelligence, 78,* 507–545. doi:10.1016/0004-3702(95)00025-9

Zaharescu, A., Rothenstein, A. L., & Tsotsos, J. K. (2005). Towards a biologically plausible active visual search model. *Proceedings of International Workshop on Attention and Performance in Computational Vision* WAPCV-2004, (pp. 133-147).

Chapter 22
Artificial Visual Attention Using Combinatorial Pyramids

E. Antúnez
Universidad de Málaga, Spain

R. Marfil
Universidad de Málaga, Spain

Y. Haxhimusa
Vienna University of Technology, Austria

W. G. Kropatsch
Vienna University of Technology, Austria

A. Bandera
Universidad de Málaga, Spain

ABSTRACT

Computer vision systems have to deal with thousands, sometimes millions of pixel values from each frame, and the computational complexity of many problems related to the interpretation of image data is very high. The task becomes especially difficult if a system has to operate in real-time. Within the Combinatorial Pyramid framework, the proposed computational model of attention integrates bottom-up and top-down factors for attention. Neurophysiologic studies have shown that, in humans, these two factors are the main responsible ones to drive attention. Bottom-up factors emanate from the scene and focus attention on regions whose features are sufficiently discriminative with respect to the features of their surroundings. On the other hand, top-down factors are derived from cognitive issues, such as knowledge about the current task. Specifically, the authors only consider in this model the knowledge of a given target to drive attention to specific regions of the image. With respect to previous approaches, their model takes into consideration not only geometrical properties and appearance information, but also internal topological layout. Once the focus of attention has been fixed to a region of the scene, the model evaluates if the focus is correctly located over the desired target. This recognition algorithm considers topological features provided by the pre-attentive stage. Thus, attention and recognition are tied together, sharing the same image descriptors.

DOI: 10.4018/978-1-4666-2672-0.ch022

INTRODUCTION

Attention in humans defines the cognitive ability to select stimuli, responses, memories or thoughts that are behaviorally relevant among the many others that are irrelevant. Thus, attention has been often compared to a virtual spotlight through which our brain perceives the world. Based on concepts that emanate from the human perception system, computational attention models aim to develop this ability in artificial systems. Humans and animals are able to delineate, detect and recognize objects in complex scenes 'at a blink of an eye'. One of the most valuable and critical resources in human visual processing is time (Evolution conditioned the usage of this resource sparsely, because of survival necessity), therefore a highly parallel model is the biological answer dealing satisfactorily with this resource, since 'all complex behaviors are carried in less than 100 steps' (Feldman et al, 1982) (called the 100 step rule). That is, since neurons have a computational speed of a few milliseconds and each perceptual phenomenon occurs in a few hundreds of milliseconds yield that biologically motivated algorithms must be carried out in less than 100 steps. Tsotsos (1988, 1990, 1992) performed complexity analysis to show that hierarchical internal representation and hierarchical processing are the credible approach to deal with space and performance constrains, observed in human visual systems.

In the last years mobile robots have begun to address complex tasks that require them to obtain a detailed description of the environment. Human-robot interaction and object recognition are two examples of tasks that could be hardly achieved using range sensors and that usually need the use of vision. In these cases, the broad amount of information provided by vision systems makes its use more computationally expensive, a problem that can be solved by dealing only with a set of image entities (regions, points or edges). Following this feature-based strategy, it is now easier to find proposals that solve the simultaneous localization and mapping problem or the human motion capture problem using vision, without employing external beacons or markers. If a mobile robot needs to solve several different tasks, we must consider that each task will need the detection of a specific set of features (local points of interest, human body parts...), so the perception system should be also changed according to the task. In this way, not only the generality of use is lost but also the robot will need to simultaneously manage different perception modules, as it will need to correctly attend to a very diverse set of situations. In biological vision systems, the attention mechanism is responsible for preselecting possible relevant information from the sensed field of view so that the complete scene can be analyzed using a sequence of rapid eye saccades. In recent years, efforts have been made to imitate such attention behavior in artificial vision systems, because it allows optimizing the computational resources as they can be focused on the processing of a set of selected regions only. Moreover, although these models can be influenced by the task to reach, they also include a bottom-up component to choose the more relevant item of the scene independently of the task. This allows to link perception and action, with perception influenced by the task to reach and the action by the perceived items.

The aim of this proposal is to present a new object-based framework of visual attention. With respect to previous approaches, our main contribution will be the representation of objects not only using appearance information, but also its internal topological configuration. This system will integrate bottom-up (data-driven) and top-down (model-driven) processing. The bottom-up component will determine salient 'pre-attentive objects' by integrating different features into the same hierarchical structure. Specifically, we propose to achieve this perception-based grouping process using a Combinatorial Pyramid (Brun and Kropatsch, 2001). Using this framework, the image topology will be preserved at upper levels; allowing correctly encoding relationships among image regions (Brun and Kropatsch, 2001). It must be noted that these 'pre-attentive objects' or 'proto-objects' (Orabona et al, 2007; Pylyshyn, 2001) will be image entities that will not necessary correspond with a recognizable object, although they will possess some of the characteristic of objects. It could be considered that they will be the result of the segmentation of each frame of the input video sequence into candidate objects (i.e. grouping together those input pixels which will be likely to correspond to parts of the same object in the real world, separately from those which are likely to belong to other objects). This process will cluster the image pixels into entities that can be considered as segmented perceptual units (Antúnez et al, 2011). The top-down component will make use of object templates to filter out data and shift the attention to objects which are relevant to accomplish the current task (e.g. human faces in a human-robot interaction framework). Generic knowledge could be used to select potential areas of attention in this component. If the knowledge is acquired before, it could lead to a hierarchy describing the structure of an articulated object with abstract properties of the entities (e.g. connectivity, articulation...). Such information can be efficiently used in the top-down search to focus quickly on the more relevant parts of the objects. Thus, our model will only consider how the a priori knowledge about the target can bias the attention.

The rest of this work is organized as follows: First, we will introduce some concepts related with artificial attention and will briefly unfold several computational models of attention. Next, we will describe our proposal. The basis of our model is the resembling of the visual ventral stream using a hierarchical grouping process that is conducted by encoding the input image into a Combinatorial Pyramid. Thus, we will firstly introduce this structure and the encoding of the image information through combinatorial maps. The bottom-up component of attention will decompose the image into regions (proto-objects) by a segmentation strategy based on the Combinatorial Pyramid. Saliency values will be associated to each image region according to color and brightness contrasts. On the other hand, the top-down component of attention looks for a specific target in the hierarchy, assigning saliency values to the image regions as a function of their similarities with the desired target. This saliency bias will be conducted by weighting the bottom-up saliency values, as suggested by Bichot et al (2005) or Wolfe (2007). Bottom-up (data-driven) and top-down (target-dependent) saliency values will be combined to determine the saliency values of image regions. Once the focus of attention has been fixed over a region of the space, enclosing a chosen proto-object, the topological and photometric properties of the proto-object will be compared to the properties of the target. This recognition task will then employ the same descriptors that drive the attention. The paper finalizes presenting several experimental results, conclusions and future work.

BACKGROUND

Cognitive vision is the research area concerned with endowing computer vision systems with cognitive capabilities in an attempt to increase their robustness and adaptability (Vernon, 2008). Although there are several quite distinct paradigms to the understanding and synthesis of cognitive systems, it is generally assumed that a good starting point for the development of such a system can be provided by looking at how nature deals with cognition. Thus, one of the trends is to exploit new knowledge gained from research in the Neurosciences or in Psychology. Specifically, one of the aspects of cognitive vision systems that have obtained more benefits from this interdisciplinary collaboration is visual attention. In human cognitive vision, attention constitutes a critical issue, which is in charge of directing the finite computational capacity of our visual cortex to relevant stimuli within the visual field while ignoring everything else (Tsotsos, 1997). In this sense, it has been often compared to a virtual spotlight through which our brain perceives the world (Navalpakkam and Itti, 2005). However, although this definition is rigorously correct, it is also very limited, and it does not take into consideration all different effects of attention. Nowadays, it is generally assumed that attention plays an important role in all aspects of visual perception including not only sensing, but also visual reasoning, recognition and visual context (Navalpakkam and Itti, 2005; Tsotsos, 2006). This assertion leads to a more general definition of attention, which considers that search, at all stages of visual perception, is not only driven by those factors that directly emanate from the scene (bottom-up, data-driven factors), but also by those derived from cognitive issues, such as knowledge of the task, gist of the scene and nature of the target (top-down, task-dependent factors).

Based on concepts that emanate from the human perception system, there exist on the literature a relatively large number of computational models whose aim is to develop some of the specific abilities of attention for searching or selection. With the aim of explaining the main functional role of visual attention as a mechanism to direct the computational resources for selective sensing, the feature integration theory proposed by Treisman and Gelade (1980) suggests that attention is used to combine (binding) different features (e.g. colour and shape) of an object during visual perception. According to this model, methods compute image features in a number of parallel channels in a pre-attentive, task-independent stage (Koch and Ullman, 1985; Itti et al, 1998). In the first implementation of this model, Koch and Ullman (1985) propose to integrate the extracted features into a single saliency map, which codes the saliency of each image pixel. The iNVT of Itti et al (1998) is one of the most popular systems, and it has obtained good results to simulate human eye movements and in applications ranging from object recognition to robotics. One problem

of these approaches is that the fusion of feature channels with per se not comparable characteristic is somewhat arbitrary (Klein and Frintrop, 2011). These approaches mainly resemble the so-called ventral and dorsal streams for attention, the more relevant visual pathways in the brain. These streams have a hierarchical architecture in which visual form information is analyzed in an increasingly complex fashion (Chikkerur et al, 2010; Tsotsos, 1990; Tsotsos, 1991, Tsotsos, 1992). Although the feature integration theory has been mainly accepted, posterior works have attempted to account specific behavioural effects of attention (e.g. modelling the influence of the scene context (Torralba, 2003) or the pop-out of salient objects (Itti et al, 1998)) or physiological evidences (e.g. the feature-based attention (Bichot et al, 2005)). Thus, the top-down bias of target features is an especially important behavioural effect, which was considered in the seminal Guided Search model proposed by Wolfe (2007). Following the scheme by Koch and Ullman (1985), this model computes and combines a set of features over the image, but in addition, it achieves feature-based biasing by weighting feature maps in a top-down manner (Wolfe, 2007). As neurobiology had showed before, bottom-up and top-down components for attention are not mutually exclusive, and nowadays, efforts in computational attention are being conducted to develop models which combine both factors (Tsotsos et al, 1995; Navalpakkam and Itti, 2005; Chikkerur et al, 2010). Having selected the focus of attention, pre-attentive features may be also used for object representation and recognition (Navalpakkam and Itti, 2005). Attention arises then as an important link connecting sensing and recognition (Tsotsos, 2006), an assertion that does not imply that recognition before attention makes no sense. In fact, it is believed that attention selects objects, object parts or groups of objects rather than spatial locations. For instance, Walther and Koch (2006) have combined the saliency model with the standard model of object recognition, considering the shape of the attended object to shape the area of attention. Proto-objects or pre-attentive objects possess some but not all the characteristics of objects, and they constitute a step above the mere localized features (Borji et al, 2010; Yu et al, 2010). In our previous proposal for computational modelling of attention (Palomino et al, 2011), we have developed an object-based model for the bottom-up processing. This model was endowed into a hierarchical structure for image grouping, where each level of abstraction is encoded as a graph with a reduced set of nodes. The whole hierarchy can be divided up into two consecutive stages. From the basic features associated to the image pixels, the first stage clusters pixels into uniform blobs (pre-segmentation stage). Then, the second stage groups the set of uniform blobs into a reduced set of pre-attentive objects, taking into account higher-level features (perceptual grouping stage). Target-based saliency maps are generated and provided as an independent input to this model, being combined with the rest of bottom-up feature maps to obtain a global, unique saliency map (Koch and Ullman, 1985).

THE PROPOSED ARTIFICIAL MODEL OF ATTENTION

The Combinatorial Pyramid

In this work, the hierarchical organization of the visual stimuli conducted by the ventral stream is encoded using an irregular pyramid. Irregular pyramids represent the input frame as a stack of graphs with decreasing number of vertices. Such hierarchies present many interesting properties within the Image Processing and Analysis framework such as: reducing the influence of noise by eliminating less important details in upper levels of the hierarchy, making the processing independent of the resolution of the regions of interest in the image, converting local features to global ones, reducing computational costs, etc (Kropatsch et al, 2005). The construction of the pyramid follows

the philosophy of reducing the amount of data between consecutive levels of the hierarchy by a reduction factor greater than one, a strategy that is also considered by other hierarchical approaches, such as the Ultrametric Contour Maps (UCM) proposed by Arbeláez (2006). As other irregular pyramids, the UCM hierarchy relies on the use of a simple graph (i.e., a region adjacency graph (RAG)) to represent each level of the hierarchy. Region adjacency graphs have two main drawbacks for image processing tasks:

1. They do not permit to know if two adjacent regions have one or more common boundaries, and
2. They do not allow differentiating an adjacency relationship between two regions from an inclusion relationship.

That is, the use of this graph encoding avoids that the topology will be preserved at upper levels of the hierarchies. Taking into account that objects are not only characterized by features or parts, but also by the spatial relationships among these features or parts, this limitation constitutes a severe disadvantage. Instead of simple graphs, each level of the hierarchy could be represented using a pair of dual graphs. Dual graphs preserve the topological information at upper levels representing each level of the pyramid as a pair of dual graphs and computing contraction and removal operations within them (Haxhimusa et al, 2003). Thus, they overcome the drawbacks of the RAG approach. The problem of this structure is the high increase of memory requirements and execution times since two data structures need now to be stored and processed. To avoid this problem, the described segmentation approach accomplishes the grouping process by means of the combinatorial pyramid (Brun and Kropatsch, 2001). A combinatorial pyramid is a hierarchical stack of combinatorial maps successively reduced by a sequence of contraction or removal operations (see (Brun and Kropatsch, 2001) for further details). Combinatorial pyramids combine the advantages of dual graph pyramids with an explicit orientation of the boundary segments of the embedded object thanks to one of the permutations which defines the map (Brun and Kropatsch, 2001). Moreover, it uses a combinatorial map at each level of the pyramid instead of a pair of dual graphs, thus reducing the memory requirements and execution times.

As aforementioned, each level of the Combinatorial Pyramid is encoded by a combinatorial map. A combinatorial map is a combinatorial representation describing the subdivision of a space. It encodes all the vertices that compound this subdivision and all the incidence and adjacency relationships among them. That is, an n-dimensional combinatorial map is an $(n+1)$-tuple $M=(D,\beta_1,\beta_2,...,\beta_n)$ such that D is the set of abstract elements called darts, β_1 is a permutation on D and the other β_i are involutions on D. An involution is a permutation whose cycle has the length of two or less. Two-dimensional (2D) combinatorial maps may be defined with the triplet $G = (D, \sigma, \alpha)$, where D is the set of darts, σ is a permutation in D encoding the set of darts encountered when turning (counter) clockwise around a vertex, and α is an involution in D connecting two darts belonging to the same arc:

$$\forall d \in D, \alpha^2 (d)=d$$

Figure 1.a shows an example of a combinatorial map. In Figure 1.b, the values of α and σ for such a combinatorial map can be found. In our approach, counter-clockwise orientation (ccw) for σ is chosen.

The symbols $\sigma*(d)$ and $\alpha*(d)$ stand the σ and α orbits of the dart d, respectively. The orbit of a permutation is obtained applying successively such a permutation over the element that is defined. In this case, the orbit $\sigma *$ encodes the set of darts encountered when turning counter-clockwise around the vertex encoded by the dart d. The orbit $\alpha*$ encodes the darts that belong to the same arc. Therefore, the orbits of σ encode the vertices of the graph and the orbits of α define the arcs of the graph. In the example of Figure 1, $\alpha*(1) = \{1,-1\}$ and $\sigma*(1) = \{1,5,2\}$. Given a combinatorial map, its dual is defined by $\sim G=(D, \varphi, \alpha)$ with $\varphi=\sigma \circ \alpha$. The orbits of φ encode the faces of the combinatorial map. Thus, the orbit $\varphi*$ can be seen as the set of darts obtained when turning-clockwise a face of the map. In Figure 1 $\varphi*(1) = \{1,-3,-2\}$. Thus, 2D combinatorial maps encode a subdivision of a 2D space into vertices ($V= \sigma*(D)$), arcs ($E=\alpha*(D)$) and faces ($F=\varphi*(D)$).

When a combinatorial map is built from an image, the vertices of such a map G could be used to represent the pixels (regions) of the image. Then, in its dual $\sim G$, instead of vertices, faces are used to represent pixels (regions). Both maps store the same information and there is not so much

Figure 1. a) Example of combinatorial map; and b) values of α and σ for the combinatorial map in a)

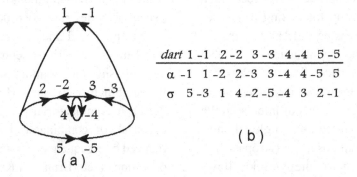

dart	1	-1	2	-2	3	-3	4	-4	5	-5
α	-1	1	-2	2	-3	3	-4	4	-5	5
σ	5	-3	1	4	-2	-5	-4	3	2	-1

(b)

(a)

difference in working with G or $\sim G$. However, as the base entity of the combinatorial map is the dart, it is not possible that this map contains only one vertex and no arcs. Therefore, if we choose to work with G, and taking into account that the map could be composed by a unique region, it is necessary to add special darts to represent the infinite region which surrounds the image (the background). Adding these darts, it is avoided that the map will contain only one vertex. On the other hand, when $\sim G$ is chosen, the background also exists but there is no need to add special darts to represent it. In this case, a map with only one region (face) would be made out of two darts related by α and σ. In our case, the base level of the pyramid will be a combinatorial map where each face represents a pixel of the image as a homogeneous region.

Bottom-Up (Data-Driven) Component of Attention

Object-based attention theories are based on the assumption that attention must be directed to an object or group of objects, instead to a generic region of the space (Orabona et al, 2007). In fact, neurophysiologic studies (Scholl, 2001) show that, in selective attention, the boundaries of segmented objects, and not just spatial position, determines what is selected and how attention is deployed. Therefore, these models will reflect the fact that the perception abilities must be optimized to interact with objects and not just with disembodied spatial locations. In the last few years, these models of visual attention have received an increasing interest in computational neuroscience and in computer vision. Thus, visual systems will segment complex scenes into objects which can be subsequently used for recognition and action. However, recent psychological research shows that, in natural vision, the pre-attentive process divides a visual input into raw or primitive objects (Olson, 2001) instead of well-defined objects. As aforementioned, some authors use the notion of proto-objects (Orabona et al, 2007; Yu et al, 2010) (pre-attentive objects or object files) to refer to these primitive objects, which are defined as units of visual information that can be bound into a coherent and stable object. The salient local regions of the image can be obtained by searching for discontinuities. For instance, Kadir and Brady (2001) use the brightness entropy of a region to measure its magnitude and scale of saliency. Escalera et al (2008) propose a model that allows to detect the most relevant image features based on their complexity. Entropy-based approaches have problems to deal with scenarios where the absence of structure makes an item salient. The Bonn Information-Theoretic Saliency model (Klein and Frintrop, 2011) is based on the feature integration theory, but centric-surround contrast is here determined in an information-theoretic way using the Kullback-Leibler Divergence. In our model, the partitioning of an image into proto-objects will be conducted by a segmentation algorithm.

The main aim of segmentation approaches is the clustering of visual information, grouping image pixels into entities of increasing size and semantic significance. Hierarchical, agglomerative approaches represent this perceptual organization by a tree of regions, ordered by inclusion (Arbeláez, 2006). In this tree, each region represents a portion of the image at a certain scale of observation. An efficient way to manage all the information in these hierarchies is to represent each level of the hierarchy as a graph. Graph-based approaches for image segmentation consider the input image as a graph in which pixels are usually vertices and the local dissimilarity between pixels sets the arc weights. Then, they attempt to merge nodes into larger components (Felzenszwalb and Huttenlocher, 2004) or to partition this image graph into a set of regions (Ren and Malik, 2003). In the graph-based merging algorithm proposed by Felzenszwalb and Huttenlocher (2004), the largest weight in the minimum spanning tree (MST) of a region (sub-graph) of the graph image defines the internal difference of this region, and the

Figure 2. a) Original images of the BSDS300 (Arbeláez et al, 2011; Martin et al, 2001); b) perceptual segmentation; and c) their associated bottom-up saliency maps

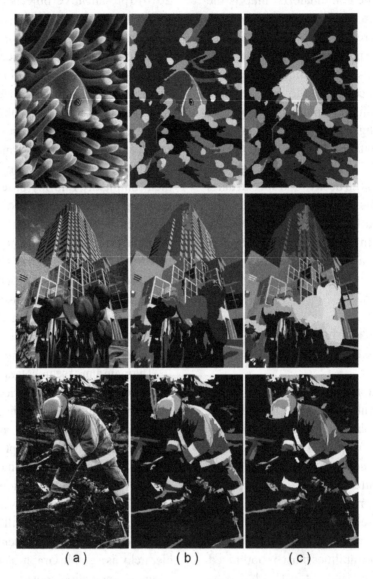

(a)   (b)   (c)

minimum weight arc connecting two regions defines the external difference between them. Each vertex is initially considered as a region of the image, after putting graph arcs into nondecreasing order by weight, the algorithm merges two regions if the external difference between them is small relative to the internal differences within at least one of the regions. A thresholding function is used to set a preference for a region size. In the hierarchy of partitions (Haxhimusa et al, 2003), the merging process based on internal and external differences is conducted through an irregular pyramid in which each level is encoded by a pair of dual graphs. Dual graph contraction (Kropatsch, 1995) is used to preserve the graph topology. If these approaches obtain global image evidences from the accumulation of local cues, the Normalized Cuts criterion has been widely used to integrate global information into the grouping procedure (Shi and Malik, 2000). In our proposal, the grouping process is accomplished by means of the MST Combinatorial Pyramid (Haxhimusa et al, 2006; Ion et al, 2006). The MST pyramid takes as input an image graph and obtains a hierarchy of partitions by using the MST algorithm and the region internal/external differences (Felzenszwalb and Huttenlocher, 2004). Specifically, the internal difference (contrast) of a region is defined as the largest weight of the arcs on its MST. As afore-mentioned, the approach was initially proposed in the dual graph-based irregular pyramid framework (Haxhimusa et al, 2003), and subsequently adapted to the combinatorial pyramid framework (Haxhi-musa et al, 2006; Haxhimusa et al, 2005; Ion et al, 2006). Our segmentation algorithm (Antúnez et al, 2011) generalizes the cited previous work:

- It employs contour and region properties, encoded in the darts and faces of the com-binatorial maps, respectively; and
- Two different measures to conduct the seg-mentation at the different levels of the hi-erarchy are used.

Thus, at the low levels of the hierarchy, a dis-tance based on color is used to divide the image into a set of regions whose spatial distribution is physically representative of the image content. The aim of this pre-segmentation stage is to represent the image by means of a set of superpixels whose number will be commonly very much less than the original number of image pixels. Besides, these superpixels will preserve the image geometric structure as each significant feature contains at least one superpixel. Next, a perceptual grouping stage groups this set of homogeneous superpixels into a smaller set of regions taking into account not only the internal visual coherence of the ob-tained regions but mainly the external relationships among them, encoded as boundary evidences on the arcs of the combinatorial maps. Following this algorithm, bottom-up attention will organize

visual stimuli in a hierarchy of levels of abstrac-tion. At the upper level of the hierarchy, the image is decomposed into pre-attentive objects, whose saliency values will be obtained using color and brightness contrasts (Marfil et al, 2009). Figure 2 shows some examples of images and their as-sociated bottom-up saliency maps.

Top-Down (Target Dependant) Component of Attention

Studies of eye movements, physiology and psy-chophysics show that, in the human visual system, the nature of the target plays an important role in selecting the focus of attention. The hypothesis that our visual system biases the attention system according to the known target representation is suggested by the fact that prior knowledge of this target accelerates its detection in visual search tasks (Navalpakkam and Itti, 2005). This knowledge can be combined with the bottom-up stream. Thus, the low-level visual system will be influenced by the known features of the target, e.g. weighting the different feature maps that determine the bottom-up salience to give more importance to those features presented on the target (Bichot et al, 2005). In our model, the feature-based atten-tion mechanism will be conducted by modeling the target object by means of two ellipses. One of them corresponds to the most salient object region, according to color contrast, and the other one covers the entire object. Both ellipses have the same first and second order parameters that the region(s) they enclose. The target model is then composed by the shape e_t and mean color c_t of the ellipse that covers the most salient region, the shape E_t and color histogram H_t associated to the ellipse that covers the whole target, and the geometric relationships between the two com-puted ellipses (relative rotation r_t and scaling s_t). Let I be the segmented image which represents the perceived scene and $\{R\}_{i=1...n}$ the set of image regions. Once the target has been modeled, the algorithm performs the following steps:

1. Determine the subset of *{R}* whose mean color is close to c_t.

2. Compute the set of ellipses *{e_s}* corresponding to the regions obtained in step 1.

3. Given the matrix *A* that encodes the transformation between the e_t and E_t, each $e_{\{s\}i}$ shape is covariantly transformed according to *A* obtaining a set of ellipses *{E_s}*. This transformation is not unique, and there will be two possible locations for each E_s. In fact, if $e_{\{s\}i}$ is really a circle, then it will be not possible to determine the position of $E_{\{s\}i}$ (i.e. the number of possible locations will be infinity).

4. Compute a color histogram $H_{\{s\}i}$ for each $E_{\{s\}i}$ and the color difference between each of them and H_t. The bottom-up saliency of each R_i will be weighted according to this color difference, associating higher values to those regions more similar to the target.

As aforementioned, a pyramidal algorithm for segmentation is employed (Antúnez et al, 2011). This allows a fast searching of the subset of *{R}*, which will be the receptive fields of a subset of vertices at upper levels of the hierarchy. The set of ellipses *{e_s}* will be the simplified description of these receptive fields, having the same first and second moments as the originally arbitrarily shaped regions.

RECOGNITION STAGE

Once the focus of attention is fixed over a given proto-object, our model will evaluate if this corresponds to the searched target. In our model, this problem is stated as a region correspondence task. A widely used approach for finding region correspondences is to use region features such as color or texture to match regions (Li et al, 2000; Greenspan et al, 2000). The problem is that these algorithms do not take into account the neighborhood or context of the two regions being matched.

In our case, the target object as well as the scene image can be segmented using the Combinatorial Pyramid. Then, we will use the information encoded in the combinatorial maps at the top of both pyramids to solve the region correspondence problem as a graph matching problem. Although there are other solutions to this problem (e.g., see Pelillo et al (1999)), it has been solved in our system by finding a maximum clique in an association or correspondence graph. This matching process will not consider the whole image, but only the areas of the scene where the object is more probably located (focus of attention). As aforementioned, these areas have been obtained according to the bottom-up salience map. The corresponding sub-combinatorial map associated to this proto-object and the combinatorial map of the target are both used to perform the graph matching procedure. Each sub-combinatorial map is derived from the combinatorial map of the top of the pyramid obtained in the segmentation of the scene. Therefore, each sub-combinatorial map includes only the regions that are associated to a specific proto-object. In this way, recognition and attention shares the same features (Navalpakkam and Itti, 2005).

The fundamental step in the graph matching process is to build the association graph which represents valid associations between the two sets of regions to be matched. The construction of the association graph is performed through the application of relative and absolute constraints. The nodes of the graph indicate individual association compatibility and they are determined by absolute constraints. On the other hand, the edges of the correspondence graph indicate joint compatibility of the connected nodes and they are determined by a relative constraint. The method used to calculate the association graph has three major stages (Antúnez et al, 2011):

1. Computation of the nodes of the association graph. In the proposed method, being I_1 and I_2 the images whose regions want to be

Figure 3. a) Target and its more relevant constitutive part P; b) segmented original image showing those regions that resemble the part P of the target; c) ellipse enclosing the whole target; d) segmented original image showing those regions that resemble the whole target according to P; and e) the six regions on the image whose shape and photometric properties resemble the whole target

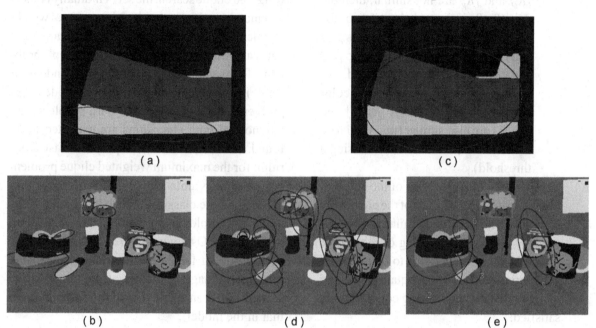

(a) (c)

(b) (d) (e)

matched, the nodes of the association graph are pairs of regions $R_i \in I_1$ and $R_j \in I_2$ that are candidate to be matched. To be matched, two regions must be similar in color (this will be set using a color threshold). However, this will be a necessary, but not a sufficient condition. To take into account topological information, the inside, contains and meets constraints (Brun and Kropatsch, 2006) will be also considered. That is, being $R_i \in I_1$ and $R_j \in I_2$ similar in color, there exist several possibilities:

a.   If $R_i \subset R_k$ and $R_j \subset R_p$ and the color difference between R_k and R_i is under a specific threshold, then R_i and R_j are candidate matches. In other case, if $R_i \subset R_k$ but R_j is not inside any region, or vice versa, the set of neighbors $\{R_n\}$ of R_j are studied. If all the regions in $\{R_n\}$ are similar in color, then R_i and R_j are

potential matches. In other case, they are discarded as candidate matches.

b.   Let $\{R_k\}$ be the set of regions inside R_i and $\{R_p\}$ the set of regions inside R_j. These two sets $\{R_k\}$ and $\{R_p\}$ are similar if, given $\{R_m\} \subset \{R_k\}$ and $\{R_n\} \subset \{R_p\}$, $\forall R_p \in \{R_m\}$ and $\forall R_q \in \{R_n\}$, R_p and R_q are similar in color. Then, R_i and R_j are candidate matches if R_i contains $\{R_k\}$ and R_j contains $\{R_p\}$, and the sets $\{R_k\}$ and $\{R_p\}$ are similar. In other case, if $\{R_k\}$ and $\{R_p\}$ are not similar, then the pair (R_i, R_j) is discarded as node of the association graph. If R_i contains $\{R_k\}$ but R_j does not contain any region, or vice versa, R_i and R_j are not considered as candidate matches.

c.   Let $\{R_k\}$ be the set of neighboring regions of R_i and $\{R_p\}$ the set of neighboring regions inside R_j. If R_i meets $\{R_k\}$

and R_j *meets* $\{R_l\}$, and the sets $\{R_k\}$ and $\{R_l\}$ are similar, then R_i and R_j are candidate matches. In other case, if $\{R_k\}$ and $\{R_l\}$ are not similar, then the pair (R_i, R_j) is discarded as node of the association graph. In this case, sets $\{R_k\}$ and $\{R_l\}$ are similar if, given $\{R_m\} \subset \{R_k\}$ and $\{R_n\} \subset \{R_l\}$, $\forall R_p \in \{R_m\}$ and $\forall R_q \in \{R_n\}$, R_p and R_q are similar in color and moreover, the ratios $|R_i|/|R_p|$ and $|R_j|/|R_q|$ are also similar (this similarity in relative size will be also set using a threshold).

2. Computation of the weights of the nodes of the association graph. Each of the previously computed pair of candidate matches has associated a weight depending of the degree of similarity of the regions to be matched. It is initialized to a value equal to one and incremented if the following conditions are satisfied:

   a. Taking into account the pair of candidate matches (R_i, R_j), the weight is incremented by 1 if $d_c(R_i, R_j) < d_c(R_i, R_m)$, being R_m the set of all possible matches of R_i.

   b. If (R_i, R_j) satisfies the contain relationship (constraint 2) and they contain the same number of regions, that is, $|R_k| = |R_l|$, where $\{R_k\}$ is the set of regions inside R_i and $\{R_l\}$ is the set of regions inside R_j, then the weight is incremented in 1.

   c. If (R_i, R_j) satisfies the meet relationship (constraint 3) and they have the same number of neighbors, that is, $|R_k| = |R_l|$, where $\{R_k\}$ is the set of neighboring regions of R_i and $\{R_l\}$ is the set of neighboring regions of R_j, then the weight is incremented in 1.

   d. Definition of the arcs of the association graph. An arc is set between two pairs of candidate matches $n_1 = (R_i, R_j)$ and $n_2 = (R_k, R_l)$ if $n_1 \neq n_2$ and $i \neq k$ and $j \neq l$.

Complete subgraphs or cliques within the association graph indicate mutual association compatibility and, by performing a maximum weighted clique search, the set of mutually consistent matches which provides a largest total weight is calculated. This problem is computationally equivalent to some other important graph problems, for example, the maximum independent (or stable) set problem and the minimum node cover problem. Since these are NP-hard problems, no polynomial time algorithms are expected to be found. In this work, we employ the fast algorithm for the maximum weighted clique problem proposed by Kumlander (2008). This algorithm is based on the classical branch and bound technique, but employing the backtracking algorithm proposed by Ostergard (2002). Figure 4 shows an example of target and the final result of the recognition stage (the parts of the target in the scene have been marked with the same indexes that in the model).

Experimental Results

To evaluate the ability of the bottom-up component of attention to extract salient regions, we compared the discriminant saliency maps obtained from a collection of natural images to the eye fixation locations recorded from human subjects, in a free-viewing task. Specifically, we have employed the human fixation database from Bruce and Tsotsos (2006). This data set was obtained from eye tracking experiments performed while subjects observed 120 different color images (see (Bruce and Tsotsos, 2006) for further details). The color and brightness contrast measures are employed as the feature to define our saliency map (Marfil et al, 2009) and, to measure the performance of the approach, obtained saliency maps are first quantized into a binary image: pixels with larger saliency values than a threshold are classified as fixated while the rest of the pixels in that image are classified as non-fixated (Tatler et al, 2005; Gao et al, 2008). Human fixations are then used as

Figure 4. a) Target; b) scene; and c) regions (1-4) associated to the target

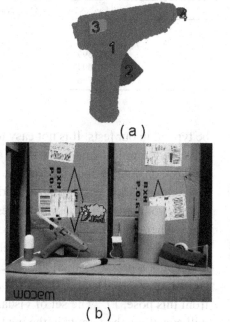

(a)

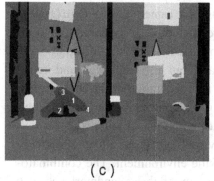

(b)   (c)

ground truth and a receiver operator characteristic (ROC) curve is drawn by varying the threshold value. The area under the curve indicates how well the saliency map predicts actual human eye fixations.

The quantitative performance of the proposed approach is shown in Table 1. In this table we also summarized the results obtained using the algorithms of Itti and Koch (2000), obtained using the Matlab saliency toolbox (Walther and Koch, 2006), Bruce and Tsotsos (2006) and Gao et al (2008). The proposal of Gao et al (2008) codifies the bottom-up saliency as a result of an optimal decision making, under constraints of computational parsimony. Thus, they derive optimal saliency detectors for intensity, color, orientation, and motion. On the other hand, Bruce and Tsotsos (2006) propose to estimate saliencies using an optimal implementation of the maximization of self-information. Moreover, as an absolute benchmark, the 'inter-subject' ROC area is also included (Gao et al, 2008; Harel et al, 2007). Although previous approaches use a more

complex set of features, it can be noted that the results obtained by our approach are similar to the ones provided by these detectors.

The influence of the top-down component of attention can be appreciated in Figure 5. The bottom-up saliency values associated to the seven objects in the scene are very high, as all of them present colors that are very different from their surroundings (Figure 5c). In fact, there are also high saliency values associated to regions of the background (e.g., the letters of the box that have not been included on the background). As Figure 5d shows, the knowledge of the target biases the attention towards those areas of the image that present a specific shape and distribution of color (see the differences between Figures 5c and 5d). In Figure 5d, we have also marked the focus of attention and scanpath.

Finally, we show an experiment where the proposed framework has been employed for visual landmark detection and recognition. The evaluation has been conducted on a Pioneer 2AT platform from ActivMedia. The image acquisition

Table 1. ROC areas for different saliency models with respect to all human fixations (see text for details)

Saliency model	ROC area
Itti and Koch (2000)	0.7277
Bruce and Tsotsos (2006)	0.7547
Proposed bottom-up algorithm	0.7599
Gao et al (2008)	0.7694
Inter-subject	0.8766

system used in the experiments employs a STH-MDCS stereoscopic camera from Videre Design. Images were restricted to 320x240 pixels. Bottom-up component drives the focus of attention to visual landmarks but also to background items. In a first trial, the robot moves in the environment and asks the user human for helping in the labeling of the detected regions. Figure 6 shows several of these visual landmarks and a top view of the layout of the environment. The combinatorial representation of these landmarks is also presented. It can be noted that some landmarks (e.g. the poster landmark) are associated to a unique face. Due to the resolution of the detected targets and the office-like environment, this will

be typical in our tests. It is not easy to find topologically complex landmarks such as the large window or double door landmarks. A set of 16 visual landmarks was initialized and located on the environment (i.e., the map stores the places from where these visual landmarks were observed). Each observation mark O_i is characterized by a robot's pose (position and orientation), $\mathbf{p}_i$, and the set of visual landmarks that were perceived from this pose, VL_i. The set of visual landmarks will constitute the targets in the next trials.

After this training trial, the robot autonomously travels through the environment in the subsequent trials. The bottom-up component of attention is triggered. However, the robot has now

Figure 5. a) Target, b) input scene, c) bottom-up saliency values, and d) combination of top-down factors with the bottom-up component of attention. This last figure shows the three first regions where the focus will be driven and the scanpath.

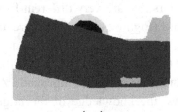

(a)

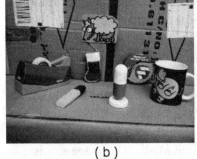

(b)

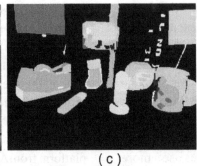

(c)

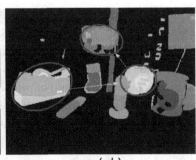

(d)

Figure 6. a) Layout of the environment, and b) examples of visual landmarks on this environment (appearance and combinatorial representations) (see text for details)

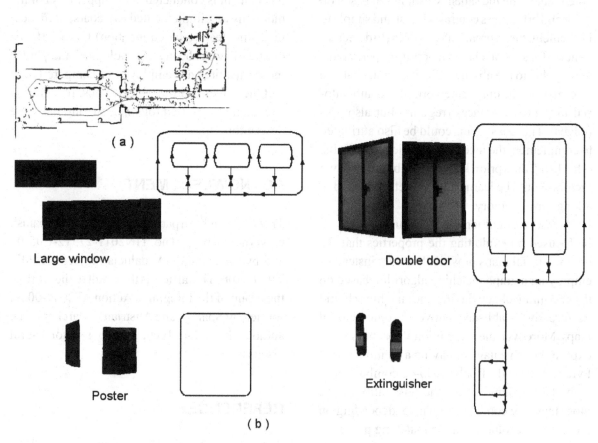

a map of the environment, which includes the observation marks. Thus, when the robot estimates that it is close to an observation mark O_i, it uses the templates of the visual landmark VL_i to implement the feature-based searching. It should be noted that, in this experimental setting, the geometric pose of the robot cannot be modified according to the perceived landmarks. Visual landmarks are not characterized by a precise 3-dimensional position on the outer world. But the experimental setting allows to evaluate the ability of the robot for efficiently searching of a predefined set of landmarks. For instance, in one of the trials (528 frames), we manually labeled 435 occurrences of the set of 16 landmarks. The robot correctly detects 396 observations (91%). It must be also noted that only 12 false observa-

tions occur. Three fixations were only allowed for each image.

FUTURE RESEARCH DIRECTIONS

Scenes are dynamic, and the attention mechanism should be able to deal with situations where the objects and visual system can be both in motion. To deal with this scenario, the computational model of attention should include a mechanism to avoid that the attention will be always focused over the same proto-objects (i.e., an implementation of the Inhibition of Return (IOR) (Palomino et al, 2010)) and that it can track the motion of recently attended objects. In the proposed framework, this could be performed by including a tracking algorithm that works inside the Combinatorial Pyramid and a short-term (working) memory. It should be also noted that while the saliency estimation based on color and brightness contrasts is fast and simple to implement, the approach has problem to detect saliencies for several classical pop-up experiments, mainly due to the absence of orientation or shape information. In our framework, the combinatorial map provides faces (regions) but also arcs (edges). These last items could be also attributed to complement the face-related saliencies. On the other hand, the a priori knowledge about top-down factors should be learnt, and correctly encoded in a long-term memory.

At the recognition stage, future work will be focused on exploiting the properties that the combinatorial maps provide. For that, instead of employing a graph matching algorithm based on the maximum clique finding, the matching should be directly established between combinatorial maps. Moreover, the recognition algorithm should exploit the advantage of having a Combinatorial Pyramid instead of only a single combinatorial map. The pyramid offers the possibility to use more than one combinatorial map, depending on the needed resolution, in the matching process.

CONCLUSION

This paper proposes a visual attention model that integrates bottom-up and top-down processing. Following an object-based strategy for attention, the bottom-up process is conducted by dividing the visual scene into perceptually uniform blobs using color and edge information. Saliency is estimated using color and brightness contrasts. Thus, the model can drive attention to proto-objects, similarly to the behavior observed in humans. Experimental results have shown that the bottom-up component of attention provides results that are similar or outperform similar other approaches. Moreover, this bottom-up mechanism is integrated with a top-down component. This top-down factors bias the attention to those regions of the scene whose color distribution and

global shape resembles the target's properties. This mechanism is conducted at the upper level of the hierarchy and implemented as a coarse statistical matching (feature-based attention). Once the focus of attention has been set, topological and photometric features are employed to compare target and the selected proto-object. Thus, recognition and attention are tied together, sharing the same descriptors.

ACKNOWLEDGMENT

This work has been partially granted by the Spanish Government project no. TIN2011-27512-C05-01 and by the Junta de Andalucía project no. P07-TIC-03106. This article is the result of the work of the group of the Integrated Action AT2009-0026, formed by Spanish and Austrian researchers. The authors thank Dr. Tech. Adrian Ion for useful discussions.

REFERENCES

Antúnez, E., Marfil, R., & Bandera, A. (2011). Region correspondence using combinatorial pyramids. In Bandera, A., Dias, J., & Escolano, F. (Eds.), *Recognition and action for scene understanding* (pp. 13–24). Málaga, Spain: SPICUM.

Arbeláez, P. (2006). Boundary extraction in natural images using ultrametric contour maps. In *Proceedings 5th IEEE Workshop Perceptual Organization in Computer Vision*, (pp. 182-189).

Arbeláez, P., Maire, M., Fowlkes, C., & Malik, J. (2011). Contour detection and hierarchical image segmentation. *IEEE Transactions on Pattern Analysis and Machine Intelligence*, (n.d), 33.

Bichot, N., Rossi, A., & Desimone, R. (2005). Parallel and serial neural mechanisms for visual search in macaque area V4. *Science*, *308*, 529–534. doi:10.1126/science.1109676

Borji, A., Ahmadabadi, M., Araabi, B., & Hamidi, M. (2010). Online learning of task-driven object-based visual attention control. *Image and Vision Computing, 28*, 1130–1145. doi:10.1016/j.imavis.2009.10.006

Bruce, N., & Tsotsos, J. (2006). In Weiss, Y., Schölkopf, B., & Platt, J. (Eds.), *Saliency based on information maximization* (*Vol. 18*, pp. 155–162). Advances in neural information processing systems Cambridge, MA: MIT Press.

Brun, L., & Kropatsch, W.G. (2001). Introduction to combinatorial pyramids. *Lecture Notes in Computer Science, 2243*, 108–128. doi:10.1007/3-540-45576-0_7

Chikkerur, S., Serre, T., Tan, C., & Poggio, T. (2010). What and where: A Bayesian inference theory of attention. *Vision Research, 50*, 2233–2247. doi:10.1016/j.visres.2010.05.013

Escalera, S., Pujol, O., & Radeva, P. (2008). Detection of complex salient regions. *EURASIP Journal on Advances in Signal Processing, 2008*. doi:10.1155/2008/451389

Feldman, J. A., & Ballard, D. H. (1982). Connectionist models and their properties. *Cognitive Science, 6*, 205–254. doi:10.1207/s15516709cog0603_1

Felzenszwalb, P., & Huttenlocher, D. (2004). Efficient graph-based image segmentation. *International Journal of Computer Vision, 59*, 167–181. doi:10.1023/B:VISI.0000022288.19776.77

Gao, D., Mahadevan, V., & Vasconcelos, N. (2008). On the plausibility of the discriminant center-surround hypothesis for visual saliency. *Journal of Vision (Charlottesville, Va.), 8*(7), 1–18. doi:10.1167/8.7.13

Greenspan, H., Dvir, G., & Rubner, Y. (2000). Region correspondence for image matching via EMD flow. *Proceedings of the IEEE Workshop on Content-based Access of Image and Video Libraries*, (pp. 27-31).

Harel, J., Koch, C., & Perona, P. (2007). In Schölkopf, B., Platt, J., & Hoffman, T. (Eds.), *Graph-based visual saliency* (*Vol. 19*, pp. 545–552). Advances in neural information processing systems Cambridge, MA: MIT Press.

Haxhimusa, Y., Ion, A., & Kropatsch, W.G. (2006). *Evaluating graph-based segmentation algorithms.*

Haxhimusa, Y., Ion, A., Kropatsch, W.G., & Brun, L. (2005). Hierarchical image partitioning using Combinatorial Maps. *Proceedings of Joint Hungarian-Austrian Conference* (OAGM), (pp. 179-186).

Haxhimusa, Y., & Kropatsch, W.G. (2003). Hierarchical image partitioning with dual graph contraction. *Lecture Notes in Computer Science, 2781*, 338–345. doi:10.1007/978-3-540-45243-0_44

Ion, A., Kropatsch, W.G., & Haxhimusa, Y. (2006). Considerations regarding the minimum spanning tree pyramid segmentation method. *Lecture Notes in Computer Science, 4109*, 182–190. doi:10.1007/11815921_19

Itti, L., & Koch, C. (2000). A saliency-based search mechanism for overt and covert shifts of visual attention. *Vision Research, 40*, 1489–1506. doi:10.1016/S0042-6989(99)00163-7

Itti, L., Koch, C., & Niebur, E. (1998). A model of saliency-based visual attention for rapid scene analysis. *IEEE Transactions on Pattern Analysis and Machine Intelligence, 20*(11), 1254–1259. doi:10.1109/34.730558

Kadir, T., & Brady, M. (2001). Saliency, scale and image description. *International Journal of Computer Vision, 45*(2), 83–105. doi:10.1023/A:1012460413855

Klein, D., & Frintrop, S. (2011). *Center-surround divergence of feature statistics for salient object detection.* International Conference on Computer Vision.

Koch, C., & Ullman, S. (1985). Shifts in selective visual attention: Towards the underlying neural circuitry. *Human Neurobiology, 4*, 219–227.

Kropatsch, W. G.(1995). Building irregular pyramids by dual graph contraction. *IEE Proceedings. Vision Image and Signal Processing, 142*(6), 366–374. doi:10.1049/ip-vis:19952115

Kropatsch, W.G., Haxhimusa, Y., Pizlo, Z., & Langs, G. (2005). Vision pyramids that do not grow too high. *Pattern Recognition Letters, 26,* 319–337. doi:10.1016/j.patrec.2004.10.026

Kumlander, D. (2008). On importance of a special sorting in the maximum-weight clique algorithm based on colour classes. *Proceedings of Second International Conference Modelling, Computation and Optimization in Information Systems,* (pp. 165-174).

Li, J., Wang, Z., & Wiederhold, G. (2000). IRM: Integrated region matching for image retrieval. *Proc. of ACM Multimedia,* 147-156.

Marfil, R., Bandera, A., Rodríguez, J. A., & Sandoval, F. (2009). A novel hierarchical framework for object-based visual attention. *Lecture Notes in Artificial Intelligence, 5395,* 27–40.

Martin, D., Fowlkes, C., Tal, D., & Malik, J. (2001). A database of human segmented natural images and its application to evaluating segmentation algorithms and measuring ecological statistics. *Proceedings of the 8th International Conference on Computer Vision,* Vol. 2, (pp. 416-423).

Navalpakkam, V., & Itti, L. (2005). Modeling the influence of task on attention. *Vision Research, 45,* 205–231. doi:10.1016/j.visres.2004.07.042

Olson, C. R. (2001). Object-based vision and attention in primates. *Current Opinion in Neurobiology, 11,* 171–179. doi:10.1016/S0959-4388(00)00193-8

Orabona, F., Metta, G., & Sandini, G. (2007). A proto-object based visual attention model. *Lecture Notes in Artificial Intelligence, 4840,* 198–215.

Ostergard, P. (2002). A fast algorithm for the maximum clique problem. *Discrete Applied Mathematics, 120,* 197–207. doi:10.1016/S0166-218X(01)00290-6

Palomino, A., Marfil, R., Bandera, J., & Bandera, A. (2011). A novel biologically inspired attention mechanism for a social robot. *EURASIP Journal on Advances in Signal Processing,* (n.d), 2011.

Pelillo, M., Siddiqi, K., & Zucker, S. (1999). *Continuous-based heuristics for graph and tree isomorphism, with application to computer vision.* Conference on Approximation and Complexity in Numerical Optimization: Continuous and Discrete Problems.

Pylyshyn, Z. (2001). Visual indexes, preconceptual objects, and situated vision. *Cognition, 80,* 127–158. doi:10.1016/S0010-0277(00)00156-6

Ren, X., & Malik, J. (2003). Learning a classification model for segmentation. *Proceedings of IEEE International Conference on Computer Vision,* (pp. 10-17).

Scholl, B. (2001). Objects and attention: the state of the art. *Cognition, 80,* 1–46. doi:10.1016/S0010-0277(00)00152-9

Shi, J., & Malik, J. (2000). Normalized cuts and image segmentation. *IEEE Transactions on Pattern Analysis and Machine Intelligence, 22,* 888–905. doi:10.1109/34.868688

Sun, Y., & Fisher, R. (2003). Object-based visual attention for computer vision. *Artificial Intelligence, 146,* 77–123. doi:10.1016/S0004-3702(02)00399-5

Tatler, B. W., Baddeley, R. J., & Gilchrist, I. D. M. (2005). Visual correlates of fixation selection: Effects of scale and time. *Vision Research, 45,* 643–659. doi:10.1016/j.visres.2004.09.017

Torralba, A. (2003). Modeling global scene factors in attention. *Journal of the Optical Society of America, 20*(7), 1407–1418. doi:10.1364/JOSAA.20.001407

Treisman, A., & Gelade, G. (1980). A feature-integration theory of attention. *Cognitive Psychology*, *12*, 97–136. doi:10.1016/0010-0285(80)90005-5

Tsotsos, J. (1997). Limited capacity of any realizable perceptual system is a sufficient reason for attentive behavior. *Consciousness and Cognition*, *6*, 429–436. doi:10.1006/ccog.1997.0302

Tsotsos, J. (2006). Cognitive vision needs attention to link sensing with recognition. In Christensen, H. I., & Nagel, H. (Eds.), *Cognitive vision systems* (pp. 25–35). Berlin, Germany: Springer. doi:10.1007/11414353_3

Tsotsos, J., Culhane, S., Wai, W., Lai, Y., Davis, N., & Nuflo, F. (1995). Modeling visual attention via selective tuning. *Artificial Intelligence*, *78*, 507–545. doi:10.1016/0004-3702(95)00025-9

Tsotsos, J. K. (1988). A complexity level analysis of immediate vision. *International Journal of Computer Vision*, *2*(1), 303–320. doi:10.1007/BF00133569

Tsotsos, J. K. (1990). Analyzing vision at the complexity level. *The Behavioral and Brain Sciences*, *13*(3), 423–469. doi:10.1017/S0140525X00079577

Tsotsos, J. K. (1992). On the relative complexity of passive vs active visual search. *International Journal of Computer Vision*, *7*(2), 127–141. doi:10.1007/BF00128132

Vernon, D. (2008). Cognitive vision: The case of embodied perception. *Image and Vision Computing*, *26*, 1–4. doi:10.1016/j.imavis.2007.09.003

Walther, D., & Koch, C. (2006). Modeling attention to salient proto-objects. *Neural Networks*, *19*, 1395–1407. doi:10.1016/j.neunet.2006.10.001

Wolfe, J. M. (2007). Guided search 4.0: Current progress with a model of visual search. In Gray, W. (Ed.), *Integrated models of cognitive system* (pp. 99–119). doi:10.1093/acprof:oso/9780195189193.003.0008

Yu, Y., Mann, G., & Gosine, R. (2010). An Object-based visual attention model for robotic applications. *IEEE Transactions on Systems, Man, and Cybernetics, B 40*, 1–15.

KEY TERMS AND DEFINITIONS

Bottom-Up Attention: Bottom-up component of attention includes all those factors that are thought to be driven by the features of the objects themselves (data-driven).

Combinatorial Map: It is a mathematical model describing the subdivision of a space. It encodes all the vertices which compound this subdivision and all the incidence and adjacency relationships among them.

Irregular Pyramid: It is a hierarchical structure that represents the input frame as a stack of graphs with decreasing number of vertices. They are constructed sequentially in a bottom-up manner using only local operations.

Minimum Spanning Tree: Given a connected, undirected graph, a minimum spanning tree (MST) is the spanning tree with a weight less than or equal to the weight of every other spanning tree. A spanning tree of a graph is a subgraph that is a tree connecting all the vertices of the graph together.

Region Adjacency Graph (RAG): If the image content is segmented into a set of non-intersected regions, the RAG is a graph where nodes are regions and an arc is set between those regions that are in contact. Then, the RAG defines the adjacency relationships among image regions.

Top-Down Attention: Top-down component of attention encloses all factors that are not under the control of the sensed scene (e.g., task or context information).

Visual Attention: It defines the ability to select visual stimuli that are behaviorally relevant among the many others that are irrelevant.

Compilation of References

AAAI. (2012). *AAAI Spring Symposium on Designing Intelligent Robots: Reintegrating AI* (SS02). Retrieved October 2011 from http://www.aaai.org/Symposia/Spring/sss12symposia.php#ss02

Aboutalib, S., & Veloso, M. (2010, October). Multiple-cue object recognition in outside datasets. *IEEE International Conference on Intelligent Robots and Systems (IROS)* (pp. 4554-4559). Taipei.

Agarwal, A., & Triggs, B. (2006). Recovering 3D human pose from monocular images. *IEEE Transactions on Pattern Analysis and Machine Intelligence, 28*, 44–58. doi:10.1109/TPAMI.2006.21

Alajlan, N., Rube, I. E., Kamel, M., & Freeman, G. (2007). Shape retrieval using triangle area representation and dynamic space warping. *Pattern Recognition, 40*, 1911–1920. doi:10.1016/j.patcog.2006.12.005

Alissandrakis, A., Nehaniv, C., & Dautenhahn, K. (2007). Correspondence mapping induced state and action metrics for robotic imitation. *IEEE Transactions on Systems, Man, and Cybernetics - Part B: Special Issue on Robot Learning by Observation. Demonstration and Imitation, 37*(2), 299–307.

Alkurdi, L. (2011). *Fuzzy logic control of a robotic blimp.* Master's thesis, University of Edinburgh, U.K.

Allili, M., & Ziou, D. (2008). Object tracking in videos using adaptive mixture models and active contours. *Neurocomputing, 71*(10-12), 2001–2011. doi:10.1016/j.neucom.2007.10.019

Allport, A. (1987). *Selection for action: Some behavioural and neurophysiological considerations of attention and action.* Erlbaum.

Almhdie, A., Léger, C., Deriche, M., & Lédée, R. (2007). 3D registration using a new implementation of the ICP algorithm based on a comprehensive lookup matrix: Application to medical imaging. *Pattern Recognition Letters, 28*(12), 1523–1533. doi:10.1016/j.patrec.2007.03.005

Althaus, P., Ishiguro, H., Kanda, T., Miyashita, T., & Christensen, H. (2004). Navigation for human-robot interaction tasks. *2004 IEEE International Conference on Robotics and Automation, ICRA'04*, Vol. 2, (pp. 1894-1900).

Amari, S.-I. (1977). Dynamics of pattern formation in lateral inhibition type neural field. *Biological Cybernetics, 27*, 77–87. doi:10.1007/BF00337259

Amtrak. (2004). Retrieved from http://www.ntsb.gov/events/2002/bourbonnais/amtrak59_anim.htm

Anderson, C., & Van Essen, D. (1987). Shifter circuits: A computational strategy for dynamic aspects of visual processing. In *Proceedings of the National Academy of Sciences USA.*

Anderson, J. R. (2004). *Cognitive psychology and its implications.* Worth Publishers.

Anderson, J. R., Bothell, D., Byrne, M. D., Douglass, S., Lebiere, C., & Qin, Y. (2004). An integrated theory of the mind. *Psychological Review, 111*(4), 1036–1060. doi:10.1037/0033-295X.111.4.1036

Angelopoulou, A., Psarrou, A., García-Rodríguez, J., & Revett, K. (2005). Automatic landmarking of 2D medical shapes using the growing neural gas network. *ICCV 2005 Workshop CVBIA* (pp. 210-219).

Angelopoulou, A., Psarrou, A., & García-Rodríguez, J. (2007). *Robust modeling and 43 tracking of non-rigid objects using Active-GNG* (pp. 1–7). ICCV.

Angulo, J., & Serra, J. (2007, Apr). Modelling and segmentation of colour images in polar representations. *Image and Vision Computing, 25*(4), 475–495. doi:10.1016/j.imavis.2006.07.018

Antman, S. (1995). *Nonlinear problems of elasticity.* Maryland, MD: Springer-Verlag.

Antúnez, E., Marfil, R., & Bandera, A. (2011). Region correspondence using combinatorial pyramids. In Bandera, A., Dias, J., & Escolano, F. (Eds.), *Recognition and action for scene understanding* (pp. 13–24). Málaga, Spain: SPICUM.

Aptoula, E. (2007). A comparative study on multivariate mathematical morphology. *Pattern Recognition, 40*(11), 2914–2929. doi:10.1016/j.patcog.2007.02.004

Arashloo, S. R., & Kittler, J. (2011). Energy normalization for pose-invariant face recognition based on MRF model image matching. *IEEE Transactions on Pattern Analysis and Machine Intelligence, 33*, 1274–1280. doi:10.1109/TPAMI.2010.209

Arbeláez, P. (2006). Boundary extraction in natural images using ultrametric contour maps. In *Proceedings 5th IEEE Workshop Perceptual Organization in Computer Vision*, (pp. 182-189).

Arbelaez, P., Maire, M., Fowlkes, C., & Malik, J. (2009). *From contours to regions: An empirical evaluation* (pp. 2294–2301). IEEE CSC on CVPR. doi:10.1109/CVPR.2009.5206707

Arbeláez, P., Maire, M., Fowlkes, C., & Malik, J. (2011). Contour detection and hierarchical image segmentation. *IEEE Transactions on Pattern Analysis and Machine Intelligence*, (n.d), 33.

Arbel, T., & Ferrie, F. (2001). Entropy-based gaze planning. *Image and Vision Computing, 19*(11), 779–786. doi:10.1016/S0262-8856(00)00103-7

Argall, B. D., & Billard, A. G. (2010). A survey of tactile human-robot interactions. *Robotics and Autonomous Systems, 58*(10), 1159–1176. doi:10.1016/j.robot.2010.07.002

Argall, B., Chernova, S., Veloso, M., & Browning, B. (2009). A survey of robot learning from demonstration. *Robotics and Autonomous Systems, 57*(5), 469–483. doi:10.1016/j.robot.2008.10.024

Arkin, R. C. (1998). *Behavior-based robotics.* Cambridge, MA: MIT press.

Arkin, R. C., Fujita, M., Takagi, T., & Hasegawa, R. (2003). An ethological and emotional basis for human–robot interaction. *Robotics and Autonomous Systems, 42*(3–4), 191–201. doi:10.1016/S0921-8890(02)00375-5

Asfour, T., Gyarfas, F., Azad, P., & Dillmann, R. (2006). Imitation learning of dual-arm manipulation tasks in humanoid robots. In *6th IEEE RAS International Conference on Humanoid Robots* (pp. 40-47). Genoa, Italy: IEEE Press.

Aurenhammer, F. (1991). *Voronoi diagrams: A survey of a fundamental geometric data structure.* ACM. doi:10.1145/116873.116880

Avenant, G. C. (2010). *Autonomous flight control system for an airship.* Master's thesis, Stellenbosch University, South Africa.

Avidan, S., & Shashua, A. (2000). Trajectory triangulation: 3D reconstruction of moving points from a monocular image sequence. *IEEE Transactions on Pattern Analysis and Machine Intelligence, 22*(4), 348–357. doi:10.1109/34.845377

Ayache, N., & Faugeras, O. (1988). Building, registrating and fusing noisy visual maps. *The International Journal of Robotics Research, 7*(6), 45–65. doi:10.1177/027836498800700605

Azad, P., Ude, A., Asfour, T., & Dillmann, R. (2010). Stereo-based markerless human motion capture for humanoid robot systems. In *IEEE International Conference on Robotics and Automation (ICRA 2007)* (pp. 3951-3956). IEEE Press.

Azarbayerjani, A., & Wren, C. (1997). Real-time 3D tracking of the human body. *Proceedings of Image'com.*

Azinheira, J. R., de Paivab, E. C., Ramosb, J. J. G., & Buenob, S. S. (2000). Hovering control of an autonomous unmanned airship. In *Proceedings of the 4th Portuguese Conference on Automatic Control*, (pp. 430–435).

Azinheira, J. R., Rives, P., Carvalho, J. R. H., Silveira, G. F., de Paiva, E. C., & Bueno, S. S. (2002). Visual servo control for the hovering of an outdoor robotic airship. In *Proceedings of the IEEE International Conference on Robotics and Automation ICRA '02*, Vol. 3, (pp. 2787–2792).

Ba, S., & Odobez, J. (2004). A probabilistic framework for joint head tracking and pose estimation. *17th International Conference on Pattern Recognition, ICPR 2004*, Vol. 4, (pp. 264-267).

Backer, G., & Mertsching, B. (2003). Two selection stages provide efficient object-based attentional control for dynamic vision. In International Workshop on Attention and Performance in Computer Vision.

Backer, G., Mertsching, B., & Bollmann, M. (2001). Data and model driven gaze control for an active vision system. *IEEE Transactions on Pattern Analysis and Machine Intelligence*, 23, 1415–1429. doi:10.1109/34.977565

Badal, S., Ravela, S., Draper, B., & Hanson, A. (1994). A practical obstacle detection and avoidance system. Proceedings of 2nd IEEE Workshop on Applications of Computer Vision, (pp. 97-104).

Bahadori, S., Iocchi, L., Leone, G., Nardi, D., & Scozzafava, L. (2007). Real-time people localization and tracking through fixed stereo vision. *Applied Intelligence*, 26, 83–97. doi:10.1007/s10489-006-0013-3

Bailey, T., Nieto, J., Guivant, J., Stevents, M., & Nebot, E. (2006). Consistency of the ekf-slam algorithm. *Proceedings of the IEEE/RSJ Conference on Intelligent Robots and Systems (IROS)*.

Bajcsy, R. (1986). *Passive perception vs. active perception*. In IEEE Workshop on Computer Vision.

Baker, S., Matthews, I., & Schneider, J. (2004). Automatic construction of active appearance models as an image coding problem. *IEEE Transactions on Pattern Analysis and Machine Intelligence*, 26(10), 1380–1384. doi:10.1109/TPAMI.2004.77

Ballard, D. H. (1990). *Animate vision, technical report 329. Technical report*. Computer Science Department, University of Rochester.

Ballard, D. H. (1991). Animate vision. *Artificial Intelligence*, 48, 57–86. doi:10.1016/0004-3702(91)90080-4

Bandera, J. P. (2010). *Vision-based gesture recognition in a robot learning by imitation framework*. Málaga, Spain: University of Málaga Press.

Bandera, J. P., Marfil, R., Bandera, A., Rodríguez, J. A., Molina-Tanco, L., & Sandoval, F. (2009). Fast gesture recognition based on a two-level representation. *Pattern Recognition Letters*, 30, 1181–1189. doi:10.1016/j.patrec.2009.05.017

Bandera, J. P., Marfil, R., Molina-Tanco, L., Rodríguez, J. A., Bandera, A., & Sandoval, F. (2007). Robot learning by active imitation. In Hackel, M. (Ed.), *Humanoid robots: Human-like machines* (pp. 519–544). Vienna, Austria: I-Tech Education and Publishing.

Bandura, A. (1969). Social learning theory of identificatory processes. In *Handbook of socialization theory and research* (pp. 213–262). Chicago, IL: Rand-McNally.

Barron, J. L., Fleet, D. J., & Beauchemin, S. S. (1994). Performance of optical flow techniques. *International Journal of Computer Vision*, 12, 43-77. Retrieved from http://dx.doi.org/10.1007/BF01420984

Barron, J., & Thacker, N. (2005). *Tutorial: Computing 2D and 3D optical flow*. (2004-012). Tina memo.

Bartoli, A., Pizarro, D., & Loog, M. (2010, August). *Stratified generalized procrustes analysis*. In British Machine and Vision Conference, Aberystwyth, UK.

Baudoin, Y., Doroftei, D., De Cubber, G., Berrabah, S. A., Pinzon, C., Warlet, F., et al. (2009). View-finder: Robotics assistance to fire-fighting services and crisis management. In *IEEE International Workshop on Safety, Security, and Rescue Robotics* (pp. 1–6). Denver, Colorado, USA.

Bauer, H.-U., Hermann, M., & Villmann, T. (1999). Neural maps and topographic vector quantization. *Neural Networks*, 12(4–5), 659–676. doi:10.1016/S0893-6080(99)00027-1

Bay, H., Tuytelaars, T., & Gool, L. V. (2008). Surf: Speeded up robust features. *Computer Vision and Image Understanding*, 110(3), 346–359. doi:10.1016/j.cviu.2007.09.014

Bay, H., Tuytelaars, T., & Van Gool, L. (2006). Surf: Speeded up robust features. In Leonardis, A., Bischof, H., & Pinz, A. (Eds.), *Computer Vision - ECCV 2006* (Vol. 3951, pp. 404–417). Lecture Notes in Computer Science. doi:10.1007/11744023_32

Baylis, G., & Driver, J. (1993). Visual attention and objects: Evidence for hierarchical coding of locations. *Journal of Experimental Psychology. Human Perception and Performance, 19*, 451–470. doi:10.1037/0096-1523.19.3.451

Belhumeur, P., Hespanha, J., & Kriegman, D. (1997). Eigenfaces vs. fisherfaces: recognition using class specific linear projection. *IEEE Transactions on Pattern Analysis and Machine Intelligence, 19*(7), 711–720. doi:10.1109/34.598228

Belongie, S., Malik, J., & Puzicha, J. (2001). Matching shapes. *Proceedings Eighth IEEE International Conference on Computer Vision, ICCV 2001,* Vol.1, (pp. 454-461).

Benezeth, Y., Jodoin, P. M., Emile, B., Laurent, H., & Rosenberger, C. (2010). Comparative study of background subtraction algorithms. *Journal of Electronic Imaging, 19*(3). doi:10.1117/1.3456695

Benjemaa, R., & Schmitt, F. (1999). Fast global registration of 3D sampled surfaces using a multi-z-buffer technique. *Image and Vision Computing, 17*(2), 113–123. doi:10.1016/S0262-8856(98)00115-2

Benoit, A., Le Callet, P., Campisi, P., & Cousseau, R. (2008). Using disparity for quality assessment of stereoscopic images. In *Proceedings International Conference on Image Processing,* (pp. 389-392). IEEE Computer Society.

Bernardin, K., Gehrig, T., & Stiefelhagen, R. (2008). Multi-level particle filter fusion of features and cues for audio-visual person tracking. In Stiefelhagen, R., Bowers, R., & Fiscus, J. (Eds.), *Multimodal Technologies for Perception of Humans* (*Vol. 4625*, pp. 70–81). Lecture Notes in Computer Science. doi:10.1007/978-3-540-68585-2_5

Besl, P. J., & McKay, N. D. (1992). A method for registration of 3-D shapes. *IEEE Transactions on Pattern Analysis and Machine Intelligence, 14*, 239–256. doi:10.1109/34.121791

Beucher, S., & Lantuejoul, C. (1979, September). *Use of watersheds in contour detection.* In International workshop on image processing: Real-time edge and motion detection/estimation, Rennes, France.

Bhattacharyya, S. (1978). Observer design for linear systems with unknown inputs. *IEEE Transactions on Automatic Control, 23*(3), 483–484. doi:10.1109/TAC.1978.1101758

Bicho, E. (1999). *Dynamic approach to behavior-based robotics.* PhD thesis, University of Minho.

Bichot, N., Rossi, A., & Desimone, R. (2005). Parallel and serial neural mechanisms for visual search in macaque area V4. *Science, 308*, 529–534. doi:10.1126/science.1109676

Bickmore, T., Schulman, D., & Yiu, L. (2010). Maintaining engagement in long-term interventions with relational agents. *Journal of Applied Artificial Intelligence; special issue on Intelligent Virtual Agents, 24*(6), 648-666.

Bio, I. D. GmbH. (2010). *The BioID face database.* Retrieved November 30, 2011, from http://support.bioid.com/downloads/facedb/index.php

Blais, G., & Levine, M. D. (1995). Registering multiview range data to create 3D computer objects. *IEEE Transactions on Pattern Analysis and Machine Intelligence, 17*(8), 820–824. doi:10.1109/34.400574

Blanc, N., Oggier, T., Gruener, G., Weingarten, J., Codourey, A., & Seitz, P. (2004). Miniaturized smart cameras for 3d-imaging in real-time. In *IEEE Sensors* (pp. 471–474). Vienna, Austria: IEEE Press. doi:10.1109/ICSENS.2004.1426202

Blanz, V., & Vetter, T. (1999, August). *A morphable model for the synthesis of 3D faces.* In International Conference and Exhibition on Computer Graphics and Interactive Techniques, Los Angeles, CA.

Blaser, E., Pylshyn, Z., & Holcombe, A. (2000). Tracking an object through feature space. *Nature, 408*(6809), 196–199. doi:10.1038/35041567

Bleyer, M., & Chambon, S. (2010). *Does color really help in dense stereo matching?* Paper presented at International Symposium 3D Data Processing, Visualization and Transmission.

Bleyer, M., & Gelautz, M. (2004). A layered stereo algorithm using image segmentation and global visibility constraints. In *Proceedings International Conference Image Processing,* (pp. 2997-3000). IEEE Computer Society.

Bleyer, M., Rhemann, C., & Rother, C. (2011). PatchMatch stereo- stereo matching with slanted support. In *Proceedings of the British Machine Vision Conference* (pp. 1-11).

Blockeel, H., & De Raedt, L. (1998). Top-down induction of first-order logical decision trees. *Artificial Intelligence, 101*, 285–297. doi:10.1016/S0004-3702(98)00034-4

Bobick, A., Intille, S., Davis, J., Baird, F., Pinhanez, C., & Campbell, L. (1999). The KidsRoom: A perceptually based interactive and immersive story environment. *Teleoperators and Virtual Environment, 8*(4), 367–391.

Boehme, H.-J., Brakensiek, A., Braumann, U.-D., Krabbes, M., & Gross, H.-M. (1998). Neural networks for gesture-based remote control of a mobile robot. *Proceedings of the IEEE World Congress Computational Intelligence*, Vol. 1, (pp. 372–377).

Bookstein, F. L. (1997). Landmark methods for forms without landmarks: Localizing group differences in outline shape. *Medical Image Analysis, 1*(3), 225–244. doi:10.1016/S1361-8415(97)85012-8

Borenstein, J., & Koren, Y. (1990). Real-time obstacle avoidance for fast mobile robots in cluttered environments. *IEEE Transactions on Systems, Man, and Cybernetics, 19*(5), 1179–1187. doi:10.1109/21.44033

Borenstein, J., & Koren, Y. (1991). The vector field histogram-fast obstacle avoidance for mobile robot. *IEEE Transactions on Robotics and Automation, 7*(3), 278–288. doi:10.1109/70.88137

Borji, A., Ahmadabadi, M., Araabi, B., & Hamidi, M. (2010). Online learning of task-driven object-based visual attention control. *Image and Vision Computing, 28*, 1130–1145. doi:10.1016/j.imavis.2009.10.006

Boyd, S., & Vandenberghe, L. (2004). *Convex optimization*. New York, NY: Cambridge University Press.

Breazeal, C., & Scassellati, B. (1999). A context-dependent attention system for a social robot. In T. Dean (Ed.), *Proceedings of the Sixteenth International Joint Conference on Artificial Intelligence (IJCAI '99)* (pp. 1146-1153). San Francisco, CA: Morgan Kaufmann Publishers Inc.

Breazeal, C. (2002). *Designing sociable robots*. Cambridge, MA: MIT Press.

Breazeal, C., Brooks, A., Gray, J., Hancher, M., McBean, J., Stiehl, D., & Strickon, J. (2003). Interactive robot theatre. *Communications of the ACM, 46*, 76–84. doi:10.1145/792704.792733

Breazeal, C., Brooks, A., Gray, J., Hoffman, G., Kidd, C., & Lee, H. (2004). Humanoid robots as cooperative partners for people. *International Journal of Humanoid Robots, 1*(2), 1–34.

Brenner, M., & Nebel, B. (2009). Continual planning and acting in dynamic multiagent environments. *Journal of Autonomous Agents and Multiagent Systems, 19*(3), 297–331. doi:10.1007/s10458-009-9081-1

Brick, T., & Scheutz, M. (2007, March). Incremental natural language processing for HRI. In *ACM/IEEE International Conference on Human-Robot Interaction (HRI)* (p. 263-270). Washington DC, USA.

Broadbent, D. E. (1958). *Perception and communication*. New York, NY: Pergamos Press. doi:10.1037/10037-000

Brook, P., Ciocarlie, M., & Hsiao, K. (2011). Collaborative grasp planning with multiple object representations. In *International Conference on Robotics and Automation (ICRA)*.

Brooks, A. G., & Arkin, R. C. (2006). Behavioral overlays for non-verbal communication expression on a humanoid robot. *Autonomous Robots, 22*, 55–74. doi:10.1007/s10514-006-9005-8

Brooks, R., & Meltzoff, A. N. (2005). The development of gaze following and its relation to language. *Developmental Science, 8*(6), 535–543. doi:10.1111/j.1467-7687.2005.00445.x

Bruce, J., Balch, T., & Veloso, M. (2000). Fast and inexpensive color image segmentation for interactive robots. *Proceedings of the IEEE International Conference on Intelligent Robots and Systems* (pp. 2061-2066). Takamatsu, Japan.

Bruce, N., & Tsotsos, J. (2006). In Weiss, Y., Schölkopf, B., & Platt, J. (Eds.), *Saliency based on information maximization* (*Vol. 18*, pp. 155–162). Advances in neural information processing systemsCambridge, MA: MIT Press.

Brun, L., & Kropatsch, W. (2001). Introduction to combinatorial pyramids. *Lecture Notes in Computer Science, 2243*, 108–128. doi:10.1007/3-540-45576-0_7

Brunnstrom, K., & Stoddart, A. J. (1996), Genetic algorithms for free-form surface matching. *Proceedings of the 13th International Conference on Pattern Recognition*, Vol. 4, (pp. 689-693).

Brunnström, K., Eklundh, J.-O., & Uhlin, T. (1994). Active fixation for scene exploration. *International Journal of Computer Vision, 17*, 137–162. doi:10.1007/BF00058749

Buonomano, D. V., Bramen, J., & Khodadadifar, M. (2009). Influence of the interstimulus interval on temporal processing and learning: Testing the state-dependent network model. *Philosophical Transactions of The Royal Society B. Biological Sciences, 364*, 1865–1873. doi:10.1098/rstb.2009.0019

Burgard, W., & Fox, D. (1997). Active mobile robot localization by entropy minimization. *Proceedings of the Euromicro Workshop on Advanced Mobile Robots*, Los Alamitos, CA, (pp. 155-162).

Burgard, W., Fox, D., Hennig, D., & Schmidt, T. (1996). Estimating the absolute position of a mobile robot using position probability grids. In *Proceedings of the National Conference on Artificial Intelligence.*

Burt, P. (1988). Attention mechanisms for vision in a dynamic world. *Proceedings Ninth International Conference on Pattern Recognition*, Beijing, China.

Buschka, P., Saffiotti, A., & Wasik, Z. (2000). Fuzzy landmark-based localization for a legged robot. In *Intelligent Robots and Systems* (pp. 1205-1210). Takamatsu, Japan.

Butko, N. J., & Movellan, J. R. (2008). *I-POMDP: An infomax model of eye movement*. In International Conference on Development and Learning (ICDL).

Cabezas, I., & Trujillo, M. (2011). A non-linear quantitative evaluation approach for disparity estimation. In *Proceedings of the International Conference on Computer Vision. Theory and Applications* (pp. 704–709).

Cabezas, I., Padilla, V., Trujillo, M., & Florian, M. (2012). A non-linear quantitative evaluation approach for disparity estimation. In *Proceedings of the World Congress on Automation*, in press.

Cabezas, I., Trujillo, M., & Florian, M. (2012). A non-linear quantitative evaluation approach for disparity estimation. In *Proceedings of the International Conference on Computer Vision, Theory and Applications* (pp. 154–163).

Cabezas, I., Padilla, V., & Trujillo, M. (2011). A measure for accuracy disparity maps evaluation. In *Proceedings of the Iberoamerican Congress on Pattern Recognition. Lecture Notes in Computer Science, 7042*, 223–231. doi:10.1007/978-3-642-25085-9_26

Cakmak, M., DePalma, N., Arriaga, R., & Thomaz, A. (2010). Exploiting social partners in robot learning. *Autonomous Robots, 29*, 309–329. doi:10.1007/s10514-010-9197-9

Calinon, S. (2007). *Continuous extraction of task constraints in a robot programming by demonstration framework*. Unpublished doctoral dissertation, École Polytechnique Fédérale de Lausanne, Lausanne, EPFL.

Cañas, J. M., Martínez de la Casa, M., & González, T. (2008). Overt visual attention inside JDE control architecture. *International Journal of Intelligent Computing in Medical Sciences and Image Procesing, 2*(2), 93-100. ISSN: 1931-308X

Canemero, L. (2010). *The HUMAINE Project*. Retrieved October 2010 from http://emotion-research.net/

Canny, J. (1986). A computational approach to edge detection. *IEEE Transactions on Pattern Analysis and Machine Intelligence*, (n.d), 679–698. doi:10.1109/TPAMI.1986.4767851

Cantrell, R., Scheutz, M., Schermerhorn, P., & Wu, X. (2010). *Robust spoken instruction understanding for HRI*. In ACM/IEEE International Conference on Human-Robot Interaction (HRI).

Cao, X., & Suganthan, P. N. (2003). Video shot motion characterization based on hierachical overlapped growing neural gas networks. *Multimedia Systems, 9*, 378–385. doi:10.1007/s00530-003-0107-2

Carcassoni, M., & Hancock, E. R. (2002). Alignment using spectral clusters. *Proceedings of the 13th British Machine Vision Conference* (pp. 213-222).

Carcassoni, M., & Hancock, E. R. (2000). *Point pattern matching with robust spectral correspondence* (pp. 1649–1655). Computer Vision and Pattern Recognition. doi:10.1109/CVPR.2000.855881

Carrera, G., Angeli, A., & Davison, A. J. (2011). Lightweight SLAM and navigation with a multi-camera rig. In *Proceedings of the 5th European Conference on Mobile Robots ECMR 2011*, (pp. 77-82).

Caselles, V., Kimmel, R., & Shapiro, G. (1997). Geodesic active contours. *International Journal of Computer Vision, 22*(1), 61–79. doi:10.1023/A:1007979827043

Catania, A. C. (2006). *Learning* (4th ed.). Santa Rosa, CA: Sloan.

CCTV Info – UK. (2005). Retrieved from http://www.cctv-information.co.uk/constant3/anpr.html

Chatila, R., & Laumond, J. (1985). Position referencing and consistent world modeling for mobile robots. In *Proceedings of the IEEE International Conference on Robotics and Automation,* (pp. 138-143).

Chen, C., Yang, P., Zhou, X., & Dong, D. (2008). A quantum-inspired qlearning algorithm for indoor robot navigation. In *IEEE International Conference on Networking, Sensing and Control (ICNSC 2008)* (pp. 1599–1603).

Chen, L., Tamer, M., & Oria, V. (2005). Robust and fast similarity search for moving object trajectories. In *Special Interest Group on Management of Data (SIGMOD 2005)* (pp. 491-502). New York, USA.

Chen, Q., Sun, Q., Ann, P. & Xia, D. (2010). Two-stage object tracking method based on kernel and active contour. *IEEE Transactions on Circuits and System for Video Technology, 20.*

Chen, W., & Saif, M. (2006). Fault detection and isolation based on novel unknown input observer design. *Proceedings of the American Control Conference,* Minneapolis, MN.

Chen, W., & Saif, M. (2006). Unknown input observer design for a class of nonlinear systems: An LMI approach. *Proceedings of the American Control Conference,* Minneapolis, MN.

Chen, X., Ji, J., Jiang, J., Jin, G., Wang, F., & Xie, J. (2010, May 10-14). *Developing high-level cognitive functions for service robots.* In International Conference on Autonomous Agents and Multiagent Systems (AAMAS). Toronto, Canada.

Chen, Y., & Medioni, G. (1991). Object modeling by registration of multiple range images. *Proceedings of the IEEE Conference on Robotics and Automation.*

Chetverikov, D., Stepanov, D., & Krsek, P. (2005). Robust Euclidean alignment of 3D point sets: The trimmed iterative closest point algorithm. *Image and Vision Computing, 23*(5), 299–309. doi:10.1016/j.imavis.2004.05.007

Chikkerur, S., Serre, T., Tan, C., & Poggio, T. (2010). What and where: A Bayesian inference theory of attention. *Vision Research, 50,* 2233–2247. doi:10.1016/j.visres.2010.05.013

Chitrakaran, V., Dawson, D. M., Dixon, W. E., & Chen, J. (2005). Identification of a moving object's velocity with a fixed camera. *Automatica, 41*(3), 553–562. doi:10.1016/j.automatica.2004.11.020

Chli, M. (2009). *Applying information theory to efficient SLAM.* Ph.D. thesis, Imperial College London.

Choset, H. (1996). *Sensor based motion planning: The hierarchical generalized Voronoi graph.* PhD thesis, California Institute of Technology.

Circadian. (2004). *Rail employee fatigue.* Retrieved from http://www.circadian.com/expert/fatigue_inattention.html

Clarkson, E., & Arkin, R. C. (2006). *Applying heuristic evaluation to human-robot interaction systems* (Tech. Rep.). Georgia Institute of Technology GIT-GVU-06-08.

Cog, X. (2011). *Cognitive systems that self-understand and self-extend.* Retrieved October 2011 from http://cogx.eu/

Comaniciu, D., Ramesh, V., & Meer, P. (2000). Real-time tracking of non-rigid objects using mean shift. *IEEE Computer Society Conference on Computer Vision and Pattern Recognition,* Vol. 2, (pp. 2142-2150).

Comaniciu, D., & Meer, P. (2002). Mean shift: A robust approach towards feature space analysis. *IEEE Transactions on Pattern Analysis and Machine Intelligence, 24*(5), 603–619. doi:10.1109/34.1000236

Comaniciu, D., Ramesh, V., & Meer, P. (2003). Kernel-based object tracking. *IEEE Transactions on Pattern Analysis and Machine Intelligence, 25,* 564–577. doi:10.1109/TPAMI.2003.1195991

Cooper, J. O., Heron, T. E., & Heward, W. L. (2007). *Applied behavior analysis* (2nd ed.). Prentice Hall.

Cootes, T. F. (2004). *Statistical models of appearance for computer vision*. Online Technical Report. Retrieved December 3, 2011, from http://www.isbe.man.ac.uk/~bim/refs.html

Cootes, T. F., & Kittipanyangam, P. (2002). Comparing variations on the active appearance model algorithm. *British Machine Vision Conference,* Vol. 2, (pp. 837-846).

Cootes, T. F., & Taylor, C. J. (2001). *Statistical models of appearance for computer vision*. Retrieved from http://www.itu.dk/stud/projects_f2004/handtracking/referencer/Cootes%20den%20lange%20-%20app_model.pdf

Cootes, T. F., Edwards, G. J., & Taylor, C. J. (1998). A comparative evaluation of active appearance models algorithms. *British Machine Vision Conference,* Vol. 2, (pp. 680-689). BMVA Press.

Cootes, T. F., Edwards, G. J., & Taylor, C. J. (1998). Active appearance models. In H. Burkhardt & B. Neumann (Ed.), *European Conference on Computer Vision,* Vol. 2, (pp. 484-498). Springer.

Cootes, T. F. (1995). Active shape models - Their training and application. *Computer Vision and Image Understanding, 61*(1), 38–59. doi:10.1006/cviu.1995.1004

Cootes, T., Edwards, G., & Taylor, C. (2001). Active appearance models. *IEEE Transactions on Pattern Analysis and Machine Intelligence, 23*(6), 681–685. doi:10.1109/34.927467

Craig, J. (1986). *Introduction to robotics: Mechanics and control*. Boston, MA: Addison-Wesley.

Croitoru, A., Agouris, P., & Stefanidis, A. (2005). 3D trajectory matching by pose normalization. In *13th ACM International Symposium on Advances in Geographic Information Systems (ACM-GIS'05)* (pp. 153-162).

Cross, A. D. J., & Hancock, E. R. (1997). *Recovering perspective pose with a dual step EM algorithm*. Eighteenth Annual Conference on Neural Information Processing Systems.

Cruz, A., Bandera, J. P., & Sandoval, F. (2009). Torso pose estimator for a robot imitation framework. In *12th International Conference on Climbing and Walking Robots and the Support Technologies for Mobile Machines (CLAWAR 2009)* (pp. 901-908). Istanbul, Turkey.

Cucchiara, R., Grana, C., & Vezzani, R. (2000). Probabilistic people tracking for occlusion handling. *International Conference on Production Research, 39,* 57-71.

Dagher, I., & Tom, K. E. (2008, Jul). WaterBalloons: A hybrid watershed balloon snake segmentation. *Image and Vision Computing, 26*(7), 905–912. doi:10.1016/j.imavis.2007.10.010

Dahl, O., Nyberg, F., & Heyden, A. (2007). Nonlinear and adaptive observers for perspective dynamic systems. *Proceedings of the American Control Conference,* (pp. 966–971). New York City, NY.

Dalley, G., & Flynn, P. (2002). Pair-wise range image registration: A study in outlier classification. *Computer Vision and Image Understanding, 87*(1-3), 104–115. doi:10.1006/cviu.2002.0986

Dani, A. P., Kan, Z., Fischer, N., & Dixon, W. E. (2010). Structure and motion estimation of a moving object using a moving camera. *Proceedings of the American Controls Conference,* (pp. 6962-6967). Baltimore, MD.

Dani, A. P., Kan, Z., Fischer, N., & Dixon, W. E. (2011). Estimating structure of a moving object using a moving camera: An unknown input observer approach. *IEEE Conference on Decision and Controls,* (pp. 5005-5010), Orlando, FL.

Dani, A. P., & Dixon, W. E. (2010). Single Camera Structure and Motion Estimation. In Chesi, G., & Hashimoto, K. (Eds.), *Visual servoing via advanced numerical methods* (pp. 209–229). Springer Lecture Notes in Control and Information Sciences. doi:10.1007/978-1-84996-089-2_12

Dani, A. P., Fischer, N., & Dixon, W. E. (2012). Single camera structure and motion. *IEEE Transactions on Automatic Control, 57*(1), 241–246. doi:10.1109/TAC.2011.2162890

Dani, A. P., Kan, Z., Fischer, N., & Dixon, W. E. (2012). Globally exponentially convergent robust observer for vision-based range estimation. *IFAC Mechatronics. Special Issue on Visual Servoing, 22*(4), 381–389.

Dani, A. P., Rifai, K., & Dixon, W. E. (2010). *Globally exponentially convergent observer for vision-based range estimation* (pp. 801–806). Yokohama, Japan: IEEE Multi-Conference on System and Control. doi:10.1109/ISIC.2010.5612878

Darouach, M., Zasadzinski, M., & Xu, S. (1994). Full-order observers for linear systems with unknown inputs. *IEEE Transactions on Automatic Control, 39*(3), 606–609. doi:10.1109/9.280770

Dautenhahn, K. W., Koay, K., Nehaniv, C., Sisbot, A., Alami, R., & Siméon, T. (2006). How may I serve you? A robot companion approaching a seated person in a helping context. *1st ACM SIGCHI/SIGART Conference on Human-Robot Interaction*, (pp. 172-179).

Dautenhahn, K., & Billard, A. (1999). Bringing up robots or -The psychology of socially intelligent robots: From theory to implementation. In *Third Annual Conference on Autonomous Agents* (pp. 366-367). Seattle, Washington, USA.

Davison, A. (2003). Real-time simultaneous localisation and mapping with a single camera. In *Proceedings of the International Conference on Computer Vision*, Nice.

Davison, A. J., Reid, I. D., Morton, N. D., & Stasse, O. (2007). MonoSLAM: Real-time single camera SLAM. *IEEE Transactions on Pattern Analysis and Machine Intelligence, 29*(6), 1052–1067. doi:10.1109/TPAMI.2007.1049

De Cubber, G., Doroftei, D., Nalpantidis, L., Sirakoulis, G. C., & Gasteratos, A. (2009). *Stereo- based terrain traversability analysis for robot navigation*. In IARP/EURON Workshop on Robotics for Risky Interventions and Environmental Surveillance. Brussels, Belgium.

De la Torre, F., & Nguyen, M. (2008, June). *Parameterized kernel principal component analysis: Theory and applications to supervised and unsupervised image alignment*. In IEEE Computer Vision and Pattern Recognition, Anchorage, AK.

De Persis, C., & Isidori, A. (2001). A geometric approach to nonlinear fault detection and isolation. *IEEE Transactions on Automatic Control, 46*, 853–865. doi:10.1109/9.928586

Deák, G. O., Fasel, I., & Movellan, J. (2001). The emergence of shared attention: Using robots to test developmental theories. In C. Balkenius, et al. (Eds.), *Proceedings 1st International Workshop on Epigenetic Robotics:* Vol. 85. (pp, 95-104). Lund University Cognitive Studies.

Deák, G. O., & Triesch, J. (2006). The emergence of attention-sharing skills in human infants. In Fujita, K., & Itakura, S. (Eds.), *Diversity of cognition*. University of Kyoto Press.

Deco, G. (2000). *A neurodynamical model of visual attention: Feedback enhancement of spatial resolution in a hierarchical system*.

Defense Advanced Research Projects Agency (DARPA). (2005). *DARPA grand challenge*. Retrieved March 14, 2005, from http://www.darpa.mil/grandchallenge

Dellaert, F., Burgard, W., Fox, D., & Thrun, S. (1999). Using the CONDENSATION algorithm for robust, vision-based mobile robot localization. In *Proceedings of the Conference on Computer Vision and Pattern Recognition, CVPR-99*, Vol. 2, (pp. 588-594). Fort Collins, CO: IEEE Computer Society.

Dellaert, F., Fox, D., Burgard, W., & Thrun, S. (1998). *Monte Carlo localization for mobile robots*. In IROS.

Demirdjian, D., Ko, T., & Darrell, T. (2005). Untethered gesture acquisition and recognition for virtual world manipulation. *Virtual Reality (Waltham Cross), 8*, 222–230. doi:10.1007/s10055-005-0155-3

Demiris, J., & Hayes, G. (2002). *Imitation as a dual-route process featuring predictive and learning components: A biologically plausible computational model*. Cambridge, MA: MIT Press.

Deng, Y., & Manjunath, B. S. (2001). Unsupervised segmentation of color-texture regions in images and video. *IEEE Transactions on Pattern Analysis and Machine Intelligence*, (n.d), 800–810. doi:10.1109/34.946985

Department for Transport – UK. (2004). Retrieved in 2004 from http://www.dft.gov.uk/stellent/groups/dft_control/documents/homepage/dft_home_page.hcsp

Desimone, R., & Duncan, J. (1995). Neural mechanisms of selective visual attention. *Annual Review of Neuroscience, 18*, 193–222. doi:10.1146/annurev.ne.18.030195.001205

Di Stefano, L., Marchionni, M., & Mattoccia, S. (2004). A fast area-based stereo matching algorithm. *Image and Vision Computing, 22*(12), 983–1005. doi:10.1016/j.imavis.2004.03.009

Diaconis, P., & Shahshahani, M. (1987). The subgroup algorithm for generating uniform random variables. *Probability in the Engineering and Informational Sciences*, *1*(1), 15–32. doi:10.1017/S0269964800000255

Dima, C. S., & Hebert, M. (1994). Classifier fusion for outdoor obstacle detection. In *IEEE International Conference on Robotics and Automation* (pp. 665–671).

Dissanayake, M., Dissanayake, M., Newman, P., Clark, S., Durrant-Whyte, H., & Csorba, M. (2001). A solution to the simultaneous localization and map building (SLAM) problem. *IEEE Transactions on Robotics and Automation*, *17*(3), 229–241. doi:10.1109/70.938381

Dixon, W. E., Fang, Y., Dawson, D. M., & Flynn, T. J. (2003). Range identification for perspective vision systems. *IEEE Transactions on Automatic Control*, *48*(12), 2232–2238. doi:10.1109/TAC.2003.820151

Dockstader, S., & Tekalp, A. (2001). Multiple camera tracking of interacting and occluded human motion. *Proceedings of the IEEE*, *89*, 1441–1455. doi:10.1109/5.959340

Domínguez, S., Zalama, E., García-Bermejo, J., & Pulido, J. (2006). Robot learning in a social robot. In S. Nolfi, G. Baldassarre, R. Calabretta, J. Hallam, D. Marocco, J.A. Meyer, O. Miglino, & D. Parisi (Eds.), *From Animals to Animats 9, Vol. 4095 of LNCS* (pp. 691-702). Berlin, Germany: Springer.

Dorai, C., Wang, G., Jain, A. K., & Mercer, C. (1998). Registration and integration of multiple object views for 3D model construction. *IEEE Transactions on Pattern Analysis and Machine Intelligence*, *20*(1), 83–89. doi:10.1109/34.655652

Doucet, A., de Freitas, J., Murphy, K., & Russel, S. (2000). Raoblackwellized particle filtering for dynamic Bayesian networks. *Proceedings of the Conference on Neural Information Processing Systems (NIPS)*, (pp. 176-183).

Doucet, A. (1998). *Technical report, Signal Processing Group*. Dept. of Engineering, University of Cambridge.

Dougherty, E. R., & Lotufo, R. A. (2003). *Hands-on morphological image processing (Vol. 59)*. Bellingham, WA: The International Society for Optical Engineering, ETATS-UNIS. doi:10.1117/3.501104

Driessens, K. (2004). *Relational reinforcement learning*. Unpublished doctoral dissertation, Katholieke Universiteit Leuven, Leuven.

Drive and Safety Alive, Inc. (2006). *Key annual statistics for the USA, 2006*. Retrieved from http://www.driveand-stayalive.com/info%20section/statistics/stats-usa.htm

Dryden, I. L., & Mardia, K. V. (1998). *Statistical shape analysis*. London, UK: John Wiley & Sons.

Duan, Y., & Hexu, X. (2005). Fuzzy reinforcement learning and its application in robot navigation. In *Proceedings of 2005 International Conference on Machine Learning and Cybernetics*, (pp. 899 –904)

Duda, R., & Hart, P. (1973). *Classification and scene analysis*. John Wiley and Sons.

Duffy, B. R., Dragone, M., & O'Hare, G. M. P. (2005). *Social robot architecture: A framework for explicit social interaction android science*. Towards Social Mechanisms, CogSci 2005 Workshop Stresa.

Duro, R., Graña, M., & de Lope, J. (2010). On the potential contributions of hybrid intelligent approaches to multicomponent robotic system development. *Information Sciences*, *180*(14), 2635–2648. doi:10.1016/j.ins.2010.02.005

Durrant-Whyte, H., Rye, D., & Nebot, E. (1996). *Localization of automatic guided vehicles*. The 7th International Symposium (ISRR).

Durrant-Whyte, H. (1987). Uncertain geometry in robotics. *IEEE Transactions on Robotics and Automation*, *4*(1), 23–31.

Durrant-Whyte, H., & Bailey, T. (2006). Simultaneous localization and mapping (slam): Part i. *Robotics and Automation Magazine, IEEE*, *13*(2), 99–110. doi:10.1109/MRA.2006.1638022

Du, S., Zheng, N., Meng, G., & Yuan, Z. (2008). Affine registration of point sets using ICP and ICA. *Signal Processing Letters*, *15*, 689–692. doi:10.1109/LSP.2008.2001823

Du, S., Zheng, N., Xiong, L., Ying, S., & Xue, J. (2010). Scaling iterative closest point algorithm for registration of m–D point sets. *Journal of Visual Communication and Image Representation*, *21*(5-6), 442–452. doi:10.1016/j.jvcir.2010.02.005

Eade, E., & Drummond, T. (2006). Scalable monocular slam. *Computer Vision and Pattern Recognition*, *1*, 469–476.

Echegoyen, Z. (2009). *Contributions to visual servoing for legged and linked multicomponent robots.* Ph.D. dissertation. San Sebastian, Spain: UPV/EHU.

Echegoyen, Z., Villaverde, I., Moreno, R., Graña, M., & d'Anjou, A. (2010). Linked multi-component mobile robots: Modeling, simulation and control. *Robotics and Autonomous Systems, 58*(12), 1292–1305. doi:10.1016/j.robot.2010.08.008

Edwards, G. J., Taylor, C. J., & Cootes, T. F. (1998). Interpreting face images using active appearance models. *International Conference on Automatic Face and Gesture Recognition,* (pp. 300-305).

Eggert, D., Lorusso, A., & Fisher, R. (1997). Estimating 3-D rigid body transformations: A comparison of four major algorithms. *Machine Vision and Applications, 9*(5), 272–290. doi:10.1007/s001380050048

Ekman, P. (1999). Basic emotions. In Dagleish, T., & Power, M. (Eds.), *Handbook of cognition and emotion.* Sussex, UK: John Wiley & Sons.

El Baf, F., Bouwmans, T., & Vachon, B. (2007). Comparison of background subtraction methods for a multimedia application. In *Systems, Signals and Image Processing, 2007 and 6th EURASIP Conference focused on Speech and Image Processing, Multimedia Communications and Services* (pp. 385-388).

Elfes, A., Montgomery, J. F., Hall, J. L., Joshi, S. S., Payne, J., & Bergh, C. F. (2005, September). *Autonomous flight control for a titan exploration aerobot.* Paper presented at the 8th International Symposium on Artificial Intelligence, Robotics and Automation in Space. Munich, Germany.

Elfes, A. (1987). Sonar based real world mapping and navigation. *IEEE Journal on Robotics and Automation, 3*(3), 249–265. doi:10.1109/JRA.1987.1087096

Elgammal, A., & Davis, L. (2001). Probabilistic framework for segmenting people under occlusion. *Eighth International Conference on Computer Vision,* Vol. 2, (pp. 145-152).

Elgammal, A., Duraiswami, R., Harwood, D., & Anddavis, L. (2002). Background and foreground modeling using nonparametric kernel density estimation for visual surveillance. *Institute of Electrical and Electronics Engineers, 7*, 1151–1163.

Elwaseif, M., & Slater, L. (2010, Jul). Quantifying MB geometries in resistivity images using watershed algorithms. *Journal of Archaeological Science, 37*(7), 1424–1436. doi:10.1016/j.jas.2010.01.002

Eng, H., Wang, J., Kam, A., & Yau, W. (2004). A bayesian framework for robust human detection and occlusion handling using human shape model. *International Conference on Production Research,* Vol. 2, (pp. 257-260).

Escalera, S., Pujol, O., & Radeva, P. (2008). Detection of complex salient regions. *EURASIP Journal on Advances in Signal Processing, 2008.* doi:10.1155/2008/451389

Estrada, C., & Tardós, J. D. (2005). Hierarchical slam: real-time accurate mapping of large environments. *IEEE Transactions on Robotics, 21*(4), 588–596. doi:10.1109/TRO.2005.844673

Falahpour, M., Moradi, H., Refai, H., Atiquzzaman, M., Rubin, R., & Lo Presti, P. G. (2009, October). *Performance comparison of classic and fuzzy logic controllers for communication airships.* Paper presented at the IEEE/AIAA 28th Digital Avionics Systems Conference DASC '09, Orlando, FL, U.S.A.

Fasola, J., & Mataric, M. (2010, August). *Robot motivator: Increasing user enjoyment and performance on a physical/cognitive task.* In International Conference on Development and Learning (ICDL). Ann Arbor, USA.

Feldman, J. A., & Ballard, D. H. (1982). Connectionist models and their properties. *Cognitive Science, 6*, 205–254. doi:10.1207/s15516709cog0603_1

Felzenszwalb, P., & Huttenlocher, D. (2004). Efficient belief propagation for early vision. In *Proceedings of Computer Vision and Pattern Recognition* (pp. 261–268). IEEE Computer Society.

Felzenszwalb, P., & Huttenlocher, D. (2004). Efficient graph-based image segmentation. *International Journal of Computer Vision, 59*, 167–181. doi:10.1023/B:VISI.0000022288.19776.77

Fergus, R., Perona, P., & Zisserman, A. (2003). *Object class recognition by unsupervised scale-invariant learning.* In International Conference on Computer Vision and Pattern Recognition (CVPR).

Fernandez-Gauna, B., Lopez-Guede, J. M., Zulueta, E., & Graña, M. (2010). Learning hose transport control with q-learning. *Neural Network World, 20*(7), 913–923.

Fidler, S., Boben, M., & Leonardis, A. (2008). *Similarity-based cross-layered hierarchical representation for object categorization*. In International Conference on Computer Vision and Pattern Recognition (CVPR).

Filliata, D., & Meyer, J. (2003). Map-based navigation in mobile robots: I. A review of localization strategies. *Cognitive Systems Research, 4*(4), 243–282. doi:10.1016/S1389-0417(03)00008-1

Finlayson, G., Hordley, S., & Hubel, P. (2001, November). Color by correlation: A simple, unifying framework for color constancy. *IEEE Transactions on Pattern Analysis and Machine Intelligence, 23*(11), 1209–1221. doi:10.1109/34.969113

Fischler, M. A., & Bolles, R. C. (1981). Random sample consensus: A paradigm for model fitting with applications to image analysis and automated cartography. *Communications of the ACM, 24*, 381–395. doi:10.1145/358669.358692

Fisher, R. B. (2004). PETS04 surveillance ground truth data set. In *Sixth IEEE International Workshop on Performance Evaluation of Tracking and Surveillance* (pp. 1-5).

Flórez, F., García, J. M., García, J., & Hernández, A. (2002). hand gesture recognition following the dynamics of a topology-preserving network. In *Proceedings of the 5th IEEE International Conference on Automatic Face and Gesture Recognition,* (pp. 318-323). Washington, DC: IEEE.

Fong, T., Nourbakhsh, I., & Dautenhahn, K. (2003). A survey of socially interactive robots. *Robotics and Autonomous Systems, 42*(3-4), 143–166. doi:10.1016/S0921-8890(02)00372-X

Freund, Y., & Schapire, R. (1995). A desicion-theoretic generalization of on-line learning and an application to boosting. In Vitányi, P. (Ed.), *Computational Learning Theory (Vol. 904,* pp. 23–37). Lecture Notes in Computer Science. doi:10.1007/3-540-59119-2_166

Frezza-Buet, H. (2008). Following non-stationary distributions by controlling the vector quatization accuracy of a growing neural gas network. *Neurocomputing, 71*, 1191–1202. doi:10.1016/j.neucom.2007.12.024

Fridman, L., Shtessel, Y., Edwards, C., & Yan, G. (2007). Higher-order sliding-mode observer for state estimation and input reconstruction in nonlinear systems. *International Journal of Robust and Nonlinear Control, 18*, 399–412. doi:10.1002/rnc.1198

Frintrop, S., Backer, G., & Rome, E. (2005). Goal-directed search with a top-down modulated computational attention system. *Lecture Notes in Computer Science, 3663*, 117–124. doi:10.1007/11550518_15

Frintrop, S., Königs, A., Hoeller, F., & Schulz, D. (2010). A component-based approach to visual person tracking from a mobile platform. *International Journal of Social Robotics, 2*, 53–62. doi:10.1007/s12369-009-0035-1

Fritzke, B. (1993). *Growing cell structures – A self-organising network for unsupervised and supervised learning*. Technical Report TR-93-026, International Computer Science Institute, Berkeley, California.

Fritzke, B. (1997). A self-organizing network that can follow non-stationary distributions. *Proceedings of the International Conference on Artificial Neural Networks '97,* (pp. 613-618). Springer.

Fritzke, B. (1995). In Tesauro, G., Touretzky, D. S., & Leen, T. K. (Eds.). Advances in Neural Information Processing Systems: *Vol. 7. A growing neural gas network learns topologies*. Cambridge, MA: MIT Press.

Fusiello, A., Trucco, E., & Verri, A. (2000). A compact algorithm for rectification of stereo pairs. *Machine Vision and Applications, 12*(1), 16–22. doi:10.1007/s001380050120

Gahinet, P., Nemirovskii, A., Laub, A., & Chilali, M. (1994). The LMI control toolbox. *IEEE Conference on Decision and Control,* (pp. 2038–2041).

Galindo, C., Fernandez-Madrigal, J.-A., Gonzalez, J., & Saffioti, A. (2008). Robot task planning using semantic maps. *Robotics and Autonomous Systems, 56*(11), 955–966. doi:10.1016/j.robot.2008.08.007

Gallup, D., Frahm, J., Mordohai, P., & Pollefeys, M. (2008). Variable baseline/resolution stereo. In *Proceedings of Computer Vision and Pattern Recognition* (pp. 1–8). IEEE Computer Society.

Gao, D., Mahadevan, V., & Vasconcelos, N. (2008). On the plausibility of the discriminant center-surround hypothesis for visual saliency. *Journal of Vision (Charlottesville, Va.)*, *8*(7), 1–18. doi:10.1167/8.7.13

Garcia, J. F. (2011). *Contributions to visual attention computation and application to the control of a humanoid robot*. PhD thesis, Universidad de Leon.

Garcia-Rodriguez, J., Angelopoulou, A., Garcia-Chamizo, J. M., & Psarrou, A. (2010). GNG based surveillance system. In International Joint Conference on Neural Networks (pp. 1-8).

García-Rodríguez, J., Flórez-Revuelta, F., & García-Chamizo, J. M. (2007). Image compression using growing neural gas. *In Proceedings of International Joint Conference on Artificial Neural Networks*, (pp. 366-370).

Garrido, S., Moreno, L., & Blanco, D. (2009). Exploration of 2D and 3D environments using Voronoi transform and fast marching method. *Journal of Intelligent & Robotic Systems*, *55*(1), 55–80. doi:10.1007/s10846-008-9293-7

Gartshore, R., Aguado, A., & Galambos, C. (2002). Incremental map buildig using an occupancy grid for an autonomous monocular robot. *Proceedings of Seventh International Conference on Control, Automation, Robotics and Vision ICARCV*, (pp. 613–618).

Geusebroek, J., van den Boomgaard, R., Smeulders, A. W. M., & Gevers, T. (2003, July). Color constancy from physical principles. *Pattern Recognition Letters*, *24*(11), 1653–1662. doi:10.1016/S0167-8655(02)00322-7

Ghallab, M., Nau, D., & Traverso, P. (2004). *Automated planning: Theory and practice*. San Francisco, CA: Morgan Kaufmann.

Ghiasi, S., Nguyen, K., & Sarrafzadeh, M. (2003). Profiling accuracy-latency characteristics of collaborative object tracking applications. *International Conference on Parallel and Distributed Computing and Systems*, (pp. 694-701).

Ghosh, B. K., Polpitiya, A. D., & Wang, W. (2007). Bio-inspired networks of visual sensors, neurons, and oscillators. *Bio-Inspired Networks of Visual Sensors, Neurons, and Oscillators*, *95*, 188–214.

Giefing, G.-J., Janûen, H., & Mallot, H. (1992). *Saccadic object recognition with an active vision system*. In 10th European Conf. Artificial Intelligence.

Gleicher, M. (1998). Retargetting motion to new characters. In *25th Annual Conference on Computer Graphics and Interactive Techniques (SIGGRAPH '98)* (pp. 33-42). New York, NY: ACM.

Głomb, P. (2009). *Detection of interest points on 3D data: Extending the Harris operator. Computer Recognition Systems 3* (pp. 103–111). Advances in Intelligent and Soft Computing.

Gobelbecker, M., Gretton, C., & Dearden, R. (2011). *A switching planner for combined task and observation planning*. In National Conference on Artificial Intelligence (AAAI). San Francisco, USA.

Godin, G., Rioux, M., & Baribeau, R. (1994). Three-dimensional registration using range and intensity information. *Proceedings of SPIE*, *2350*, (p. 279).

Goldberg, S. B., Maimone, M. W., & Matthies, L. (2002). Stereo vision and rover navigation software for planetary exploration. *Proceedings of IEEE Aerospace Conference*, (pp. 2025–2036).

Gomatam, A. M. (2004). *Non-invasive multimodal biometric recognition techniques*. Unpublished MS Thesis from Gannon University, Erie, PA, USA.

Gomatam, A. M., & Sasi, S. (2004). Enhanced gait recognition using HMM and VH techniques. *IEEE International Workshop on Imaging Systems and Techniques*, (pp. 144-147). 14 May 2004, Stresa - Lago Maggiore, Italy. DOI: 10.1109/IST.2004.1397302

Gomatam, A. M., & Sasi, S. (2004). Multimodal gait recognition based on stereo vision and 3D template matching. *Proceedings of the International Conference on Imaging Science, Systems and Technology* (CISST'04) (pp. 405-410). Las Vegas, Nevada, USA, June 21-24, 2004, CSREA Press.

Gomatam, A. M., & Sasi, S. (2005). Gait recognition based on isoluminance line and 3D template matching. *International Conference on Intelligent Sensing and Information Processing -ICISIP'05*, (pp. 156-160). January 04-07, 2005, IIT Chennai, India. DOI: 10.1109/ICISIP.2005.1529440

Gomes, S. B. V., & Ramos, J. G., Jr. (1998). Airship dynamic modeling for autonomous operation. In *Proceedings of the IEEE International Robotics and Automation Conference,* Vol. 4, (pp. 3462–3467). Leuven, Belgium.

Gong, S., McKenna, S. J., & Psarrou, A. (Eds.). (2000). *Dynamic vision: From images to face recognition.* London, UK: Imperial College Press. doi:10.1142/p155

Gonzalez, P., Burgard, W., Sanz, R., & Fernandez, J. L. (2009). Developing a low-cost autonomous indoor blimp. *Journal of Physical Agents, 3*(1), 43–52.

Goodall, C. (1991). Procrustes methods in the statistical analysis of shape. *Journal of the Royal Statistical Society. Series A, (Statistics in Society), 53*(2), 285–339.

Goodall, C. (1991). Procrustes methods in the statistical analysis of shape. *Royal Statistical Society, Series B, 53,* 285–339.

Goodrich, M. A., & Schultz, A. C. (2007). Human-robot interaction: A survey. *Foundations and Trends in Human-Computer Interaction, 1*(3), 203–275. doi:10.1561/1100000005

Gourier, N., Hall, D., & Crowley, J. (2004). Facial features detection robust to pose, illumination and identity. *2004 IEEE International Conference on Systems, Man and Cybernetics,* Vol. 1, (pp. 617-622).

Greenspan, H., Belongie, S., Goodman, R., Perona, P., Rakshit, S., & Anderson, C. H. (1994). Overcomplete steerable pyramid filters and rotation invariance. *Proceedings of IEEE Computer Vision and Pattern Recognition,* (pp. 222–228).

Greenspan, H., Dvir, G., & Rubner, Y. (2000). Region correspondence for image matching via EMD flow. *Proceedings of the IEEE Workshop on Content-based Access of Image and Video Libraries,* (pp. 27-31).

Gregoire, M., & Schomer, E. (2007). Interactive simulation of one-dimensional flexible parts. *Computer Aided Design, 39*(8), 694–707. doi:10.1016/j.cad.2007.05.005

Grisetti, G., Stachniss, C., & Burgard, W. (2006). Improved techniques for grid mapping with Rao-Blackwellized particle filters. *IEEE Transactions on Robotics, 23*(1), 34–46. doi:10.1109/TRO.2006.889486

Grisetti, G., Stachniss, C., & Burgard, W. (2007). Improved techniques for grid mapping with Raoblackwellized particle filters. *IEEE Transactions on Robotics, 23*(1), 34–46. doi:10.1109/TRO.2006.889486

Grollman, D. (2010). *Teaching old dogs new tricks: Incremental multimap regression for interactive robot learning from demonstration.* Unpublished doctoral dissertation, Department of Computer Science, Brown University.

Gryn, J. M., Wildes, R. P., & Tsotsos, J. K. (2009). Detecting motion patterns via direction maps with application to surveillance. *Computer Vision and Image Understanding, 113,* 291–307. doi:10.1016/j.cviu.2008.10.006

Guan, Y., & Saif, M. (1991). A novel approach to the design of unknown input observers. *IEEE Transactions on Automatic Control, 36*(5), 632–635. doi:10.1109/9.76372

Guedon, Y. (2005). Hybrid Markov/semi-Markov chains. *Computational Statistics & Data Analysis, 49,* 663–688. doi:10.1016/j.csda.2004.05.033

Guivant, J., & Nebot, E. (2001). Optimization of the simultaneous localization and map-building algorithm for real-time implementation. *IEEE Transactions on Robotics and Automation, 17*(3), 242–257. doi:10.1109/70.938382

Haeusler, R., & Klette, R. (2010). Benchmarking stereo data (not the matching algorithms). In M. Goesele, S. Roth, A. Kuijper, B. Schiele, & K. Schindler (Eds.), *Proceedings of the 32nd DAGM Conference on Pattern Recognition,* (pp. 383-392). Springer-Verlag.

Hall, D., Nascimento, J., Ribeiro, P., Andrade, E., Moreno, P., & Pesnel, S. … Crowley, J. (2005). Comparison of target detection algorithms using adaptive background models. *Joint IEEE International Workshop on Visual Surveillance and Performance Evaluation of Tracking and Surveillance,* (pp. 113-120).

Hall, E. (1966). *The hidden dimension.* New York, NY: Doubleday.

Hamilton, W. R. S. (Ed.). (1853). *Lectures on quaternions.* Dublin, Ireland: Hodges and Smith.

Hamker, F. H. (2000). *Distributed competition in directed attention.*

Hammouri, H., Kabore, P., & Kinnaert, M. (2001). Geometric approach to fault detection and isolation for bilinear systems. *IEEE Transactions on Automatic Control, 46*(9), 1451–1455. doi:10.1109/9.948476

Hammouri, H., Kinnaert, M., & El Yaagoubi, E. H. (1999). Observer based approach to fault detection and isolation for nonlinear systems. *IEEE Transactions on Automatic Control, 44*, 1879–1884. doi:10.1109/9.793728

Hammouri, H., & Tmar, Z. (2010). Unknown input observer for state affine systems: A necessary and sufficient condition. *Automatica, 46*, 271–278. doi:10.1016/j.automatica.2009.11.004

Hanbury, A., & Serra, J. (2001). Mathematical morphology in the HLS colour space. *Proceedings of the 12th British Machine Vision Conference.*

Han, M., & Kanade, T. (2004). Reconstruction of a scene with multiple linearly moving objects. *International Journal of Computer Vision, 59*(3), 285–300. doi:10.1023/B:VISI.0000025801.70038.c7

Happold, M., Ollis, M., & Johnson, N. (2006, August). *Enhancing supervised terrain classification with predictive unsupervised learning.* In Robotics: Science and systems. Philadelphia, USA.

Harel, J., Koch, C., & Perona, P. (2007). In Schölkopf, B., Platt, J., & Hoffman, T. (Eds.), *Graph-based visual saliency* (*Vol. 19*, pp. 545–552). Advances in neural information processing systemsCambridge, MA: MIT Press.

Haritao-Glue, I., & Flickner, M. (2001). Detection and tracking of shopping groups in stores. *IEEE Computer Vision and Pattern Recognition, 1*, 431–438.

Haritaoglu, I., Harwood, D., & Davis, L. S. (2000). W4: Real-time surveillance of people and their activities. *IEEE Transactions on Pattern Analysis and Machine Intelligence, 22*, 809–830. doi:10.1109/34.868683

Harris, C., & Stephens, M. (1988). A combined corner and edge detector. *Alvey Vision Conference*, Vol. 15, (p. 50). Manchester, UK.

Hartley, R., & Zisserman, A. (2003). *Multiple view geometry in computer vision*. Cambridge, UK: Cambridge University Press.

Hartley, R., & Zisserman, A. (2004). *Multiple view geometry in computer vision* (2nd ed.). Cambridge University Press. doi:10.1017/CBO9780511811685

Hautus, M. (1983). Strong detectability and observers. *Linear Algebra and Its Applications, 50*, 353–368. doi:10.1016/0024-3795(83)90061-7

Haxhimusa, Y., Ion, A., & Kropatsch, W. (2006). *Evaluating graph-based segmentation algorithms.*

Haxhimusa, Y., Ion, A., Kropatsch, W., & Brun, L. (2005). Hierarchical image partitioning using Combinatorial Maps. *Proceedings of Joint Hungarian-Austrian Conference* (OAGM), (pp. 179-186).

Haxhimusa, Y., & Kropatsch, W. (2003). Hierarchical image partitioning with dual graph contraction. *Lecture Notes in Computer Science, 2781*, 338–345. doi:10.1007/978-3-540-45243-0_44

Haykin, S. (1999). *Neural networks—A comprehensive foundation*. Englewood Cliffs, NJ: Prentice Hall.

Hecht, F., Azad, P., & Dillmann, R. (2009). Markerless human motion tracking with a flexible model and appearance learning. In *IEEE International Conference on Robotics and Automation (ICRA 2009)* (pp. 3173-3179). Kōbe, Japan: IEEE Press.

Henderson, J. M. (2003). Human gaze control during real-world scene perception. *Trends in Cognitive Sciences, 7*, 498–504. doi:10.1016/j.tics.2003.09.006

Hergenrother, E., & Dhne, P. (2000) Real-time virtual cables based on kinematic simulation. In Proceedings of the International Conference in Central Europe on Computer Graphics, Visualization and Computer Vision.

Herrero, S., & Bescós, J. (2009). Background subtraction techniques: Systematic evaluation and comparative analysis. In *Proceedings of the 11th International Conference on Advanced Concepts for Intelligent Vision Systems* (pp. 33-42). Springer-Verlag.

Herrero-Pérez, D., Martínez-Barberá, H., & Saffiotti, A. (2004). Fuzzy self-localization using natural features in the four-legged league. In *Proceedings of the International RoboCup Symposium*, Lisbon, Portugal.

Herrero-Pérez, D., & Martínez-Barberá, H. (2011). Fuzzy mobile-robot positioning in intelligent spaces using wireless sensor networks. *Sensors (Basel, Switzerland)*, *11*, 10820–10839. doi:10.3390/s111110820

Herrero-Pérez, D., Martínez-Barberá, H., LeBlanc, K., & Saffiotti, A. (2010). Fuzzy uncertainty modeling for grid based localization of mobile robots. *International Journal of Approximate Reasoning*, *51*, 912–932. doi:10.1016/j.ijar.2010.06.001

Hirschmüller, H., & Scharstein, D. (2009). Evaluation of stereo matching costs on images with radiometric differences. *IEEE Transactions on Pattern Analysis and Machine Intelligence*, *31*(9), 1582–1599. doi:10.1109/TPAMI.2008.221

Ho, J. Ming-Hsuan, Yang, Rangarajan, A., & Vemuri, B. (2007). *A new affine registration algorithm for matching 2D point sets*. IEEE Workshop on Applications of Computer Vision, WACV '07.

Hoey, J., Poupart, P., Bertoldi, A., Craig, T., Boutilier, C., & Mihailidis, A. (2010). Automated handwashing assistance for persons with dementia using video and a partially observable Markov decision process. *Computer Vision and Image Understanding*, *114*(5), 503–519. doi:10.1016/j.cviu.2009.06.008

Hoiem, D., Efros, A., & Hebert, M. (2007). Recovering surface layout from an image. *International Journal of Computer Vision*, *75*(1), 151–172. doi:10.1007/s11263-006-0031-y

Holdstein, Y., & Fischer, A. (2008). Three-dimensional surface reconstruction using meshing growing neural gas (MGNG). *The Visual Computer*, *24*, 295–302. doi:10.1007/s00371-007-0202-z

Hootman, J., & Helmick, C. (2006). Projections of US prevalence of arthritis and associated activity limitations. *Arthritis and Rheumatism*, *54*(1), 226–229. doi:10.1002/art.21562

Horn, B. K. P. (1987). Closed-form solution of absolute orientation using unit quaternions. *Journal of the Optical Society of America*, *4*(4), 629–642. doi:10.1364/JOSAA.4.000629

Horswill, I. (1993). Polly: A vision-based artificial agent. In *National Conference on Artificial Intelligence (AAAI)* (pp. 824-829).

Hou, M., & Muller, P. (1992). Design of observers for linear systems with unknown inputs. *IEEE Transactions on Automatic Control*, *37*(6), 871–875. doi:10.1109/9.256351

Howard, A., Turmon, M., Matthies, L., Tang, B., Angelova, A., & Mjolsness, E. (2006). Towards learned traversability for robot navigation: From underfoot to the far field. *Journal of Field Robotics*, *23*(11–12), 1005–1017. doi:10.1002/rob.20168

Hoyos, A., Congote, J., Barandiaran, I., Acosta, D., & Ruiz, O. (2011). Statistical tuning of adaptive-weight depth map algorithm. In A. Berciano, D. Diaz-Pernil, H. Molina-Abril, P. Real, & W. Kropatsch (Eds.), *Proceedings of the 14th International Conference on Computer Analysis of Images and Patterns: Part II* (pp. 563-572). Springer-Verlag.

Hübner, R., & Backer, G. (1999). Perceiving spatially inseparable objects: Evidence for feature-based object selection not mediated by location. *Journal of Experimental Psychology. Human Perception and Performance*, *11*, 583–597.

Hulse, M., McBride, S., & Lee, M. (2009). Implementing inhibition of return; embodied visual memory for robotic systems. *Proceedings of the 9th International Conference on Epigenetic Robotics: Modeling Cognitive Development in Robotic Systems*, Lund University Cognitive Studies, Vol. 146, (pp. 189–190).

Hu, Q., Li, S., He, K., & Lin, H. (2010). A robust fusion method for vehicle detection in road traffic surveillance. In Huang, D.-S., Zhang, X., Reyes Garcia, C., & Zhang, L. (Eds.), *Advanced Intelligent Computing Theories and Applications with Aspects of Artificial Intelligence* (*Vol. 6216*, pp. 180–187). Berlin, Germany: Springer. doi:10.1007/978-3-642-14932-0_23

Hu, W., Tan, T., Wang, L., & Maybank, S. (2004). A survey on visual surveillance of object motion and behaviors. *Pattern Recognition*, *34*, 334–352.

Hygounenc, E., & Soueres, P. (2003). Lateral path following GPS-based control of a small-size unmanned blimp. In *Proceedings of the IEEE International Conference of Robotics and Automation ICRA '03,* Vol. 1, (pp. 540–545).

Hygounenc, E., Jung, I., Soueres, P., & Lacroix, S. (2004). The autonomous blimp project of LAAS-CNRS: Achievements in flight control and terrain mapping. *The International Journal of Robotics Research, 23,* 473–511. doi:10.1177/0278364904042200

Hyvärinen, A., & Oja, E. (2000). Independent component analysis: Algorithms and applications. *Neural Networks, 13*(4-5), 411–430. doi:10.1016/S0893-6080(00)00026-5

Iddan, G. J., & Yahav, G. (2001). 3D imaging in the studio (and elsewhere...). *Proceedings of the Society for Photo-Instrumentation Engineers, 4298,* 48–55. doi:10.1117/12.424913

Igual, L., & De la Torre, F. (2010, June). *Continuous Procrustes analysis to learn 2D Shape models from 3D objects.* In Computer Vision and Pattern Recognition Workshops, San Francisco, CA.

Inoue, H., Tachi, S., Nakamura, Y., Hirai, K., Ohyu, N., & Hirai, S. Tanie, K., Yokoi, K., & Hirukawa, H. (2001). Overview of humanoid robotics project of meti. In *32nd International Symposium on Robotics* (pp. 1478-1482).

Ion, A., Kropatsch, W., & Haxhimusa, Y. (2006). Considerations regarding the minimum spanning tree pyramid segmentation method. *Lecture Notes in Computer Science, 4109,* 182–190. doi:10.1007/11815921_19

Itti, L. (2000). *Models of bottom-up and top-down visual attention.* Unpublished doctoral dissertation, California Institute of Technology, Pasadena, California.

Itti, L., & Koch, C. (2000). A saliency-based search mechanism for overt and covert shifts of visual attention. *Vision Research, 40,* 1489–1506. doi:10.1016/S0042-6989(99)00163-7

Itti, L., & Koch, C. (2001). Computational modelling of visual attention. *Nature Reviews. Neuroscience, 2,* 194–203. doi:10.1038/35058500

Itti, L., & Koch, C. (2005). A model of saliency-based visual attention for rapid scene analysis. *IEEE Transactions on Pattern Analysis and Machine Intelligence, 20*(11), 1254–1259. doi:10.1109/34.730558

Itti, L., Koch, C., & Niebur, E. (1998). A model of saliency-based visual attention for rapid scene analysis. *IEEE Transactions on Pattern Analysis and Machine Intelligence, 20,* 1254–1259. doi:10.1109/34.730558

Itti, L., Koch, C., & Niebur, E. (1998). Attentive mechanisms for dynamic and static scene analysis. *IEEE Transactions on Pattern Analysis and Machine Intelligence, 20,* 1254–1259. doi:10.1109/34.730558

James Clerk Maxwell, B. (1885). Experiments on colour, as perceived by the eye, with remarks on colour blindness. *Transactions of the Royal Society of Edinburgh, 21*(2), 274–299.

Jankovic, M., & Ghosh, B. (1995). Visually guided ranging from observations points, lines and curves via an identifier based nonlinear observer. *Systems & Control Letters, 25*(1), 63–73. doi:10.1016/0167-6911(94)00053-X

Jenning, C., & Murray, D. (1997). Gesture recognition for robot control. In *Proceedings of the Conference on Computer Vision and Pattern Recognition.*

Jensfelt, P., & Kristensen, S. (2001). Active global localization for a mobile robot using multiple hypothesis tracking. *IEEE Transactions on Robotics and Automation, 17*(5), 748–760. doi:10.1109/70.964673

Jiang, Z., Huynh, D. Q., Morany, W., Challay, S., & Spadaccini, N. (2010). Multiple pedestrian tracking using colour and motion models. *Digital Image Computing: Techniques and Applications,* (pp. 328-334).

Jian-guo, G., & Jun, Z. (2008, December). *Altitude control system of autonomous airship based on fuzzy logic.* Paper presented at the 2nd International Symposium on Systems and Control in Aerospace and Astronautics, Shenzhen, China.

Jolliffe, I. (2005). Principal component analysis. In Everitt, B., & Howell, D. (Eds.), *Encyclopedia of statistics in behavioral science.* New York, NY: John Wiley & Sons, Ltd.doi:10.1002/0470013192.bsa501

Jolliffe, I. T. (Ed.). (2010). *Principal component analysis*. New York, NY: Springer-Verlag.

Jones, M. J., & Poggio, T. (1989). Multidimensional morphable models. In *International Conference on Computer Vision*, Vol. 29, (pp. 683-688). Springer.

Jones, M. J., Sinha, P., Vetter, T., & Poggio, T. (1997). Top–down learning of low-level vision tasks. *Current Biology*, *12*(7), 991–994. doi:10.1016/S0960-9822(06)00419-2

Julesz, B. (1991). Early vision and focal attention. *Reviews of Modern Physics*, *66*(3), 735–772. doi:10.1103/RevModPhys.63.735

Julier, S., & Uhlmann, J. (2001). A counter example to the theory of simultaneous localization and map building. In *Proceedings of the IEEE International Conference on Robotics and Automation*, (pp. 4238-4243).

Junger, M., Liebling, T., Naddef, D., Newhauser, G., & Wolsey, L. (2010). *50 years of integer programming, 1958-2008*. Berlin, Germany: Springer-Verlag. doi:10.1007/978-3-540-68279-0

KadewTraKuPong. P., & Bowden, R. (2001). An improved adaptive background mixture model for real-time tracking with shadow detection. *Proceedings of the 2nd European Workshop on Advanced Video-Based Surveillance Systems*.

Kadir, T., & Brady, M. (2001). Saliency, scale and image description. *International Journal of Computer Vision*, *45*, 83–105. doi:10.1023/A:1012460413855

Kaelbling, L., & Lozano-Perez, T. (2011). *Domain and plan representation for task and motion planning in uncertain domains*. In IROS 2011 Workshop on Knowledge Representation for Autonomous Robots.

Kaelbling, L., Littman, M., & Cassandra, A. (1998). Planning and acting in partially observable stochastic domains. *Artificial Intelligence*, *101*(1-2), 99–134. doi:10.1016/S0004-3702(98)00023-X

Kahl, F., & Hartley, R. (2008). Multiple-view geometry under the L1-norm. *IEEE Transactions on Pattern Analysis and Machine Intelligence*, *30*(9), 1603–1617. doi:10.1109/TPAMI.2007.70824

Kahn, J. P., Kanda, T., Ishiguro, H., Gill, B. T., Ruckert, J. H., Shen, S., et al. (2012). Do people hold a humanoid robot morally accountable for the harm it causes? *The Seventh Annual ACM/IEEE International Conference on Human-Robot Interaction* (pp. 33-40). New York, NY: ACM.

Kalal, Z., Matas, J., & Mikolajczyk, K. (2010). *P-N learning: Bootstrapping binary classifiers by structural constraints* (pp. 49–56). IEEE Computer Vision and Pattern Recognition. doi:10.1109/CVPR.2010.5540231

Kalman, R. E. (1960). A new approach to linear filtering and prediction problems. *Journal of Basic Engineering*, *82*, 35–45. doi:10.1115/1.3662552

Kaminski, J., & Teicher, M. (2004). A general framework for trajectory triangulation. *Journal of Mathematical Imaging and Visualization*, *21*(1), 27–41. doi:10.1023/B:JMIV.0000026555.79056.b8

Kampke, T., & Elfes, A. (2003). Optimal aerobot trajectory planning for wind-based opportunistic flight control. *IEEE/RSJ International Conference on Intelligent Robots and Systems (IROS 2003)*, Vol. 1, (pp. 67–74).

Kanekazi, A., Nakayama, H., Harada, T., & Kuniyoshi, Y. (2010). High-speed 3D object recognition using additive features in a linear subspace. In *IEEE International Conference on Robotics and Automation (ICRA 2010)* (pp. 3128-3134). Anchorage, AK: IEEE Press.

Kaplan, F., & Hafner, V. (2004). The challenges of joint attention. *Interaction Studies: Social Behaviour and Communication in Biological and Artificial Systems*, *7*(2), 67–74.

Kaplan, F., & Hafner, V. V. (2006). The challenges of joint attention. *Interaction Studies: Social Behaviour and Communication in Biological and Artificial Systems*, *7*, 135–169. doi:10.1075/is.7.2.04kap

Kass, M., Witkin, A., & Terzopoulos, D. (1988). Snakes: Active contour models. *International Journal of Computer Vision*, *8*(2), 321–331. doi:10.1007/BF00133570

Kawai, Y., Hirano, H., Azuma, T., & Fujita, M. (2004). Visual feedback control of an unmanned planar blimp system with self-scheduling parameter via receding horizon approach. In *Proceedings of the IEEE International Conference on Control Applications, Vol. 2*, (pp. 1063-1069).

Kawai, Y., Kitagawa, S., Izoe, S., & Fujita, M. (2003, August). *An unmanned planar blimp on visual feedback control: Experimental results.* Paper presented at the SICE 2003 Annual Conference, Fukui, Japan.

Kelly, P., O'Connor, N., & Smeaton, A. (2007). A framework for evaluating stereo-based pedestrian detection techniques. *IEEE Transactions on Circuits and Systems for Video Technology, 18*(8), 1163–1167. doi:10.1109/TCSVT.2008.928228

Kemmotsu, K., Koketsua, Y., & Iehara, M. (2008). Human behavior recognition using unconscious cameras and a visible robot in a network robot system. *Robotics and Autonomous Systems, 56*(10), 857–864. doi:10.1016/j.robot.2008.06.004

Kendon, A. (1967). Some functions of gaze-direction in social interaction. *Acta Psychologica, 26*, 22–63. doi:10.1016/0001-6918(67)90005-4

Keni, B., & Rainer, S. (2008). Evaluating multiple object tracking performance: The CLEAR MOT metrics. *EURASIP Journal on Image and Video Processing, 2008*, 10.

Kennedy, W., Bugajska, M., Marge, M., Adams, W., Fransen, B., Perzanowski, D., et al. (2007). Spatial representation and reasoning for human-robot interaction. In *Twenty-Second Conference on Artificial Intelligence (AAAI)* (pp. 1554-1559). Toronto, Canada.

Khan, S., & Shah, M. (2000). Tracking people in presence of occlusion. *Asian Conference on Computer Vision.*

Khatib, O. (1996). Motion coordination and reactive control of autonomous multi-manipulator system. *Journal of Robotic Systems, 15*(4), 300–319.

Khatib, O. (1999). Robot in human environments: Basic autonomous capabilities. *The International Journal of Robotics Research, 18*(7), 684–696. doi:10.1177/02783649922066501

Khoury, G. A., & Gillett, J. D. (2002). *Airship technology.* Cambridge, UK: Cambridge University.

Kim, D., Sun, J., Min, S., James, O., Rehg, M., & Bobick, A. F. (2006). *Traversability classification using unsupervised on-line visual learning for outdoor robot navigation.* In IEEE International Conference on Robotics and Automation.

Kim, H., Jasso, H., Deak, G., & Triesch, J. (2008). A robotic model of the development of gaze following. *7th IEEE International Conference on Development and Learning* (ICDL 2008) (pp. 238-243).

Kim, W., & Kweon, I. (2011). Moving object detection and tracking from moving camera, *The 8th International Conference on Ubiquitous Robots and Ambient Intelligence*, Nov. 23-26.

Kim, J., & Sukkarieh, S. (2007). Real-time implementation of airborne inertial-SLAM. *Robotics and Autonomous Systems, 55*, 62–71. doi:10.1016/j.robot.2006.06.006

Kim, M. (2008). Face tracking and recognition with visual constraints in real-world videos. *Computer Vision and Pattern Recognition, CVPR, 2008*, 1–8.

Kirby, R. (2010). *Social robot navigation.* PhD Thesis, Robotics Institute, Carnegie Mellon University, Pittsburgh, PA.

Kitt, W. W., Chekima, A., & Dhargam, J. A. (2011). Design and verification of an ANFIS based control system for an autonomous airship on a simulation environment. *Global Journal of Computer Application and Technology, 1*, 119–128.

Klein, D., & Frintrop, S. (2011). *Center-surround divergence of feature statistics for salient object detection.* International Conference on Computer Vision.

Kleinberg, J., & Tardos, E. (2008, May 17-20). *Balanced outcomes in social exchange networks.* In ACM Symposium on Theory of Computing.

Kleinke, C. (1986). Gaze and eye contact: A research review. *Psychological Bulletin, 100*(1), 78. doi:10.1037/0033-2909.100.1.78

Knox, W., & Stone, P. (2010, May). *Combining manual feedback with subsequent MDP reward signals for reinforcement learning.* In International Conference on Autonomous Agents and Multiagent Systems (AAMAS).

Koch, C., & Ullman, S. (1985). Shifts in selective visual attention: Towards the underlying neural circuitry. *Human Neurobiology, 4*, 219–227.

Koenig, D., & Mammar, S. (2001). Design of a class of reduced order unknown inputs nonlinear observer for fault diagnosis. *Proceedings of the American Control Conference*, (pp. 2143–2147). Arlington, VA.

Kojo, N., Inamura, T., Okada, K., & Inaba, M. (2006). Gesture recognition for humanoids using proto-symbol space. In *6th IEEE RAS International Conference on Humanoid Robots* (pp. 76-81). Genoa, Italy: IEEE Press.

Kolmogorov, V., & Zabih, R. (2001). Computing visual correspondence with occlusion using graph cuts. In *Proceedings of Eight International Conference on Computer Vision*, (pp. 508-515). IEEE Computer Society.

Kolmogorov, V., & Zabih, R. (2001). Computing visual correspondence with occlusion using graph cuts. In *Proceedings of European Conference on Computer Vision*, (pp. 82-96). Springer-Verlag.

Kornienko, A. (2006). *System identification approach for determining flight dynamical characteristics of an airship from flight data*. PhD thesis, University of Stuttgart, Germany.

Kostavelis, I., Boukas, E., Nalpantidis, L., Gasteratos, A., & Aviles Rodrigalvarez, M. (2011, November). *SPARTAN system: Towards a low-cost and high-performance vision architecture for space exploratory rovers*. In 2nd International Workshop on Computer Vision in Vehicle Technology: From Earth to Mars, in Conjunction with ICCV 2011. Barcelona, Spain.

Kostavelis, I., Nalpantidis, L., & Gasteratos, A. (2011). Supervised traversability learning for robot navigation. In *12th Conference towards Autonomous Robotic Systems* (Vol. 6856, pp. 289–298). Sheffield, UK: Springer-Verlag.

Kostlivá, J., Cech, J., & Sara, R. (2003). Dense stereo matching algorithm performance for view prediction and structure reconstruction. In J. Bigun & T. Gustavsson (Eds.), *Proceedings of the 13th Scandinavian Conference on Image Analysis*, (pp. 101-107). Springer-Verlag.

Kostlivá, J., Cech, J., & Sara, R. (2007). *Feasibility boundary in dense and semi-dense stereo matching. Computer Vision and Pattern Recognition* (pp. 1–8). IEEE Computer Society.

Krause, A., Singh, A., & Guestrin, C. (2008). Near-optimal sensor placements in Gaussian processes: Theory, efficient algorithms and empirical studies. *Journal of Machine Learning Research, 9*, 235–284.

Kreucher, C., Kastella, K., & Hero, A. (2005). Sensor management using an active sensing approach. *IEEE Transactions on Signal Processing, 85*(3), 607–624.

Kropatsch, W. (1995). Building irregular pyramids by dual graph contraction. *IEE Proceedings. Vision Image and Signal Processing, 142*(6), 366–374. doi:10.1049/ip-vis:19952115

Kropatsch, W., Haxhimusa, Y., Pizlo, Z., & Langs, G. (2005). Vision pyramids that do not grow too high. *Pattern Recognition Letters, 26*, 319–337. doi:10.1016/j.patrec.2004.10.026

Krotkov, E. (1987). *Exploratory visual sensing for determining spacial layout with an agile camera system*. PhD thesis, University of Pennsylvenia.

Kuffner, J. J. (2004). Effective sampling and distance metrics for 3D rigid body path planning. In *Proceedings of IEEE International Conference on Robotics and Automation, Vol. 4*, (pp. 3993-3998).

Kuipers, J. B. (Ed.). (1999). *Quaternions and rotation sequences: A primer with applications to orbits, aerospace and virtual reality*. Princeton, NJ: Princeton University Press.

Kulczycki, E. A., Joshi, S. S., & Hess, R. A. (2006). *Towards controller design for autonomous airships using SLC and LQR methods*. In AIAA Guidance Navigation and Control Conference and Exhibit.

Kulic, D., Lee, D., & Nakamura, Y. (2009). Whole body motion primitive segmentation from monocular video. In *IEEE International Conference on Robotics and Automation (ICRA 2009)* (pp. 3166-3172). Kōbe, Japan: IEEE Press.

Kumlander, D. (2008). On importance of a special sorting in the maximum-weight clique algorithm based on colour classes. *Proceedings of Second International Conference Modelling, Computation and Optimization in Information Systems*, (pp. 165-174).

Kunchev, V., Jain, L., Ivancevic, V., & Finn, A. (2006). Path planning and obstacle avoidance for autonomous mobile robots: A review. In *International Conference on Knowledge-Based and Intelligent Information and Engineering Systems* (Vol. 4252, pp. 537–544). Springer-Verlag.

Kungl, P., Schlenker, M., Wimmer, D. A., & Kröplin, B. H. (2004). Instrumentation of remote controlled airship "LOTTE" for in-flight measurements. *Aerospace Science and Technology*, 8(7), 599–610. doi:10.1016/j.ast.2004.06.004

Kyung Hyun, C., Minh Ngoc, N., & Asif Ali, R. (2008). A real time collision avoidance algorithm for mobile robot based on elastic force. *International Journal of Mechanical. Industrial and Aerospace Engineering*, 2(4), 230–233.

Labayrade, R., Aubert, D., & Tarel, J.-P. (2002). Real time obstacle detection in stereovision on non flat road geometry through "v-disparity" representation. In IEEE Intelligent Vehicle Symposium (Vol. 2, pp. 646–651). Versailles, France.

LaBerge, D. (1995). *Attentional processing*. Harvard University Press.

LaBerge, D., Carlson, R. L., Williams, J. K., & Bunney, B. G. (1997). Shifting attention in visual space: Tests of moving-spotlight models versus an activity-distribution model. *Journal of Experimental Psychology. Human Perception and Performance*, 23, 1380–1392. doi:10.1037/0096-1523.23.5.1380

Lacroix, S., Jung, I.-K., Soueres, P., Hygounenc, E., & Berry, J.-P. (2003). The autonomous blimp project of LASS/CNRS: Current status. *Experimental Robotics VIII*, 5, 487–496. doi:10.1007/3-540-36268-1_44

Lai, K., Bo, L., Ren, X., & Fox, D. (2011, May 9-13). *Sparse distance learning for object recognition combining RGB and depth information*. In International Conference on Robotics and Automation (ICRA). Shanghai, China.

Lambert, D. (2004). *Body language*. London, UK: Harper Collins.

Lambert, J. (1760). *Photometria sive de mensure de gratibus luminis. Colorum umbrae*. Eberhard Klett.

Lammens, J. M. G. (1994). *A computational model of color perception and color naming*. Doctoral dissertation, Computer Science Department, State University of New York at Buffalo, NY.

Land, M. F., & Hayhoe, M. (2001). In what ways do eye movements contribute to everyday activities? *Vision Research*, 41, 3559–3565. doi:10.1016/S0042-6989(01)00102-X

Langer, D. (1994). A behavior-based system for off-road navigation. *IEEE Transactions on Robotics and Automation*, 10(6), 776–783. doi:10.1109/70.338532

Lanitis, A., Taylor, C. J., & Cootes, T. F. (1995). Automatic face identification system using flexible appearance models. *Image and Vision Computing*, 13(5), 393–401. doi:10.1016/0262-8856(95)99726-H

Lanitis, A., Taylor, C. J., & Cootes, T. F. (1997). Automatic interpretation and coding of face images using flexible models. *IEEE Transactions on Pattern Analysis and Machine Intelligence*, 19(7), 742–756. doi:10.1109/34.598231

Lanz, O., & Brunelli, R. (2008). An appearance-based particle filter for visual tracking in smart rooms. In Stiefelhagen, R., Bowers, R., & Fiscus, J. E. (Eds.), *Multimodal Technologies for Perception of Humans* (Vol. 4625, pp. 57–69). Lecture Notes in Computer Science. doi:10.1007/978-3-540-68585-2_4

Lawler, E. J. (2001). An affect theory of social exchange. *American Journal of Sociology*, 107(2), 321–352. doi:10.1086/324071

Lawrence, S., Giles, C., Tsoi, A., & Back, A. (1997). Face recognition: A convolutional neural-network approach. *IEEE Transactions on Neural Networks*, 8(1), 98–113. doi:10.1109/72.554195

Leclerc, Y. G., Luong, Q., & Fua, P. (2000). Measuring the self-consistency of stereo algorithms. In *European Conference on Computer Vision*, (pp. 282–298). Springer-Verlag.

Lerios, A. (1995). *Rotations and quaternions*. Technical Report. Retrieved from http://server2.phys.uniroma1.it/doc/giansanti/COMP_BIOPHYS_2008/LECTURES/LECT_2/Lerios1995.pdf

Li, C., Parikh, D., & Chen, T. (2011, November 6-13). *Extracting adaptive contextual cues from unlabeled regions*. In International Conference on Computer Vision (ICCV). Barcelona, Spain.

Li, C., Xue, J., Du, S., & Zheng, N. (2010). A fast multi-resolution iterative closest point algorithm. *Chinese Conference on Pattern Recognition (CCPR)*, (pp. 1-5, 21-23).

Li, J., Wang, Z., & Wiederhold, G. (2000). IRM: Integrated region matching for image retrieval. *Proc. of ACM Multimedia*, 147-156.

Li, L., Bulitko, V., Greiner, R., & Levner, I. (2003). *Improving an adaptive image interpretation system by leveraging*. In Australian and New Zealand Conference on Intelligent Information Systems.

Li, X., & Sridharan, M. (2012, June 5). *Vision-based autonomous learning of object models on a mobile robots*. In Autonomous Robots and Multirobot Systems (ARMS) Workshop at the International Conference on Autonomous Agents and Multiagent Systems (AAMAS), Valencia, Spain.

Li, X., Sridharan, M., & Zhang, S. (2011, May 9-13). *Autonomous learning of vision-based layered object models on mobile robots*. In International Conference on Robotics and Automation (ICRA). Shanghai, China.

Lienhart, R., & Maydt, J. (2002). An extended set of Haar-like features for rapid object detection. *2002 International Conference on Image Processing*, Vol. 1, (pp. 900-903).

Lipton, A., Fujiyoshi, H., & Patil, R. (1998). *Moving target classification and tracking from real-time video*. DARPA Image Understanding Workshop.

Liu, F., Farza, M., & M'Saad, M. (2006). *Unknown input observers design for a class of nonlinear systems - application to biochemical process*. IFAC Symposium on Robust Control Design.

Liu, Y., Pan, Z., Stirling, D., & Naghdy, F. (2009). Control of autonomous airship. In *Proceedings of the IEEE International Conference on Robotics and Biomimetics* (pp. 2457–2462).

Lopes, M., & Santos-Victor, J. (2005). Visual learning by imitation with motor representations. *IEEE Transactions on Systems, Man, and Cybernetics. Part B, Cybernetics*, *35*, 438–449. doi:10.1109/TSMCB.2005.846654

Lopez-Guede, J. M., Fernandez-Gauna, B., Graña, M., & Zulueta, E. (2011). Empirical study of q-learning based elemental hose transport control. In Corchado, E., Kurzynski, M., & Wozniak, M. (Eds.), *Hybrid Artificial Intelligent Systems* (*Vol. 6679*, pp. 455–462). Berlin, Germany: Springer. doi:10.1007/978-3-642-21222-2_55

Lopez, M. T., Fernandez-Caballero, A., Fernandez, M. A., Mira, J., & Delgado, A. E. (2006). Motion features to enhance scene segmentation in active visual attention. *Pattern Recognition Letters*, *27*, 469–478. doi:10.1016/j.patrec.2005.09.010

Lopez, M. T., Fernandez, M. A., Fernandez-Caballero, A., Mira, J., & Delgado, A. E. (2007). Dynamic visual attention model in image sequences. *Image and Vision Computing*, *25*, 597–613. doi:10.1016/j.imavis.2006.05.004

Lotufo, R. A., Morgan, A. D., & Johnson, A. S. (1990). *Automatic number plate recognition*. IEE Colloquium on Image Analysis for Transport Applications, February 1990, London, INSPEC Accession Number: 3649590.

Louloudi, A., Mosallam, A., Marturi, N., Janse, P., & Hernandez, V. (2010). *Integration of the humanoid robot nao inside a smart home: A case study*. The Swedish AI Society Workshop.

Lowe, D. G. (2004). Distinctive image features from scale-invariant keypoints. *International Journal of Computer Vision*, *60*(2), 91–110. doi:10.1023/B:VISI.0000029664.99615.94

Lü, C., Huang, J., & Shen, Y. (2011), Subjective assessment of noised stereo images. *International Conference on Multimedia Technology*, (pp.783-785).

Lutz, T., Funk, P., Jakobi, A., & Wagner, S. (2002, July). *Summary of Aerodynamic Studies on the LOTTE Airship*. Paper presented at the 4th International Airship Convention and Exhibition. Cambridge, UK.

Maccormick, J., & Blake, A. (2000). Probabilistic exclusion and partitioned sampling for multiple object tracking. *International Journal of Computer Vision*, *39*, 57–71. doi:10.1023/A:1008122218374

Majecka, B. (2009). *Statistical models of pedestrian behavior in the Forum*. Master's thesis, University of Edinburgh, U.K.

Maki, A., Nordlund, P., & Eklundh, J.-O. (1996). A computational model of depth-based attention. In *Proceedings of the 13th International Conference on Pattern Recognition*.

Maki, A., Nordlund, P., & Eklundh, J.-O. (2000). Attentional scene segmentation: Integrating depth and motion. *Computer Vision and Image Understanding, 78*, 351–373. doi:10.1006/cviu.2000.0840

Mallick, S., Zickler, T., Kriegman, D., & Belhumeur, P. (2005). Beyond Lambert: Reconstructing specular surfaces using color. In *IEEE Computer Society Conference on Computer Vision and Pattern Recognition, CVPR 2005* (Vol. 2, pp. 619-626).

Malpica, W. S., & Bovik, A. C. (2009). Range image quality assessment by structural similarity. *IEEE International Conference on Acoustics, Speech and Signal Processing*, (pp. 1149-1152).

Manduchi, R. (2004). Learning outdoor color classification from just one training image. In *European Conference on Computer Vision* (Vol. 4, pp. 402–413).

Manz, A., Liscano, R., & Green, D. (1993). A comparison of realtime obstacle avoidance methods for mobile robots. In *Experimental Robotics II* (pp. 299–316). Springer-Verlag. doi:10.1007/BFb0036147

Marfil, R., Bandera, A., Rodríguez, J. A., & Sandoval, F. (2004). Real-time template-based tracking of non-rigid objects using bounded irregular pyramids. In *IEEE/RSJ International Conference on Intelligent Robotics and Systems*, Vol. 1 (pp. 301-306). Sendai, Japan: IEEE Press.

Marfil, R., Bandera, A., Rodríguez, J.A., & Sandoval, F. (2008). A novel hierarchical framework for object-based visual attention. *Attention in Cognitive Systems, LNCS 5395*.

Marfil, R., Bandera, A., Rodríguez, J. A., & Sandoval, F. (2009). A novel hierarchical framework for object-based visual attention. *Lecture Notes in Artificial Intelligence, 5395*, 27–40.

Mariottini, G. L., & Roumeliotis, S. I. (2011). Active vision-based robot localization and navigation in a visual memory. *Proceedings of International Conference on Robots and Automation*.

Marocco, D., & Floreano, D. (2002). Active vision and feature selection in evolutionary behavioral systems. *Proceedings of International Conference on Simulation of Adaptive Behavior* (SAB-7), (pp. 247-255).

Marthi, B., Russell, S., & Wolfe, J. (2009, August). *Angelic hierarchical planning: Optimal and online algorithms*. Technical Report UCB/EECE-2009-122, EECS Department, University of California Berkeley.

Martin, D., Fowlkes, C., Tal, D., & Malik, J. (2001, July). A database of human segmented natural images and its application to evaluating segmentation algorithms and measuring ecological statistics. In *Proceedings of the 8th International Conference on Computer Vision* (Vol. 2, pp. 416–423).

Martinetz, T., Berkovich, S. G., & Schulten, K. J. (1993). Neural-gas network for vector quantization and its application to time-series prediction. *IEEE Transactions on Neural Networks, 4*(4), 558–569. doi:10.1109/72.238311

Martinetz, T., & Schulten, K. (1994). Topology representing networks. *Neural Networks, 7*(3), 507–522. doi:10.1016/0893-6080(94)90109-0

Martínez-Barberá, H., & Saffiotti, A. (2000). *Thinking-Cap-II architecture*. Retrieved November 20, 2011, from http://robolab.dif.um.es/tc2/

Martínez, J., Romero-Garcés, A., Bandera, J. P., Marfil, R., & Bandera, A. (2012). (in press). A DDS-based middleware for quality-of-service and high-performance networked robotics. *Concurrency and Computation*. doi:doi:10.1002/cpe.2816

Masuda, T., Sakaue, K., & Yokoya, N. (1996). Registration and integration of multiple range images for 3-D model construction. *Proceedings of the 13th International Conference on Pattern Recognition*, Vol. 1, (pp. 879-883).

Matthews, I., & Baker, S. (2004). Active appearance models revisited. *International Journal of Computer Vision, 60*(2), 135-164. Retrieved December 1, 2011, from http://www.ri.cmu.edu/publication_view.html?pub_id=4601

Ma, Y., Soatto, S., Kosecka, J., & Sastry, S. (2004). *An invitation to 3-D vision*. Springer.

McKenna, S., Jabri, J., Duran, Z., & Wechsler, H. (2000). Tracking interacting people. *International Workshop on Face and Gesture Recognition*, (pp. 348-353).

McKenna, S., Raja, Y., & Gong, S. (1999). Tracking colour objects using adaptive mixture models. *Image and Vision Computing*, (n.d), 225–231. doi:10.1016/S0262-8856(98)00104-8

Melo, F., & Ribeiro, M. (2008). Reinforcement learning with function approximation for cooperative navigation tasks. In *IEEE International Conference on Robotics and Automation*, (pp. 3321 –3327)

Meltzoff, A., Brooks, R., Shon, A., & Rao, R. (2010, October-November). Social robots are psychological agents for infants: A test of gaze following. *Neural Networks*, *23*(8-9), 966–972. doi:10.1016/j.neunet.2010.09.005

Meltzoff, A., & Moore, M. (1989). Imitation in newborn infants: Exploring the range of gestures imitated and the underlying mechanisms. *Developmental Psychology*, *25*, 954–962. doi:10.1037/0012-1649.25.6.954

Mertsching, B., Bollmann, M., Hoischen, R., & Schmalz, S. (1999). The neural active vision system navis. *Handbook of Computer Vision and Applications*, *3*, 543–568.

Miao, X., & Rao, R. (2007). Learning the lie groups of visual invariance. *Neural Computation*, *19*, 2665–2693. doi:10.1162/neco.2007.19.10.2665

Mikolajczyk, K., & Schmid, C. (2004). Scale and affine invariant interest point detectors. *International Journal of Computer Vision*, *60*(1), 63–86. doi:10.1023/B:VISI.0000027790.02288.f2

Milanese, R. (1993). *Detecting salient regions in an image: From biological evidence to computer implementation.* PhD thesis, University of Geneva.

Milanese, R. (1990). *Detection of salient features for focus of attention. Technical report.* Computing Science Center, University of Geneva.

Milanese, R., Gil, S., & Pun, T. (1995). Attentive mechanisms for dynamic and static scene analysis. *Optical Engineering (Redondo Beach, Calif.)*, *34*, 2428–2434. doi:10.1117/12.205668

Mileva, Y., Bruhn, A., & Weickert, J. (2007). Illumination-robust variational optical flow with photometric invariants. In *Pattern Recognition* (pp. 152–162).

Mita, T., Kaneko, T., & Hori, O. (2005). Joint haar-like features for face detection. *10th IEEE International Conference on Computer Vision*, Vol. 2, (pp. 1619-1626).

Mittal, A., Moorthy, A. K., Ghosh, J., & Bovik, A. C. (2011). *Algorithmic assessment of 3D quality of experience for images and videos. Digital Signal Processing Workshop and IEEE Signal Processing Education Workshop (DSP/SPE)* (pp. 338–343). IEEE Computer Society.

Miyata, S., Nakamura, H., Yanou, A., & Takehara, S. (2009). Automatic path search for roving robot using reinforcement learning. In *Fourth International Conference on Innovative Computing, Information and Control*, (pp. 169 –172).

MoBo. (2004). *The CMU motion of body (MoBo) database.* Retrieved from http://www.ri.cmu.edu/publication_view.html?pub_id=3904

Moeslund, T., Hilton, A., & Krüger, V. (2006). A survey of advances in vision-based human motion capture and analysis. *Vision and Image Understanding*, *104*, 90–126. doi:10.1016/j.cviu.2006.08.002

Mohammad, Y., & Nishida, T. (2009). Interactive perception for amplification of intended behavior in complex noisy environments. *AI & Society*, *23*(2), 167–186. doi:10.1007/s00146-007-0137-y

Monteiro, S., & Bicho, E. (2010). Attractor dynamics approach to formation control: Theory and application. *Autonomous Robots*, *29*, 331–355. doi:10.1007/s10514-010-9198-8

Montemerlo, M., Thrun, S., Koller, D., & Wegbreit, B. (2003). Fastslam 2.0: An improved particle filtering algorithm for simultaneous localization and mapping that provably converges. *In Proceedings of International Conference on Artificial Intelligence (IJCAI)*.

Morales, S., & Klette, R. (2009). A third eye for performance evaluation in stereo sequence analysis. In X. Jiang & N. Petkov (Eds.), *Proceedings of the 13th International Conference on Computer Analysis of Images and Patterns* (pp. 1078-1086). Springer-Verlag.

Morales, S., & Klette, R. (2010). Ground truth evaluation of stereo algorithms for real world applications. In R. Koch & F. Huang (Eds.), *Proceedings of the 2010 International Conference on Computer Vision: Part II*, (pp. 152-162). Springer-Verlag.

Moravec, H. (1980). *Obstacle avoidance and navigation in the real world by a seeing robot rover.* Technical Report CMU-RI-TR-3, Carnegie-Mellon University.

Moravec, P. (1987). Certainty grids for mobile robots. In *NASA/JPL Space Telerobotics Workshop* (Vol. 3, pp. 307–312).

Morency, L.-P., Sundberg, P., & Darrell, T. (2003). Pose estimation using 3D view-based Eigenspaces. ICCV Workshop on Analysis and Modeling of Face and Gesture, (pp. 45-52). Nice, France, October.

Moreno, J. A. (2000). Unknown input observers for SISO nonlinear systems. *IEEE Conference on Decision and Control*, (pp. 790–801). Sydney, NSW Australia.

Moreno, R. Graña, & d'Anjou, A. (2010, July). An image color gradient preserving color constancy. *In Fuzz-IEEE 2010* (p. 710-714).

Moreno, R., Graña, M., & d'Anjou, A. (2010). A color transformation for robust detection of color landmarks in robotic contexts. *In Trends in Practical Applications of Agents and Multiagent Systems* (pp. 665–672).

Moreno, R., Lopez-Guede, J., & d'Anjou, A. (2010). Hybrid color space transformation to visualize color constancy. *In Hybrid Artificial Intelligence Systems* (pp. 241–247).

Moreno, J. A., & Dochain, D. (2008). Global observability and detectability analysis of uncertain reaction systems and observer design. *International Journal of Control, 81,* 1062–1070. doi:10.1080/00207170701636534

Moreno, R., Graña, M., & Zulueta, E. (2010, June). RGB colour gradient following colour constancy preservation. *Electronics Letters, 46*(13), 908–910. doi:10.1049/el.2010.0553

Morik, K., Wrobel, S., Kietz, J.-U., & Emde, W. (1993). *Knowledge acquisition and machine learning: Theory, methods, and applications.* San Francisco, CA: Academic.

Motorola Solutions for Government. (2005). *Government and enterprise North America.* Retrieved from http://www.motorola.com/governmentandenterprise/northamerica/en-us/solution.aspx?navigationpath=id_801i/id_826i/id_2694i/id_2695i

Mühlig, M., Gienger, M., Hellbach, S., Steil, J. J., & Goerick, C. (2009). Task-level imitation learning using variance-based movement optimization. In *IEEE International Conference on Robotics and Automation (ICRA 2009)* (pp. 1177-1184). Kōbe, Japan: IEEE Press.

Muhlmann, K., Maier, D., Hesser, J., & Manner, R. (2002). Calculating dense disparity maps from color stereo images, an efficient implementation. *International Journal of Computer Vision, 47*(1–3), 79–88. doi:10.1023/A:1014581421794

Mumford, D., & Shah, J. (1989). Optimal approximations by piecewise smooth functions and associated variational problems. *Communications on Pure and Applied Mathematics, 42*(5), 577–685. doi:10.1002/cpa.3160420503

Muñoz-Salinas, R., Aguirre, E., & Garcá-Silvente, M. (2007). People detection and tracking using stereo vision and color. *Image and Vision Computing, 25*(6), 995–1007. doi:10.1016/j.imavis.2006.07.012

Murarka, A., Sridharan, M., & Kuipers, B. (2008). *Detecting obstacles and drop-offs using stereo and motion cues for safe local motion.* In International Conference on Intelligent Robots and Systems (IROS), Nice, France.

Murphy, K. (1999). Bayesian map learning in dynamic environments. *Proceedings of the Conference on Neural Information Processing Systems (NIPS),* (pp. 1015-1021).

Murphy-Chutorian, E., & Trivedi, M. (2009). Head pose estimation in computer vision: A survey. *IEEE Transactions on Pattern Analysis and Machine Intelligence, 31*(4), 607–626. doi:10.1109/TPAMI.2008.106

Musa, Z., & Watada, J. (2008). Video tracking system: A survey. *An International Journal of Research and Surveys, 2,* 65–72.

Mutlu, B., & Forlizzi, J. (2008). Robots in organizations: The role of workflow, social, and environmental factors in human-robot interaction. *The 3rd ACM/IEEE International Conference on Human Robot Interaction* (pp. 287-294). New York, NY: ACM.

Mutlu, B., Forlizzi, J., & Hodgins, J. (2006). A storytelling robot: Modeling and evaluation of human-like gaze behavior. *6th IEEE-RAS International Conference on Humanoid Robots,* (pp. 518 -523).

Mutlu, B., Shiwa, T., Kanda, T., Ishiguro, H., & Hagita, N. (2009). Footing in human-robot conversations: how robots might shape participant roles using gaze cues. In *HRI'09: Proceedings of the 4th ACM/IEEE International Conference on Human Robot Interaction* (pp. 61–68)

Nagai, Y. (2004). *Understanding the development of joint attention from a viewpoint of cognitive developmental robotic*. Unpublished doctoral dissertation, Osaka University, Japan.

Nagai, Y. (2005). The role of motion information in learning human-robot joint attention. In *Proceedings of the 2005 IEEE International Conference on Robotics and Automation* (ICRA), (pp. 2069–2074)

Nagai, Y., Hosoda, A., & Asada, M. (2003). A constructive model for the development of joint attention. *Connection Science*, *15*(4), 211–229. doi:10.1080/0954009 0310001655101

Naimark, M. A. (Ed.). (1964). *Linear representation of the Lorentz Group*. New York, NY: Macmillan.

Nakauchi, Y., & Simmons, R. (2002). A social robot that stands in line. *Autonomous Robots*, *12*, 313–324. doi:10.1023/A:1015273816637

Nalpantidis, L., & Gasteratos, A. (2010). Stereo vision for robotic applications in the presence of non-ideal lighting conditions. *Image and Vision Computing*, *28*(6), 940–951. doi:10.1016/j.imavis.2009.11.011

Nalpantidis, L., & Gasteratos, A. (2011). Stereovision-based fuzzy obstacle avoidance method. *International Journal of Humanoid Robotics*, *8*(1), 169–183. doi:10.1142/S0219843611002381

Nalpantidis, L., Sirakoulis, G. C., & Gasteratos, A. (2008). Review of stereo vision algorithms: from software to hardware. *International Journal of Optomechatronics*, *2*(4), 435–462. doi:10.1080/15599610802438680

National Transportation Safety Board. (2004). *Symposium Proceedings 2004*. Retrieved from http://www.ntsb.gov/ Events/symp_rec/proceedings/authors/scaman.htm

Navalpakkam, V., & Itti, L. (2005). Modeling the influence of task on attention. *Vision Research*, *45*, 205–231. doi:10.1016/j.visres.2004.07.042

Navalpakkan, V., & Itti, L. (2006). An integrated model of top-down and bottom-up attention for optimizing detection speed. In *Proceedings of the 2006 IEEE Computer Society Conference on Computer Vision and Pattern Recognition*.

Nehaniv, C. L., & Dautenhahn, K. (2002). The correspondence problem. In Dautenhahn, K., & Nehaniv, C. L. (Eds.), *Imitation in animals and artifacts* (pp. 41–61). Cambridge, MA: MIT Press.

Neilson, D., & Yang, Y. (2008). Evaluation of constructable match cost measures for stereo correspondence using cluster ranking. In *Computer Vision and Pattern Recognition* (pp. 1–8). IEEE Computer Society. doi:10.1109/ CVPR.2008.4587692

Neira, J., & Tardos, J. (2001). Data association in stochastic mapping using the joint compatibility test. *IEEE Transactions on Robotics and Automation*, *17*(6), 890–897. doi:10.1109/70.976019

Neugebauer, P. J. (1997), Geometrical cloning of 3D objects via simultaneous registration of multiple range images. *Proceedings 1997 International Conference on Shape Modeling and Applications*, (pp. 130-139).

Neumann, O., van der Hejiden, A. H. C., & Allport, A. (1986). Visual selective attention: Introductory remarks. *Psychological Research*, *48*, 185–188. doi:10.1007/ BF00309082

Newcombe, R. A., & Davison, A. J. (2010). Live dense reconstruction with a single moving camera. In *Proceedings of the IEEE Conference on Computer Vision and Pattern Recognition (CVPR)*, 2010, (pp. 1498-1505). San Francisco, California (USA), June 2010.

News, C. N. N. (2005). *Train collision near Los Angeles kills 11*. Retrieved from http://www.cnn.com/2005/ US/01/26/train.derailment/

Noreils, F. R. (1993). Toward a robot architecture integrating cooperation between mobile robots: Application to indoor environment. *The International Journal of Robotics Research*, *12*(1), 79–98. doi:10.1177/027836499301200106

Ntelidakis, A. (2010). *Using a blimp to build an interior model of the forum*. Master's thesis, University of Edinburgh, U.K.

Nuance. (2001). *Nuance: Introduction to the Nuance System*. South Yarra, Australia: Nuance Communications Inc.

OECD. (2007). *OECD demographic and labour force database*. Organisation for Economic Co-operation and Development.

Office of Law Enforcement Technology Commercialization. (2005). *License plate recognition*. Retrieved from http://www.oletc.org/oletctoday/0415_licplate. pdf#search=%22automatic%20license%20plate%20 recognition%22

O'Grady, M., Muldoon, C., Dragone, M., Tynan, R., & O'Hare, G. (2010). Towards evolutionary ambient assisted living systems. *Journal of Ambient Intelligence and Humanized Computing, 1*, 15–29. doi:10.1007/s12652-009-0003-5

Ohya, A., Kosaka, A., & Kak, A. (1998). Vision-based navigation of mobile robot with obstacle avoidance by single camera vision and ultrasonic sensing. *IEEE Transactions on Robotics and Automation, 14*(6), 969–978. doi:10.1109/70.736780

Oliensis, J. (2000). A critique of structure-from-motion algorithms. *Computer Vision and Image Understanding, 80*, 172–214. doi:10.1006/cviu.2000.0869

Ollero, A., & Merino, L. (2004). Control and perception techniques for aerial robotics. *Annual Reviews in Control, 28*(2), 167–178. doi:10.1016/j.arcontrol.2004.05.003

Olshausen, B., Anderson, C., & Essen, C. V. (1995). A multiscale dynamic routing circuit for forming size- and position-invariant object representations. *Journal of Computational Neuroscience, 2*, 45–62. doi:10.1007/BF00962707

Olson, C. R. (2001). Object-based vision and attention in primates. *Current Opinion in Neurobiology, 11*, 171–179. doi:10.1016/S0959-4388(00)00193-8

Ong, S. C., Png, S. W., Hsu, D., & Lee, W. S. (2010, July). Planning under uncertainty for robotic tasks with mixed observability. *The International Journal of Robotics Research, 29*(8), 1053–1068. doi:10.1177/0278364910369861

Orabona, F., Metta, G., & Sandini, G. (2007). A proto-object based visual attention model. *Lecture Notes in Artificial Intelligence, 4840*, 198–215.

Ortigosa, N., Morillas, S., & Peris-Fajarns, G. (2011). Obstacle-free pathway detection by means of depth maps. *Journal of Intelligent & Robotic Systems, 63*, 115–129. doi:10.1007/s10846-010-9498-4

Osher, S., & Paragios, N. (Eds.). (2003). *Geometric level set methods in imaging, vision, and graphics*. New York, NY: Springer.

Oskoei, A., Walters, M., & Dautenhahn, K. (2010). *An autonomous proxemic system for a mobile companion robot*. AISB.

Ostergard, P. (2002). A fast algorithm for the maximum clique problem. *Discrete Applied Mathematics, 120*, 197–207. doi:10.1016/S0166-218X(01)00290-6

Osuna, E., Freund, R., & Girosi, F. (1997). Training support vector machines: An application to face detection. *IEEE Computer Society Conference on Computer Vision and Pattern Recognition*, (p. 130).

Otterlo, V. (2005). *A survey of reinforcement learning in relational domains*. CTIT Technical Report, TRCTIT-05-31.

Pacchierotti, E., Christensen, H., & Jensfelt, P. (2007). Evaluation of passing distance for social robots. *The 15th IEEE International Symposium on Robot and Human Interactive Communication* (pp. 315-320). IEEE.

Pai, D. (2002). Strands: Interactive simulation of thin solids using cosserat models. *Computer Graphics Forum, 21*(3), 347–352. doi:10.1111/1467-8659.00594

Palma-Amestoy, R., Guerrero, P., Vallejos, P., & Ruiz-del-Solar, J. (2006). *Context-dependent color segmentation for Aibo robots*. In 3rd IEEE Latin American Robotics Symposium, Santiago de Chile, Chile.

Palmer, P. L., Dabis, H., & Kittler, J. (1996). Performance measure for boundary detection algorithms. *Computer Vision and Image Understanding, 63*(3), 476–494. doi:10.1006/cviu.1996.0036

Palomino, A., Marfil, R., Bandera, J., & Bandera, A. (2011). A novel biologically inspired attention mechanism for a social robot. *EURASIP Journal on Advances in Signal Processing*, (n.d), 2011.

Papadakis, N., & Caselles, V. (2010). Multi-label depth estimation for graph cuts stereo problems. *Journal of Mathematical Imaging and Vision, 38*(1), 70–82. doi:10.1007/s10851-010-0212-8

Paragios, N., & Deriche, R. (2005). Geodesic active regions and level set methods for motion estimation and tracking. *Computer Vision and Image Understanding, 97,* 259–282. doi:10.1016/j.cviu.2003.04.001

Park, H., Shiratori, T., Matthews, I., & Sheikh, Y. (2010). 3D reconstruction of a moving point from a series of 2D projections. *European Conference on Computer Vision, Vol. 6313,* (pp. 158–171).

Parker, J. R. (1994). *Practical computer vision using C.* New York, NY: Wiley.

Parker, L. E. (2003). The effect of heterogeneity in teams of 100+ mobile robots. In Schultz, A., Parker, L. E., & Schneider, F. (Eds.), *Multi-robot systems: From swarms to intelligent automata* (Vol. 2, pp. 205–215).

Park, M., Makhmalbaf, A., & Brilakis, I. (2011). Comparative study of vision tracking methods for tracking of construction site resources. *Automation in Construction, 20,* 905–915. doi:10.1016/j.autcon.2011.03.007

Pashler, H., & Badgio, P. (1985). Visual attention and stimulus identification. *Journal of Experimental Psychology, 11,* 105–121.

Patton, R. J., & Chen, J. (1993). Optimal selection of unknown input distribution matrix in the design of robust observers for fault diagnosis. *Automatica, 29,* 837–841. doi:10.1016/0005-1098(93)90089-C

Payne, J., & Joshi, S. S. (2004). *6 degree-of-freedom non-linear robotic airship model for autonomous control.* Unpublished technical report, University of California. Oakland, California, U.S.

Pelillo, M., Siddiqi, K., & Zucker, S. (1999). *Continuous-based heuristics for graph and tree isomorphism, with application to computer vision.* Conference on Approximation and Complexity in Numerical Optimization: Continuous and Discrete Problems.

Pertew, A., Marquez, H., & Zhao, Q. (2005). Design of unknown input observers for Lipschitz nonlinear systems. *Proceedings of American Control Conference,* (pp. 4198–4203).

Petrick, R., & Bacchus, F. (2004). *Extending the knowledge-based approach to planning with incomplete information and sensing.* In International Conference on Automated Planning and Scheduling (ICAPS).

Piccardi, M. (2004). Background subtraction techniques: A review. *2004 IEEE International Conference on Systems, Man and Cybernetics,* Vol. 4, (pp. 3099-3104).

Pierce, W. D., & Cheney, C. D. (2008). *Learning* (4th ed.). Hove, UK: Psychology Press.

Pineau, J., Montemerlo, M., Pollack, M., Roy, N., & Thrun, S. (2003). Towards robotic assistants in nursing homes: Challenges and results. *Robotics and Autonomous Systems. Special Issue on Socially Interactive Robots, 42*(3-4), 271–281.

Pizarro, D., & Bartoli, A. (2011). Global optimization for optimal generalized Procrustes analysis. In *IEEE Conference on Computer Vision and Pattern Recognition.* (pp. 2409-2415).

Policastro, C. A., Romero, R. A. F., Zuliani, G., & Pizzolato, E. (2009). Learning of shared attention in sociable robotics. *Journal of Algorithms, 64*(4), 139–151. doi:10.1016/j.jalgor.2009.04.005

Posner, M., Snyder, C., & Davidson, B. (1980). Attention and the detection of signals. *Journal of Experimental Psychology, 109,* 160–174.

Postma, E. O., van den Herik, H. J., & Hudson, P. T. W. (1997). Scan: A scalable model of attentional selection. *Neural Networks, 10,* 993–1015. doi:10.1016/S0893-6080(97)00034-8

Puig, D., Garcia, M. A., & Wu, L. (2011). A new global optimization strategy for coordinated multi-robot exploration: Development and comparative evaluation. *Robotics and Autonomous Systems, 59*(9), 635–653. doi:10.1016/j.robot.2011.05.004

Pulli, K. (1999). Multiview registration for large data sets. *Proceedings Second International Conference on 3-D Digital Imaging and Modeling,* (pp. 160-168).

Pulli, K., & Shapiro, L. G. (1997). Surface reconstruction and display from range and color data. *Graphical Models, 62*(3), 165–201. doi:10.1006/gmod.1999.0519

Pylyshyn, Z. (2001). Visual indexes, preconceptual objects, and situated vision. *Cognition, 80,* 127–158. doi:10.1016/S0010-0277(00)00156-6

Qin, H., & Terzopoulos, D. (1996). *D-nurbs: A physics-based framework for geometric design.* Los Alamitos, CA. USA, Tech. Rep.

Quigley, M., Conley, K., Gerkey, B., Faust, J., Foote, T., Leibs, J., et al. (2009). *ROS: An open-source robot operating system*. In ICRA Workshop on Open Source Software.

Rabiner, L. R. (1989). A tutorial on hidden Markov models and selected applications in speech recognition. *Proceedings of the IEEE*, *77*(2), 257–286. doi:10.1109/5.18626

Rajamani, R. (1998). Observers for Lipschitz nonlinear systems. *IEEE Transactions on Automatic Control*, *43*(3), 397–401. doi:10.1109/9.661604

Rajko, S., Qian, G., Ingalls, T., & James, J. (2007). Real-time gesture recognition with minimal training requirements and on-line learning. In *IEEE Conference on Computer Vision and Pattern Recognition (CVPR'07)* (pp. 1-8). IEEE Press.

Ramanan, D., Forsyth, D., & Zisserman, A. (2007). Tracking people by learning their appearance. *IEEE Transactions on Pattern Analysis and Machine Intelligence*, *29*, 65–81. doi:10.1109/TPAMI.2007.250600

Ramos, J. J. G., de Paiva, E. C., Azinheira, J. R., Bueno, S. S., Bergermana, M., Ferreira, P. A. V., & Carvalho, J. R. H. (2001). Lateral/directional control for an autonomous unmanned airship. *Aircraft Engineering and Aerospace Technology*, *73*(5), 453–458. doi:10.1108/EUM0000000005880

Reddy, V. (2012). Object tracking based on pattern matching. *International Journal of Advanced Research in Computer Science and Software Engineering, 2*.

Rehman, A., & Wang, Z. (2010). Reduced-reference SSIM estimation. *IEEE International Conference on Image Processing (ICIP)* (pp. 289-292).

Reignier, P. (1994). Fuzzy logic techniques for mobile robot obstacle avoidance. *Robotics and Autonomous Systems*, *12*(3-4), 143–153. doi:10.1016/0921-8890(94)90021-3

Remazeilles, A., Chaumette, F., & Gros, P. (2006). 3D navigation based on a visual memory. *Proceedings of Robotics and Automation*.

Ren, X., & Malik, J. (2003). Learning a classification model for segmentation. *Proceedings of IEEE International Conference on Computer Vision*, (pp. 10-17).

Richard, H., & Zisserman, A. (2003). *Multiple view geometry in computer vision*. Cambridge University Press.

Rittscher, J., Kato, J., Joga, S., & Blake, A. (2000). A probabilistic background model for tracking. *European Conference on Computer Vision*, Vol. 2, (pp. 336-350).

Rizzo, A., Parsons, T., Buckwalter, G., & Kenny, P. (2010, March 21). *A new generation of intelligent virtual patients for clinical training*. In IEEE Virtual Reality Conference. Waltham, USA.

Robins, B., Dautenhahn, K., Boekhorst, R., Billard, A., Keates, S., & Clarkson, J. (2004). *Effects of repeated exposure of a humanoid robot on children with autism*. Springer-Verlag. doi:10.1007/978-0-85729-372-5_23

Rodríguez, W., Last, M., Kandel, A., & Bunke, H. (2004). 3-dimensional curve similarity using string matching. *Robotics and Autonomous Systems*, *49*, 165–172. doi:10.1016/j.robot.2004.09.004

Roh, H., Kang, S., & Lee, S. (2000). Multiple people tracking using an appearance model based on temporal color. *International Conference on Pattern Recognition*, (pp. 643-646).

Rosenthal, S., Veloso, M., & Dey, A. (2011, August). *Learning accuracy and availability of humans who help mobile robots*. Twenty-Fifth Conference on Artificial Intelligence (AAAI), San Francisco, USA.

Rowley, H., Baluja, S., & Kanade, T. (1995). *Human face detection in visual scenes. Technical report*. Carnegie Mellon University.

Rubin, M. (2000). *Cosserat theories: Shells, rods and points*. Norwell, MA: Kluwer.

Rusinkiewicz, S., & Levoy, M. (2001). Efficient variants of the ICP algorithm. In *Proceedings of the Third International Conference on 3D Digital Imaging and Modeling*, (pp. 145-152).

Rusu, R. B., Blodow, N., & Beetz, M. (2009). Fast point feature histograms (FPFH) for 3D registration. *IEEE International Conference on Robotics and Automation, ICRA '09*, (pp. 3212-3217).

Sabe, K., Fukuchi, M., Gutmann, J.-S., Ohashi, T., Kawamoto, K., & Yoshigahara, T. (2004). Obstacle avoidance and path planning for humanoid robots using stereo vision. In *IEEE International Conference on Robotics and Automation* (Vol. 1, pp. 592–597).

Saffiotti, A., & LeBlanc, K. (2000). Active perceptual anchoring of robot behavior in a dynamic environment. *In Proceedings of the IEEE International Conference on Robotics and Automation* (pp. 3796-3802). San Francisco, USA.

Saffiotti, A., & Wesley, L. P. (1996). Perception-based self-localization using fuzzy locations. In *Proceedings of the 1st Workshop on Reasoning with Uncertainty in Robotics* (pp. 368-385). Amsterdam, NL.

Saffiotti, A. (1997). The uses of fuzzy logic in autonomous robot navigation. *Soft Computing, 1*(4), 180–197. doi:10.1007/s005000050020

Sahoo, P. K., Soltani, S., Wong, A. K. C., & Chen, Y. C. (1988). A survey of thresholding techniques. *Computer Vision Graphics and Image Processing, 41*(2), 233–260. doi:10.1016/0734-189X(88)90022-9

Saito, M., et al. (2011). Semantic object search in large-scale indoor environments. *Proceedings of IROS, 2011* San Francisco, CA, USA.

Salah, A., Morros, R., Luque, J., Segura, C., Hernando, J., & Ambekar, O. (2008). Multimodal identification and localization of users in a smart environment. *Journal on Multimodal User Interfaces, 2*, 75–91. doi:10.1007/s12193-008-0008-y

Salichs, M. A., Barber, R., Khamis, A. M., Malfaz, M., Gorostiza, J. F., & Pacheco, R. … Garcia, D. (2006). Maggie: A robotic platform for human-robot social interaction. *IEEE Conference on Robotics, Automation and Mechatronics,* (pp. 1-7).

Sandhu, R., Dambreville, S., & Tannenbaum, A. (2008). Particle filtering for registration of 2D and 3D point sets with stochastic dynamics. *IEEE Conference on Computer Vision and Pattern Recognition, CVPR 2008,* (pp. 1-8, 23-28).

Scassellati, B. (1996). *Mechanisms of shared attention for a humanoid robot* (pp. 102–106).

Scassellati, B. (1999). *Imitation and mechanisms of joint attention: A developmental structure for building social skills on a humanoid robot* (pp. 176–195). New York, NY: Springer. doi:10.1007/3-540-48834-0_11

Schaal, S. (1999). Is imitation learning the route to humanoid robots? *Trends in Cognitive Sciences, 3*(6), 233–242. doi:10.1016/S1364-6613(99)01327-3

Scharstein, D., & Szeliski, R. (2012). *Middlebury stereo evaluation - Version 2*. Retrieved April 30th, 2012, from http://vision.middlebury.edu/stereo/eval/

Scharstein, D., & Szeliski, R. (2002). A taxonomy and evaluation of dense two-frame stereo correspondence algorithms. *International Journal of Computer Vision, 47*, 7–42. doi:10.1023/A:1014573219977

Scharstein, D., & Szeliski, R. (2003). High-accuracy stereo depth maps using structured light. In *Computer Vision and Pattern Recognition* (pp. I–195–I–202). IEEE Computer Society. doi:10.1109/CVPR.2003.1211354

Scheutz, M., Schermerhorn, P., Kramer, J., & Anderson, D. (2007). First steps towards natural human-like HRI. *Autonomous Robots, 22*(4), 411–423. doi:10.1007/s10514-006-9018-3

Schilbach, L., Wilms, M., Eickhoff, S. B., Romanzetti, S., Tepest, R., & Bente, G. (2010). Minds made for sharing: Initiating joint attention recruits reward-related neurocircuitry. *Journal of Cognitive Neuroscience, 22*(12), 2702–2715. doi:10.1162/jocn.2009.21401

Scholl, B. (2001). Objects and attention: the state of the art. *Cognition, 80*, 1–46. doi:10.1016/S0010-0277(00)00152-9

Schöner, G., Dose, M., & Engels, C. (1995). Dynamics of behavior: Theory and applications for autonomous robot architectures. *Robotics and Autonomous Systems, 16*(2-4), 213–245. doi:10.1016/0921-8890(95)00049-6

Schreer, O. (1998). Stereo vision-based navigation in unknown indoor environment. In *5th European Conference on Computer Vision* (Vol. 1, pp. 203–217).

Sclaroff, S., & Isidoro, J. (1998). Active blobs. *IEEE International Conference on Computer Vision,* (pp. 1146-1153).

Sclaroff, S., & Isidoro, J. (2003). Active blobs: Region-based, deformable appearance models. *Computer Vision and Image Understanding, 89*(2/3), 197–225. doi:10.1016/S1077-3142(03)00003-1

Senior, A. (2002). Tracking with probabilistic appearance models. *ECCV Workshop on Performance Evaluation of Tracking and Surveillance Systems*, (pp. 48-55).

Senior, A., Hampapur, A., Tian, Y., Brown, L., Pankanti, S., & Bolle, R. (2006). Appearance models for occlusion handling. *Image and Vision Computing, 24*(11). doi:10.1016/j.imavis.2005.06.007

Serre, T., Wolf, L., Bileschi, S., Riesenhuber, M., & Poggio, T. (2007, March). Robust object recognition with cortex-like mechanisms. *IEEE Transactions on Pattern Analysis and Machine Intelligence, 29*(3). doi:10.1109/TPAMI.2007.56

Shafer, S. A. (1984, April). Using color to separate reflection components. *Color Research and Application, 10*, 43–51.

Shan, Y., Sawhney, H., Matei, B., & Kumar, R. (2006). Shapeme histogram projection and matching for partial object recognition. *IEEE Transactions on Pattern Analysis and Machine Intelligence, 28*, 568–577. doi:10.1109/TPAMI.2006.83

Sharma, R., & Sasi, S. (2007). *Vision-based monitoring system for detecting red signal crossing. Innovations and Advanced Techniques in Computer and Information Sciences and Engineering* (pp. 29–33). Springer.

Sheikh, Y., Javed, O., & Kanade, T. (2009). Background subtraction for freely moving cameras. *IEEE 12th International Conference on Computer Vision* (pp. 1219-1225).

Shen, H., Zhang, H., Shao, S., & Xin, J. H. (2008, August). Chromaticity-based separation of reflection components in a single image. *Pattern Recognition, 41*, 2461–2469. doi:10.1016/j.patcog.2008.01.026

Shi, J., & Malik, J. (2000). Normalized cuts and image segmentation. *IEEE Transactions on Pattern Analysis and Machine Intelligence, 22*, 888–905. doi:10.1109/34.868688

Shi, J., & Tomasi, C. (1994). Good features to track. *IEEE Computer Vision and Pattern Recognition, 1*, 593–600.

Shin, H. J., Lee, J., Shin, S. Y., & Gleicher, M. (2001). Computer pupettry: An importance based approach. *ACM Transactions on Graphics, 20*(2), 67–94. doi:10.1145/502122.502123

Shneier, M. O., Shackleford, W. P., Hong, T. H., & Chang, T. Y. (2006). Performance evaluation of a terrain traversability learning algorithm in the darpa lagr program performance evaluation of a terrain traversability learning algorithm in the DARPA LAGR program. In *Performance Metrics for Intelligent Systems Workshop* (pp. 103–110). Gaithersburg, MD, USA.

Shoemake, K. (1991). *Quaternions*. Tech Report. Retrieved from http://campar.in.tum.de/twiki/pub/Chair/DwarfTutorial/quatut.pdf

Shoemake, K. (1992). Uniform random rotations. In Kirk, D. (Ed.), *Graphics Gems III* (pp. 124–132). San Francisco, CA: Morgan Kaufmann.

Shon, A. P., Grimes, D. B., Baker, C. L., Hoffman, M. W., Zhou, S., & Rao, R. P. N. (2005). Probabilistic gaze imitation and saliency learning in a robotic head. *Proceedings of the 2005 IEEE International Conference on Robotics and Automation (ICRA 2005)*, (pp. 2865-2870)

Siino, R. M., & Hinds, P. (2004). *Making sense of new technology as a lead-in to structuring: The case of an autonomous mobile robot* (pp. E1–E6). Academy of Management Proceedings.

Silva, R. R., Policastro, C. A., Zuliani, G., Pizzolato, E., & Romero, R. A. F. (2008). Concept learning by human tutelage for social robots. *Learning and Nonlinear Models, 6*, 44–67.

Silva, R. R., & Romero, R. A. F. (2012). Modelling shared attention through relational reinforcement learning. *Journal of Intelligent & Robotic Systems, 66*(1), 167–182. doi:10.1007/s10846-011-9624-y

Simon, D. (1996). *Fast and accurate shape-based registration*. Ph.D. Dissertation, tech. Report CMU-RI-TR-96-45, Robotics Institute, Carnegie Mellon University.

Singh, S., Simmons, R., Smith, T., Stentz, A., Verma, I., Yahja, A., et al. (2000). Recent progress in local and global traversability for planetary rovers. In *IEEE International Conference on Robotics and Automation* (Vol. 2, pp. 1194–1200).

Sisbot, E. A., Marin-Urias, L. F., Broquère, X., Sidobre, D., & Alami, R. (2010). Synthesizing robot motions adapted to human presence. *International Journal of Social Robotics, 2*, 329–343. doi:10.1007/s12369-010-0059-6

Smith, R., & Cheesman, P. (1987). On the representation and estimation of spatial uncertainty. *The International Journal of Robotics Research, 5*(4), 56–68. doi:10.1177/027836498600500404

Smith, S. M., & Brady, J. M. (1997). SUSAN - A new approach to low level image processing. *International Journal of Computer Vision, 23*(1), 45–78. doi:10.1023/A:1007963824710

Smyth, M. M., & Pendleton, L. R. (1990). Space and movement in working memory. *The Quarterly Journal of Experimental Psychology Section A, 42,* 291–304. doi:10.1080/14640749008401223

Soatto, S., & Perona, P. (1998). Reducing "structure from motion": A general framework for dynamic vision, part 1: Modeling. *IEEE Transactions on Pattern Analysis and Machine Intelligence, 20*(9), 933–942. doi:10.1109/34.713360

Sola, J., Monin, A., Devy, M., & Lemaire, T. (2005). *Undelayed initialization in bearing only slam.* IEEE International Conference on Intelligent Robots and Systems.

Solanas, A., & Garcia, M. A. (2004). Coordinated multi-robot exploration through unsupervised clustering of unknown space. In *Proceedings of IEEE/RSJ International Conference on Intelligent Robots and Systems,* (pp. 717-721).

Solis, A., Nayak, A., Stojmenovic, M., & Zaguia, N. (2009). Robust line extraction based on repeated segment directions on image contours. *Proceedings of the IEEE Symposium on CISDA.*

Sonka, M., Hlavac, V., & Boyle, R. (1996). *Image processing analysis and machine vision.* International Thomson Computer Press.

Soquet, N., Aubert, D., & Hautiere, N. (2007). Road segmentation supervised by an extended V-disparity algorithm for autonomous navigation. In *IEEE Intelligent Vehicles Symposium* (pp. 160–165). Istanbul, Turkey.

Sridharan, M. (2011, December 18-21). *Augmented reinforcement learning for interaction with non-expert humans in agent domains.* In International Conference on Machine Learning Applications (ICMLA), Honolulu, Hawaii.

Sridharan, M., & Li, X. (2009). *Learning sensor models for autonomous information fusion on a humanoid robot.* In IEEE-RAS International Conference on Humanoid Robots. Paris, France.

Sridharan, M., & Stone, P. (2007). *Global action selection for illumination invariant color modeling.* In International Conference on Intelligent Robots and Systems (IROS). San Diego, USA.

Sridharan, M., Wyatt, J., & Dearden, R. (2008, September). *HiPPo: Hierarchical POMDPs for planning information processing and sensing actions on a robot.* In International Conference on Automated Planning and Scheduling (ICAPS). Sydney, Australia.

Sridharan, M., & Stone, P. (2009). Color learning and illumination invariance on mobile robots: A survey. *Robotics and Autonomous Systems, 57*(6-7), 629–644. doi:10.1016/j.robot.2009.01.004

Sridharan, M., Wyatt, J., & Dearden, R. (2010). Planning to see: A hierarchical approach to planning visual actions on a robot using POMDPs. *Artificial Intelligence, 174,* 704–725. doi:10.1016/j.artint.2010.04.022

Srinivasan, V., Thurrowgood, S., & Soccol, D. (2006). An optical system for guidance of terrain following in UAVs. *Proceedings of the IEEE International Conference on Video and Signal Based Surveillance* (AVSS), (pp. 51–56).

Stachniss, C. (2009). *Robotic mapping and exploration,* Vol. 55. STAR Springer tracts in advanced robotics. Springer.

Staudte, M., & Crocker, M. W. (2009). Visual attention in spoken human-robot interaction. In *HRI '09: Proceedings of the 4th ACM/IEEE International Conference on Human Robot Interaction,* (pp. 77–84).

Stauffer, C., & Grimson, W. (2000). Learning patterns of activity using real time tracking. *IEEE Transactions on Pattern Analysis and Machine Intelligence, 22,* 747–767. doi:10.1109/34.868677

Steg, H., Strese, H., Loroff, C., Hull, J., & Schmidt, S. (2006). *Europe is facing a demographic challenge Ambient Assisted Living offers solutions.*

Stone, P., Sutton, R., & Kuhlmann, G. (2005). Reinforcement learning for RoboCup soccer keepaway. *Adaptive Behavior, 13*, 165–188. doi:10.1177/105971230501300301

Strecha, C., von Hansen, W., Van Gool, L., Fua, P., & Thoennessen, U. (2008). On benchmarking camera calibration and multi-view stereo for high resolution imagery. In *Proceedings Conference on Computer Vision and Pattern Recognition,* (pp. 1-8). IEEE Computer Society.

Sturm, P., & Triggs, B. (1996). A factorization based algorithm for multi-image projective structure and motion. *Lecture Notes in Computer Science, 1065*, 709–720. doi:10.1007/3-540-61123-1_183

Sumioka, H., Yoshikawa, Y., & Asada, M. (2010). Reproducing interaction contingency toward open-ended development of social actions: Case study on joint attention. *IEEE Transactions in Autonomous Mental Development, 2*(1), 40–50. doi:10.1109/TAMD.2010.2042167

Sun, S., Haynor, D., & Kim, Y. (2003). Semiautomatic video object segmentation using V Snakes. *IEEE Transactions on Circuits and Systems for Video Technology, 13*, 75–82. doi:10.1109/TCSVT.2002.808089

Sun, Y., & Fisher, R. (2003). Object-based visual attention for computer vision. *Artificial Intelligence, 146*, 77–123. doi:10.1016/S0004-3702(02)00399-5

Sutton, R., & Barto, A. (1998). *Reinforcement learning: An introduction.* Cambridge, MA: MIT Press.

Swaminathan, R., & Sridharan, M. (2011, December). *Towards Natural human-robot interaction using multimodal cues.* Technical Report, Department of Computer Science, Texas Tech University.

Syrdal, D. S., Lee Koay, K., & Walters, M. L. (2007). A personalized robot companion? - The role of individual differences on spatial preferences in HRI scenarios. *RO-MAN 2007 - The 16th IEEE International Symposium on Robot and Human Interactive Communication* (pp. 1143-1148). IEEE.

Szeliski, R. (1999). Prediction error as a quality metric for motion and stereo. In *International Conference on Computer Vision*, Vol. 2, (pp. 781–788). IEEE Computer Society.

Szeliski, R., & Zabih, R. (2000). An experimental comparison of stereo algorithms. In *Proceedings of the International Workshop on Vision Algorithms,* (pp. 1–19). Springer-Verlag.

Takaya, T., Kawamura, H., Minagawa, Y., Yamamoto, M., & Ohuchi, A. (2006). PID landing orbit motion controller for an indoor blimp robot. *Artificial Life and Robotics, 10*(2), 177–184. doi:10.1007/s10015-006-0385-9

Talamadupula, K., Benton, J., Kambhampati, S., Schermerhorn, P., & Scheutz, M. (2010). Planning for human-robot teaming in open worlds. *ACM Transactions on Intelligent Systems and Technology, 1*(2), 14:1-14:24.

Tan, T., Nishino, K., & Ikeuchi, K. (2003, 18-20 June). Illumination chromaticity estimation using inverse-intensity chromaticity space. In *Proceedings 2003 IEEE Computer Society Conference on Computer Vision and Pattern Recognition,* (Vol. 1, pp. 673-680).

Tan, P.-N., Steinbach, M., & Kumar, V. (2005). *Introduction to data mining.* Boston, MA: Addison-Wesley Longman.

Tan, R. T., Nishino, K., & Ikeuchi, K. (2004, Oct). Separating reflection components based on chromaticity and noise analysis. *IEEE Transactions on Pattern Analysis and Machine Intelligence, 26*(10), 1373–1379. doi:10.1109/TPAMI.2004.90

Tapus, A., Mataric, M., & Scassellati, B. (2007, March). The grand challenges in socially assistive robotics. *Robotics and Automation Magazine. Special Issue on Grand Challenges in Robotics, 14*(1), 35–42.

Tarabalka, Y., Chanussot, J., & Benediktsson, J. A. (2010). Segmentation and classification of hyperspectral images using watershed transformation. *Pattern Recognition, 43*(7), 2367–2379. doi:10.1016/j.patcog.2010.01.016

Tatler, B. W., Baddeley, R. J., & Gilchrist, I. D. M. (2005). Visual correlates of fixation selection: Effects of scale and time. *Vision Research, 45*, 643–659. doi:10.1016/j.visres.2004.09.017

Theetten, A., Grisoni, L., Andriot, C., & Barsky, B. (2008). Geometrically exact dynamic splines. *Computer Aided Design, 40*(1), 35–48. doi:10.1016/j.cad.2007.05.008

Theocharous, G., Murphy, K., & Kaelbling, L. P. (2004). Representing hierarchical POMDPs as DBNs for multi-scale robot localization. In *International Conference on Robotics and Automation (ICRA)*.

Theodoridis, S., & Koutroumbas, K. (Eds.). (2009). *Pattern recognition*. Burlington, MA: Academic Press, Elsevier.

Thrun, S., & Bucken, A. (1996). Integrating grid-based and topological maps for mobile robot navigation. *Proceedings of the Thirteenth National Conference on Artificial Intelligence*, Portland, Oregon.

Thrun, S., & Bucken, A. (1997). Learning maps for indoor mobile robot navigation. *Technical report*, 1-60.

Thrun, S. (2004). Toward a framework for human-robot interaction. *Human-Computer Interaction, 19*(1), 9–24. doi:10.1207/s15327051hci1901&2_2

Thrun, S. (2006). Stanley: The robot that won the DARPA grand challenge. *Field Robotics, 23*(9), 661–692. doi:10.1002/rob.20147

Thrun, S., Burgard, W., & Fox, D. (2006). *Probabilistic robotics*. Cambridge, MA: The MIT Press.

Thrun, S., Montemerlo, M., Dahlkamp, H., Stavens, D., Aron, A., & Diebel, J. (2006, September). Stanley: The robot that won the DARPA grand challenge: Research articles. *Journal of Robotic Systems, 23*, 661–692.

Tistarelli, M., & Sandini, G. (1991). Direct estimation of time-to-impact from optical flow. *IEEE Workshop on Visual Motion*, (pp. 52–60).

Tombari, F., Di Stefano, L., Mattoccia, S., & Mainetti, A. (2010). A 3D reconstruction system based on improved spacetime stereo. In *Proceedings of International Conference on Control, Automation, Robotics and Vision*, (pp. 1886-1896). IEEE Computer Society.

Tombari, F., Mattoccia, S., & Di Stefano, L. (2010). Stereo for robots: Quantitative evaluation of efficient and low-memory dense stereo algorithms. In *Proceedings of International Conference Control Automation Robotics and Vision*, (pp. 1231–1238). IEEE Computer Society.

Tombari, F., Mattoccia, S., Di Stefano, L., & Addimanda, E. (2008). Classification and evaluation of cost aggregation methods for stereo correspondence. In *Proceedings Conference on Computer Vision and Pattern Recognition*, (pp. 1-8). IEEE Computer Society.

Toro, J. (2008). Dichromatic illumination estimation without pre-segmentation. *Pattern Recognition Letters, 29*, 871–877. doi:10.1016/j.patrec.2008.01.004

Torralba, A. (2003). Modeling global scene factors in attention. *Journal of the Optical Society of America, 20*(7), 1407–1418. doi:10.1364/JOSAA.20.001407

Torralba, A., Oliva, A., Castellanos, M. S., & Henderson, J. M. (2006). Contextual guidance of eyes movements and attention in real-world scenes: The role of the global features in object research. *Psychological Review, 113*, 766–786. doi:10.1037/0033-295X.113.4.766

Torra, P. H. S., & Criminisi, A. (2004). Dense stereo using pivoted dynamic programming. *Image and Vision Computing, 22*(10), 795–806. doi:10.1016/j.imavis.2004.02.012

Torta, E., Cuijpers, R., Juola, J., & van der Pol, D. (2011). Design of robust robotic proxemic behaviour. In Mutlu, B. A., Ham, J., Evers, V., & Kanda, T. (Eds.), *Social Robotics* (*Vol. 7072*, pp. 21–30). Lecture Notes in Computer Science Berlin, Germany: Springer. doi:10.1007/978-3-642-25504-5_3

Toussaint, M., Charlin, L., & Poupart, P. (2008). Hierarchical POMDP controller optimization by likelihood maximization. In *Proceedings of the 24th Annual Conference on Uncertainty in AI (UAI)*.

Train Accident Report. (2004). Retrieved from http://www.visualexpert.com/Resoures/trainaccidents.html

Trax. (2004). Retrieved from http://www.avtangeltrax.com/digital.htm

Treisman, A., & Gelade, G. (1980). A feature-integration theory of attention. *Cognitive Psychology, 12*, 97–136. doi:10.1016/0010-0285(80)90005-5

Triesch, J., & Malsburg, C. (2001). Democratic integration: Self-organized integration of adaptive cues. *Neural Computation, 13*(9), 2049–2074. doi:10.1162/089976601750399308

Triesch, J., Teuscher, C., Deak, G. O., & Carlson, E. (2006). Gaze following: Why (not) learn it? *Developmental Science*, *9*(2), 125–147. doi:10.1111/j.1467-7687.2006.00470.x

Tsotsos, J. (1987). A complexity level analysis of vision. In *Proceedings International Conference on Computer Vision: Human and Machine Vision Workshop,* London, England.

Tsotsos, J. (1997). Limited capacity of any realizable perceptual system is a sufficient reason for attentive behavior. *Consciousness and Cognition*, *6*, 429–436. doi:10.1006/ccog.1997.0302

Tsotsos, J. (2006). Cognitive vision needs attention to link sensing with recognition. In Christensen, H. I., & Nagel, H. (Eds.), *Cognitive vision systems* (pp. 25–35). Berlin, Germany: Springer. doi:10.1007/11414353_3

Tsotsos, J. K. (1988). A complexity level analysis of immediate vision. *International Journal of Computer Vision*, *2*(1), 303–320. doi:10.1007/BF00133569

Tsotsos, J. K. (1990). Analyzing vision at the complexity level. *The Behavioral and Brain Sciences*, *13*(3), 423–469. doi:10.1017/S0140525X00079577

Tsotsos, J. K. (1992). On the relative complexity of passive vs active visual search. *International Journal of Computer Vision*, *7*(2), 127–141. doi:10.1007/BF00128132

Tsotsos, J., Culhane, S., Wai, W., Lai, Y., Davis, N., & Nuflo, F. (1995). Modeling visual attention via selective tuning. *Artificial Intelligence*, *78*, 507–545. doi:10.1016/0004-3702(95)00025-9

Tsotsos, J., Liu, Y., Martinez-Trujillo, J. C., Pomplun, M., Simine, E., & Zhou, K. (2005). Attending to visual motion. *Computer Vision and Image Understanding*, *100*, 3–40. doi:10.1016/j.cviu.2004.10.011

Tsui, C.-C. (1996). A new design approach to unknown input observers. *IEEE Transactions on Automatic Control*, *41*(3), 464–468. doi:10.1109/9.486653

Turk, G., & Levoy, M. (1994). Zippered polygon meshes from range images. In *Proceedings of the 21st annual conference on Computer graphics and interactive techniques* (SIGGRAPH '94) (pp. 311-318). New York, NY: ACM.

Uchizawa, K., Douglas, R., & Maass, W. (2006). On the computational power of threshold circuits with sparse activity. *Neural Computation*, *28*, 2994–3008. doi:10.1162/neco.2006.18.12.2994

Uhr, L. (1972). Layered recognition cone networks that preprocess, classify and describe. *IEEE Transactions on Computers*, *21*, 758–768. doi:10.1109/T-C.1972.223579

Ullman, S., & Basri, R. (1991). Recognition by linear combinations of models. *IEEE Transactions on Pattern Analysis and Machine Intelligence*, *13*(10), 992–1006. doi:10.1109/34.99234

University of Florida website. (2007). *Neural network training.* Retrieved from http://www.math.ufl.edu/help/matlab/ReferenceTOC.html

van de Weijer, J., & Gevers, T. (2004). Robust optical flow from photometric invariants In *ICIP: 2004 International Conference on Image Processing,* Vols. 1- 5 (pp. 1835-1838).

Van der Mark, W., & Gavrila, D. (2006). Real-time dense stereo for intelligent vehicles. *IEEE Transactions on Intelligent Transportation Systems*, *7*(1), 38–50. doi:10.1109/TITS.2006.869625

van der Pol, D., Cuijpers, R., & Juola, J. (2011). Head pose estimation for a domestic robot. *The 6th International Conference on Human-Robot Interaction, HRI '11*, (pp. 277-278).

van der Spiegel, J., Kreider, G., Claeys, C., Debusschere, I., Sandini, G., Dario, P., … Soncini, G. (1989). *A foveated retina like sensor using ccd technology.* Kluwe.

Van der Zwaan, S., Bernardino, A., & Santos-Victor, J. (2000). Vision based station keeping and docking for an aerial blimp. In *Proceedings of the International Conference on Intelligent Robots and Systems,* Vol. 1, (pp. 614-619). Takamatsu, Japan.

Van Zwynsvoorde, D., Simeon, T., & Alami, R. (2001). Building topological models for navigation in large scale environments. *Proceedings of ICRA*, *4*, 4256–4261.

Vandapel, N., Huber, D., Kapuria, A., & Hebert, M. (2004). Natural terrain classification using 3-D ladar data. In *IEEE International Conference on Robotics and Automation* (Vol. 5, pp. 5117–5122).

Vandorpe, J., Van Brussel, H., & Xu, H. (1996). Exact dynamic map building for a mobile robot using geometrical primitives produced by a 2d range finder. In *IEEE International Conference on Robotics and Automation* (pp. 901–908). Minneapolis, USA.

Vapnik, V. N. (1995). *The nature of statistical learning theory*. New York, NY: Springer.

Velastin, S. A., Boghossian, B. A., Ping, B., Lo, L., Sun, J., & Vicencio-Silva, M. A. (2005). PRISMATICA: Toward ambient intelligence in public transport environments. In *Good Practice for the Management and Operation of Town Centre CCTV, European Conference on Security and Detection,* Vol. 35 (pp. 164-182)

VenuGopal, T., & Naik, P. P. S. (2011). Image segmentation and comparative analysis of edge detection algorithms. *International Journal of Electrical. Electronics & Computing Technology*, *1*(3), 38–42.

Vernon, D. (2008). Cognitive vision: The case of embodied perception. *Image and Vision Computing*, *26*, 1–4. doi:10.1016/j.imavis.2007.09.003

Vidal, R., Ma, Y., Soatto, S., & Sastry, S. (2006). Two-view multibody structure from motion. *International Journal of Computer Vision*, *68*(1), 7–25. doi:10.1007/s11263-005-4839-7

Viejo, D., & Cazorla, M. (2007). 3D plane-based ego-motion for slam on semi-structured environment. *IEEE/RSJ International Conference on Intelligent Robots and Systems, IROS 2007,* (pp. 2761-2766).

Viejo, D., & Cazorla, M. (2008). *3D model based map building*. International Symposium on Robotics, ISR 2008, Seoul, Korea.

Vieville, T., Clergue, E., Enriso, R., & Mathieu, H. (1995). Experimenting with 3-d vision on a robotic head. *Robotics and Autonomous Systems*, *14*, 1–27. doi:10.1016/0921-8890(94)00019-X

Vig, L., & Adams, J. A. (2005). A framework for multi-robot coalition formation. *Proceedings of the 2nd Indian International Conference on Artificial Intelligence.*

Viola, P., & Jones, M. (2002). Robust real-time face detection. *International Journal of Computer Vision*, *57*(2), 137–154. doi:10.1023/B:VISI.0000013087.49260.fb

Voit, M., Nickel, K., & Stiefelhagen, R. (2007). Neural network-based head pose estimation and multi-view fusion. In Stiefelhagen, R., & Garofolo, J. (Eds.), *Multimodal Technologies for Perception of Humans* (*Vol. 4122*, pp. 291–298). Lecture Notes in Computer Science. doi:10.1007/978-3-540-69568-4_26

Voit, M., Nickel, K., & Stiefelhagen, R. (2008). Head pose estimation in single- and multi-view environments - results on the clear'07 benchmarks. In Stiefelhagen, R., Bowers, R., & Fiscus, J. (Eds.), *Multimodal Technologies for Perception of Humans* (*Vol. 4625*, pp. 307–316). Lecture Notes in Computer Science. doi:10.1007/978-3-540-68585-2_29

Wagner, A., & Arkin, R. (2008). Analyzing social situations for human-robot interaction. *Interaction Studies: Social Behaviour and Communication in Biological and Artificial Systems*, *9*(2), 277–300. doi:10.1075/is.9.2.07wag

Waldherr, S., Thrun, S., & Romero, R. (2000). A gesture-based interface for human-robot interaction. *Autonomous Robots*, *9*, 151–173. doi:10.1023/A:1008918401478

Walters, M., Dautenhahn, K., Boekhorst, R., Koay, K., Syrdal, D., & Nehaniv, C. (2009). An empirical framework for human-robot proxemics. In Dautenhahn, K. (Ed.), *New frontiers in human-robot interaction* (pp. 144–149).

Walther, D. (2006). *Interactions of visual attention and object recognition: Computational modeling, algorithms, and psychophysics*. PhD thesis, California Institute of Technology, 2006.

Walther, D., & Koch, C. (2006). Modeling attention to salient proto-objects. *Neural Networks*, *19*, 1395–1407. doi:10.1016/j.neunet.2006.10.001

Wang, Z., & Bovik, A. C. (2009). Mean squared error: Love it or leave it? A new look at signal fidelity measures. *Signal Processing Magazine*, *26*(1), 98–117. doi:10.1109/MSP.2008.930649

Wasik, Z., & Safiotti, A. (2002). Robust color segmentation for the RoboCup domain. *Proceedings of the International Conference on Pattern Recognition*, Quebec City, Quebec, CA.

Watson. (n.d.). *WATSON: A real-time head tracking and gesture recognition*. Retrieved from http://projects.ict.usc.edu/vision/watson/

Weber, C., & Wermter, S. (2007). A self-organizing map of sigma-pi units. *Neurocomputing, 70*(13-15), 2552–2560. doi:10.1016/j.neucom.2006.05.014

Weik, S. (1997). Registration of 3-D partial surface models using luminance and depth information. *Proceedings of the International Conference on Recent Advances in 3-D Digital Imaging and Modeling,* (pp. 93-100).

West, G., Newman, C., & Greenhill, S. (2005). Using a camera to implement virtual sensors in a smart house. *From Smart Homes to Smart Care: International Conference on Smart Homes and Health Telematics,* Vol. 15, (pp. 83-90).

Westelius, C. J. (1995). *Focus of attention and gaze control for robot vision.* PhD thesis, Department of Electrical Engineering, Linköping University, Sweden.

Whalen, C., & Schreibman, L. (2003). Joint attention training for children with autism using behavior modification procedures. *Journal of Child Psychology and Psychiatry, and Allied Disciplines, 44*(3). doi:10.1111/1469-7610.00135

Wikipedia. (2005). *Automatic number plate recognition.* Retrieved from http://en.wikipedia.org/wiki/Automatic_number_plate_recognition

Wikipedia. (2007). *Interlacing.* Retrieved from http://en.wikipedia.org/wiki/Interlacing

Wimmer, D., Bildstein, M., Well, K. H., Schlenker, M., Kungl, P., & Kröplin, B.-H. (2002, October) *Research airship "lotte" development and operation of controllers for autonomous flight phases.* Paper presented at the International Conference on Intelligent Robots and Systems. Lausanne, Switzerland.

Wittenbauma, G., Stasser, G., & Merry, C. (1996). Tacit coordination in anticipation of small group task completion. *Journal of Experimental Social Psychology, 32*(2), 129–152. doi:10.1006/jesp.1996.0006

Wolfe, J. M. (1994). Guided search 2.0: A revised model of visual search. *Psychonomic Bulletin & Review, 1,* 202–238. doi:10.3758/BF03200774

Wolfe, J. M. (2007). *Guided search 4.0: Current progress with a model of visual search.* Brigham and Womens Hospital and Harvard Medical School.

Wolfe, J. M. (2007). Guided search 4.0: Current progress with a model of visual search. In Gray, W. (Ed.), *Integrated models of cognitive system* (pp. 99–119). doi:10.1093/acprof:oso/9780195189193.003.0008

Wolfe, J. M., Cave, K. R., & Franzel, S. L. (1989). Guided search: An alternative to the feature integration model for visual search. *Journal of Experimental Psychology. Human Perception and Performance, 15,* 419–433. doi:10.1037/0096-1523.15.3.419

Wren, C., Azarbayejani, A., & Pentland, A. (1997). Pfinder: Real time tracking of the human body. *IEEE Transactions on Pattern Analysis and Machine Intelligence, 19*(7), 780–785. doi:10.1109/34.598236

Wright, R. D., & Ward, L. M. (2008). *Orienting of attention.* Oxford University Press.

Wurm, K. M., Stachniss, C., & Burgard, W. (2006). Coordinated multi-robot exploration using a segmentation of the environment. *IEEE/RSJ International Conference on Intelligent Robots and Systems* (pp. 1160-1165).

Xie, L., & Khargonekar, P. P. (2010). Lyapunov-based adaptive state estimation for a class of nonlinear stochastic systems. *Proceedings of the American Controls Conference,* (pp. 6071–6076). Baltimore, MD.

Xie, S.-R., Luo, J., Rao, J.-J., & Gong, Z.-B. (2007). Computer vision-based navigation and predefined track following control of a small robotic airship. *Acta Automatica Sinica, 33*(3), 286–291. doi:10.1360/aas-007-0286

Xin, W., & Pu, J. (2010). An improved ICP algorithm for point cloud registration. *2010 International Conference on Computational and Information Sciences (ICCIS),* (pp. 565-568).

Yamaoka, F., Kanda, T., Ishiguro, H., & Hagita, N. (2010). A model of proximity control for information-presenting robots. *IEEE Transactions on Robotics, 26*(1), 187–195. doi:10.1109/TRO.2009.2035747

Yamazaki, A., Yamazaki, K., Kuno, Y., Burdelski, M., Kawashima, M., & Kuzuoka, H. (2008). Precision timing in human-robot interaction: Coordination of head movement and utterance. In *CHI '08: Proceeding of the Twenty-Sixth Annual SIGCHI Conference on Human Factors in Computing Systems,* (pp. 131–140).

Yanco, H. A., Drury, J. L., & Scholtz, J. (2004). Beyond usability evaluation: Analysis of human-robot interaction at a major robotics competition. *Human-Computer Interaction*, *19*(1), 117–149. doi:10.1207/s15327051hci1901&2_6

Yang, Q., Wang, L., Yang, R., Stewénius, H., & Nister, D. (2008). Stereo matching with color-weigthed correlation, hierarchical belief propagation and occlusion handling. In *Transactions on Pattern Analysis and machine. Intelligence*, *31*(3), 492–504.

Yan, T. (2009). Automatic facial landmark labeling with minimal supervision. *Computer Vision and Pattern Recognition, CVPR, 2009*, 2097–2104.

Yan, W., Weber, C., & Wermter, S. (2011). A hybrid probabilistic neural model for person tracking based on a ceiling-mounted camera. *Journal of Ambient Intelligence and Smart Environments*, *3*(3), 237–252.

Yarbus, A. L. (1969). *Eye movements and vision*. New York, NY: Plenum.

Yaz, E., & Azemi, A. (1993). Observer design for discrete and continuous nonlinear stochastic systems. *International Journal of Systems Science*, *24*(12), 2289–2302. doi:10.1080/00207729308949629

Yelal, M. R., Sasi, S., Shaffer, G. R., & Kumar, A. K. (2006). *Color-based signal light tracking in real-time video*. IEEE International Conference on Advanced Video and Signal Based Surveillance, November 22-24, 2006, Sydney, Australia. DOI: 10.1109/AVSS.2006.34

Yershova, A., Jain, S., LaValle, S. M., & Mitchell, J. C. (2010). Generating uniform incremental grids on SO (3) using the Hopf fibration. *The International Journal of Robotics Research*, *29*(7), 801–812. doi:10.1177/0278364909352700

Yilmaz, A., Li, X., & Shah, M. (2004). Contour-based object tracking with occlusion handling in video acquired using mobile cameras. *IEEE Transactions on Pattern Analysis and Machine Intelligence*, *26*(11). doi:10.1109/TPAMI.2004.96

Yoon, K. J., Chofi, Y. J., & Kweon, I. S. (2005, 11-14 Sept.). Dichromatic-based color constancy using dichromatic slope and dichromatic line space. In *IEEE International Conference on Image Processing, ICIP 2005* (Vol. 3, pp. 960-3).

Yoon, K., & Kweon, I. (2007). Stereo matching with the distinctive similarity measure. *Proceedings of International Conference on Computer Vision*, (pp. 1-7). IEEE Computer Society.

Yoon, K.-J., & Kweon, I. S. (2006). Adaptive support-weight approach for correspondence search. *IEEE Transactions on Pattern Analysis and Machine Intelligence*, *28*(4), 650–656. doi:10.1109/TPAMI.2006.70

Yoon, S., Park, S.-K., Kang, S., & Kwak, Y. K. (2005). Fast correlation-based stereo matching with the reduction of systematic errors. *Pattern Recognition Letters*, *26*(14), 2221–2231. doi:10.1016/j.patrec.2005.03.037

Yu, C., Scheutz, M., & Schermerhorn, P. (2010). Investigating multimodal real-time patterns of joint attention in an HRI word learning task. In *HRI '10: Proceeding of the 5th ACM/IEEE International Conference on Human–Robot interaction*, (pp. 309–316).

Yu, Y., Mann, G., & Gosine, R. (2010). An Object-based visual attention model for robotic applications. *IEEE Transactions on Systems, Man, and Cybernetics, B 40*, 1–15.

Yuan, C., & Medioni, G. (2006). 3D reconstruction of background and objects moving on ground plane viewed from a moving camera. *Computer Vision and Pattern Recognition*, *2*, 2261–2268.

Yücel, Z., Salah, A., Merigli, C., & Mericli, T. (2009). Joint visual attention modeling for naturally interacting robotic agents. *24th International Symposium on Computer and Information Sciences*, (pp. 242-247).

Zach, C., Karner, K., & Bischof, H. (2004). Hierarchical disparity estimation with programmable 3D hardware. In *International Conference in Central Europe on Computer Graphics, Visualization and Computer Vision* (pp. 275–282).

Zadeh, L. A. (1978). Fuzzy sets as a basis for a theory of possibility. *Fuzzy Sets and Systems*, *1*, 3–28. doi:10.1016/0165-0114(78)90029-5

Zaharescu, A., Rothenstein, A. L., & Tsotsos, J. K. (2005). Towards a biologically plausible active visual search model. *Proceedings of International Workshop on Attention and Performance in Computational Vision WAPCV-2004*, (pp. 133-147).

Zang, P., Irani, A., Zhou, P., Isbell, C., & Thomaz, A. (2010). Using training regimens to teach expanding function approximators. In *International Joint Conference on Autonomous Agents and Multiagent Systems (AAMAS)* (pp. 341-348).

Zangenmeister, W. (1996). *Designing an anthropomorphic head-eye system.*

Zhang, L., Choi, S.-I., & Park, S.-Y. (2011). Robust ICP registration using biunique correspondence. *2011 International Conference on 3D Imaging, Modeling, Processing, Visualization and Transmission (3DIMPVT),* (pp. 80-85).

Zhang, S., & Sridharan, M. (2011, August 7-8). *Visual search and multirobot collaboration on mobile robots.* In International Workshop on Automated Action Planning for Autonomous Mobile Robots. San Francisco, USA.

Zhang, S., & Sridharan, M. (2012, June 4-8). *Active visual search and collaboration on mobile robots.* In International Conference on Autonomous Agents and Multiagent Systems (AAMAS). Valencia, Spain.

Zhang, S., Bao, F. S., & Sridharan, M. (2012, June 5). *ASP-POMDP: Integrating non-monotonic logical reasoning and probabilistic planning on mobile robots.* In Autonomous Robots and Multirobot Systems (ARMS) Workshop at the International Conference on Autonomous Agents and Multiagent Systems (AAMAS), Valencia, Spain.

Zhang, Z., Hou, C., Shen, L., & Yang, J. (2009). An objective evaluation for disparity map based on the disparity gradient and disparity acceleration. In *Proceedings of International Conference on Information Technology and Computer Science* (pp. 452-455).

Zhang, T., & Freedman, D. (2005). Improving performance of distribution tracking through background mismatch. *IEEE Transactions on Pattern Analysis and Machine Intelligence, 27*(2), 282–287. doi:10.1109/TPAMI.2005.31

Zhang, Y. (1997). Evaluation and comparison of different segmentation algorithms. *Pattern Recognition Letters, 18*(10), 963–974. doi:10.1016/S0167-8655(97)00083-4

Zhao, J., Katupitiya, J., & Ward, J. (2007). Global correlation based ground plane estimation using V-disparity image. In *IEEE International Conference on Robotics and Automation* (pp. 529–534). Rome, Italy.

Zhao, W., & Chellappa, R. (2006). *Face processing: Advanced modeling and methods.* Burlington, MA: Academic Press, Elsevier.

Zhao, W., Chellappa, R., & Rosenfeld, A. (2003). Face recognition: A literature survey. *ACM Computing Surveys, 35*(4), 399–458. doi:10.1145/954339.954342

Zhou, H., Yuan, Y., & Shi, C. (2008). Object tracking using SIFT features and mean shift. *International Journal of Computer Vision, 113*(3).

Zickler, T., Mallick, S., Kriegman, D., & Belhumeur, P. (2006). Color subspaces as photometric invariants. In *2006 IEEE Computer Society Conference on Computer Vision and Pattern Recognition,* (Vol. 2, pp. 2000–2010).

Zinsser, T., Schmidt, J., & Niemann, H. (2003). A refined ICP algorithm for robust 3-D correspondence estimation. *Proceedings 2003 International Conference on Image Processing, ICIP 2003,* Vol. 2, (pp. 695-8).

Zivkovic, Z., & Krose, B. (2004). An EM-like algorithm for color-histogram-based object tracking. *IEEE Computer Society Conference on Computer Vision and Pattern Recognition,* Vol. 1, (pp. 798-803).

About the Contributors

José García Rodríguez received his BSc, MSc, and PhD in Computer Science from the University of Alicante (Spain) in 1994, 1996, and 2009, respectively. He is currently Associate Professor in the Department of Computer Technology at the University of Alicante. His research interests are focused on computer vision, neural networks, man-machine interaction, ambient intelligence, robotics, and algorithms parallelization and acceleration.

Miguel Cazorla received a BS degree in Computer Science from the University of Alicante (Spain) in 1995 and a PhD in Computer Science from the same University in 2000. He is currently Associate Professor in the Department of Computer Science and Artificial Intelligence at the University of Alicante. He has done several postdocs stays: ACFR at University of Sydney with Eduardo Nebot, IPAB at University of Edinburgh with Robert Fisher, CMU with Sebastian Thrun and SKERI with Alan Yuille. He has published several papers on robotics and computer vision. His research interests are focused on computer vision and mobile robotics (mainly using vision to implement robotics tasks).

* * *

Juan José Alcaraz-Jiménez received his Telecommunications Engineering Degree in 2006, and two years later, obtained his Master's in Information and Communication Technologies in the Technical University of Cartagena. He is studying now for his PhD at the Faculty of Informatics in the University of Murcia, Spain. He has participated in the Robocup Nao Standard Platform League from 2008 to 2011 with different teams. His research interests are biped locomotion, balance control, and machine learning.

Raed Almomani is a Ph.D. student in the department of Computer Science at Wayne State University. He received his B.S. degree in Computer Science from Al alBayt University, Mafraq, Jordan, in 1999. He received his first M.S. degree in Computer Science from Al alBayt University, Mafraq, Jordan in 2002 and his second M.S. degree in Computer Science from Wayne State University. He worked as a Lecturer in the department of Computer Science at Al alBayt University between 2003 and 2009. He worked as research and teaching assistant in the department of Computer Science at Wayne State University. His research interests include computer vision, pattern recognition, and data mining.

Laith Alkurdi received his BSc from the University of Jordan in Mechatronics Engineering, 2010 and his MSc in Intelligent Robotics with distinction from the University of Edinburgh, 2011. He is currently a PhD candidate at the Technical University of Munich investigating the application of machine learning and intelligent control to intention recognition and human robot interaction.

Anastassia Angelopoulou is currently working towards the PhD degree with the Harrow School of Computer Science, University of Westminster, London, United Kingdom. In 2003, she joined the Computer Vision Research Lab at Harrow School of Computer Science and she is currently working in the field of mathematical modelling for computer vision applications. Her research interests include medical image processing, computer vision, and shape modelling (mainly using statistical methods for automatic model building).

Esther Antúnez received her title of Telecommunication Engineering from the University of Málaga, Spain, in 2008. During 2009, she worked at the PRIP group of the Technical University of Vienna, under the supervision of Dr. Walter G. Kropatsch. Her research is focused on graph and irregular pyramids, and their applications in image segmentation and artificial attention.

Jorge Azorin-Lopez completed a degree in Computer Science Engineering in 2001 and a Ph.D. degree in Computer Science at the University of Alicante (Spain) in 2007. Since 2001, he has been a faculty member of the Department of Computer Science Technology and Computation at the same university, where he is currently an Associate Professor. His current areas of research interest include vision systems to perceive under adverse conditions, real scenes segmentation and labelling, automated visual inspection, and reconfigurable hardware. At present, he is working on behavior analyses, 3D modeling, and free-form 3D changes modeling.

Antonio Bandera received his M.S. and Ph. D. degrees in Electronic Engineering from the University of Málaga (Spain) in 1995 and 2000, respectively. From 2001, he is a Lecturer in the Department of Electronic Technology, University of Málaga (Spain). Dr. Antonio Bandera has developed his research activities in the fields of computer vision, robotics, and pattern recognition. He has published more than 50 papers in international journals (29 cited in the Journal Citation Report), and has got more than 60 contributions on national or international conferences. Nowadays, he is the main researcher of two projects funded by the Spanish Government, and of an integrated action with the PRIP of TU Vienna. Finally, he is the academic coordinator of a Master course on Electronic Technologies for Smart Environments, organized by the University of Málaga (Spain).

Juan Pedro Bandera received his M.S. and Ph. D. degrees from the University of Málaga, in 2003 and 2010, respectively. From 2003 to 2006 he held a Research Grant at ISIS group, at the University of Málaga. From 2006 to 2010 he was Assistant Professor, and since 2010 he is PhD Assistant Professor at the University of Málaga. Juan Pedro Bandera is author of over 25 technical publications, proceedings and books. His research interests include social robotics, learning by imitation, and visual perception.

Humberto Martínez Barberá received the M.S. and Ph.D. degrees in Computer Science in 1995 and 2001, respectively. He is an Associate Professor at the Department of Information and Communications Engineering, University of Murcia, Spain. He has been visiting Researcher and visiting Professor at Örebro University, Sweden, and visiting Professor at University of Versailles, France. His current research interests include autonomous mobile robots, robot control architectures, and software. He is the head of the Applied Engineering Research Group at University of Murcia, where he leads technology transfer projects related to robotics, vision, and automation.

Lourdes Mattos Brasil Electrical Engineer, Federal University of Santa Catarina (1984), Master of Science in Electrical Engineering/Biomedical Engineering, Federal University of Santa Catarina (1994), Sandwich Doctorate in Mathematical Applied from Facultes Universitaires Notre_Dame de La Paix (FUNDP), Belgium (1997–1998), and D.Sc. in Electrical Engineering/Biomedical Engineering (1999). She is currently professor/researcher and head of Electronic Course and Lato Sensu in Clinical Engineering Course at the University of Brasilia, Gama Faculty. She is interested in: artificial intelligence, data mining, machine learning, knowledge acquisition, knowledge based systems, hybrid expert systems, virtual reality, intelligent tutoring systems, informatics in health, and e-learning.

Ivan Cabezas received the Diploma degree in Systems Engineering from the Universidad del Valle, Cali, Colombia in 2004. Currently, he is pursuing the PhD Degree in Computer Engineering from the Universidad del Valle, by researching on Stereo Vision. In 2004 he joined the School of Computer and Systems Engineering as an Auxiliary Professor, where he is currently a Research Assistant, member of the Multimedia and Vision Research Lab. His current research interests include optimisation applied to inverse problems, evolutionary computing, computer vision, pattern recognition, and multimedia signal processing. He has evaluated, published, and led multidisciplinary projects on these areas.

Jose María Cañas got his Ph.D. at the Technical University of Madrid (2003). He is Associate Professor at the Rey Juan Carlos University, where he leads the Robotics Group. He was member of Robot Learning Lab at Carnegie Mellon University (advised by Reid Simmons), worked five years at Instituto de Automática Industrial (CSIC) and did a postdoc stay at Georgia Institute of Technology with Ron Arkin. He has published several papers on robotics and computer vision. He is interested in the perception and control architecture for intelligent behaviors in mobile robots, and in artificial vision systems.

Juan Manuel García Chamizo received the BSc in Physics from the University of Granada (Spain) in 1980 and PhD from the University of Alicante (Spain) in 1994. He is currently Professor with the Department of Computer Technology of the University of Alicante and head of the Industrial Informatics and Computer Nets research group. His research interest areas are computer vision, neural networks, industrial informatics, and biomedicine.

Raymond H. Cuijpers received his Ph.D. degree from Utrecht University in 2000. During his PhD he became an expert on visual perception and its relation to computer vision. He did a post-doc on the role of shape perception on human visuomotor control at Erasmus MC Rotterdam. In 2004 he did a second post-doc at the Radboud University Nijmegen in the context of the European FP6 project called Joint Action Science and Technology (JAST), where he studied cognitive models of joint action. Since 2008 he is Assistant Professor at the Eindhoven University of Technology. Currently, he is project coordinator of the European FP7 project Knowledgeable SErvice Robots for Aging (KSERA). In this context, he studies how robots can autonomously interact with humans in a natural and fluent way, so that they can help improve the quality of life of older persons. Raymond Cuijpers is author of over 22 publications in journals and conference proceedings on human perception, human motor control, artificial intelligence and cognitive robotics.

Ashwin Dani received his B.S. degree in Mechanical Engineering from India in 2005, and Ph.D. degree in Mechanical Engineering from University of Florida in 2011. He is currently pursuing his post-doctoral research at the University of Illinois at Urbana-Champaign. Dr. Dani's main research interest is in the interdisciplinary area of computer vision, nonlinear controls, and robotics. Specifically, his research interests include vision-based estimation and control, autonomous navigation, estimation and control of multi-agent systems. His dissertation work won the Best Dissertation Award in Dynamics, Systems and Control from the Department of Mechanical and Aerospace Engineering at the University of Florida.

Warren Dixon received his PhD degree in 2000 from the Department of Electrical and Computer Engineering from Clemson University. After completing his Doctoral studies he was selected as an Eugene P. Wigner Fellow at Oak Ridge National Laboratory (ORNL). In 2004, Dr. Dixon joined the faculty of the University of Florida in the Mechanical and Aerospace Engineering Department. Dr. Dixon's main research interest has been the development and application of Lyapunov-based control techniques for uncertain nonlinear systems. He has published 3 books, an edited collection, 9 chapters, and over 250 refereed journal and conference papers. He has won several awards for his work including 2011 American Society of Mechanical Engineers (ASME) Dynamics Systems and Control Division Outstanding Young Investigator Award, 2009 American Automatic Control Council (AACC) O. Hugo Schuck Award, and 2006 IEEE Robotics and Automation Society (RAS) Early Academic Career Award. Dr. Dixon is a senior member of IEEE.

Ming Dong received his B. S. degrees in Electrical Engineering and Industrial Management Engineering from Shanghai Jiao Tong University, Shanghai, China, in 1995. He received his Ph. D degree in Electrical Engineering from University of Cincinnati in 2001. He joined the faculty of Wayne State University in 2002 and is currently an Associate Professor in the Department Computer Science and the director of Machine Vision and Pattern Recognition Laboratory, part of the Center for Visual Informatics and Intelligence. Dr. Dong's areas of research include pattern recognition, data mining, and multimedia content analysis. His research is funded by National Science Foundation, State of Michigan, and Industries. He has published over 90 technical articles in premium journals and conferences in related fields, e.g., TPAMI, TKDE, TNN, TVCG, CVPR, ACM MM and WWW. He was as an associate editor of *IEEE Transactions on Neural Networks* (2008-2011) and *Pattern Analysis and Applications* (2007-2010), and served in many conference program committees.

Sergio Orts Escolano received a BSc and MSc in Computer Science from the University of Alicante (Spain) in 2008 and 2010, respectively. He is currently a PhD student and researcher in the Department of Computer Technology at the University of Alicante. His research interests are focused on computer vision, robotics, and neural networks.

Robert B. Fisher received a B.S. with Honors (Mathematics) from California Institute of Technology (1974) and a M.S. (Computer Science) from Stanford University (1978). He received his PhD from University of Edinburgh (1987), investigating computer vision. Since then, Bob has been an academic at Edinburgh University, now in the School of Informatics, where helped found the Institute of Perception, Action and Behaviour. He is currently the Dean of Research in the College of Science and Engineering. His research covers topics in high level and 3D computer vision, medical imaging, and video sequence analysis. He has published or edited 11 books and about 260 peer-reviewed scientific articles, including 47 journal papers. He is a Fellow of the Int. Association for Pattern Recognition (2008) and the British Machine Vision Association (2010).

Nic Fischer received his B.S. degree and M.S. degree in Mechanical Engineering from the University of Florida in 2008 and 2010. He is currently pursuing a PhD degree in the Department of Mechanical and Aerospace Engineering at the University of Florida in the field of Dynamics, Systems and Control under the advisement of Dr. Warren Dixon. Nic's research interests include Lyapunov-based control of input-delayed nonlinear systems, saturated feedback control, autonomous underwater vehicles, and vision-based estimation and control.

Juan F. García received his PhD. in Computer Science from the University of León, Spain, in 2011. He had a four years research grant at Mathematics Department, University of León and he has researched in National University of Ireland, Galway. He currently works for Indra, a global technology, innovation, and talent company. His research interests include robotics, attention control, and artificial vision.

Antonios Gasteratos is an Assistant Professor at the Department of Production and Management, Democritus University of Thrace, Greece. He teaches the courses of Robotics, Automatic Control Systems, Measurements Technology, and Electronics. He holds a B.Eng. and a Ph.D. from the Department of Electrical and Computer Engineering, DUTH, Greece. During 1999-2000 he was a visiting Researcher at the Laboratory of Integrated Advanced Robotics (LIRA-Lab), DIST, University of Genoa, Italy, with a TMR grant from VIRGO and SMART networks. He has served as a reviewer to numerous scientific journals and international conferences. His research interests are mainly in mechatronics and in robot vision. He has published more than 60 papers in books, journals, and conferences. He is a member of the IEEE, IAPR, ECAI, and the Technical Chamber of Greece (TEE). Prof. Gasteratos is a member of EURON, euCognition, and I*Proms networks.

Andrés Fuster Guilló received his B.S. degree in Computer Science Engineering from the Polytechnic University of Valencia (Spain) in 1995 and his PhD in Computer Science at the University of Alicante (Spain) in 2003. He has been a member of the Department of IT and Computation at the University of Alicante since 1997, where he is currently Assistant Professor. His research interests are in the area of computer vision.

Yll Haxhimusa is an Assistant Professor at Vienna University of Technology since 2008. He finished his PhD at the Vienna University of Technology, under the supervision of Walter G. Kropatsch. As post doc Dr. Haxhimusa has worked with Prof. Zygmunt Pizlo at Purdue University, studying graph and combinatorial pyramids in the human problem solving. His research is focused on graph and combinatorial pyramids, and their properties and applications in image segmentation and human problem solving.

David Herrero-Pérez received the M.E. degree in Electrical and Electronic Engineering from the Technical University of Cartagena, Spain, in 2002, and the Ph.D. degree in Computer Science from the University of Murcia, Spain, in 2007. From 2007 to 2009, he was a Visiting Professor at the Department of Systems Engineering and Automation, University Carlos III, Madrid, Spain. From 2009 to 2010, he held a research position at the University of Murcia, Spain. Since November 2010, he is an Assistant Professor at Technical University of Cartagena, Spain. He was a Visiting Researcher at the Center of Applied Autonomous Sensor Systems (AASS) of Örebro University, Örebro, Sweden, and at the Instituto de Sistemas e Robótica of Instituto Superior Tecnico (ISR/IST), Lisbon, Portugal. His research interests are in the field of real-time systems, fault-tolerant distributed systems, cooperative robotics, and multi-sensor fusion, with a special focus on industrial applications.

Zhen Kan received his Bachelor of Engineering degree in Mechanical Engineering in 2005, and a Master of Science degree in Mechatronic Engineering in 2007 from Hefei University of Technology (HFUT), China. He then joined the Nonlinear Controls and Robotics (NCR) research group at the University of Florida (UF) and completed his Ph.D. in December 2011. After completing his Doctoral studies, he is working as a postdoctoral research fellow with the Air Force Research Laboratory (AFRL) at Eglin AFB and University of Florida REEF. Dr. Kan's current research interests include networked control systems, cooperative control of autonomous agents, vision-based estimation and control, and Lyapunov-based nonlinear control.

Patrycia Barros de Lima Klavdianos Graduated in Computer Science at the Catholic University of Brasilia (UCB). She postgraduated in Software Engineering at UCB, and earned her Master's in Computer Vision at Université de Bourgogne. With more than 17 years of experience as architect and developer of solutions built in Java and C/C ++, she has been working with robotics for constructing cognitive and interactive systems. She cooperates in the LNCC Project for building a 3D intelligent environment to simulate surgical procedures using computer graphics and robotic interfaces (Phantom and Force Dimension). She has been using AAM for hand tracking and object shape analysis.

Ioannis Kostavelis was born in Thessaloniki, Greece, in 1987. He received the diploma degree in Production and Management Engineering from the Democritus University of Thrace and the M.Sc. degree (with Honors) in Informatics from the Aristotle University of Thessaloniki in 2009 and 2011, respectively. He is currently with the Laboratory of Robotics and Automation, Department of Production and Management Engineering, Democritus University of Thrace, where he is pursuing the Ph.D. degree in the field of Robotic Vision. He has been involved in different research projects funded by the European Space Agency and the Greek state. His current research interests include vision systems for robotic applications and machine learning techniques.

Walter G. Kropatsch is full Professor at Vienna University of Technology since 1990. His interest in image pyramids goes back to 1984 when he spent one year at the Center for Automation Research of the University of Maryland on invitation of Prof. Azriel Rosenfeld. Since then he explored different variants of hierarchical data and processing structures. In collaboration with Peter Meer, Annick Montanvert, and Jean-Michel Jolion, he extended the scope of pyramids to graphs and then to plane graphs for which the first proof of topology preservation could be shown. His current interest are in the extension of the pyramidal concept towards dynamical hierarchical description of the 3D moving objects in their environment and its derivation from one or more 2D image sequences. In this high dimensional context, topology receives more importance. Prof. Kropatsch served the IAPR in many positions and was its president from 2004 to 2006. Together with J.-M. Jolion, he initiated the IAPR TC15 on graph-based representations in 1996.

Kurosh Madani is graduated in fundamental physics in 1985 from PARIS 7 – Jussieu University, Kurosh Madani received his MSc. in Microelectronics from University PARIS 11 (PARIS-SUD), Orsay, in 1986. He received his Ph.D. of Sciences from University PARIS 11 in 1990. In 1995, he received the DHDR Doctor Hab. degree from University PARIS-EST Creteil (UPEC). Professor in Senart-FB Institute of Technology of UPEC, he is Vice-director of LISSI lab and head of Intelligent Machines & Systems research team of this laboratory.

Rebeca Marfil, M.S. Degree in Telecommunication Engineering and PhD Degree from the University of Málaga, in 2002 and 2006, respectively. Since 2002, she has worked as Research Assistant at the Department of Tecnología Electrónica of the University of Málaga. Her research is focused on artificial vision and hierarchical image processing.

Vicente Matellán got his PhD. in Computer Science from Technical University of Madrid, Spain. He joined the University of León in February 2008 as Associate Professor of Computer Science. He had been working at the Rey Juan Carlos University since1999, where he founded and leaded the Robotics Group (Móstoles-Madrid, Spain), and previously as Researcher and Assistant Professor at the Carlos III University of Madrid from 1993 till 1999. He is member of IEEE Technical Committee on Software Engineering for Robotics, part of the Steering Committee of the CompCog research programme of the European Science Foundation, editor of the *Journal of Physical Agents* (http://jopha.net), is leading the national network on physical agents, and is also involved in several research projects. He has published over 100 papers in journals, books, and conferences in the robotics area.

Nils Meins received his Diploma degree in Computer Science with a focus on computer vision and machine learning from the University of Hamburg in 2010. Since 2011, he is a Research Associate of the Knowledge Technology group in the Department of Computer Science at the University of Hamburg and works toward his Ph.D. degree. His research interests are computer vision, machine learning, and bio-inspired cognitive and intelligence systems with a focus on face perception.

Jairo Simão Santana Melo graduated in Computer Science at the Catholic University of Brasilia (UCB). He postgraduated in Object Oriented Systems at UCB, and earned a Master's in Knowledge Management and Technology at UCB. He was approved in 2009 for the Ph.D. in Electrical Engineering - Subarea Automation at the University of Brasilia (UnB) to develop projects related to robotics and simulation in a 3D environment. He has been an architect and developer of solutions in Java, C and C ++ for the area of electronic transactions. He was a LNCC Project developer in the area of navigation, interaction and visualization, and anatomical structures in a three dimensional environment in order to provide an interface for medical training using the resources of computer graphics and robotic interfaces (phantom and force dimension) simulating small surgical procedures.

Luis Molina-Tanco is a Lecturer at University of Málaga. He studied Telecommunication Engineering at Universidad Politécnica de Madrid, worked in the Telecommunications Industry and then went back to Academia to undertake a Ph. D. in Human Motion Synthesis from Captured Data at the Centre for Vision, Speech and Signal Processing at University of Surrey (UK). In 2003 he joined the Department of Electronic Technology (DTE) of the University of Málaga, as Researcher and Lecturing Assistant, and as Lecturer since 2007.

Vicente Morell received a BSc and MSc in Computer Science from the University of Alicante (Spain) in 2008 and 2010, respectively. He is currently a PhD student and researcher in the Artificial Intelligence Department at the University of Alicante. His research interests are focused on computer vision, robotics, and neural networks.

Ramón Moreno received de degree of Engineer in Computer Sciences on 2006 at University of the Basque Country, afterwards he has been working for 6 years as PhD student within the Computational Intelligence Group of this university. Relevant works are in image segmentation, reflectance analysis, robot vision, and hyperspectral image analysis. He got a (formacion de personal Investigador - FPI) grant from the Basque Governement and he will get de degree of PhD in June 2012 in the same university. He has been in LSSI lab of University of Creteil in 2010 working in robot vision systems, and in 2011 in INPE (National Intitute for Space Research) Brasil working with hyperspectral images.

Lazaros Nalpantidis was born in Thessaloniki, Greece in 1980. He is currently a post-doctoral researcher at the Centre for Autonomous Systems, Computer Vision & Active Perception Lab. of the Royal Institute of Technology (KTH) in Stockholm, Sweden. He holds a PhD (2010) from the Department of Production and Management Engineering, Democritus University of Thrace, Greece in the field of Robot Vision. He holds a BSc degree (2003) in Physics and the MSc degree (2005) (with Honors) in Electronic Engineering both from the Aristotle University of Thessaloniki, Greece. He has been involved in numerous research projects funded by the European Commission, European Space Agency, Greek state, and Swedish state. His current research interests and involvement in international research programs include active and cognitive vision systems for robots in scene understanding, depth perception, obstacle avoidance, and SLAM applications.

Pedro Cavestany Olivares received the MSc degree in Industrial Engineering from Polytechnic University of Cartagena in 2005. He also holds a MSc degree in Information Technology and Advanced Telematics from University of Murcia. Pedro works in the research project "Collective Action and Perception in Multi-agent System on Humanoid Robots" in University of Murcia, and has participated in several Robocup championships. Pedro's research field is computer vision in mobile robots and projective geometry. Pedro is currently studying a joint PhD between University of Murcia and Cranfield University in distributed scene understanding from multiple mobile platforms.

Eduardo Perdices got his Bachelor's Degree in Computer Science at the Rey Juan Carlos University in 2009 and his Master's Degree in Telematic and Information Systems at the same university in 2010. He is currently a full time research staff member and his main research fields are computer vision and localization methods for mobile robots.

Xavier Perez-Sala received the B.S. degree in Industrial Electronics at Universitat Politècnica de Catalunya (UPC) in 2008 and the MS degree in Artificial Intelligence at UPC and Universitat de Barcelona (UB) in 2010. He did his Master's thesis at Technical Research Centre for Dependency Care and Autonomous Living (CETpD), combining computer vision and reinforcement learning for social robots navigation. He is currently pursuing a PhD degree in Artificial Intelligence at the CETpD and Computer Vision Center (UAB), focused on computer vision techniques for human pose recovery and behavior analysis, in the field of social robotics.

Alexandra Psarrou is Head of the Artificial Intelligence and Interactive Multimedia Department, at Harrow School of Computer Science, University of Westminster. Dr. Psarrou was born in 1963 in Athens, Greece and received her BSc (1987), MSc (1988) in Computer Science, and PhD (1996) in Computer Vision from Queen Mary and Westfield College (QMW), London. Her research interest include dynamic image understanding, modelling and prediction of motion using artificial neural networks, 2D and 3D shape indexing for content based search in image and video databases, and modelling intelligent man-machine interfaces.

Domenec Puig received the MS and PhD degrees in Computer Science from Polytechnic University of Catalonia, Barcelona, Spain, in 1992 and 2004, respectively. In 1992, he joined the Department of Computer Science and Mathematics at Rovira i Virgili University, Tarragona, Spain, where he is currently Associate Professor. Since July 2006, he is the Head of the Intelligent Robotics and Computer Vision group at the same university. His research interests include image processing, texture analysis, perceptual models for image analysis, scene analysis, and mobile robotics.

Francisco J. Rodríguez received his Computing Engineering degree in Computer Science from University of León, Spain, in 2006. Currently, he is a PhD student at Mechanical, Information Technology, and Aerospace Engineering Department, University of León. His current research interests include human-robot interaction, cognitive robotics, and image processing.

Juan A. Rodríguez is a Lecturer at University of Malaga since 2002. He studied Telecommunication Engineering at University of Málaga, worked for a year in a Telecommunications Company developing software and finally joined the Department of Electronic Technology (DTE) in 1996 as Lecturing Assistant. He received the PhD in Image Hierarchical Processing in 2001.

Manuel Graña Romay received the M.Sc. degree in Computer Science in 1982 and the Ph.D. degree in Computer Sciences in 1989, both from Universidad del Pais Vasco, Spain. His current position is Full Professor (Catedratico de Universidad) in the Computer Science and Artificial Intelligence Department of the Universidad del Pais Vasco, in San Sebastian. He is the head of the Computational Intelligence Group (Grupo de Inteligencia Computational) which has been awarded funding as a high performance university research group since the year 2001. Current research interests are in applications of computational intelligence to multicomponent robotic systems, specifically linked multicomponent robotic systems, medical image in the neurosciences, multimodal human computer interaction, remote sensing image processing, content based image retrieval, lattice computing, semantic modelling, data processing, classification, and data mining. Until 2010 he has coauthored more than seventy journal papers, over two hundred conference papers.

Roseli Aparecida Francelin Romero received her Ph.D. degree in electrical engineering from the University of Campinas, Brazil, in 1993. She is an Associate Professor in Department of Computer Science at ICMC of the University of Sao Paulo (USP), since 1988. From 1996 to 1998, she was a Visiting Scientist at Carnegie Mellon's Robot Learning Lab, USA. Her research interests include artificial neural networks, machine learning techniques, fuzzy logic, robot learning and computational vision. She has been reviewer for important journals and for several International and National Conferences of her area. She has already organized Special Sessions in important Conferences and organized important events in Brazil. Dr. Romero is a member of INNS – International Neural Networks Society and Computer Brazilian Society (SBC). She is a vice coordinator of the Center for Robotics of São Carlos – CRob-SC/USP and President of the Research Development Committee of ICMC-USP.

Sreela Sasi is currently working as Professor in the Department of Computer and Information Science at Gannon University, Erie, PA, USA. She has done PhD in Computer Engineering from Wayne State University, Michigan, MS in Electrical Engineering from University of Idaho, Idaho, and BSc (Engg.) in Electronics and Communication Engineering from College of Engineering - Trivandrum, University of Kerala, India. Research interests include computer vision, intelligent system design using fuzzy logic, neural networking and wavelet transforms, web mining, digital watermarking, modeling of electrophysiology cardiac mapping, and VLSI design. She is a Senior member of IEEE, Eta Kappa Nu honor society, IEEE Women in Engineering, Fellow IETE (L) - India, Member ISTE (L) – India.

Marcelo Saval received the BS degree in 2010 and MS degree in 2011 in Computer Engineering from the University of Alicante. Currently, he is pursuing his PhD in the Industrial Computing and Computer Networks (IIRC, Spanish acronym) research group within the topic of 3D computer vision. His interests include computer vision in behavior analyses, 3D modeling, and free-form 3D changes modeling.

José Antonio Serra received a BSc in Telecommunications from Lasalle University in Barcelona (Spain) in 1996 and MSc in Computer Science from University of Alicante in 2012. He is currently a PhD student and researcher in the Computer Technology Department at the University of Alicante. His research interests are focused on computer vision, robotics and neural networks.

Renato Ramos da Silva received his B.S. degree in Computer Science from Federal University of Lavras in 2005, and his M.S. in Computer Science at Institute of Mathematics and Computer Science at University of São Paulo in 2009, where he is currently a Ph.D. candidate. His research focuses in the area of Human-Robot Interaction where the robot is designed to interact with people as a socially intelligent partner. His current research involves study the social robotic architecture and learning mechanism to improve the interaction between a robot and a caregiver using shared attention mechanism, supported by FAPESP.

Mohan Sridharan is an Assistant Professor of Computer Science at Texas Tech University. He was a Research Fellow in the School of Computer Science at University of Birmingham (UK), working on the EU Cognitive Systems (CoSy) project, between August 2007 and October 2008. He received his PhD (Aug 2007) and Masters (May 2004) in Electrical and Computer Engineering at The University of Texas at Austin. Dr. Sridharan's doctoral research enabled mobile robots to autonomously learn models of color distributions, and use the learned models to detect and adapt to illumination changes. His post-doctoral research developed novel hierarchical decompositions in partially observable Markov decision processes to enable a mobile robot to automatically tailor visual sensing and information processing to the task at hand. Dr. Sridharan's current research interests include autonomous robots, stochastic machine learning, computer vision, human-robot interaction, and applied cognitive systems.

Elena Torta received her M.S. in Industrial Automation Engineering from the Universitá Politecnica delle Marche, Ancona (IT), in 2009. She is currently working toward the Ph.D. degree in the Department of Industrial Engineering and Innovation Sciences, Human Technology Interaction group, at the Eindhoven University of Technology (NL). Her main research interests include control system design, ambient assisted living, and human-technology interaction.

Maria Trujillo received the Diploma degree in Statistics from the Universidad del Valle, Cali, Colombia, in 1991, the M.Sc. degree in Statistics from the Colegio de Posgraduados, Montecillo, Mexico in 1994, and the Ph.D. degree in Electronic Engineering from the University of London, United Kingdom, 2005. In 1999 she joined the Department of Industrial Engineering and Statistics at Universidad del Valle, Colombia; and in 2006 moved into the School of Computer and Systems Engineering, where she is currently Associate Professor. She is also the Head of the Multimedia and Vision Research Lab. Her current research interests include multimedia signal processing, computer vision, pattern recognition, and medical image processing. She has published in these areas. Prof. Trujillo has served the IEEE and other professional Societies in many capacities. She is currently a member of the Peer Review College of the IET Technical Committee on Computer Vision, the Textile Research Journal and DYNA, among others. Also, she is in the International Program Committee of various international conferences.

David van der Pol received his Bachelor's degree in Electrical Engineering from the Avans Hogeschool in Breda in 2005. After his studies in Breda, he studied Human Technology Interaction at Eindhoven University of Technology where he received his Master's degree in 2009. His research focused on the perception of eye-contact in pictorial space (Mona Lisa Effect). For two years he worked for the University of Eindhoven on the E.U. funded KSERA project as a Human Robot Interaction researcher focusing on eye-contact and joint attention. David currently works for DVC Machine Vision and focuses on applications in computer vision.

Julio Vega was born in Badajoz. He got his Technical Engineering in Computer Systems at University of Extremadura in 2005. He achieved a Seneca Grant while pursuing his Bachelor's Engineering Degree at Politechnical University of Madrid. And finally, he moved to Rey Juan Carlos University, where he got a Collaboration Fellowship on Robotics Group in 2006. He got his Bachelor's Engineering Degree in Computer Science at Rey Juan Carlos University in 2008, when he joined the Doctoral program through a Research Fellowship and is currently a full time research staff member, supervised by Dr. José María Cañas.

Diego Viejo obtained his BSc and MSc in Computer Science in 2002 and his PhD in 2008 both from the University of Alicante. Since 2004, he is a Lecturer and a Researcher in the Department of Computer Science and Artificial Intelligence (DCCIA) at the University of Alicante. His research interests are focused on 3D vision applied to mobile robotics.

Cornelius Weber graduated in physics, Bielefeld, Germany, and received his PhD in Computer Science at the Technische Universität Berlin in 2000. Then he was a postdoctoral fellow in Brain and Cognitive Sciences, University of Rochester, USA. From 2002 to 2005 he was a research scientist in Hybrid Intelligent Systems, University of Sunderland, UK. Then, Junior Fellow at the Frankfurt Institute for Advanced Studies, Frankfurt am Main, Germany, until 2010. He is now Lab Manager at the Knowledge Technology group at the University of Hamburg. His core interest is computational neuroscience. Scope is on the development of feature detectors, neural models of image segmentation and frame of reference transformations in the visual system, furthermore, reinforcement learning and motor system models for robot control and related applications.

Stefan Wermter is Full Professor at the University of Hamburg and Director of the Centre for Knowledge Technology. He holds an MSc from the University of Massachusetts in Computer Science, and a PhD and Habilitation in Computer Science from the University of Hamburg. He has been a research scientist at the International Computer Science Institute in Berkeley before leading the Chair in Intelligent Systems at the University of Sunderland. His main research interests are in the fields of neural networks, hybrid systems, cognitive neuroscience, cognitive robotics, and natural language processing. Professor Wermter has written or edited five books and published more than 150 reviewed articles. He is on the board of the European Neural Network Society, an associate editor of the journals *Connection Science, International Journal for Hybrid Intelligent Systems,* and *Knowledge and Information Systems.* He is on the editorial board of the journals *Neural Networks, Cognitive Systems Research,* and *Journal of Computational Intelligence.*

Wenjie Yan received his Bachelor's degree in Mechanical Engineering from the joint college of the Hamburg University of Applied Science and the University of Shanghai for Science and Technology in 2002, and his diploma degree in Mechatronics Engineering from the Karlsruhe Institute of Technology in 2009. He is now a Research Associate of the Knowledge Technology group in the Department of Computer Science at the University of Hamburg in Germany. His research interests are in the area of artificial intelligence, cognitive systems, computer vision, and robotics.

Index